AF412175

HIDDEN WORLDS IN
QUANTUM PHYSICS

HIDDEN WORLDS IN QUANTUM PHYSICS

Gérard Gouesbet

CNRS – Université et INSA de Rouen, France

World Scientific

NEW JERSEY · LONDON · SINGAPORE · BEIJING · SHANGHAI · HONG KONG · TAIPEI · CHENNAI

Published by

World Scientific Publishing Co. Pte. Ltd.

5 Toh Tuck Link, Singapore 596224

USA office: 27 Warren Street, Suite 401-402, Hackensack, NJ 07601

UK office: 57 Shelton Street, Covent Garden, London WC2H 9HE

Library of Congress Control Number: 2024950372

British Library Cataloguing-in-Publication Data
A catalogue record for this book is available from the British Library.

HIDDEN WORLDS IN QUANTUM PHYSICS

ISBN 978-981-98-0648-5 (hardcover)
ISBN 978-981-98-0649-2 (ebook for institutions)
ISBN 978-981-98-0650-8 (ebook for individuals)

For any available supplementary material, please visit
https://www.worldscientific.com/worldscibooks/10.1142/14138#t=suppl

Typeset by Stallion Press
Email: enquiries@stallionpress.com

Preface

This book by Gérard Gouesbet gives an extraordinarily encyclopedic overview of the "hidden worlds" of quantum mechanics, based on his vast knowledge of both physics and philosophy. He covers almost every possible problem posed by quantum mechanics and almost every possible interpretation on alternative view of quantum mechanics that has been proposed.

Ever since its inception in the 1920s, quantum mechanics has provoked heated debates between physicists as well as between philosophers. On the one hand, there is an "orthodox" view, often called the Copenhagen interpretation [1], whose name refers to the Danish physicist Niels Bohr who, together with Werner Heisenberg, Wolfgang Pauli, Max Born, and many other eminent physicists developed this view. On the other hand, there have always been critics of this view, also including eminent physicists — Albert Einstein, Erwin Schrödinger, Louis de Broglie, David Bohm, and John Bell, among others.

In this preface, I would like to outline what are the real problems raised by quantum mechanics and what could be a possible solution to those problems. What follows is of course based on the work of the critics of the orthodox interpretation, who are discussed in detail by Gérard Gouesbet.

In quantum mechanics, one associates to any physical system an abstract mathematical object, called the wave function or the state vector. This state changes with time according to precise rules that, in practice, are nonambiguous. Thanks to those rules, one can make a great number of predictions that are spectacularly well confirmed experimentally. Moreover, quantum mechanics is at the basis of all modern electronics and therefore of all informatics and telecommunications.

The problem is that it is difficult to say what the state vector means. Indeed, in many situations, it is in a "superposed" state, which is often explained by saying that the system possesses simultaneously mutually exclusive properties. For example, it could be both black *and* white — not gray, but simultaneously completely black and completely white. Of course, we are not talking here literally about colors but about "properties" such as spin, or energy, or polarization, or angular momentum. And, of course, when we observe something, we never find both of these mutually exclusive properties but only one of them. The state is then supposed to "collapse" or be "reduced" to a state that is no longer a superposition of states with different properties but a state with only one of those properties.

It has to be emphasized that a superposed state is not supposed to be understood only in the sense of reflecting our ignorance, as if we simply did not know whether the particle *is* black or white, until we look. If that were the case, it would be like a coin that is either heads or tails; we may not know in which state it is until we look, but we know that it is in one or the other and, by looking, we simply learn which one it is. In ordinary quantum mechanics, there is no sense attached to the idea that the system *is* either black or white before one looks, and that makes the situation radically different from the one of "classical" probabilities, such as coin tossing.

Why do physicists use such a language? Because if one did not use the language of superposed states, if instead, one attributed a state corresponding to only one of the two superposed properties to the system, many empirically correct results could not be derived.

An example of such a use of superposed states is given by the double-slit experiment. A set of particles is sent toward a wall in which two slits can be opened and the particles are detected on another wall at a certain distance behind the first wall. If only one slit is open, we find a certain distribution of particles on the second wall, centered around the slit, as one would expect in a "random" experiment, meaning that not all particles passing through the slit follow exactly the same path. If only the other slit is open, one finds a similar distribution centered around the second slit. If both slits are open, one would naively expect the distribution of particles to be the "sum" of the distributions obtained when each slit is open. But that is not the case: The particles distribute themselves as if a wave had been going through both slits, displaying an interference pattern, with places where there are even fewer particles arriving than when only one slit is open.

One sometimes describes that experiment by saying that the "particle" sometimes behaves as a wave (when both slits ate open) and sometimes as a particle (when only one slit is open); hence, one speaks of wave-particle duality. One also says that the particle goes through both slits (when they are both open), which is an example of mutually incompatible properties (2).

But all that does not answer the question of what the state vector means nor does it allow us to understand how to think about superposed states.

To dramatize the problem, Schrödinger invented his famous example of the cat [3]. Imagine a cat in a sealed room together with a capsule containing a poison that can kill the cat if the capsule is open. Imagine also a mechanism that, when one measures whether a system is black or white (to follow our metaphor), breaks the capsule if the result is white and does not break it if it is black. Then, if one applies the rules of quantum mechanics to that entire system (including the capsule and the cat), the cat is *both* dead and alive — not in between life and death but completely alive *and* completely dead.

Schrödinger viewed this situation, which he called "quite ridiculous," as a *reductio ad absurdum* of the traditional interpretation of quantum mechanics, i.e., of the idea that the quantum state gives a complete description of the system. In the case of the cat, the description of it as "both dead and alive" cannot possibly be complete.

Of course, in the usual interpretation, the state vector is applied to microscopic objects, not to macroscopic ones, like measuring devices or a cat, but it is never stated where one puts the dividing line between the two. If one applied the logic of quantum mechanics to every object, large or small, then, as Schrödinger showed, one easily arrives at the cat problem. And, if quantum mechanics applies to one atom, or two atoms, or ten, where does one put the limit between that and a cat?

One possible answer to that puzzle is to say that the state vector does not *mean* anything physically, but it only enters into an algorithm through which "results of measurements" can be predicted, and nothing else. But the following question then arises: What is the point of such predictions? In science, the role of experiments is to test theories, but the theories themselves are about the world, not *only* about the experiments. Bones of dinosaurs constitute evidence for the past existence of dinosaurs, but, when we speak of dinosaurs, we do not think of them as being reduced to their bones, i.e., to what is directly observed.

When Einstein asked his famous question to Pais, "Do you really believe that the moon is not there when nobody looks?", he was thinking exactly about that problem. A physical theory that does not speak about the world but *only* about what happens in our laboratories suffers from a serious deficiency.

Actually, Einstein also invented an example, Einstein's boxes [4], that illustrates another problem raised by the idea that quantum mechanics "deals only with measurements." Imagine a box that contains a *single* particle. (Of course, it is unrealistic, but we are talking here about a thought experiment. [5]) Assume that there is a mechanism that allows the box to be divided into two equal and airtight parts. Assume also that the box is initially in Paris and that one sends one half box to New York and the other half box to Tokyo. If one follows quantum mechanics, the "complete description" of the state of the particle is that it is in a superposed state, "the particle is in the half box in New York" *and* "the particle is in the half box in Tokyo." Remember, this does not mean that it is in one half box *or* the other but in one *and* the other. Now, open one of the boxes, say the one in Tokyo. If one finds the particle there, one can predict that no particle will be found in New York if the box there is opened afterward. But, if one does not find the particle in Tokyo, one can, with equal certainty, predict that one would find the particle in New

York if the box there is opened. Indeed, opening the empty box in Tokyo qualifies as a "measurement" and therefore modifies the state of the system.

If one were to think in terms of "classical" probabilities, then there would be nothing mysterious in that situation: We would not know, before opening one of the boxes, in which one the particle is (but we would be sure that it is in one of them), and we would just *learn* in which it is after opening either of the boxes. But that is *not* the way one is supposed to think in quantum mechanics, because the quantum state is supposed to be the complete description of the system, meaning that there is no fact of the matter as to which half box the particle is in (or maybe, one would say, in orthodox quantum mechanics, that the sentence "the particle is in one box or the other" is meaningless).

However, since the two boxes are far apart from each other (and, in principle, they could be light years apart), a new problem arises: Is an action in Tokyo, i.e., opening a box, affecting in any way the physical situation in New York? That is, does it affect not just our *knowledge* of the situation but the situation itself? If it does, then one would have to admit the existence of actions at a distance, which moreover are instantaneous and extend arbitrarily far, since, in this thought experiment, there is no limit to the distance between the boxes or to the time elapsed between the opening of the first box and of the second one. Now, one problem with simultaneous actions at a distance is that simultaneity is a notion that depends on the frame of the reference, at least if one takes the theory of relativity into account, so that actions at a distance also imply, in principle, actions into one's own past [6], and *that* is certainly hard to swallow, to say the least.

To examine further this notion of action at a distance, consider the story, popularized by John Bell of Bertlmann's socks [7]. Mr. Bertlmann is a scientist who always wears socks of different colors. If one looks at one of his socks, one knows that the other sock is of another color. Putting aside matters of taste, there is nothing mysterious here. But assume that the socks are, like Schrödinger's cat, in a superposed state. And, to pursue this metaphor, it is better to switch back to the ordinary situation where both socks are of the same color. Assume that the state is a superposition between a state where both socks are black and a state where both socks are white. They are, as one says, correlated — neither of them is purely white or black, since they are in a superposition, but if one of them is "observed" to be black, then a later observation on the other sock will find it black too. And the same holds for "white".

Now, one may also imagine that the socks, instead of being attached to Mr. Bertlmann's feet are sent away, one to New York and the other to Tokyo, or even sent light years away, while remaining correlated. The same question raised about Einstein's boxes can be asked here: If the colors do not exist before being measured, how come that they are correlated? Does a measurement in one place directly (and instantaneously) affect the situation in another place?

Of course, the role played by the socks here is purely metaphorical, but there are experiments done on pairs of photons that display similar features, with the colors being replaced by photon polarizations along certain directions. One could also think of doing experiments with electrons, with the spin (again, along certain directions) replacing the colors.

This argument is actually the one given in 1935 by Einstein, Podolsky, and Rosen [8] to demonstrate the incomplete character of quantum mechanics: If there is no action at a distance, said Einstein, Podolski, and Rosen, then the perfect correlations can only be accounted for by assuming that the colors (or the polarization or the spin values) exist prior to the measurements being taken. But ordinary quantum mechanics does not include those values in its description of the world — hence it is incomplete.

Unlike Einstein's boxes, the example of Bertlmann's socks can be analyzed in more detail. In the case of the boxes, one is simply perplexed: How can it be that the particle is in neither box, since that implies a form of action at a distance? Maybe quantum mechanics is not the whole story, and so, maybe there is no action at a distance after all. But John Bell invented an argument that shows that a form of action at a distance does exist, irrespective of whether or not quantum mechanics is complete.

For the socks, unlike for the boxes, one can think of a more complicated state, which metaphorically would be both socks "in a superposition of black and white" but also "in a superposition of both red and green" and also "in a superposition of both yellow and blue" (which, in real life, corresponds to values of the spin or of the polarization along different directions). Now, by looking at or by measuring one of the socks, one would have a choice: See whether it is (after measurement) black or white or red or green or yellow or blue. But one could perform one type of observation for one property (say, black-white) on one sock and another (say, yellow-blue) on the other sock. Now, a simple argument shows that the statistics of those measurements (when they are applied to the real quantum mechanical systems, photons instead of socks and polarization instead of colors), where one looks at different properties of the two different socks, imply that it is simply *inconsistent* to assume that the colors exist prior to their measurement [9].

This implies that the colors are "created" by the measurements but, since, when one measures the same pair of properties on both sides, one always obtains perfect correlations, one has to conclude that this "creation," although caused by a "measurement" on one side only, affects the physical situation on the other side.

The strength of Bell's argument is that it demonstrates the existence of actions at a distance [10] directly, by merely looking at the statistical distributions of the colors, and not because of quantum mechanics or any interpretation of it.

What does all that imply concerning the status of quantum mechanics? Do we have to admit that the cat is both dead and alive, that the particle is both in the box in New York and in the one in Tokyo, and that measurements "create reality"? Not really. There exists a theory, introduced by Louis de Broglie in 1927, developed by David Bohm in 1952, and popularized by John Bell, that avoids all these paradoxes. In that theory, the quantum state is not the complete description of the system, because particles have also positions that evolve in time, and therefore they also have trajectories.

The quantum state evolves in the usual way in the theory, through Schrödinger's equation, and it guides the evolution of the positions of the particles. It also defines a probability distribution for the positions of the particles that implies the usual statistical predictions of quantum mechanics [11]. The problem of the boxes has a clear solution: The particle is always in one box or the other. There is no cat problem either, because, if one determines all the positions of the particles composing the cat, then one determines also whether the cat is dead or alive (e.g., consider the motion of the particles constituting the blood of the cat).

The de Broglie–Bohm theory also explains why the properties that have been metaphorically described as colors above do not exist prior to their "measurements", so that, in this theory, the so-called measurements do not really measure any preexisting property of the system, but they should be viewed as interactions between the measuring device and the system [12]. Of course, that interaction must involve, because of Bell's result, a form of action at a distance. But the de Broglie–Bohm does include such an action. In fact, it is because Bell wanted to see whether this property could possibly be avoided in another theory that he was led to derive his result, which showed that the action at a distance present in the de Broglie–Bohm theory is not an artificial feature of that theory but is unavoidable.

The de Broglie–Bohm theory is not a theory about "results of measurements" but a theory about the world: about particles and their motion. In that theory, there are no paradoxes about measurements; the latter are just physical processes, described by the theory itself, and not some *deus ex machina*, as in ordinary quantum mechanics.

One sometimes asks how to compare ordinary quantum mechanics and the de Broglie–Bohm theory, but this is not the right question, because ordinary quantum mechanics is not really a theory: It gives an algorithm that predicts results of measurements, but it does not tell us anything about the world. In fact, one should consider ordinary quantum mechanics simply as a mutilated form of the de

Broglie–Bohm theory (which is a theory of the world, viewed as made of moving particles): Ordinary quantum mechanics amounts to forgetting the particles and their trajectories, and using the de Broglie–Bohm theory to predict the statistics of measurements, which are of course the usual ones. But, if one does that, one creates all kinds of artificial puzzles, such as those raised by Schrödinger's cat or by Einstein's boxes.

I must say that I do not agree with the objections to the de Broglie–Bohm theory raised by Gérard Gouesbet in Chapter 17. First, the underdetermination thesis is a general skeptical argument that applies to all our knowledge and that is simply to be ignored when doing science [13]. Besides, as explained in the previous paragraph, because ordinary quantum mechanics and the de Broglie–Bohm theory cannot be put on the same footing, there is no issue of underdetermination between the two. Of course, point particles are some sort of idealization (another of Gouesbet's objections), both in classical mechanics and in the de Broglie–Bohm theory, an idealization to be replaced by something else in a deeper theory. But if the de Broglie–Bohm theory reaches the degree of clarity of classical mechanics, then it is difficult to see what more could one ask from it. All physical theories will involve idealizations and approximations and that can hardly be considered as an argument against any particular theory.

It is quite unfortunate that the de Broglie–Bohm theory (also called pilot wave theory) is largely ignored in the physics community. As John Bell puts it, "Why is the pilot wave picture ignored in text books? Should it not be taught, not as the only way, but as an antidote to the prevailing complacency? To show that vagueness, subjectivity, and indeterminism, are not forced on us by experimental facts, but by deliberate theoretical choice?" [14].

Jean Bricmont
August, 2012

[1] As discussed in Chapter 3, it is not easy to define precisely this interpretation. Here, we will follow what is presented under that name in textbooks.

[2] For a visual demonstration of the effect, see: http://www.youtube.com/watch?v=DfPeprQ7oGc.

[3] E. Schrödinger, The present situation in quantum mechanics, translated from the German and originally published in *Proceedings of the American Philosophical Society,* 124, 323–338, also Section I.11 of Part I of *Quantum Theory and Measurement,* edited by J.A. Wheeler and W.H. Zurek, (Princeton University Press, Princeton, New Jersey, 1983).

[4] See T. Norsen, Einstein's boxes, *Am. J. Phys.,* 73 (2), February 2005, pp. 164–176. The original idea was proposed by Einstein at the Solvay Congress in 1927.

[5] In the article mentioned in the previous note, one discusses experiments performed already in Hungary in the 1950s with photons that are reflected or not by mirrors; that situation is very similar to the one of Einstein's boxes.

[6] Indeed, if one could, for example, send a message instantaneously, then one could send it to someone in a moving reference frame, who would be in our present, but whose equal time slice would include part of our past. Then, that person could forward the message instantaneously into our past.

[7] See Bertlmann's socks and the nature of reality, in J.S. Bell, *Speakable and Unspeakable in Quantum Mechanics,* 2nd edition (Cambridge University Press, Cambridge, 2004).

[8] See A. Einstein, B. Podolsky, and N. Rosen. Can quantum mechanical description of reality be considered complete?, *Phys. Rev.* 47, 777–780 (1935). The version of that argument given here is due to David Bohm.

[9] See D. Dürr, S. Goldstein, R. Tumulka, and N. Zanghi, John Bell and Bell's theorem in the *Encyclopedia of Philosophy,* 2nd edition, edited by D.M. Borchert (Macmillan Reference, New York, 2006) for a simple proof of Bell's theorem along the line of this example. (One can replace the three choices of pairs of colors by three different directions of the spin and associate one of each color in each pair to a plus one and the other color to a minus one.)

[10] Of course, one has to worry about the problem of backward causation acting on one's own past! This is not easy to solve, but the sort of actions at a distance considered here are rather special and do not permit the transmission of signals, which means that the worst paradoxes are at least avoided. But this does not mean that the presence of these actions at a distance does not cause a serious problem if one wants to combine quantum mechanics with relativity.

[11] For a more detailed exposition of that theory, see D. Dürr, and S. Teufel, *Bohmian Mechanics: The Physics and Mathematics of Quantum Theory* (Springer, Berlin, 2009) or D. Dürr, S. Goldstein, and N. Zanghi, Quantum equilibrium and the origin of absolute uncertainty, *J. Stat. Phys.*, 67, 843–907 (1992).

[12] See M. Daumer, D. Dürr, S. Goldstein and N. Zanghi, Naive realism about operators, *Erkenntniss*, 45, 379–397 (1996) and D. Albert, *Quantum Mechanics and Experience* (Harvard University Press, Cambridge, Massachusetts, 1992).

[13] For a more detailed discussion of underdetermination and other such arguments, see J. Bricmont and A. Sokal, Science and sociology of science: Beyond war and peace, in *The One Culture?: A Conversation about Science*, edited by Jay Labinger and Harry Collins (University of Chicago Press, Chicago, 2001), pp. 27–47.

[14] J.S. Bell, *Speakable and Unspeakable in Quantum Mechanics* (Cambridge University Press, Cambridge, 1993), p. 160.

Foreword

There are plenty of books on quantum mechanics displayed in nearly endless sets of shelves in university and personal libraries, laboratories, and in many other places. Surely, I have straight away to apologize for committing the offence of adding one more entry to an already huge list. I need to ask the reader to show leniency and, in any case, I have to justify myself of any possible accusation.

The excuse is very simple indeed: I always found quantum mechanics fascinating but, at the same time, incomprehensible. Approaching the end of my scientific life, I just decided that the time had come to discuss the lack of understanding attached to quantum mechanics, and to examine whether and how we could escape from such an unsatisfactory situation. There are two issues to be discussed with respect to this issue, two issues that I call: Hidden worlds.

The first one (the main one actually which will run all along this book) concerns hidden variables which have been introduced to (tentatively) drive back quantum mechanics to a more classical vision of physics, and to gain, if possible, a good and fair enough understanding of quantum events. As a result, we possess hidden variables theories which provide alternatives to the usual quantum mechanics, the one of the textbooks. It is sometimes believed that the debate concerning the existence of hidden variables is definitely closed after Bell's theorems and their associated experimental investigations. As we shall see, such is however not the case and, in particular, one of these hidden variables theories, namely the pilot wave version of Bohm, is still defended by some scientists and/or philosophers. Furthermore, the opposition, or even the conflict, between defenders of hidden variables and their opponents, may be sometimes sharp, or even harsh, as I shall illustrate with various quotations. Actually, it has sometimes been harsh in the past, and it is sometimes still harsh nowadays. As a result, statements pertaining to the framework of the quantum mechanics of the textbooks may be found dubious (or even worst) by the defenders of hidden variables and, conversely, statements pertaining to the framework of hidden variables theories may be found dubious (or even worst) by the defenders of the quantum mechanics of the textbooks. We are then facing a controversial matter, involving many controversial issues, for which I propose what I believe to be a well-tempered account, in which arguments, formulations, and quotations, from all camps, will be honestly reproduced, *even when I do not agree with them.* Similarly, the reader is kindly requested to accept to sympathetically read arguments, formulations, and quotations, from all camps, *even if he (she) does not agree with them*: He (she) just have to honestly listen the voices of other interpretations which may be different from the one he (she) believes in. Furthermore, in 1973, Belinfante published a survey of hidden-variables theories [1]. About forty years later, it was certainly the right time to publish an up-to-dated account of the topic.

The second hidden world is the one of first principles, that is to say of basic principles which satisfy the intuition and, therefore, open the way to some kind of understanding. This issue is discussed in the two last chapters of the book (namely Chapters 16 and 17), and is actually connected with the issue of hidden variables. A non-singularity principle is put forward and its consequences are drawn. One of these consequences is that Bohmian mechanics must be rejected (if we hold fast on the non-singularity principle, and its consequences). There is however a loophole for those who defend Bohmian mechanics, namely, they can reject the non-singularity principle, and/or its consequences, leading to a rejection of the rejection.

In these two last chapters, the ingredients associated with the non-singularity principle are fairly speculative, although these chapters also contain some amount of non speculative material which is useful to deepen our understanding of hidden-variables proposals. Therefore, whatever the final choice, if any, made by the reader, whether he (she) rejects or not hidden variables, their contents provide complementary aspects to the issue.

This book is not intended to some defenders of usual quantum mechanics, more precisely the ones who decide to hold fast on it and want to avoid loosing time with what they feel to be side- or even blind-alleys. But it is intended to scientists, whether defenders of or opponents to usual quantum mechanics, who are interested by alternative approaches and believe that they would not loose their energy if they spend a bit of time with the issues involved. But it is not only a book devoted to scientific matters. It is also devoted to philosophy of sciences and to history of sciences in such a way that a fairly large audience of scientists, philosophers, and historians is expected to be possibly motivated enough to read this book.

However, some parts of the track are decorated with mathematics that some readers might find heavy. Therefore, this book contains two threads, one for everyday PEDESTRIANS, such as the ones randomly picked up in the streets of Paris, where they are lazily strolling, and the one for tough mountaineers crossing the Himalaya chain. Bifurcations to the path for PEDESTRIANS, with no mathematics or only with easy mathematics, or warnings for PEDESTRIANS, will be indicated by using the word "PEDESTRIAN" (or the plural "PEDESTRIANS") followed by a tag, namely $\sim\sim\sim\sim$, a wavy tag which just means: *Fluctuat nec mergitur*, that is to say: You are entering an easy path, or the path is not too difficult, there is no risk to sink. Furthermore, information will be given to the PEDESTRIANS for them to know where they have to glue back to the bulk. The implementation of a path for PEDESTRIANS may generate a bit of redundancy, but redundancy is a useful ingredient for educational skills.

Now, mountaineers or PEDESTRIANS, have a nice trip.

Du même auteur

G. Gouesbet. Morphôses. Éditions Saint-Germain-des-Prés, Paris. 1980.

G. Gouesbet. Manifêtes. Éditions Saint-Germain-des-Prés, Paris. 1984.

G. Gouesbet, S. Meunier-Guttin-Cluzel, O. Ménard. Chaos and its reconstruction, supplemented with contributions from L.A. Aguirre (Brazil), Kevin Judd (Australia), Boris Bezruchko *et al.* (Russia), H.U. Voss *et al.* (Germany) and Natalia B. Janson *et al.* (UK and Russia). Nova Science Publishers. 2003.

G. Gouesbet, G. Gréhan. *Generalized Lorenz-Mie Theories.* Springer-Verlag, Berlin, Heidelberg. 2011. With a preface by James A. Lock (CSU, Cleveland State University).

G. Gouesbet. *Hidden Worlds in Quantum Physics.* Dover Publications. 2013. With a preface by J. Bricmont (Université catholique de Louvain).

G. Gouesbet. Violences de la Nature. L'Harmattan, Paris. 2016.

G. Gouesbet. Violences des Hommes. L'Harmattan, Paris. 2016.

G. Gouesbet, G. Gréhan. *Generalized Lorenz-Mie Theories.* Second Edition. Springer International Publishing, 2017.

G. Gouesbet. Violences des Dieux. L'Harmattan, Paris. 2019. Avec une préface de Keith Moser (Professor of French and Francophone Studies, Mississippi State University) et une postface de Thierry Murcia (Docteur en histoire des religions, Aix-Marseille Université, CNRS).

G. Gouesbet. Violences des Idoles. L'Harmattan, Paris. 2021. Avec une préface d'Issa Asgarally (Docteur en Linguistique de l'université Paris V-René Descartes, Professeur à l'Institut de l'Éducation de Maurice) et une postface de Keith Moser (Professor of French and Francophone Studies, Mississippi State University).

G. Gouesbet, G. Gréhan. *Generalized Lorenz-Mie Theories.* Third Edition. Springer Nature Switzerland, 2023.

Contents

Chapter 1

Introduction

Relativity versus quantum mechanics

It is nowadays a cliché to state that relativity and wave mechanics, more generally quantum mechanics (I essentially restrict myself in this book to the non-relativistic version), resulting from the condensation of the famous two overshadowing Lord Kelvin clouds, pointed out during a 1900-lecture at the Royal Institution of Great Britain, constitute the brightest achievements of human kind in the last century. They form the gleaming pillars of modern physics and irrevocably tell us something most significant concerning the plumbing of our observable universe (such a beautiful plumbing which however is not really and readily able to tell us something significant concerning awareness). Relativity and quantum mechanics are however of different natures, in many respects.

Special relativity is easy to teach, and can be introduced to first year students in the university, relying only on elementary mathematics. General relativity is more demanding, requiring *a prior* extensive mastering of this most attractive calculus named tensor calculus, and therefore appears much more abstract than special relativity. Yet the main ideas, such as the idea of material objects following the geodesics of the space-time, can be conveyed only using plain words, although with a bit of popularization. With a sufficient amount of training, we can even make our intuition comfortable with the relativity realm. When hiking in mountains, you can now definitely feel why you cannot grasp this summit in front of you, although it is flattened on two-dimensional retinal images: It is far in space, for sure, but it is also far in time. After having taught such theories to student for many years, I remember, on the contrary, how I felt most uncomfortable when I had to return to classical mechanical lectures, with these most unnatural, and evanescent, absolute space and time of Newton.

One reason, for this ease to handle relativity, is that it can be deduced by starting from a basic principle (or postulate) which is clear to the mind, or in other words: Clear and distinct to the intuition. Such basic principles, which are clear and distinct to the intuition, I call them first principles, a notion that we shall have many opportunities to refine. For instance, the special principle of relativity states that the laws of nature have to be the same in all inertial reference frames. Its extension in general relativity states that the invariance of the laws of nature should remain true in all reference frames, whether inertial or non-inertial. These are rather technical accounts. But, if we accept the anthropological point of view to attach a human observer to each frame, the meaning is incredibly clear and distinct. It is simply a principle of mediocrity. It means that the laws of nature should be the same whether they are expressed by Einstein or by Bohr, by Alice or by Bob. Although it took a long time since *homo faber* and the beginning of *homo sapiens* to express it clearly, the principle of relativity, which nowadays possesses the taste of a Kantian *a priori*, is something that we would like most reluctantly to challenge. It is a basic principle of physics in the strongest sense, that is to say a first principle. Concerning the second basic principle of special relativity, the one for the constancy of the speed of light in vacuum (I am not going to discuss the equivalence principle of general relativity), it is a postulate whose roots are of an empirical nature (Maxwell's equations gathering together many experimental facts, or Michelson–Morley experiments) and therefore, at least up to now, possibly waiting for an enlightening derivation or a complementary discussion, it must be viewed as an *a posteriori* principle. For this reason, although it is a basic principle of relativity, I do not consider it as a first principle of physics, for the time being. Let us however give it a name: The second postulate, for further use.

A second reason, for our easiness to cope with relativity theories, is that we are dealing with chains of events occurring causally in a space-time arena. Our common sense concepts of space, time, simultaneity are altered, and space and time had to be amalgamated in a four-dimensional space-time continuum, but this does not drive us too far away from our everyday experience. We just have to adapt ourselves a bit. It is therefore not surprising that some authors consider that relativities still pertain to the realm of (although somewhat extended) classical physics. This is the point of view that I shall adopt too in this book.

In deep contrast, with quantum mechanics, we are indeed abruptly jumping in another kingdom, the quantum kingdom. It does not match our intuition of everyday experience, even extended, and as a result lacks of intelligibility. Experimental observational features, through so-called measurements, are not interpreted in a deterministic framework, but rather exhibit an intrinsic indeterminacy, and occur outside of the categories of space and time. What I shall call the classical trio (space, time, causality) is ruled out from the quantum kingdom, in a way that we shall clarify later. In particular, non-locality (and/or non-separability), outside of our classical space-time pictures, is predicted by the quantum mechanical formulation, in agreement with experimental results. Even if we later dismiss the quantum mechanical formulation, non-separability will remain an indestructible experimental feature of the world that, us, human beings, may observe.

Also, may be more important (at least in my mind), there is the fact that the quantum mechanical formulation may be introduced by using basic principles, in an axiomatic way, but none of these principles possesses any *a priori* flavour. Therefore, they do not clearly satisfy our intuition or may even be counter-intuitive. They are actually not first principles in the sense that I introduced previously when I commented the principle of relativity. They are all inspired by experimental data, and therefore they are all *a posteriori*. They all contradict Einstein's belief that *pure thought is competent to comprehend the real, as the ancients dreamed* [2].

I am not going to dispute in this section the fact to know whether science should eventually be always *a posteriori*, meaning that *a priori* statements would be illusory. But the lack of clear and distinct *a priori* principles also, unquestionably, participates to the lack of intelligibility of quantum mechanics. The quantum mechanical formulation is genuinely a mathematical structure, to be applied as a recipe, or if you prefer as an algorithm, with the only aim to *save the appearances* by correctly predicting observational data. It is genuinely an empirical theory. On one hand, you have physical facts (the observations), and, on the other hand, you have mathematics, which are assumed to be able to mimic the physical facts. In the absence of genuine first principles, this is just a juxtaposition of logically unrelated heterogeneous structures. Such an epistemological status is not without any danger, as we are going to see below by analyzing what might be, in the history of human kind, the first attempt to produce a complete, and presumably final scientific explanation of the structure and of the texture of the world.

The beginning of theoretical physics

It is often difficult to trace back the origin of something, in practice because origins are often not reported, or even not reportable, or lost, and in principle because actually all we have is a somewhat continuous flow of events in which some particular occurrences are distinguished by us, possibly in an arbitrary (but convenient) way. Yet, I am taking the risk of assigning the beginning of theoretical physics (in our modern sense) to my old cherished pre-socratic Greeks [3, 4], followed by the acme of Plato, in what we may call the theory of elements. Rather than Plato, Rovelli would have given a privilege to Anaximander [5], but, within a filiation, the most representative individual is sometimes difficult to univocally identify.

The idea that the world is composed of basic elements is often granted to the Persians (on the present territory of Iran), the center from which this idea would have spread to India, China, Greece and to all the Ancient World. According to some sources, the word element would have been introduced by the Ionian school of Miletus, represented by Thales, Anaximander or Anaximenes, but the discovery of elements would be Phenician. The four elements of Empedocles, soon discussed below, are also present in Egyptian cosmology. The existence of elements is however attested in antique Buddhist texts and, more explicitly, by about 600 BC., Lao Tzu, in China, put forward the idea that matter would be composed of five elements (metal, wood, water, fire and earth, usually to be taken in that order), animated by two forces, the yin and the yang. These elements had to play a significant influence in many aspects of the Chinese culture, including in Chinese gastronomy where they are associated with five flavours, namely: Hot, bitter, sour, sweet and salty. This is however not yet theoretical physics, but only a zoological description insofar as the association between the observations and mathematics is lacking, although there is already some amount of interpretation in the description of the observations.

At this stage, in which I mentioned both Eastern and Western cultures, I have to apologize straight away for what is going to go on later in this book. I shall indeed give a privilege to Western references. In such an attitude, there will be no contempt to Eastern cultures. I am actually fond of these cultures. I can manage rather well with Japanese language (fluent for everyday life) and a bit, although in a rudimentary way, with Chinese language, and I am convinced that a synthesis between both kinds of culture "might save the world". However, due to my Western education (I am contingently born in West), I intend to avoid any mistake which would occur in referring to a culture not enough familiar to me. And therefore I am just asking my Eastern friends to possibly, if they like, adapt my arguments to their cultures, feelings, and histories. Considering the deep unity and uniqueness of the human kind, this should not be too difficult. Apologizing again.

With this proviso stated, I may now say that the mathematical elements lacking in the Chinese system (as far as I know) were to be found in what are now called the Platonic solids (although this is a misnomer, as I shall comment). These solids, also known as perfect solids, are convex regular polyhedrons, which can be inscribed in the sphere. There are five of them, and only five of them, a fact whose demonstration is available from the Elements of Euclid (from 300 BC). They are named according to the number of their faces, which are all identical for a given solid. We then have

(i) the tetrahedron, with four faces, also called the pyramid (ii) the hexahedron, with six faces, also called the cube (iii) the octahedron, with eight faces (iv) the dodecahedron, with twelve faces and (v) the icosahedron, with twenty faces.

It seems that these perfect figures were already known to Pythagoras (about 580–490 BC), although this statement is somewhat controversial. The issue is not convincingly attested (except may be for expert historians), an obvious drawback being that no text from Pythagoras reached us. But the statement seems to be reasonably true in view of subsequent testimonies and historical evidences from the literature. Furthermore, Pythagoras would have made a first connection between a series of physical elements and a corresponding series of mathematical perfect solids. For him, the pyramid would have produced the fire, the cube the earth, the octahedron the air, the icosahedron the water and the dodecahedron the sphere of the universe. Here is the beginning of theoretical physics, a beginning afterward discussed again, and eventually extended, by successors.

For Parmenides (born somewhere in the range 520–510 BC, that is to say before the death of Pythagoras, and indeed a Pythagorean), there are only two elements, the fire and the earth and, in his doctrine, the air and the water would be formed by mixing fire and earth. This might be viewed as a progress insofar as the reduction of the number of basic elements could possibly manifest a desire and a trend towards unification (a backbone of physics since at least Newton). But it could also be viewed as a regression since the mathematical contact with perfect solids is lost. The Parmenides justification for using only two elements is that there are only two entities (or two concepts): Being and Non-Being. Zeno of Elea (from about 495 to about 430 BC), a member of Parmenides school, distinguished four "elements", changing continuously from one to the other, namely: Hot, cold, dry and humid, but these "elements", in a modern language, are more properties than objects.

Empedocles (about 484 to 424 BC), another successor of Parmenides, the weird man with sandals made out from bronze, has been much influenced by Pythagoreanism, and has even been accused of having plundered Pythagoras. However, in any case, probably because he has been well documented, even by contemporaries, he will remain the man of the theory of the four elements, actually called four roots by him. The four elements are the fire, the air, the water and the earth but, in contrast with those of Zeno of Elea, they are eternal. The changes observed in the world are then the result of the attraction and of the repulsion of the elements, under the action of two forces: The Love which explains the attraction of the elements, and the Strife which accounts for their separation. Similarly as for the Chinese who attached five flavours to their five elements, Empedocles attached four colors to his four elements, namely white, black, red and ochre. We know that, following the Pythagorean tradition, he also associated the cube with the earth, but it is likely that he actually also associated the other polyhedrons with the other elements, although this is not attested. Philaolaos (presumably born in 470 BC, dead in 400 BC), a Pythagoras follower, from Pythagorean Italy, assumed that the sphere of the world was made of five elements, namely again the four elements (fire, air, water and earth) embedded inside the sphere, plus the fifth element representing the shell of the sphere.

We now leave the fragmented world of the pre-socratic era and arrive to Plato (probably 427–348 BC) whose dialogues reached us, basically as uncorrupted as the works of W.V. Quine in the last century. Plato provided a refined account of the theory of elements, and of the first theory of mathematical physics, based on the perfect solids. He did not discover the existence of these solids (that is why the expression Platonic solids is a misnomer) but, very likely, learnt of them during an about ten years stay he spent in Sicily and southern Italy, possibly from a friend of him, Archytas, who was a senior member of Pythagoras school (to better fix the chronology in mind, just remark that only about 60 years elapsed between the death of Pythagoras and the birth of Plato). In his Timaeus [6], he pushed forward the physical theory of elements, associated with the mathematics of the perfect solids, to some kind of *logical conclusion* (*logical conclusion*: An expression that we shall later encounter again in a slightly different context).

The Timaeus is a theoretical treatise which explains the creation of the world and its constitution. The sensible world (the one which is unreliable, viewed as a degraded copy of the world of Ideals) is created by the demiurge, the blacksmith of the world, by using four elements. This may be proved (at least, Plato believed he did). Indeed, in order to be sensible, that is to say perceptible by senses, the world must be in relation with the two senses most used by human beings, namely the vision which (obviously!) implies the fire and the sense of touch which (obviously!) implies the earth. This being understood, mathematical arguments may afterward conclude to the necessity of the two other elements, the air and the fire.

The mathematical arguments used by Plato again rely on the existence of perfect solids associated with elements (let us recall, the pyramid for the fire, the cube for the earth, the octahedron for the air and the icosahedron for the water). The analysis of these solids is refined by considering triangles forming their faces. The faces of the pyramid, octahedron and icosahedron are formed by equilateral triangles (with four, eight and twenty triangles respectively). The faces of the cube are squares which can be reduced to four isosceles rectangle triangles. Let us also note that the faces of the dodecahedron (which is not associated with any of the four elements) are regular pentagons (in the number of 12). Analyzing further the triangles forming the faces of the four polyhedrons associated with the four elements, noting that equilateral triangles can be built from six scalene rectangle triangles (i.e. with all sides having different lengths), we then see that the mathematical structure associated with the four elements can be built from two kinds of elementary rectangle triangles: Isosceles and scalene. We then may say that these triangles are letters (that we may like today to designate by 0 and 1), that these letters form syllables (the faces), and that syllables form words (the polyhedrons). With such words, we may write the book of the world and of its changes (see notes of Luc Brisson in [6]). Therefore, for Plato, as much later on for Galileo Galilei, the book of nature is written in a mathematical language.

To understand how changes with respect to time may occur, let us set down a principle of conservation: The two kinds of elementary rectangle triangles cannot be created nor destroyed. Therefore, the fire (the pyramid), the air (the octahedron) and the water (the icosahedron), all built on scalene triangles, can transform one into the other. But the earth, the only one to be built from isosceles triangles, can only experience decomposition and recombination processes.

We still need to understand which guide, beyond tradition, justifies that such or such polyhedron is associated with such or such element. As an example, let us consider the case which is the simplest one to discuss, namely the one concerning the association between the cube and the earth. The earth is the element which is the most difficult to move and therefore the most stable. And, among the polyhedrons, the cube is also the most stable, since it can safely rest on one of its flat faces. Therefore, the mathematical counterpart of the earth must be the cube (this is a summary of Plato argument which, in the original text, is a bit more lengthy and elaborate). Similar considerations can be applied to the other elements.

But there is a fifth polyhedron, the dodecahedron. Therefore, there must be a *quinta essentia*, a quintessence (a fifth element), to solve the discrepancy between experiments and theory. Since the four elements already correctly deal with the sensible world, the fifth element must correspond to something which is not familiar on earth. This is what we would call today a theoretical prediction which must have a counterpart in the reality. Hence, concluded Plato, *the god used it for the universe*. In other words, the cosmos itself is built from the fifth element, whose existence is, in a sense, predicted by the theory.

While Plato is pointing a finger up toward the sky, his student, Aristotle (384–322 BC) lowers his hand down to the earth, as in the famous painting "School of Athens" by Raphaël. In his chief cosmological treatise [7], what is on earth, in the sublunary world, is again made out from the four elements. Then there is the realm of the non-perishable heaven, peacefully rotating in perfect constant circular motion around the earth, to which the fifth element, or ether, is attached.

The theory of elements afterward survived all along the Middle Age (remember for instance its use by Kepler, to build his first System of the World [8, 9]), up to modern chemistry where the number of elements increased from five to more than one hundred (nothwithstanding the zoo of elementary particles), with the concept of forces (like Love and Strife) remaining a must. Actually, the end of the theory of the ancient elements is due to Lavoisier who empirically reduced the fire (and its avatar, the phlogiston) to a combustion process, and empirically demonstrated that the air and water, these ancient elements, were made out from sub-elements, which are actually modern elements of modern chemistry.

Nevertheless, the strong analogy between the antique theory of elements and our current scientific knowledge is striking. In both cases, we have physical elements (*elements of reality*) put in parallel with a mathematical structure which is supposed to sustain the empirical world. The economy of the mathematical structure is as strong as it was possible: All the huge, somehow incredible, diversity of the world of appearances, is reduced to only two kinds of triangles in the world of Ideals. Symmetries, so essential to our current physics, are present in the various symmetries of the perfect solids. And we also have basic principles, such as a law of conservation of two basic kinds of triangles which helped to understand the evolution of the sensible world, this embodiment of the out of time eternity.

The theory of elements has been discussed by Heisenberg [10]. His first reaction was that this theory was absurd. He was worrying why a philosopher like Plato could get lost in such speculations, invoking however the excuse that the old Greeks did not possess the detailed empirical knowledge which is available to us nowadays. However, in parallel, the idea that we should eventually reach mathematical forms to describe the elementary parts of matter exerted on him a certain fascination (Chapter 1 of [10]). But eventually (Chapter 20), he admitted that elementary particles can be compared to the regular bodies of Plato's *Timaeus*. And, elsewhere [11], he claimed that he reached the conviction that it was not really possible to deal with the modern atomic physics without knowing the Greek philosophy of nature. Another relevant reference from Heisenberg is [12].

Anyway, as we know, Plato's theory suffers from essential limitations. On the physical side, the identification of elements, in the absence of sophisticated enough chemical or physical techniques to analyze the constitution of matter, was poor. On the theoretical side, the mathematics available in that epoch were underdeveloped. This last feature however may be viewed as fortunate because, underdeveloped as they were, they could adapt themselves to an underdeveloped level of physical knowledge. There was no mismatch between two kinds of underdevelopments, the physical one and the mathematical one, so that my old Greeks have shown a way, a track to follow.

But, in my opinion, the most striking limitation is of an epistemological nature. The theory of elements uses basic principles, which are somehow *ad hoc*, put forward to permit the construction of a correspondence between physical and mathematical ingredients. They are not first principles in the sense discussed in the previous section, and their acceptance is conditioned by the mesmeric bedazzlement provoked by the beauty of mathematics. I shall call this the Greek syndrome. To be more specific, we are in danger of suffering of the Greek syndrome whenever we are content when we find a correspondence between physical observations and mathematical structures, and stop asking questions when the correspondence just works. To be very provocative, I would say that this is no more physics, but something rather akin to numerology.

Therefore, from such an epistemological point of view, the theory of elements is similar to quantum mechanics (no real understanding, but just a complacency, at least from some people, and for some people, once a satisfactory correspondence has been exhibited), but is not similar to relativity where we can exhibit a first principle satisfying our intuition, not only a basic principle. I claim that this might be a weakness of our present quantum mechanics, which could possibly be infected by the Greek syndrome. This position may be found extreme by some readers (after all, quantum mechanics is operationally much more efficient than the theory of elements) but, whether we can possibly escape from it or not, I do not dare to say in this section: This is an issue which will run throughout this book.

The shared perception of the unintelligibility of quantum mechanics

As we may see, the lack of intelligibility of quantum mechanics ultimately originates from two sources (i) the strangeness of quantum events, so remote from our everyday experience, making our physical intuition hopeless and (ii) correspondingly, the lack of first principles, making our intellectual intuition disarmed. Complaining students, professors, researchers, could after all reassure themselves by assuming that they just do not know enough, that their lack of knowledge keep them erring in doubt and in darkness, and that, most surely, there must exist clever enlightened people to which the truth has been clearly and distinctly revealed. This is however contradicted by many testimonies from such clever people, including some Founding Fathers.

Writing to a friend, Einstein declared that the more success the quantum theory had, the sillier it looks (e.g. [13]). Schrödinger expressed his despair by saying that, if these damned quantum jumps indeed existed, he would regret having been ever involved in quantum theory. Niels Bohr compassionately replied: But, us, we are very grateful to you having dealt with it, for your wave mechanics represents, due to its clarity and mathematical simplicity, a tremendous improvement with respect to the previous formulation of quantum mechanics [10].

Laments from Einstein or Schrödinger, which have been opponents to quantum mechanics, at least to what is called the Copenhagen interpretation, may however be found of little value, just because they indeed have been opponents. After all, what opponents cannot understand and accept could be understandable and acceptable for proponents. Such proponents should just express themselves clearly enough. And, as we know, arguments of opponents can be biased, even in a vicious way, as we are used to by watching and listening some intellectually dishonest people.

But we also have Feynman, a defender, who stated that no one could explain more than that had been, up to now, explained [14], and also that he could *safely say that nobody understands quantum mechanics*, a famous statement indeed [15–18] adding that if someone pretends he understands quantum mechanics, then it means that he actually did not understand anything of it (no archival reference provided for this last part of the quotation that I merely picked up from my personal central neuronal system). As another example, we have Gell-Mann speaking of *quantum mechanics, that mysterious, confusing discipline, which none of us really understands but which we know to use*[19]. We also have Niels Bohr, may be the most emblematic defender, telling that anyone who would pretend possibly thinking on quantum physics without feeling any vertigo just demonstrate that he did not understand it [20]. And Van Fraassen acknowledged that, *to a traditional mind, quantum theory is perplexing — and we all start with traditional minds* [21]. But, even after having been learning, studying, thinking, and after having spent a lot of time to escape from tradition, the perplexity of the mind is still there, may be not for a few inspired individuals, but at least for most of us.

The many-interpretations of the quantum kingdom

In face of such a state of affair, it is not a surprise that so many interpretations, often conflicting, flourished. Although there is a dominant interpretation, the Copenhagen one (itself however with various sub-interpretations), other interpretations appeared from the very beginning, some of them proposing alterations of the mathematical formulation already available from the late twenties of the last century, others basically accepting this formulation, but managing to make it speaking another physical language. Such discussions on the interpretative aspects of the theory never ceased and are still active nowadays, even if we remain in the nonrelativistic framework (small enough energies to avoid creation and annihilation processes of particles) to which most of this book is devoted. They produced indeed a huge literature in parallel with more routine, albeit much sophisticated, computational and experimental works. It is likely correct and fair to state that no theory, even relativity theories, ever produced such an amount of commentary exegeses (at least among scientists, skipping the existence of various kinds of laymen). As a consequence, we have to deal with a many-interpretations quantum world (which is not the same as a quantum many-worlds interpretation, like the one of Everett, to be mentioned later).

Most often, in university lectures and most praised textbooks, the problems of interpretation (or the problem of interpretations) are swept under the carpet, in favour of a training to axiomatic basic principles and to the mastering of associated computational techniques. This is fair for the sake of efficiency (predictions of the quantum mechanics of the textbooks have not yet ever been challenged by any experiment) but possibly dangerous as far as research is concerned (our physics is far from being achieved, and alternative proposals might contain, at least, a germ of truth). In particular, as we shall see, we can learn a lot by examining hidden-variables theories.

Positivism and interpretations

A seemingly good reason to dismiss the issue of interpretations is to convoke a philosophical posture, known as positivism. We shall have to discuss positivism more extensively when appropriate but, for the time being, just recall that, according to it, the aim of science should be to achieve correct predictions, even of a statistical nature, rather than discovering some kind of ontology, i.e. what things really are. We are here facing an alternative which, after asking the question: *What is science?*, is expressed by Van Fraassen [21] by two other questions: *What is happening?*, and: *What is really going on?*. The first question concerning what is happening, or more generally concerning what is going to happen, is meaningful for positivists. The second question is meaningless for them.

We can and shall draw many objections against positivism but, in this section, I shall be content to tease the proponents of positivism, many quantum mechanists indeed, by remarking that, positivism, according to the above cursory definition, tells us that we should be satisfied if we succeed to *save the appearances.* A variant is the constructive empiricism advocated by Van Fraassen [21] according to which *the aim of science is not truth as such but only empirical adequacy, that is, truth with respect to the observable phenomena.* Indeed, saving the appearances is empirically adequate.

From this restricted point of view, the success of quantum mechanics to describe the systems in the world is on the same footing that Ptolemy astronomy to describe the System of the World. As we know, Ptolemy used a complicated architecture of epicycles and deferents, what has been called an astronomy of eccentrics and epicycles by Duhem [22], to describe the celestial motions and, to better approach the observations, it is sometimes believed that it has been necessary to add epicycles on epicycles; at least this legendary way of speaking is well established, even if it is erroneous. It is less known to a significant number of scientists that the heliocentric system of Copernicus relies on the same machinery than the geocentric system of Ptolemy, and that it produced, as a result, a system which is just as complicated than Ptolemy one, or even more complicated [8]. Adding epicycles on epicycles is certainly not a good idea to make good science, and I must acknowledge that quantum mechanics did not proceed in that way, but, from a positivistic point of view, it should not matter: Empirical adequacy should be sufficient for us to be satisfied...

Hence, if we hold fast on a positivistic attitude, considering all the predictive successes of quantum mechanics, we have to be satisfied with its mathematical formalism, and with its use in the physical world. Nothing more is required and nothing more has to be demanded. The formalism contains the essence and the quintessence of the theory. Any concern for statements directed to the issue of interpretations leads to consequences which have to be viewed as superfluous appendages, meaningless superstructures or even, if we are rude enough to take the risk of offending, as metaphysical digressions. We may then charitably, or possibly condescendingly, accept that some alien individuals go on dealing with the issue of interpretation, the poor ones who are philosophically inclined. But, of course, since philosophers never agreed, they should arrive to different choices motivated by personal philosophical biases and/or prejudices. Surely, this should not be a way of making science.

We may conversely feel that the attitude we just described, the one concerning what is the right way of making science, might be a bit extreme, or even dogmatic. We may like to state that it is reductionist and that, by claiming that some questions are meaningless, it could be dangerous for the future of science. We may express our reluctance to it insofar as it manifests a renunciation to the desire, most legitimate for human beings, to understand what is going on (and why), not simply how it is going on. Expressing the idea that we have to be content, for practical use, to deal with formalisms and computational algorithms, that we indeed have solely to compute, make experiments, and to shut up on the rest, is also a renunciation to this other idea that the human mind could be able to grasp the most inner secrets of nature, at least some pieces of them, even if it is overwhelming difficult to achieve it.

Admittedly, the positivistic renunciations are well fitted with a certain spirit of the time, a mood opposed to any question which would not be of an immanent nature, a process of dissolution of the metaphysics, which after the excessive commitments of Descartes, was basically initiated by Locke (1632–1704) and its empiricism, followed by Berkeley (1685–1753) and Hume (1711–1776), influenced by Locke, and then also by Kant (1724–1804), influenced by Hume. This venerable tradition may be pursued with Nietzsche (1844–1889, the last date being the one of his mental collapse), and the dramatic development of the neo-positivist school of the Vienna Circle, without forgetting Quine and Wittgenstein.

Yet, we should still allow ourselves to ask so-described meaningless questions, if only because we have to demand for our liberty of remaining free of thinking, and speculating, as a principle of human dignity, but also as a pragmatic commitment, just because it could possibly lead us to meaningful developments. Elsasser [23] stated that, notwithstanding the silence which reigned these last years, the questions of interpretations must still occupy the physicist. Although this was enunciated more than half a century ago, it is still valid nowadays. And, may be, the best way to end this section is to quote the last sentence of a book by Jammer [24], a sentence actually borrowed from the French moralist Joseph Joubert: *It is better to debate a question without settling it than to settle a question without debating it.*

The end of determinism

Questions of interpretations had to struggle with many problems, such as the issues of determinism, objectivity, realism or non-locality, or the famous measurement problem (the most formidable problem in quantum mechanics) which, still nowadays, divides the quantum mechanics community into a Babel tower confusion.

Recurrently, the proposed answers aim to bridge the gap, at least partially, between the quantum world and the classical concepts, and to reconcile them as much as possible. In this introductory chapter, it is most convenient to restrict oneself to the issue of determinism, postponing the discussion of other issues to later chapters, just because it is the easiest, and the best known to a large audience.

Of course, this does not mean that it is the most important issue at stake. For instance, Squires [13] remarks that the measurement problem is a more powerful motivation (to think more and to invent alternative interpretations) than the desire for a restoration of determinism, this determinism which is knocked down and driven to an end by quantum mechanics.

The end of determinism has been quite a shock to many, in particular to the most emblematic one among all of them, namely Einstein who, in a celebrated statement, expressed his reluctance to a gambling God (*Der Hergott würfelt nicht*). In a letter dated May 1, 1924, to Paul Ehrenfest, in the framework of the famous Bohr-Einstein debate, he also wrote that *a final abandonment of strict causality is very hard* (for him) *to tolerate* (see [24]). Einstein's conviction that determinism is right may also be found reflected when he said that *everything is determined, the beginning as well as the end, by forces over which we have no control. It is determined for the insect as well as for the star. Human beings, vegetables or cosmic dust, we all dance to a mysterious tune, intoned in the distance by an invisible player* [2].

The reluctance to give up causality may also been detected in a quotation from Popper who considered the principle of causality as a metaphysical principle. Do not believe here that there is an accusation of metaphysicism concerning causality and that, for this reason, causality should be given up. On the contrary, in a critique addressed to quantum mechanics [25], Popper proposed that we should depart from the indeterministic metaphysics (so popular nowadays, he said). For Popper, what distinguishes this indeterministic metaphysics from the determinist metaphysics which was previously in fashion is less an increase of lucidity than an increase of sterility.

The restoration of determinism

However, is it assured that quantum mechanics tolls the bell for determinism? At the present stage, we must state that the answer is: No. So, would it be possible to propose an interpretation of quantum mechanics which would restore determinism? At the present stage again, the answer is: Yes. Before giving arguments on these questions and providing answers, let us first discuss an example, borrowed to the history of sciences, which created a well known and much commented precedent, and which is worth recalling.

For this example, let us consider the thermodynamics of a gas, a perfect gas actually, for the sake of simplicity. To macroscopically describe the state of one mole of such a gas in equilibrium, we may use three macroscopic thermodynamical variables, which are easily observable by using classical instruments, namely pressure denoted by P, temperature denoted by T, and volume denoted by V. These macroscopic variables define classical observables which are related by the state equation of perfect gases, reading as $PV = RT$, in which R is the constant of perfect gases. Due to this relation, a state of our mole of perfect gas can be defined by using only two observables, say pressure and temperature, so that, in technical terms, the space of states forms a two-dimensional manifold, spanned by our two chosen observables. At this macroscopic thermodynamical level of description, our understanding of the situation is completely deterministic. It simply relies on the deterministic character of our state equation. Apparently, this equation tells us nothing on the evolution of the system since time is not involved in it. But the fact that time is not involved indeed actually tells us something on the evolution of the system. It tells us that, if the system is in equilibrium at a certain time t, it will remain at equilibrium at any later time. This is after all what we should expect of an equilibrium. Nothing happens in equilibrium (hence time is not useful, and we might even say that it does not flow any more). But, if nothing happens in equilibrium at a certain time t, we may predict that nothing will happen at later times. In the present case study, no evolution generates no evolution.

But, there is also, below, a sub-thermodynamical level made out of atoms and molecules (say underlying entities), with a junction between the two levels brilliantly revealed by the works of Boltzmann. If we consider (or provisionally assume) that these atoms and molecules are of a classical nature, then we may say that the trajectories of the underlying entities are deterministic but, in practice, we have to consider them as random, or more elegantly stochastic (molecular chaos hypothesis). However, such a stochasticity is not of an intrinsic nature. It does not conflict with determinism, but simply points out to our ignorance and to our inability to deal with a too huge amount of data. Conversely, with a deeper point of view, namely the quantum mechanical point of view, we have to consider that the underlying entities exhibit an intrinsic indeterministic character, reflecting their quantum nature. We are then facing two levels of description of the reality, a macroscopic thermodynamical level which is deterministic, and a sub-thermodynamical level which is indeterministic. Of course, the deterministic character of the upper level is the result of averages over many lower level indeterministic processes. But one important lesson from this example is that determinism and indeterminism do not necessarily conflict. We may have an interplay between determinism and indeterminism, associated with two different layers of our description of the world.

Another lesson may be drawn from the concept of observability. At the upper level, we are dealing with observables, in the most common sense meaning of the term. When the idea of the existence of an atomic sub-level has been introduced and developed to a coherent theory, by Boltzmann, the underlying entities of the sub-level were not observables and, for many who relied on a positivistic attitude, they were of a metaphysical nature. It took a bit of time for these entities to be accepted as genuine components of the reality. The analysis of Brownian motion first provided an indirect, although convincing enough, evidence, and more evidences came later which, nowadays, with our contemporary technological arsenal, are out of doubt. But there has been a time were the variables associated with the sub-level were hidden variables. The lesson is then that hidden variables should not be necessarily rejected from our theories, and that variables which are hidden at some time may be revealed at later times.

What is the corresponding situation in quantum mechanics, the sub-level of the above example? Well, we may imagine a sub-sub-level of the above example, that is to say a sub-quantum mechanical level, made out from other hidden variables. Also, as the above example demonstrated, we should not worry whether these variables are observable or not (after all, they could later become observable). Furthermore, if we assume that the hidden variables are of a deterministic nature, then the determinism would be restored at the sub-sub-level. The connection between determinism at the bottom level and indeterminism at the medium level could then be explained by assuming that the bottom deterministic level is indeed deterministic, but that some of the quantities involved, such as initial conditions for sub-sub-trajectories are unknown. Then the indeterminacy at the medium level is not of an intrinsic nature any more but is the result of a lack of knowledge of the actual values of hidden variables at the bottom level, a classical kind of ignorance indeed, the same kind of ignorance that we meet when we describe the trajectories of the medium level in a classical way. We are facing, in such a vision, a sub-molecular chaos. The meaning of the word chaos can be the same than in the molecular chaos hypothesis but also, as we shall see, it can be given the more precise meaning involved in modern chaos theory. And we have then a picture of the world in terms of a pile of three layers, from the top deterministic level, to the medium indeterministic level, and to the bottom level where determinism is restored. Of course, we could pursue this construction game, piling layers over layers, eventually making the issue of determinism quite undecidable (such a construction game is evoked in the literature, but I must confess that I am reluctant to seriously advocate it). Anyway, this is enough to set the stage on which we shall play the hidden variables performance of the quantum world.

Life and death of hidden variables

A brief history of hidden-variables theories in quantum mechanics will help us to discuss whether the debate is closed, or still open, or at least to provide insights on this issue. The emergence of hidden-variables theories is coextensive to the development of the now usual quantum mechanics, at a time when, while trying to find a right track, researchers generated a profusion of ideas. However, in 1932, von Neumann [26] provided a proof of impossibility of hidden variables in quantum mechanics, claiming the death of hidden variables and apparently closing the debate for ever, more or less at the very beginning of the quantum mechanical story. The widespread mood generated by this proof of impossibility is well reflected by London and Bauer [27] asserting that *one can convince oneself that statistical distributions, such as are given by quantum mechanics and verified by experiment, have such a structure that they cannot be reproduced by hidden parameters.* However, later on, it has been firmly established that the so-called von Neumann proof of impossibility was invalid, or even circular, so that the mood changed. Thirty five years later, Kochen and Specker [28] could start a famous paper with the following statements: *Forty years after the advent of quantum mechanics the problem of hidden variables, that is, the possibility of imbedding quantum theory into a classical theory, remains a controversial and obscure subject. Whereas to most physicists the possibility of a classical reinterpretation of quantum mechanics remains remote and perhaps irrelevant to current problems, a minority have kept the issue alive throughout this period.* This was a matter of introduction but, in the bulk of their paper, Kochen and Specker proceeded further by establishing another proof of impossibility, a proof that, as the reader might have guessed, became eventually dismissed.

That proofs of impossibility did not actually reach their target is evident from the existence of a logically consistent hidden-variables theory due to Bohm [29, 30], in the filiation of previous efforts by Louis de Broglie, a theory which has been published 20 years after von Neumann and fifteen years before Kochen and Specker. Bohm's theory indeed gives a not yet ended life to hidden variables. Of course, such a life is made vivid by the fact that von Neumann, Kochen and Specker, and other proofs of impossibility, do not apply to Bohm's work. Although it is logically consistent, Bohm's theory is not found satisfactory enough by most of quantum mechanists, for various reasons which are not good reasons, and that we shall have the opportunity to dismiss. In 1980, Bohm [31] could write: *The question to whether there are hidden variables underlying the quantum theory was thought to have been settled definitively in the negative long ago. As a result, the majority of modern physicists no longer regard this question as relevant for physical theory. In the past few years however, a number of physicists, including myself, have developed a new approach to this problem, which raises the question of hidden variables again.*

To exemplify the confusion reigning in this domain, let us remark that, only three years later, Wheeler [32] could still write: *Is there not some underground machinery beneath the working of the world which one can ferret out to secure an advance indication of the outcome? Some secret determiner, some hidden-variables? Every attempt, theoretical or observational, to defend such a hypothesis has been struck down.* But what about Bohm's theory which was not dead, and that Wheeler failed to explain us why it was struck down? Nowadays, after the subsequent developments of the story, after Bell's inequalities, and all the non-locality stuff and associated experiments, that we shall have to examine, there is again a huge majority of physicists who would agree with Wheeler statements, made 25 years ago, and who would claim that the debate is over, that Ψ is complete, and that there is nothing hidden beyond Ψ. But, again, what about Bohm's theory, which has been indeed dismissed and rejected, although logically consistent and therefore alive, and which is still, for the time being, a provocative skeleton enclosed in the closet?

Whether eventually the debate is to be considered as close or open is not a matter for an introductory chapter, and is better suited to a conclusive chapter, after having made a comprehensive enough tour in the quantum landscape. But here is the right place to discuss both terms of the alternative. Let us first assume that the debate is closed. If such is the case, then, nevertheless, the one century examination of the issue was not a loss of time, first because concluding that the affair is over is a definite and useful result, and second because the examination of the issue itself provided many opportunities to deepen and reinforce our familiarity with the interpretative problems raised by quantum mechanics, and therefore deepened our understanding of quantum mechanics itself. Conversely, if the debate is still open, then more works are required, and a refined knowledge of the past of the story is compulsory. In both cases, wandering in what some people could call the marsh of hidden variables is at least a good exercise, even if it makes us splashing around. But it could also possibly be more than an exercise, and reveal itself as a lane to royal avenues for the future of physics.

The never-ending future of physics

Some people recurrently give way to the arrogant temptation to claim that we have reached the end of our trip in the physical world. More than 2000 year ago already, Aritotle, contemplating his encyclopedic achievement, believed that, apart from the need of some minor refinements and complements, he had told all what was to be told (see Pellegrin's introduction in [33]). It is however not sure whether Aristotle always maintained such an optimistic auto-appraisal. Indeed, it has been said that Aristotle, watching the tides, after having observed them for a long time, and unable to explain them, hurled him in despair into the sea, and voluntarily drowned himself [34]. In his principles of philosophy, Descartes [35] similarly stated that he can demonstrate, by an easy count, that there was no phenomenon in nature whose explanation had been omitted in his treatise. Also, one century ago, there was the over-optimistic assertion of Lord Kelvin, with his two clouds overshadowing the landscape, but soon to be dissipated. Nowadays, it seems that the same kind of mistake is put forward again. With the superstring theory, we should soon possess the Holy Grail of reconciliation of quantum mechanics and of the Einstein theory of gravitation, the famous still hidden "theory M", the Theory of Everything. If you believe in God, how could you imagine that, only after four centuries of use of the so-called scientific method, we would have succeeded in exhausting the Power of Its Imagination? And, if you do not believe in God, how could you imagine that, only after four centuries of use of the so-called scientific method, we would have succeeded in understanding what is the world in which we are embedded, My Enigma, and Yours?

I am not equipped to deeply discuss the superstring theory (or better said the superstring theories), but I may possibly rely, pushing a default button, on the analysis of the situation recently made by an unquestionable expert, namely Smolin [9]. He concluded that, after more than thirty five years of continuous developments, with the efforts of more than one thousand of researchers, among the brightest ones: We have failed. Defenders and most actors of the superstring theories might feel offended by such a statement, which might look as too much definitive. But there could be something true in it, at least according to Smolin. It might well also be that the superstring theories are suffering from a Greek syndrome. If we believe that there is some kind of rationality in this universe, and that there is at least some amount of rationality which can be properly expressed with mathematical structures, then we should expect (we must expect) that there is a mathematical structure, for example the group E8, which would sustain the existence and the properties of so-called elementary particles, in the same way that the perfect solids sustained the existence and properties of elements. Even if successful, the exhibition of such a mathematical structure could only possibly to be viewed as another manifestation of the Greek syndrome, that is to say the fact to mimic physical observations with mathematics, without explaining anything. This might be the ultimate fate of our science, an unsurpassable limit produced by the organization of our brain.

But, actually, it might be that we have failed from the beginning, even if this eventuality may look implausible to many. It might be that, somewhere, we followed the wrong route at some insecure and misleading bifurcation. It might be that, dazzled by the many indisputable successes of quantum mechanics, we have been running forward, computing and experimenting, without questioning any more the so beautiful messages delivered to us by the so clever Founding Fathers. Smolin again, in his book, listed a list of problems which are vital to the future developments of physics. One of them is to solve the problems of the foundations of quantum mechanics, by giving a meaning to the theory as it stands today, or by inventing instead a new theory which would have a clear meaning. In particular, as a way to attack these problems, he expressed the possibility of a new interpretation of the theory, that is to say a new way to read the equations, which would be realist, in such a way that measurement and observation would not play any more any role in the description of the fundamental reality which would run outside of us, even if we do not observe it.

In any case, we have to admit that our physics is in bad shape. The hope that we could soon have to our disposal a general enough and efficient enough theory is fading away. And we are left with much more than two clouds. We are left with dark matter, dark energy, the cosmological constant and its reconciliation with the value predicted by quantum mechanics, differing by so many orders of magnitude, and with many other problems. It is not ridiculous to believe that, by digging more, in particular by digging more the field of hidden variables, further insights might be obtained. It might be beneficial, or even a necessity, for the potential final success of our scientific enterprise. In such a situation, it is compulsory that some people take the time and the risk to revisit seemingly established results and concepts. I express the hope that this book, that I have been forced to elaborate (because I am not able to stop asking questions), both for scientific and psychological reasons, could be of some help to someone (one would already be great).

Difficulties with the literature on hidden-variable theories

Besides hopefully providing new information and results or, may be, better said: Speculative insights (in the two last chapters), this book essentially constitutes a review which should help the reader, in particular the newcomer, to better find his way in a much fractal-like ramified field. Indeed, I found the literature on hidden variables very extensive, enormously huge, very far from always being easy reading, and often rather obscure and sometimes even discouraging. I hope I have nevertheless been able to pave the way to those who would like to deepen further the matter. Clearly, I have made the choice to more extensively discuss some issues which seemed to me more relevant. But, without pretending to full exhaustiveness, I tried to direct the reader to all aspects I am aware of, even if it is only by providing entry references.

The exposition of the hidden-variables topic is furthermore made a bit more complicated because it is not solely of a scientific nature, in a strict sense, but also incorporates many other aspects, creeping below the surface, pertaining to metaphysics, religions, philosophy, epistemology, history of sciences, societal features, ... i.e. to many various ingredients centered on human beings, rather than on the world outside of us. As a single exemplifying matter of fact, Belinfante [1], in his famous survey (reviewed by Ballentine [36]), stressed the implications of the issue for religious beliefs as follows. On one hand, we can feel an harmony between the usual indeterministic interpretation of quantum mechanics and the belief to God, because we can then understand that, even by complying with the laws of nature He has erected, He still have the power of making repeated decisions concerning the development of the world when time goes on, decisions that creatures could not predict. On the other hand, in a completely deterministic world, God must seemingly be absent (a discussion to be refined later).

In the present book, such aspects of the topic cannot be omitted, because they play a significant role from the point of view of the theory of knowledge, and from the point of view of any theory of production of knowledge. They may also help us to take care of prejudices which could spoil our way of playing with concepts. And, in any case, as Einstein pointed out, *an interest in philosophy makes one a better scientist* [2].

Organization of the book

The book is organized as follows. We have started with an introductory chapter, something similar to the overture of an opera, in which the main themes, to be later developed, have been presented. Chapter 2 recalls us a background in classical mechanics, essentially the Hamilton–Jacobi's formulation of classical mechanics, a formulation having played an essential historical and conceptual part in the development of wave mechanics, and of hidden-variables theories, and which, at the end of the book, will prove valuable again for some final, albeit speculative, conclusions. Although the reader entering this book is supposed to possess some minimal knowledge in quantum mechanics, a background in this topic is recalled, for the sake of accuracy, in Chapter 3. Postulates defining the formalism are provided, and commented, and the Copenhagen interpretation associated with the formalism is discussed. Chapter 4 is devoted to various objections which have been raised against quantum mechanics and which determined the search for alternative theories and/or interpretations, namely hidden-variables theories. Chapter 5, strongly correlated with Chapter 4, serves again the same kinds of dishes, but in a more philosophically oriented framework. Being at this stage convinced (or feigning to be convinced) that the objections raised may have at least a bit of relevance, we are ready to understand the efforts of those, like Louis de Broglie, and later on Bohm, who attempted to get rid of these objections by introducing hidden variables, and developed what we shall call causal theories (double solution, and pilot wave), which are hidden-variables theories. A brief history of causal theories is then sketched in Chapter 6, giving us a general view of the chronology of the events. Chapter 7 is thereafter devoted to the double solution and Chapter 8 to the pilot wave. Other hidden-variables theories are discussed in Chapter 9. Actually, there have been proofs of impossibility and possibility proofs for the existence of hidden-variables theories, which have lead to a rather confusing situation which is reviewed in Chapter 10. Two main issues are connected with quantum mechanics, hidden-variables theories, and quantum experiments, namely contextuality and non-locality. These are discussed in Chapters 11 and 12 respectively. It is there that we shall encounter the EPR-paradox, Bell's inequalities and Bell's theorems. The amount of knowledge gained is then sufficient to propose a clear classification of hidden-variables theories provided in Chapter 13. Stochastic quantum mechanics which is akin to hidden-variables theories, but deserves a special mention, is separately studied in Chapter 14. With Chapter 15, we return to non scientific discussions, to prepare the stage for the last acts. With Chapters 16 and 17, we are entering a realm containing some ingredients which are of a speculative nature. More specifically, Chapter 16 discusses a main issue, the one to know whether and how we can gain a genuine understanding of quantum mechanics, ending with a statement which should have been obvious, namely the fact that Newtonian trajectories of matter points do not exist, and whose consequences will have to be drawn. Chapter 17 provides a coup de grâce, a coup de grâce for who or what? This is something that I am not going to tell now; but there will be a loophole to avoid this coup de grâce. Chapter 18 is a conclusion.

Chapter 2

Background in Classical Mechanics

The trekking for mountaineers

PEDESTRIANS ($\sim\sim\sim\sim$) should go to the following section.

With this brief chapter, I introduce a background in classical mechanics which shall serve us to better understand the origin of some hidden-variables theories (the ones from Louis de Broglie and Bohm), that we shall call causal theories and, much later on, to criticize them. Other ingredients regarding classical mechanics will be introduced outside of this chapter when we shall need them. We know that classical mechanics can be declined under four different formulations, which are mathematically and empirically equivalent. These are the Newton's, Lagrange's, Hamilton's and Hamilton–Jacobi's formulations. We only need to discuss Newton's and Hamilton–Jacobi's formulations, the other ones being irrelevant for most of my purpose. Discussing Newton's formulation is fast. It is just sufficient (and necessary) to recall that, using the basic law telling us that force is equal to mass multiplied by acceleration, and integrating with initial conditions, we can build trajectories of matter points.

We however need to be a bit more eloquent with Hamilton–Jacobi's formulation (see for instance Louis de Broglie [37], Blotkhintsev [38], Landau and Lifchitz [39], and Holland [40]). Hamilton–Jacobi formulation of non-relativistic classical mechanics of a matter point relies on an equation, that I shall call Hamilton–Jacobi equation, reading as

$$-\frac{\partial S}{\partial t} = \frac{1}{2m}\left(\frac{\partial S}{\partial x_j}\right)^2 + V \tag{2.1}$$

This equation allows one to study the motions of a particle of mass m in a potential $V = V(x_j, t)$. The x_j's denote Cartesian coordinates and t is the time. The field $S = S(x_j, t)$ is a real field that I shall call the Jacobi's field. Equation (2.1) has to be complemented by two other equations reading as

$$W = -\frac{\partial S}{\partial t} \tag{2.2}$$

$$p_j = \frac{\partial S}{\partial x_j} \tag{2.3}$$

in which W is the energy and p_j is the momentum. From Eq. (2.2), we see that S is an action (energy multiplied by time) and, from now on, we may call it the action. Also, inserting Eqs. (2.2) and (2.3) in Eq. (2.1), we see that we obtain $W = T + V$, which should be enough to convince us of the equivalence between Newton's and Hamilton–Jacobi's formulations.

For a conservative motion, the energy (that we denote E in that case) is constant along each particular motion, and Eq. (2.2) implies

$$S(x_j, t) = S_0(x_j) - Et \tag{2.4}$$

Inserting Eq. (2.4) into Eq. (2.1), we obtain

$$\left(\frac{\partial S_0}{\partial x_j}\right)^2 = 2m(E - V) \tag{2.5}$$

We now consider the locus of the points for which S_0 possesses a given value C_0:

$$S_0(x_j) = C_0 \tag{2.6}$$

Equation (2.6) shows that the locus is a time-independent surface. There is one surface, and only one, containing a point P of space, according to $C_0 = S_0(x_j(P))$. The whole space is therefore filled by a set of motionless surfaces forming what I call the Jacobi's static field. From Eqs. (2.3) and (2.4), we have

$$p_j = \left(\frac{\partial S}{\partial x_j}\right) = \left(\frac{\partial S_0}{\partial x_j}\right) \tag{2.7}$$

Therefore, p_j is the gradient of S (and S_0). This means that trajectories are orthogonal to the surfaces S (and to the surfaces S_0). Next, we consider the locus of the points for which the action S possesses a given value C:

$$S(x_j, t) = C \tag{2.8}$$

Equation (2.8) shows that the locus is still a surface but which now depends on time. When times goes on, the surface moves and, in general, experiences a deformation. For a given time t, the moving surface $S(x_j, t) = C$ coincides with a motionless surface $S_0(x_j) = C_0$, according to, from Eq. (2.4): $C = C_0 - Et$. Therefore, when time goes on, the moving surface $S = C$ sweeps over all motionless surfaces $S_0 = C_0$.

We now consider a fictitious point P, pertaining to the surface $S = C$, and therefore moving with it, with the constraint that its displacement remains orthogonal to the swept surfaces $S_0 = C_0$. The velocity of the moving surface may then be defined as

$$w_j = \frac{dx_j}{dt} \tag{2.9}$$

in which dx_j is an infinitesimal displacement of the point P. But, we have

$$\frac{dS}{dt} = \frac{dC}{dt} = 0 \tag{2.10}$$

that is to say

$$\frac{\partial S}{\partial x_j}\frac{dx_j}{dt} + \frac{\partial S}{\partial t} = 0 \tag{2.11}$$

leading to

$$p_j w_j = E \tag{2.12}$$

But, w_j (modulus : w) is colinear to p_j (modulus : p). Hence, with E positive, we obtain

$$w = \frac{E}{p} = \frac{E}{\sqrt{2m(E - V)}} \tag{2.13}$$

We are therefore facing two different velocities (i) the velocity $v = p/m$ of the material point and (ii) the velocity $w = E/p$ of the fictitious point P.

Finally, inserting Eq. (2.13) into Eq. (2.5), we obtain

$$\left(\frac{\partial S_0}{\partial x_j}\right)^2 = \frac{E^2}{w^2} = p^2 \tag{2.14}$$

Other elements of classical mechanics will be later introduced when appropriate, but for the time being, we do not need knowing more. The situation is pictorially represented in Fig. 1. Figure 1(a) deals with Newton's formulation, relying on the existence of trajectories (only one of them is represented), while Fig. 1(b) deals with Hamilton–Jacobi's formulation in which we have both trajectories and a field filling the space.

Since we are used to observe trajectories of macroscopic objects in our everyday experience (macroscopic objects, not matter points however), we are keen to believe that the field S of the Hamilton–Jacobi's formulation does not have any physical reality, but is simply an intermediary for computations. After all, this field does not pertain to our sense data. Anticipating much on the end of this book, let us nevertheless straight away observe that, actually, we shall have to dismiss Newtonian trajectories of matter points and, even if it is counter-intuitive at the present time, to erect S as the single significant mathematical (and physical) object of classical mechanics. Another issue which could immediately have struck the

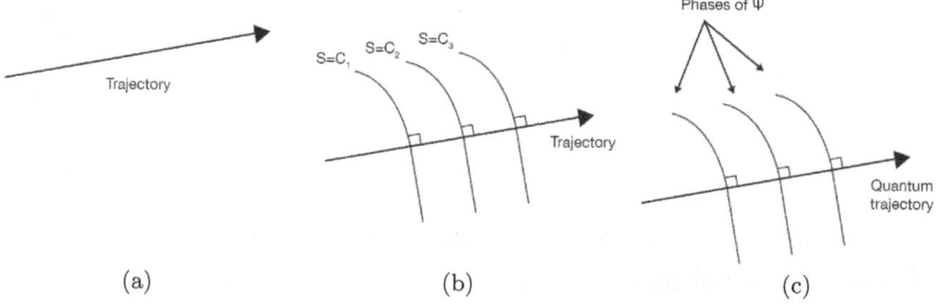

Figure 1. Pictorial representation of (a) Newton's formulation, (b) Hamilton–Jacobi's formulation, and (c) causal theories.

mind of the reader is that, although Newton's and Hamilton–Jacobi's formulations are mathematically and empirically equivalent, they are not ontologically equivalent: They do not tell us the same thing concerning the structure of the world, excepted if we just shrug our shoulders and rashly sweep S under the carpet. This is an issue that we shall have to investigate when appropriate, and whose consequences will be most significant, and will have to be drawn.

Another remark which will reveal later its importance is as follows. Newtonian formulation is a local formulation: You can move along a trajectory, but there is nothing physically significant outside of it: Even if you derive the force acting on the particle from an extended potential, the force itself is applied to the point where the particle is located. Conversely, the Hamilton–Jacobi's formulation is highly non-local. The very existence of the trajectories depends on the existence of the extended action S which, somehow, pilots the particles. As we shall see, the non-locality of this formulation anticipates the non-locality of quantum mechanics.

Figure 1(c) is for further use. It will provide a key-point to understand how causal interpretations of quantum mechanics are related to classical mechanics. We also have Fig. 2 which is most appropriately presented in this chapter. When compared with Fig. 1, we see that it tells us something else. Actually, it tells us the genuine story but, before being in position to approach this question, and to tell the genuine story, we have to make a cruise along the coasts, and between the islands, of the quantum kingdom.

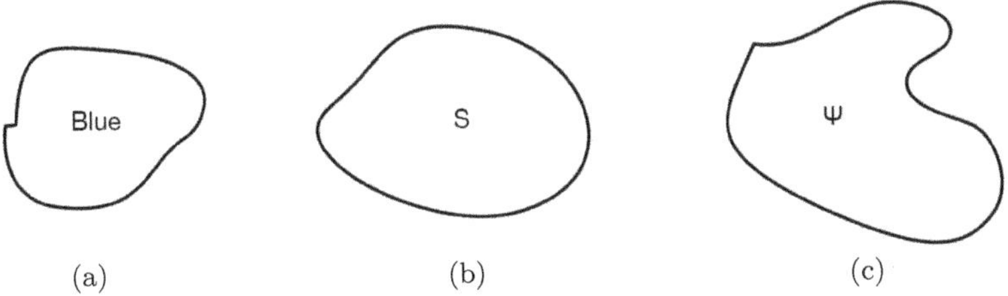

Figure 2. Pictorial representation of classical mechanics to wave mechanics: (a) Newtonian physics. (b) Hamilton–Jacobi formalism, and (c) quantum mechanics.

The path for PEDESTRIANS (∼∼∼∼)

In the formulation of classical mechanics by Newton, a matter point is a local object which, being acted upon by forces, follows trajectories, as depicted in Fig. 1(a). There exists (at least) another version, telling us something completely different, as depicted in Fig. 1(b), which is Hamilton–Jacobi's formulation: Instead of only a local object, we also have to consider an extended field S and trajectories are orthogonal to isovalued surfaces of S. Naively, we might believe that S does not exist (we cannot see it) and that it is simply a convenient conceptual intermediary for computations. Actually, it will later be extensively discussed that S is an ingredient in the famous wave function Ψ of quantum mechanics. Therefore, it possesses, in some sense a high level of reality. The reader should keep in mind that, although Newton's and Hamilton–Jacobi's formulations deliver us different ontologies, they are mathematically and empirically equivalent. Figure 1(c) is for further use.

After having followed this easy path, the PEDESTRIAN (∼∼∼∼) might like to have a look to the more difficult one, for mountaineers, and, concentrating on comments, possibly gain an extended vision of the issues discussed in this chapter.

Chapter 3

Background in Quantum Mechanics

In the same way that we needed a background in classical mechanics (which I have scrupulously restricted to a necessary minimum), we need a background in quantum mechanics. Of course, the reader who embarked himself in this book is likely to possess at least a bit of basic knowledge on the topic. But we may have different basic knowledges depending on what one has previously read and studied. Therefore, recalling a background, valid for any reader, is compulsory to properly set the stage. It will furthermore secure later expositions and argumentations.

There are two extreme ways of exposing quantum mechanics (and various hybrid ways). You may progressively and heuristically introduce the basic principles by relying on experimental results (such as from the celebrated two-slit experiment), well in agreement with the fact that quantum theory is an empirical theory, in which experimental data have been successfully put in parallel with a mathematical formulation. This is pedagogically very efficient and, for instance, well exemplified in the books by Feynman [14] or Bohm [41]. Or you may expose the mathematical formulation in itself, that is as a genuine mathematical formulation, relying on basic principles which can then be appropriately called axioms, from which you can deduce theorems. Having in mind that the axioms and theorems of the so-generated mathematical framework have to correspond to physical facts, the word "axiom" is certainly a bit restrictive and arid, and we shall better use the word "postulate". Providing a list of postulates which are basic principles (but they are not first principles) is an expedient and convenient way to expose quantum mechanics, and it is the one I shall choose. The origins of such a postulational approach may be found in the famous books by Dirac [42] and von Neumann [26]. But I shall prefer the most recent exposition by Cohen-Tannoudji *et al.* [43].

Postulates

PEDESTRIANS ($\sim\sim\sim\sim$): Please, quantum mechanically jump to the next section.

The postulates of quantum mechanics, following Cohen-Tannoudji *et al.* [43], possibly with a bit of rewording, but using the same numbering, are as follows.

Postulate 1.

At a given time t_0, the state of a physical system is defined by a ket $|\Psi(t_0)\rangle$, called a vector, or state vector, pertaining to the space of states $\mathcal{E}$. In this book, rather than using the ket $|\Psi\rangle$, I shall also very often use the wave function Ψ which, in technical terms, is simply the ket $|\Psi\rangle$ in what is called the $|\mathbf{r}\rangle$-representation.

Postulate 2.

Any measurable physical property$\mathcal{A}$ is described by a Hermitian operator A operating in $\mathcal{E}$. This operator is an observable, that is to say it generates an orthonormal set of Eigenvectors which forms a basis in the space of states. This being kept in mind, the word observable may also be used without any damage to denote the measurable physical property $\mathcal{A}$, in view of the biunivocal relationship between A and $\mathcal{A}$.

Postulate 3.

The result of a measurement of a measurable physical property $\mathcal{A}$ can only be one of the Eigenvalues of the associated observable A.

Postulate 4.

Let us consider the case of a non-degenerate discrete spectrum. Then, when one measures the measurable physical property $\mathcal{A}$ on a system described by a normed state $|\Psi\rangle$, the probability $\mathcal{P}(a_n)$ to obtain the non-degenerate Eigenvalue a_n of the associated observable A is given by:

$$\mathcal{P}(a_n) = |\langle u_n \mid \Psi \rangle|^2 \tag{3.1}$$

in which $|u_n\rangle$ is the normed Eigenvector of A associated with the Eigenvalue a_n, and $\langle u_n|$ is the bra associated with the ket $|u_n\rangle$.

For the case of a degenerated discrete spectrum, Eq. (3.1) has to be generalized to

$$\mathcal{P}(a_n) = \sum_{i=1}^{g_n} \left|\langle u_n^i \mid \Psi \rangle\right|^2 \tag{3.2}$$

in which g_n is the degree of degeneracy of the Eigenvalue a_n and the set $\{|\ u_n^i >\}, i = 1, 2, \ldots, g_n$ is a set of vectors forming an orthonormal basis in the Eigen-subspace $\mathcal{E}_n$ associated with the Eigenvalue a_n of A.

These rules can be generalized to the cases of a continuous spectrum and of a mixed spectrum (comprising both discrete and continuous components).

Postulate 5.

If the measurement of a measurable physical property $\mathcal{A}$ on a system in state $|\Psi\rangle$ provides the result a_n, then the state of the system immediately after the measurement, that is to say at a time that we may denote as $(0+)$, is the normed projection:

$$|\Psi_{0+}\rangle = \frac{P_n |\Psi\rangle}{\sqrt{\langle \Psi \mid P_n \mid \Psi \rangle}} \tag{3.3}$$

of $|\Psi\rangle$ on the proper subspace associated with a_n, in which P_n is the projector of $|\Psi\rangle$ on the proper subspace (with rather trivial generalizations to the cases of continuous and mixed spectra).

Postulate 6.

The time evolution of the state vector $|\Psi(t)\rangle$ is governed by Schrödinger's equation:

$$i\hbar\frac{d\,|\Psi(t)\rangle}{dt} = H(t)\,|\Psi(t)\rangle \tag{3.4}$$

in which $H(t)$ is the Hamiltonian observable associated with the total energy of the system and $\hbar$ is the Planck's constant h over 2π (For convenience, I shall simply call $\hbar$ the Planck's constant).

This postulate, as expressed by Cohen-Tannoudji *et al.* [43], does not specify when this equation of evolution is valid, may be to avoid a difficulty. Indeed, it is sometimes understood that it is valid only, say, between measurements, and that the measurement process itself is controlled by the other postulates devoted to measurements. In other cases, it may be understood that it must be valid in all cases, including during measurements, with then the challenge to explain how a single outcome may emerge from the evolution of a superposition of states (of the so-called apparatus). Bohmian mechanics will do that. But, in this section, we are concerned with usual postulates of the usual quantum mechanics, satisfying many people, and dissatisfying others. Bohmian mechanics has to wait a bit. Also, the demand that "measurement", ideally, as required by some persons, should not appear in the fundamental physical postulates, i.e. that the measurement interactions should be described by using the same physics as is used for everything else, is not implemented in the version of the usual quantum mechanics that we are considering in the present chapter, i.e. in the postulates listed above. This is a matter for further discussion.

Postulates are similarly exposed by Le Bellac [16] with again the process of measurement taken as requiring specific postulates, rather than being described by the same physics as is used for everything else. Concerning postulate 6 above, Le Bellac explicitly states that it is valid for *closed* quantum systems, meaning, in the words of Le Bellac, that the quantum system under consideration must not be a part of a larger quantum system. However, *"closed"* does not mean *"isolated"* because the equation of evolution 3.4 is valid if the quantum system interacts with a "classical" system, e.g. in the case of a spin (1/2) interacting with a variable magnetic field.

There are also extra-postulates such as associated with the need of symmetrization of the expression of some operators, or required for systems incorporating several identical particles, whose discussions are not compulsory for this book.

The afore-listed postulates, as they stand, even if unsatisfactory in the eyes of opponents, are aiming to a non-ambiguous definition of quantum mechanics, and define a kind of quantum mechanics which is also named as being: Pure, standard, orthodox, conventional, of the textbooks, and so on. All these qualifying names or expressions however may convey undesirable connotations. The most exemplifying case is the word *orthodox* which immediately suggests that any other alternative framework is a *heresy*. Therefore, most often (but not always), I prefer to use the more neutral appellation of usual quantum mechanics, or even better, this point of terminology being stressed enough, simply of quantum mechanics. We may also distinguish between wave mechanics (de Broglie, Schrödinger) and matrix mechanics (Heisenberg, Born, Jordan), depending on the emphasis we intend to stress. These two mechanics have been independently developed and soon after shown to be equivalent. Quantum mechanics also holds for their synthesis.

Postulates for PEDESTRIANS ($\sim\sim\sim\sim$)

Postulate 1.

A table may be described by a set of properties, e.g. the size of the table is such or such, it is painted in blue, and so on. Let us designate with the letter Ψ the description of the table. We then shall write $\Psi = $ (the size of the table is such or such, it is painted in blue, and so on). The description of a quantum mechanical system, let us call it its state, is similarly denoted by using the letter Ψ, called the wave function, which may take the form of a mathematical expression. There is a more general description, called the ket $|\Psi\rangle$, that the PEDESTRIAN ($\sim\sim\sim\sim$) may ignore or, even if it is a somewhat incorrect procedure, identify with the wave function Ψ.

Postulate 2.

To measure the length of a table, we make a physical operation by using, for instance, a comparison with another length taken as a reference. To measure a property of a quantum mechanical system, we let an operator (formally of a mathematical nature) act on the state Ψ of the system.

Postulate 3.

The length of the table obtained by the measurement is a precise number (notwithstanding usual uncertainty measurements), corresponding to the true value of the length. The result of a measurement of a property of a quantum mechanical system is not a precise number corresponding to the true value of the property. It is a precise number taken among a set of precise numbers, called Eigenvalues, depending on the measured property.

Postulate 4.

The measurement of the true value of the length of the table provides a precise number giving the true value of the length (notwithstanding usual uncertainty measurements), with certainty, i.e. with probability 1. The measurement of a property of a quantum mechanical system provides a precise number, taken among the set of Eigenvalues, which is not, in general, obtained with certainty, but with a probability depending on the property considered, the state of the system, and the result of the measurement.

Postulate 5.

After the measurement of the true value of the length of the table, the true value of the length of the table is still equal to the true value of the length of the table before the measurement. But, after the measurement of a property of a quantum mechanical system, the state of the quantum mechanical system is (in general) modified, implying a change of the results (values and probabilities) of consecutive measurements.

Postulate 6.

The true value of the length of the table does not change with respect to time (if no action, from an observer or from the environment, say measurement, is applied). The state of a quantum mechanical system evolves with time according to a deterministic equation called Schrödinger's equation. This equation is linear, i.e. if Ψ_1 and Ψ_2 are solutions, then $(\Psi_1 + \Psi_2)$ is is a solution as well. See further comments, not repeated here, after Eq. (3.4).

Comments on postulates

As a way of checking that the postulates above indeed provide a fair account of quantum mechanics, we could cross-correlate their contents with other accounts of quantum mechanics to be found in the literature. For instance, we may refer to Bohm [31] who stated that *although there are several alternative formulations of this theory (due to Heisenberg, Schrödinger, Dirac, von Neumann, and Bohr) which differ somewhat in interpretation, they all have the following basic assumptions in common*, this being pursued by the enunciation of the mentioned basic assumptions. The reader may like to return to Bohm [31] and check the quality of the cross-correlation with Cohen *et al.* [43]. This is also a way of saying that the list of postulates given by Cohen *et al.* [43] is not only a recent and modern account, but also correlates well with more ancient postulational approaches, reflecting well the history of quantum mechanics.

The first postulate associates Ψ with an individual system. It has not always be understood, or believed, that it should be so: Connecting Ψ with an individual system. To match and try to understand the occurrence of quantum probabilities, which depart so much from classical probabilities (the former ones being viewed as intrinsic and the latter ones resulting from a lack of accurate enough knowledge of the system under study), it has conversely (and alternatively) be proposed that Ψ should represent an *ensemble* of systems. In particular, one knows that Einstein favoured such an interpretation, that he called the statistical interpretation of quantum mechanics, according to which the state vector does not describe a single system, but rather a family of systems having experienced the same quantum preparation, that is to say a succession of measurements with a complete set of commuting observables. There is nowadays enough opportunities to achieve measurements on individual systems to rule out this alternative.

One of the most important comments to stress concerns however the sixth postulate which tells us that the wave equation of quantum mechanics is linear, so that solutions can superpose linearly. Such a superposition principle was already used for many kinds of classical waves (electromagnetic, acoustic or some kinds of hydrodynamical waves). But such waves are of a real nature, i.e. they can be described by real functions. If they are described by complex functions, as we usually prefer to do, we have to remember that the extension from reality to complexity is helpful for intermediary computations, but that, eventually we only retain real contributions. In contrast, the wave-function Ψ is genuinely of an irreducible complex character, as beautifully commented by Bohm [41]. Therefore, the superposition principle used in quantum mechanics is a complex superposition principle, in contrast with the one used in classical physics which is a real superposition principle. Of course, the use of the complex superposition principle is motivated by the existence of interference effects which, although of a different nature in quantum mechanics than for classical waves, were also present in classical physics, and well modeled by using the idea of linear superposition of waves. Bohm however remarks that *whether this is the only hypothesis* (i.e. the superposition principle) *which can explain interferences is not known*. Hence, the sixth postulate, from this point of view, is indeed a postulate, not the consequence of a rational necessity.

We shall have many opportunities to return to the linearity issue. But, for the time being, elaborating a bit on the previous remark, we may observe that, if we add nonlinear terms to Schrödinger's equation, and if these terms are small enough or designed intelligently enough for not spoiling available experimental results, then they could possibly be legitimate. Even the way used by Schrödinger to derive his equation is not an obstacle to the introduction of nonlinear terms. Indeed, Schrödinger relied on an analogy between Hamilton–Jacobi's mechanics and geometrical optics, afterward conveyed to the wave equation of optics (see [44, 45], and Chapter 16). The linearity of the wave equation of optics is therefore automatically transferred to the wave equation of wave mechanics. However, a derivation is not a demonstration and, therefore, some room is left for the introduction of nonlinear terms. There is actually a significant literature on nonlinear Schrödinger's equations and, whether there is eventually a good reason to dismiss nonlinear terms, this is something I do not intend to discuss now. Linearity of quantum mechanics is also present in the fact that all physical results are calculated by using Hermitian operators which operate linearly on wave functions, more generally on state vectors.

Another important comment concerns postulates 4 and 5, in connection with postulate 6. In contrast to postulate 6 which exhibits a deterministic evolution of the state vector, postulate 4 breaks down the deterministic evolution and introduces a probabilistic character which, in quantum mechanics at least, is of an intrinsic nature (it is exactly the place where the famous indeterminacy of quantum mechanics occurs). We are then facing two different kinds of evolution of quantum systems, a deterministic one when systems are not observed, and an indeterministic one in the measurement process, an unpleasant feature indeed. Furthermore, according to postulate 5, the measurement implies an abrupt modification of the state vector, a *quantum jump*, so odious to Schrödinger, often referred to the *collapse of the wave function*. These features, in essence, are ingredients to the famous measurement problem to which, again, we shall return later.

Copenhagen interpretation

From a positivist point of view, the formulation of quantum mechanics which provides a way to make correct, and up to now unchallenged, predictions of experimental results is strictly sufficient. We would not need more. But, actually, the quantum world has been found so strange, from the very beginning, that the desire and the need for a satisfactory interpretation (possibly a *superstructure)* immediately became an inescapable motivation for a parallel, often philosophically oriented, line of investigation.

The dominant interpretation, which won during the debates at the Solvay Congress of 1927 [46, 47], is called the Copenhagen interpretation, in the honour of Niels Bohr which, may be, has been his most pugnacious defender, or at least one of them. We should not however forget the utmost role of Born, working in Göttingen, who introduced the probabilistic formulation of the measurement process. Such experts as Pauli, Heisenberg, Jordan, Fermi, von Neumann, Wigner, Weisskopf, Oppenheimer or Dirac have been assistants of Born [47]. Hence, some historians of science might prefer the denomination of the Copenhagen-Göttingen interpretation. But I shall conform myself to an established tradition, dismissing the reference to Göttingen.

While the postulates we have listed above define a formalism, the Copenhagen school defined an interpretation. Surely, the Copenhagen interpretation matches very well the quantum mechanical formalism, but the difference between formalism and interpretation is something that we must always keep in mind [48]. A formalism, particularly if it is expressed in mathematical terms, is the tool to make predictions, even if only of a statistical nature. An interpretation may try to make more, like explaining with plain words how the world looks like, or even to claim what the world is. In particular, we shall later have another interpretation, called an ontological interpretation, also matching the quantum mechanical formalism, with a very small price to be paid, as far as the formulation itself is concerned, which consists essentially to a different reading of the equations (this will be the pilot wave).

Since we may have at least one alternative interpretation to quantum mechanics (the pilot wave again), it should not be too much surprising that we may also have various interpretations of the Copenhagen interpretation. And, therefore, it should not be a surprise no longer if various authors put forward various interpretations of it, even contradictory. For instance, Penrose [49] stated that the

*standard Copenhagen interpretation is based on the absurd concoction
of Schrödinger evolution of the one hand and state-vector reduction
on the other.* We might agree with the idea that there is here
an absurd concoction, although this problem pertains more to the
formalism than to the interpretation, insofar as the mentioned
concoction is explicitly implied by the postulates. The expression
"absurd concoction" furthermore points to the idea that the set of
postulates might be incoherent, an option certainly not shared by all
quantum mechanists who feel comfortable with it.

Closer to what seems to be a more correct and charitable interpretation of the Copenhagen interpretation, Vigier [50] stated that,
in the Copenhagen school, the classical notion of knowledge is a non-
sense because the aim of physics is not to describe the real behavior
of things, but only to build a mathematical formulation allowing one
to match experimental results (with however, I must add, the risk of
being victimized by the Greek syndrome). This quotation from Vigier
is reminiscent from another well known quotation from Bohr himself
saying that *there is no quantum world. There is only an abstract
quantum physical description. It is wrong to think that the task of
physics is to find out how nature is. Physics concerns what we can
say about nature* (see [51]).

Bohr and Heisenberg, both tenants of the Copenhagen interpretation, do not exactly share the same idea about what this
interpretation is, as described in Chevalley's preface of a book
by Bohr [52]. The point of view of Bohr may be found in this
lastly quoted book [52], while the point of view of Heisenberg is
available from Heisenberg himself [53]. The relationship between
the views of Bohr and Heisenberg is also discussed by Heisenberg
[10]. A discussion of the differences between the points of view
of Bohr and Heisenberg is available from Shimony [54]. Basically,
according to Shimony, *in denying that the words "physical reality" are
meaningful without reference to an experimental arrangement, Bohr
renounces any knowledge of the "thing in itself" ... Heisenberg departs
from Bohr in enunciating a metaphysical implication of quantum
mechanics. Heisenberg asserts that there is an intermediate modality*

-potentiality- between logical possibility and existence. However, *his attempts to explain the transition from potentiality to existence are not convincing.*

In trying to provide a synthetic view, Stapp [55] commits itself *to give a clear account of the logical essence of the Copenhagen interpretation. This logical essence should be distinguished from the inhomogeneous body of opinions and views that now constitute the Copenhagen interpretation itself.* Squires [13] expresses his idea that *a precise account of what the Copenhagen interpretation actually is does not exist,* in basic agreement with Jammer [24] who stated that *the Copenhagen interpretation is not a single, clear-cut, unambiguously defined set of ideas but rather a common denominator for a variety of related viewpoints.* For Van Fraassen [21], the Copenhagen interpretation is *really a roughly correlated set of attitudes expressed by members of the Copenhagen school, and not a precise interpretation.* For Cushing [48], the Copenhagen interpretation *requires complementarity (e.g. wave-particle duality), inherent indeterminism at the most fundamental level of quantum phenomena, and the impossibility of an event-by-event causal representation in a continuous space-time background.*

Various ingredients may indeed be attached to the Copenhagen interpretation including indeterminism and the statistical interpretation of the wave function by Born, the correspondence principle of Bohr, and also the statement that a measurable quantity does not possess a specific value outside of the context of measurement, i.e. that the measurement "creates the value" or, in the words of Van Fraassen [21], that a measurement *yields* a value as its outcome, but does not *reveal* a value. However, when we attempt to focus on the common denominator, the most important concept to put forward is the one of complementary, fathered by Bohr and presented for the first time at Como, in 1927, during an international congress of physics commemorating the hundredth anniversary of the death of Volta. The idea of complementarity is closely related to the Heisenberg uncertainty relations (see [52]).

An example of complementarity is the one associated with the concepts of wave and particle. For instance, in the double slit experiment, the interference pattern (wave character) is destroyed if we manage to observe through which slit the quantum object had passed (particle character). This suggests that the wave and particle characters are complementary in the sense that their observations require mutually exclusive experiments. This however is not quite correct, and even erroneous, because the interference pattern may be observably shown to consist of single dots so that wave and particle characters actually manifest themselves in a single experiment. This point is in particular discussed by Popper [25], in an appendix. It is a good reason, among others, to definitively abandon and even reject the notion of wave-particle duality. It seems that, as a consequence, Bohr eventually gave up this aspect of complementarity. At least, he stopped discussing it.

A better example concerns the attribution of position and momentum to a quantum object or, more generally, the attribution of conjugate variables, which genuinely requires mutually exclusive experiments. We say that such conjugate variables are complementary. The more you are able to specify one of them through a measurement process, the less you are allowed to specify the other. This statement is made quantitative by the virtue of Heisenberg uncertainty relations.

The last and in one sense may be the most important example of complementarity to be discussed here concerns the renunciation to a causal description of atomic phenomena in space and time, as discussed by Bohr at the Solvay Congress of 1927 [46, 47]. Bohr actually did not reject the existence of a physical reality, but questioned what we can learn of it (as for d'Espagnat, Bohr here saw the reality as a veiled reality that we shall discuss later). Indeed, Bohr pointed out the impossibility of specifying simultaneously a causal development and a spatiotemporal location. For instance, we have, on one hand, the causal development of Schrödinger's equation in a ghostly limbo between observations, a development which does not occur in space and time since it occurs in a configuration space and, on the other hand, if we need to obtain information on space and time locations, then we have to make measurements and face the intrinsic indeterminacy of quantum mechanics. In short, we may have causality but outside of space and time, and we may have space and time but outside of causality. Following the spirit of complementarity, what cannot be simultaneously reached has to be decreed as complementary. Therefore causality and space-time become complementary concepts.

As a summary, the whole Copenhagen interpretation may be viewed as an interpretational appendage to the formalism. But it is a rather ill-defined appendage so that the expressions "quantum mechanics" which refers to the postulates which are clearly stated, and "Copenhagen interpretation" which refers to a rather fuzzy (elusive in the words of Cushing [48]) interpretation, should not be considered as equivalent. Therefore, in the rest of the book, I may use one expression or the other, depending whether I intend to put the emphasis on the formalism or on its possible meaning, but I do not consider them as exactly synonymous.

Chapter 4

Objections to Quantum Mechanics

Several objections, to be discussed in this chapter, have been raised against quantum mechanics. Nowadays, likely depending on individuals, these objections may have received satisfactory answers, or partially satisfactory answers, or no answer at all. But, in any case, they are worth discussing, firstly because they will allow one to deepen our understanding of quantum mechanics, or may be better stated our lack of understanding, and secondly because they historically provided impetuses for the development of hidden-variables theories. The themes considered in this chapter are essentially the same than the ones we shall also consider in the next chapter but, there, they will be examined with a more philosophical orientation.

The issue of determinism

The principle of causality lies at the very foundations of classical physics, at least since Newton, and it somehow looked so obvious to many, so anchored in the human brain, that Kant made out from it an *a priori* form of our conceptual world, something strictly required and inescapable for a genuine understanding. We also already mentioned how much the demand of causality, or say determinism, was, for Einstein, a necessary prerequisite to build any satisfactory physical theory.

However, quantum mechanics does not satisfy this principle. When floating in the world of Heisenberg's potentialities, quantum objects indeed satisfy the deterministic, differential, Schrödinger's equation, but this deterministic evolution is disrupted by the actualizing act of observation. For Belinfante [1], the first dissast-isfaction about quantum theory comes from people who reason that, if different measurement values are associated with a same Ψ, then there must exist different microstates, at a sub-quantum mechanical level, associated with that Ψ, a concern already discussed fairly extensively by von Neumann [26].

Formally, the situation runs as follows. Let us consider classical mechanics for a system made out from a set of N matter points (e.g. [39, 56]). The geometrical description of the system, i.e. its location, requires a $3N$-dimensional manifold spanned by $3N$ degrees of freedom. These degrees of freedom are generalized (not necessarily Cartesian) coordinates q_k, $k = 1 \ldots 3N$. The dynamical description of the system, i.e. its velocity, requires another $3N$-dimensional manifold spanned by $3N$ generalized velocities, or better momenta $p_k, k = 1 \ldots 3N$. The mechanical description of the system, or state, requires a $6N$-dimensional manifold formed by the product of the geometrical and dynamical manifolds, with values (q_k, p_k). The knowledge of a point in the state manifold (say $6N$ values at a time t_0, called initial conditions) allows one, by using the Newtonian equations of motion, to deterministically predict the future for $t > t_0$ or to retrodict the past for $t < t_0$.

In contrast, the situation in quantum mechanics is quite different. The state of a quantum mechanical system built from N quantum objects may be defined by a wave function $\Psi(q_1, q_2, \ldots, q_{3N})$. In this expression, we have chosen to give a privilege to coordinates and, hence, in the quantum mechanical framework, momenta *must not* appear in the list of arguments. Also, the coordinates appearing in the expression are not coordinates denoting positions of the particles because, in the potential world outside of observations, particles *do not have* any position. We also insist on the fact that, according to postulate 1, Ψ is assumed to be complete since it completely determines the state of the system. This complete determination of the state is associated with a completely deterministic evolution driven by Schrödinger's equation, outside of any observation. Now, if we measure the positions of the particles, we cannot predict the positions that particles had before the measurements (these are meaningless because particles did not have any position), but we can predict densities of probabilities for the various possible outcomes. One of these possibilities will actually occur, but in an indeterministic (and unpredictable) way. We therefore have to be content with statistical determinism for measurement outcomes instead of determinism for any individual measurement (excepted if the wave function represents an Eigenstate of the observable). Furthermore, the measurement process exhibits an irreversible character which makes retrodiction impossible.

Well, up to now, it could be all right. After all, why should we hold fast on causality, which just could be an erroneous habit of our mind, instead of giving it up straight away? We could for instance pretend that indeterminism is an experimental fact to which we have necessarily to obey without discussing any more (however forgetting that quantum indeterminacy is not an experimental fact but already an interpretation of facts). Let us nevertheless make the case as strange as it is indeed, by considering a specific example. Let Ψ given by:

$$\Psi = \frac{1}{2}\left[\sqrt{\frac{3}{2}}Y_2^{-2} + Y_2^0 + \sqrt{\frac{3}{2}}Y_2^2\right] \tag{4.1}$$

in which Y_l^m's are spherical harmonics (PEDESTRIANS ($\sim\sim\sim\sim$), sorry for this mathematical expression, but stay with us). Let us measure the z-component of the angular momentum. The possible measurement outcomes are $m = -2, 0, 2$, and it is an exercise for students to demonstrate that the outcomes will happen with probabilities $3/8, 1/4$ and $3/8$ respectively (summing up to 1 as it should). But how is this possible?

In a classical coin tossing experiment, with head and tail occurring with equal probabilities $(1/2)$, there is nothing mysterious. We have many hidden variables: The precise way characterizing the launching of the coin, convection during the journey of the coin from its launching base to the landing table, details of the properties of the table such as roughness, and so on. We are not able to analyze all these tiny and much complicated and intertwined causes. From a conceptual point of view, we are however not reluctant to admit that the effects of such many causes will compensate, by the virtue of some averaging processes taken on the values of the hidden variables. Forgetting this sub-level kind of explanation, we may also simply rely on a law of symmetry, a case where actually the symmetry makes the law [57]. Indeed, if nothing distinguishes head and tail, the coin being assumed to be well balanced, then there must be an equiprobability between head and tail. Hence, without any doubt nor reluctance, we may predict that both head and tail will occur with equal probabilities (1/2).

However, in the above quantum mechanical example, in which Ψ is complete according to quantum mechanics, the situation is quite different because we have no hidden variables. Now, let us imagine that we make an indefinite, or better infinite, number of successive measurements on the system defined by the Ψ of Eq.4.1. Then the outcomes of the first five measurements could be $-2, 0, 0, 2, -2$ or even $0, 0, 0, 0, 0$. But, after an infinite duration required to make an infinite number of measurements, we must observe on our experiments that the results $-2, 0, 2$ indeed occur with probabilities $3/8, 1/4, 3/8$.

This is indeed magic, and more stranger than strange. How measurements made on March 24th 2008, February 5th 2572, December 13th 9897, and so on, manage to adjust themselves, in the absence of hidden variables like the ones we invoked in our analysis of coin tossing, to eventually agree with the probabilities predicted by the theory? How do measurements know what they have to exhibit? And how can *we, us,* deal with this? May be God manages with this problem, a somehow metaphysical (non-scientific) but reasonable assumption. Possibly, after having created the world and the laws of nature, He is carefully choosing the outcomes of quantum measurements in such a way that predicted probabilities are indeed observed in experiments. This might be the way He found, through intensive computations that we are not able to achieve, to allow Him to go on interacting with the universe He made. Such questions are not stupid and Newton, for whom the absolute space was the *Sensorium Dei* [58], anticipated the modern chaos theory when he realized that the motions of planets driven by his gravitation theory could be unstable. Then, God was in charge of watching and checking that the eternal harmony of the cosmic clock would be preserved. But, are modern scientists ready to accept such a metaphysical loophole? If not, there remains the question to understand how independent successive measurements manage to satisfy the quantum mechanical statistical laws.

Would not be more reasonable to propose that, actually, quantum mechanical processes are amenable to the same kind of analysis we have made for the tossing of coin, that is to say that Ψ is not complete, that Ψ alone does not provide a faithful knowledge of the quantum system? If we pursue this idea, we are invincibly driven to the conclusion that there must be a set of supplementary data, not embodied in Ψ and quantum mechanics, called again hidden variables, which are unobserved or even possibly unobservable, but which would univocally determine the outcomes of quantum measurements. The indeterminacy involved in quantum measurements would then merely be due to our lack of knowledge concerning the values of the hidden variables. That is to say, we do not precisely know the state of the system, and there is therefore no wonder that we are only able to make statistical predictions. This is one of the boosting impetus for the search of hidden-variables theories. If such a search eventually fails, then we might have to admit that there is no complete causality in nature, that our intuition is ultimately misleading and, if we do not refer to God as an alibi, that we have already the cliff of our ignorance and of our conceptual abilities facing us, dominating our mind and disabled power of understanding.

To make us immune to the indeterminism disease, or at least to reduce the suffering produced by this disease, we may recognize that, if causality looks like an *a priori* form demanded by our mind, it is actually not immediately exhibited by the experimental facts (nothwithstanding the problem to know whether it could be mediately present via an adequate interpretation). Therefore, we may develop strategies to reassure ourselves by understanding that, being not really present in the empirical world such as it is immediately observed by us, causality should not be necessary required in our physical theories. For sure, causality is a key ingredient of classical mechanics, more generally of classical physics. But classical physics pertains to our modeling of the world, abstracted from the experimental facts, but does not necessarily perfectly reflect the experimental facts themselves.

For example, if we want to experimentally check the development of a Newtonian trajectory, we need a perfectly accurate knowledge of the initial conditions. But initial conditions can never be known with infinite accuracy. Then the limitations imposed on our knowledge of initial conditions will propagate on all the future of the trajectory. The theoretical trajectory computed from the initial conditions, with relevant inaccuracies attached to them, would rather look like a kind of tube, with flesh surrounding the linear skeleton of the ideal trajectory of our Newtonian model. But even this picture of a comfortable tube may be misleading. Indeed, the situation is particularly dramatic in presence of chaotic phenomena where two near-by trajectories exponentially depart one from the other in phase space. We then have to face a horizon of predictability usually measured from the largest Lyapunov exponent [59]. Of course, a lack of predictability is not a lack of causality but it severely limits our pretension to check classical mechanics. Obviously, we may argue that we are here discussing practical limitations but that, in principle, the causality is perfectly satisfied, that if we could indeed specify initial conditions with perfect accuracy, then the above difficulties would evaporate. But this is a metaphysical belief which cannot be experimentally checked. Worst, we are simply not allowed to specify initial conditions (say a location, and a velocity) with infinite accuracy. Indeed, if we try to do so, then, soon or later, we shall collide the quantum mechanical domain where Heisenberg uncertainty relations prevent us to simultaneously define both location and velocity with an infinite precision. Not only do we have problems with initial conditions to empirically check the validity of causality, but we also have problems with the forces producing the motion (in a given frame of reference). We usually have to disregard tiny influences that we feel irrelevant and, if they are relevant and have to be incorporated in our equations of motion, they also cannot be specified with infinite precision. The search for such an infinite precision would again put us in contact with the quantum world.

In 1957, Bohm [60] could write in about the same spirit: *Thus we do not expect that any causal relationship will represent absolute truth; for do this, they would have to apply without approximation, and unconditionally. At any particular stage in the development of science, our concepts concerning causal relationships will then be true only relative to a certain approximation and to certain conditions.* Nevertheless, even if causality is not in the facts, it was in Bohm's mind since he opted for a causal interpretation of quantum experimental facts. In his seminal papers [29, 30], he could argue that quantum mechanics *requires us to give up the possibility of even conceiving precisely what might determine the behavior of an individual system at the quantum level, without providing adequate proof that such a renunciation is necessary. The usual interpretation is admittedly consistent; but the mere demonstration of such consistency does not exclude the possibility of other equally consistent interpretation, which would involve additional elements or parameters permitting a detailed causal and continuous description of all processes, and not requiring us to forego the possibility of conceiving the quantum level in precise terms.*

Therefore, if our belief in causality cannot be exactly corroborated by experimental facts due to their inherent inaccuracies, nor rejected by quantum mechanics because we may invoke the existence of additional parameters (hidden variables), then it is immune to experiments. We may have to claim that this belief is of a metaphysical nature in a precise sense, the one which allowed Popper to draw a demarcation line between science and metaphysics [25, 61]. Again causality is not in the facts, but in our mind, not only Bohm's one. Are we really ready to extirpate it without at least a bit of suffering and reluctance?

May be so, at least for some people. Indeed, in contrast with those who question indeterminism in quantum mechanics, with the desire of a restoration of determinism, there are also those who accept indeterminism without any scruple. For such an acceptation, a criticism of causality as sketched above can help. There may also be some kind of mere resignation, or renunciation, or also a sympathy or adherence with a positivistic Zeitgeist, or even a deep rejection of a deterministic conception of the world which is so contradictory with the intuitive belief to free will, this free will without which we may deeply question whether there is a sense of life or not. Nature may be ignorant about such questions, but human beings are not.

Louis de Broglie, in his preface to a book by Bohm [60] refers to positivism as follows: *The construction of purely probabilistic formulae that all theoreticians use today was thus completely justified. However, the majority of them, under the influence of preconceived ideas derived from positivist doctrine, have thought that they could go further and assert that the uncertain and incomplete character of the knowledge that experiment at its present stage gives us about what really happens in microphysics is the result of a real indeterminacy of the physical states and of their evolution. Such an extrapolation does not appear in any way to be justified.* Similar opinions and statements may be found in several places in the bulk of the book by Bohm. And also, as you may know, believe it or not, there are those people who are enthusiastically clapping to the end of determinism.

Space and time

As a systematic (and clever) opponent to quantum mechanics, we already have met Einstein in connection with the issue of determinism. Not surprisingly, we meet him again in connection with the issue of space and time. For him, *it is ... characteristic ... of physical objects that they are thought of as arranged in a space-time continuum*, hence its reluctance, expressed in several places, to accept a world of physical phenomena overflowing and transcending the categories of space and time. In particular, Einstein expressed his disapproval by using a Gedanken experiment expounded at the Solvay Congress of 1927, available from the final discussion of this Congress [46], and discussed by Louis de Broglie in many occasions such as in [37, 62–65].

This Gedanken experiment runs as follows. Let us consider a corpuscle and its associated monochromatic plane wave, normally incident on a screen pierced by a small circular hole. A photographic plate, having the form of a large radius hemisphere, is positioned behind the hole. After having passed the hole, the incident plane wave has diffracted and it essentially covers the whole angular range of the hemisphere. According to quantum mechanics, there is a certain probability (actually, a certain density of probability) for the corpuscle to manifest itself, through a photoelectric effect, at a certain point P of the plate (actually, around a certain point P of the plate). However, if this happens at point P, at time t, this cannot happen at any other point of the plate, since we assumed that there was only one (indivisible) corpuscle associated with the incident wave. If we rely on quantum mechanics, we have to think something as follows: The corpuscle is associated with an extended diverging wave behind the hole, represented by some Ψ, and not really localized in space. Rather, it is somehow present in a fuzzy and diffuse way, spreading in the Ψ after the hole. Now, once the corpuscle has manifested itself at point P, the wave function has instantaneously to become zero everywhere. This vividly illustrates how much quantum mechanics departs from our usual vision of space and was seemingly so conflicting with relativity that Einstein could not accept it. Worst, in the quantum mechanical framework, there is no known mechanism (even relativistic) able to explain such an extraordinary collapse of the wave function. This looks like a kind of *fiat*.

Here is now another Gedanken experiment attributed to Louis de Broglie, addressing the same issue [66]. It starts with a box containing one electron. The density of probability of finding the electron at any place in the box, if we make a measurement of its location, is *more or less* the same anywhere insofar as the wave function which describes it is spread in the whole volume of the box. Next, the box is splitted in two sub-boxes A and B of equal volume. Now, the wave function is distributed over the two sub-boxes and it tells us that the probability of finding the electron in box A is equal to the probability of finding the electron in box B, namely, say, 1/2. Let us leave box A in Paris and translate box B to Beijing. There, in Beijing, let us open box B and let us assume that we find the electron in it. Then, we know with certainty that the electron is not in box A located in Paris. What quantum mechanics states to describe these observations is that the opening of the box B at Beijing and the detection of the electron in this box instantaneously annihilates the wave function in box A, although it is far away from box B, another extraordinary collapse: Still a snake difficult to swallow [66].

Besides illustrating the fact that our usual conceptions of space and time are inadequate for a complete interpretation of microscopic phenomena, the above Gedanken experiments are also our first concrete encounters with the measurement problem in quantum mechanics. It has been one of the impetuses to Louis de Broglie's motivation to the development of causal theories. The analysis of the above Einstein's objection by Louis de Broglie indeed made him remarking [65] that the expression "potential presence of the corpuscle at any point of the wave", expression that we are nearly obliged to employ to expound this (Copenhagen) interpretation, is a matter of confusion: Saying that the corpuscle may be present at any point of the wave, does that mean that at any time it is located somewhere in the wave, but that we ignore which is this point, or does that mean that the corpuscle possesses some kind of mysterious "omnipresence" in the whole wave packet? And, he added: It seems that the second interpretation is the one adopted by the advocates of the present interpretation (again Copenhagen), although some of them keep on this issue a prudent silence. Obviously, while the second option points out to the Copenhagen interpretation, the first one points out to hidden-variables causal theories developed by Louis de Broglie. The word "mysterious" expresses a dissatisfaction against quantum mechanics, while the expression "prudent silence" designates an epistemological coward attitude. It might also be interpreted as denoting a resignation, or the positivist desire to avoid asking meaningless questions. An ultimate position of some experts, far beyond quantum mechanics itself, is to claim that space and time are emergent structures, resting on something bubbling at a deeper more fundamental level, some kind of coarse-grained illusion [9].

The fact is that, actually, quantum mechanics, in many respects, is not compatible with our usual conceptions of space and time (nor with the more sophisticated concept of space-time) inherited from pre-quantum physics. According to Stapp [55], *the principal difficulty in understanding quantum mechanics lies in the fact that its completeness is incompatible with external existence of the space-time continuum of classical physics.* Louis de Broglie [62] remarks that a corpuscle in quantum mechanics is no more a well defined object defined in the framework of space and time and, quoting Bohr, that quantum objects are *unsharply defined individuals within finite space-time limits.*

We are now going to be more explicit on this issue and, as a first comment, we should remember a previously discussed version of the complementarity principle with space and time on one hand, and causality on the other hand, which alone should force us to stop thinking of space and time as absolute necessary ingredients of the world. Besides, the most famous theme concerning the inadequacy of space to deal with quantum phenomena is the one of non-locality, attached to keywords such as: EPR-paradox, quantum correlations at a distance of separated objects, Bell's inequalities and theorems, teleportation ... We shall devote a whole Chapter to this issue (Chapter 12 to name it).

However, for the time being, and anticipating in a convenient way, let us simply mention a corresponding theoretical feature of quantum mechanics, which is also attested by experimental results: Two quantum objects having interacted in the past, spatially separated, manifest quantum correlations at a distance such as if one measurement is performed on one object, then the other object is instantaneously affected (or, with a bit more of caution : Looks like being instantaneously affected) by it. Theoretically, such features are attached to the existence of non factorizable wave functions (or more generally state vectors), i.e. such as we have, for the example of a two-body system: $\Psi(1,2) \neq \Psi(1)\Psi(2)$. This non factorizable form implies that particles 1 and 2 do not constitute separated quantum systems but are in contrast embodied in a single quantum system to be viewed as a whole object. The reluctance of Einstein to accept non-locality in space (which has been for him more shocking than quantum indeterminacy) made him speaking of *spooky actions at a distance (spukhafte Fernwirkung)*, when he realized the non-local implications of the quantum theory.

This issue has also been a motivation for the search of a class of hidden-variables theories, in order to avoid non-locality. But, whatever the future of quantum mechanical formalisms will be, non-locality in space is now a well confirmed experimental fact, with the consequence that the above class of hidden-variables theories (call them: Local hidden variables theories), aiming to provide a classical explanation for distant correlations, has to be dismissed. As a result, we have to claim that hidden-variables theories must be non-local. A related and akin issue, which can be here for the time being confused with the one of non-locality in space, although we shall later establish a distinction between both issues, is the one of contextuality (Chapter 11). As we shall see, there is then another condition for the possible acceptance of hidden-variables theories: They have to be contextualist. Here the reader has to be a bit patient: Contextuality and non-locality will later be discussed more extensively when appropriate.

Strongly associated with the problem of non-locality in space (discussed above by referring to a N-body problem, with $N = 2$), there is also the problem of localization in space already apparent for an one-body problem. Even if we admit the idea that a quantum system is not localized in space, and actually spreads out when time goes on, as a result of the governing Schrödinger's equation, we would be reluctant to consider that such a behavior should also be attributed to macroscopic objects. After all, the Moon looks like being there, rather than looking like a spreading conglomeration. Nowadays, we may have many arguments (some of them reserved for future discussions: GRW equation, decoherence) for not being, possibly, disturbed any more by this issue, but it provided an opportunity to Einstein to raise another objection.

This objection is discussed by Louis de Broglie [63] that I am translating as follows. Let us consider a moving sphere of mass M. In wave mechanics (recall Ehrenfest's theorem) as well as in classical mechanics, one can demonstrate that the barycenter of the system, i.e. the barycenter of the sphere, is the same as the one of a matter point of mass M. It is then represented by the propagation of a wave train satisfying the equation of propagation of waves Ψ with the value M for the mass. At time t_0, this wave train will have a certain shape, with uncertainties on the initial values of the barycenter coordinates. After a long time t, the wave train will have spread, and the uncertainties on the values of the barycenter coordinates will have increased. If, at this time, we take a stereoscopic photograph of the moving body, we will be able to deduce with a very good accuracy (however compatible with uncertainty relations) the values of the barycenter coordinates. The train wave Ψ will therefore be abruptly reduced in a dramatic way. But, the Copenhagen interpretation of this reduction of the wave packet probability is that it is the consequence of the action of the measurement process. However, in the present case, this interpretation is unacceptable because the light beams which illuminate the body when taking the photograph certainly could not exert a significant action on this body whose mass M can be considerable.

Besides non-locality in space, there is also the issue of non-locality in time which may be attached to a class of experiments, called delayed-choice experiments. In short, such experiments may be for instance interferometry experiments with photons, say in a Mach–Zehnder instrument. By introducing a certain device before the observation of the photons, we may force the photons to manifest themselves as waves or particles. The device can be furthermore introduced at the last time, once the photons have already propagated nearly all along the interferometer. A way of speaking of such experiments (not necessarily the best way however) is that the introduction of the device at the last time implies a retrodiction on the past, modifying the way in which potentialities evolved. Even if we are not allowed to tell it, it looks like this introduction retrodictively forced the photons to propagate as waves or particles. Possibly, as a source, we may have used photons received by telescopes, having travelled a long way through the cosmos, for instance since the time when our universe became transparent, about 400 000 years after the Big Bang.

Wheeler [67] expressed a possible analysis of such features by saying that *in the delayed-choice experiments we, by a decision in the here and now, have an irretrievable influence of what we will want to say about the past — a strange inversion of the normal order of time,* a point of view which is not independent of the idea of a participatory universe, cherished by Wheeler, and to which we shall return later. We may also take another point of view in which the universe deploys its potentialities, and that measurements, making potentialities actual, retrodictively modify and restrict the ranges of potentialities. The ghostly nature of unobserved quantum phenomena would progressively take consistency in a series of actualization processes. In about the same kind of spirit, Bohm and Hiley [68] stated that delayed-choice experiments *are designed to show that according to the quantum theory the choice to measure one or another of a pair of complementary variables at a given time can apparently affect the physical states of things for considerable periods of time before such a decision is made.* For d'Espagnat [69]), *when the existence of superluminal influences is denied altogether it seems difficult not to consider that the action of backward causality has some degree of pertinence.* Similar but somewhat more extraordinary experiments allow one to erase the past, or to shape it. For an easy access and a popularized account, see Greene [70]. See also Wickes *et al.* [71].

Our previous discussion would give some seemingly paradoxical characters to the concept of time. It could even look extreme to some people, but it is worth noting and even better worth thinking. What may be the best way to interpret delayed-choice experiments however is to refer to the Bohr orthodoxy of the Copenhagen interpretation. Then, the entire experiment is to be seen as a whole indivisible affair. What makes sense for us, human beings, is the irruption of the potentialities into an actual world of space and time, through measurement processes, and wave function collapses. This may be expressed by saying that *no elementary phenomenon is a phenomenon until it is a registered (observed) phenomenon* [32]. In this conception, we do not face any more any paradoxical feature of time, just because time is not any more a relevant category to analyze the situation. We might possibly argue that, between observations, quantum systems deterministically evolve according to Schrödinger's equation in which time is a parameter. But such an evolution does not occur in our ordinary space and time framework. It does occur in a configuration space. Before turning to a discussion of this configuration space, let us end this discussion of delayed-choice experiments by mentioning a recent experimental implementation of them by Jacques *et al.* [72].

Configuration space

Let us consider a wave function $\Psi(x_j)$ for a single particle. The x_j's are Cartesian coordinates of the usual 3D-Newtonian space but, as we already noticed, they are not coordinates of the position of the quantum object represented by this Ψ. In quantum mechanics, the x_j's indeed cannot be coordinates of the position of the quantum object because such a position does not pre-exist to the measurement, but is conversely the result of the measurement itself which, in a deep way, "creates" it. Therefore, already for a single particle, although t is the usual Newtonian time parameter, Ψ does not evolve in the usual space and time arena of classical physics, because the associated quantum object does not possess a trajectory in this arena. We say that Ψ evolves in a configuration space. The concept of configuration space does not exclusively pertain to quantum mechanics. It has already much used in classical physics, including classical statistical physics. But a trajectory in the classical configuration space of a single particle, in classical mechanics, identifies with the trajectory in the usual space. In contrast, such is no more the case in quantum mechanics.

The situation may look even stranger for a N-body problem. Let us discuss it for $N = 2$, without any loss of generality as far as principles and conclusions are concerned. In this case, the wave function receives the form $\Psi(x_j^{(1)}, x_j^{(2)})$ in which $x_j^{(1)}$ are Cartesian coordinates associated with particle (1) and $x_j^{(2)}$ are Cartesian coordinates associated with particle (2). The 6-dimensional space $(x_j^{(1)}, x_j^{(2)})$ forms the configuration space of the 2-body problem.

In classical physics, there would not be anything mysterious with such a configuration space. I remember the performance of two wind-surfers, dropped from a plane high in the sky, gliding on the air, whirling around each other, down to the ground. They were so close together that, although each of them was evolving in the 3D-Newtonian space, I could nearly intuitively perceive the 6-dimensional configuration space which could be used to describe their combined motion. But, nothing of that sort is possible in quantum mechanics, just because quantum wind-surfers do not follow any trajectory. Furthermore, if we carry out measurements on the positions of two quantum wind-surfers, one of them may appear here, and the other there, possibly at a far remote distance. This is also a manifestation of quantum non-locality. Hence, the configuration space of quantum mechanics is much more abstract than the one of classical mechanics.

Louis de Broglie [37] has been complaining on the fact that the quantum configuration space is fictitious and expressed the hope that it could be possible to turn it to an alternative description in which the N-body problem could be represented by N different waves, each one evolving in a 3D-space (N different wavy wind-surfers), a task that he did not succeed to complete in a satisfactory way. For him, the use of the quantum configuration space, which is required to describe Ψ in quantum mechanics, implies that Ψ does not describe any physical reality, but merely introduces a useful symbolic representation to deal with quantum probabilities.

Some philosophers of science could reply that the usual space itself is merely a representation and, nowadays, the status of space (and time) as fundamental ingredients of the stuff of the universe is legitimately questioned. These philosophers might also argue that the use of the Newtonian space and time provides a convenient and efficient representation to deal with a certain class of phenomena, those to which our mind is accustomed due to our everyday contact with the everyday experience. Along the same line, without any epistemological contradiction, relying not only on quantum mechanical formulation, but also on undeniable experimental results, the same philosophers would remark that, for microscopic processes, we simply have to accept to switch to another representation, with non-locality and all associated matters, which is also convenient and efficient to deal with a different extended class of phenomena. Therefore, with such a kind of epistemological symmetrical point of view, if we endorse the classical trio (space, time, causality) for classical phenomena, we should not be reluctant to endorse a non classical trio (non-space, non-time, non-causality) for quantum phenomena.

The subjectivity of the wave function

Furthermore, Louis de Broglie, who has been complaining about the status of the configuration space in quantum mechanics, in particular about its fictitious character, most naturally also complained about the nature of Ψ itself. Not only the configuration space was not describing a physical reality, but the wave function, for him, should receive a similar status. That is to say, the wave function is not an objective reality, but a purely subjective one. This idea is expressed by Louis de Broglie in many places, such as in references [37, 62, 63, 65], among others. For Louis de Broglie, and also for many other scientists, the wave Ψ is not a wave in a classical sense. It is not a wave pertaining to a large class of waves formed from optical waves, acoustical waves, hydrodynamical waves, or even gravitational waves, to which we may allocate an objective nature (at least in the framework of classical physics). On the contrary, it is only a tool for making predictions, and more importantly for making predictions not of events, but only of probabilities of events. Indeed, the detailed expression we can assign to a Ψ depends on prior observations having reached the knowledge or even the awareness of the observer. A complete determination of Ψ, that is to say a preparation, requires the use of a complete set of commuting observables. Surely, between two observations, the wave function evolves according to the deterministic Schrödinger's equation, but this evolution is unobservable. In this sense, it is metaphysical. But, each time an observation is made, that is to say each time some amount of information is gained by an observer, or more rigorously by *the* observer who observes, the expression of Ψ has to be modified, excepted if Ψ already represented an Eigenstate of the measured observable. The modification of Ψ required by any gained information is brutal, violent, instantaneous, generating the problem of the wave function collapse. Another important remark is in order, as pointed out again by Louis de Broglie [63]. Let us consider two normed wave functions Ψ_1 and Ψ_2. Let us build a new Ψ by the combination $(\Psi_1 + \Psi_2)$. Then the resulting Ψ must be normed again. This shows that the complex superposition principle of quantum mechanics is far from behaving like the ordinary principle of superposition of classical physics. We do not simply have a summation of amplitudes, but an extra-renormalization is required.

For Louis de Broglie, and many others, these features can only be understood if Ψ is actually a representation of our present knowledge of the state of the system, this means if it is a subjective representation, and certainly not the representation of an objective reality. Louis de Broglie [65] summarized his feeling concerning this issue under the form of a dilemma: As a representation of probabilities, the wave has to be subjective and, because it determines physical phenomena, it has to be objective, adding that some authors perpetually oscillate between these two conflicting points of view because they feel that none of them can be rejected. For him, the key-point is that it seems difficult to assign a physical reality to the probability waves since they do not only depend on the system under study, but also on the knowledge we have of it. Similar opinions concerning quantum mechanics are extensively discussed by Bohm [60] who, dissatisfied by such a state of affair, will eventually end up by giving the status of a strict objective reality to the wave function, this being however done in the framework of his causal theory to be discussed later. More pragmatically, Heisenberg asserts, with some amount of artfulness, that quantum probabilitites form a new kind of objective reality.

Schrödinger, the demiurge of Ψ, soon also became dissatisfied with what became the prevalent (Copenhagen) interpretation. He considered Ψ as possessing something of a psychological nature (picked up from [63]). He wrote [73] that it must have been given to Louis de Broglie the same blow and the same disappointment that have been given to himself, when he learnt that some kind of transcendental interpretation, nearly mystical, of the undulatory phenomenon, has been put forward, such an interpretation being very quickly welcomed by the majority of theoretician masters as the only one agreeing with experiments, and which became the orthodox dogma, accepted by nearly everyone, with however some rather remarkable exceptions. He also pursued as follows. The non physical character of the wave function (sometimes said to merely embody everything we know) is even reinforced by the fact that, in the orthodox view, its change produced by the measurement process depends on the fact that the observer informs himself of the result. Furthermore, the change only occurs for the observer who informs himself of the result. If you are present, but are not informed of the result — even if you have the most meticulous knowledge both of the wave function before the measurement, and of the experimental set-up used — the modified wave function does not concern yourself, it is so to speak non-existent; for you there is at best a wave function related to the experimental set-up augmented of the system under study, a wave function having no particular relationship with the wave function adopted by the observer who knows the result of the measurement. This Schrödinger's point of view will be soon made still more vivid when I shall introduce the reader to Wigner's friends.

Louis de Broglie, who was well aware of this implication of quantum mechanics, that the observer does not only have to observe but also have to be aware of his observation, grumbled against it. To the consequences of the structure of Ψ in the measurement process, he wanted to provide a more objective character. When compared to quantum mechanics, the opposite extreme realist point of view is that an observation should not require any observer. He expressed himself on this issue by saying [65] that he would like to insist on the role, certainly exaggerated, which is often put forward, in the analysis of the observation of microphysical entities, to the measurement set-up. Very often, he said, there is no intervening measuring set-up in a proper sense. When the arrival of a photon or of an electron on a photographic plate generates a local blackening and when one examines this behavior with the eye, where is the measuring set-up?

When a physicist sees the trajectory of an electron as given by the lining up of liquid droplets in a Wilson chamber, where is the measuring set-up? Certainly, in both cases, and in analogous cases, there is always an experimental set-up: One inserts a photographic plate on the path of a photon, one sends an electron in the Wilson chamber, etc... But, strictly speaking, there is no measuring device. Nevertheless, in certain cases, one can use a proper measuring device: For instance, one measures the darkening of the photographic plate by using an apparatus quantifying the local opacity of the plate... But, in any case, the measuring system only intervenes at the end of an observable localization process, when the chain reaction will have sufficiently amplified the phenomenon so that it became measurable by a measurement apparatus, in the ordinary sense of the word.

The content of this rather lengthy account of the opinion of Louis de Broglie seems very reasonable and, if it were really so, quantum mechanics would be easier to deal with than it is actually. The fact is that this opinion is deeply heretical. A consistent orthodox defender of quantum mechanics, like von Neumann, would remark that "observing with the eye" is not an innocuous final step of the measurement. On the contrary, it is the most important step of the process. The existence of such conflicting ways of understanding what is a measurement in quantum mechanics forces us to now examine one of the most disturbing issues in quantum mechanics: The measurement problem.

Measurement problem

Generalities for the measurement problem

All, let us say: Annoyances, associated with the aforementioned issues, related to the questioning on the classical set of the classical trio (space, time, causality) are somehow exacerbated and condensed in what is called the measurement problem, already made previously concrete by the two objections from Einstein concerning the detection of a corpuscle on a hemispherical screen and the observation of the position of a macroscopic object, as well as by an objection from Louis de Broglie in his two-box Gedanken

Experiment (see for instance the general discussion in Chevalley's preface to a book by Bohr [52]). It may be expressed by the measurement paradox of Wigner [74] as *the contradiction between the deterministic nature of the quantum-mechanical equations of motion and the probabilistic outcome of the measurements.* It is a key-point of quantum mechanics, most likely its Achilles' heel. It has been so for a long time, and it is still so nowadays. As long as it is not properly solved or understood, it is the weakest issue in quantum mechanics, making its position vulnerable in the mind of many people. Unfortunately, it is a puzzling and confusing matter, and there is no shared enough agreement concerning the interpretation it should received [41]. On the contrary, it is a matter of diverging opinions, even concerning the point to know whether a satisfactory theory of quantum measurement is even possible. For Jarrett [75], *no quantum-mechanically acceptable account of measurement interactions has yet been given. Although this is not a claim with which I would expect a great many physicists to agree, it is a claim for which ... a solid case is available.*

A consistent positivist posture might be to remark that this problem pertains to interpretation so may be to metaphysics and that we should not worry with it insofar as we possess adequate postulates to make correct predictions. But this is making *eyes wide shut* and *ears wide blocked up.*

An overall evaluation of the situation may be given by quoting Jammer [24] saying that *measurement, the scientist's ultimate appeal to nature, becomes in quantum mechanics the most problematic and controversial notion.* In the Copenhagen interpretation of quantum mechanics indeed, the act of observing, measuring, watching, looking at ... *produces* the outcome in the strongest sense. We really do not measure a property existing before the measurement (hence measurement is a misnomer) but, in a strong way, we "create" it, in compliance with probabilistic rules. So, indeed, what about the location of Einstein's corpuscle before observation and what about the relocalization of the corpuscle during the process of observation? The helplessness of Einstein on this issue is vividly illustrated when he says [76] that no one doubts that, at a given instant, the barycenter of the Moon is well located, even in the absence of any observer, actual or potential. But, if we rely on quantum mechanics, then one must admit that this location is the result of the observation, but did not exist before it. However, this conclusion, when applied to a macroscopic body instead of to an electron or an atom, is intuitively unbearable.

Indeed, we are particularly puzzled by the applications of quantum mechanical measurement processes to macroscopic objects, a puzzle made most weird by wondering where is the Moon when it is not observed or also: Where were the dinosaurs before someone, somewhere, discovered a first fossilized remnant? In the words of Bell [77] discussing quantum mechanics for the whole world: *Was the world wave function waiting to jump for thousands of millions of year until a single-celled living creature appeared? Or did it have to wait a little longer for some more highly qualifier measurer- with a Ph.D?*

The status of unobserved things, made so acute and precise in quantum mechanics, where it is performed on the scientific stage, has also previously been a philosophical or even metaphysical issue of utmost significance, all along the debate between realism and idealism. For Berkeley, the most idealist philosopher among the idealists, being is being perceived (his famous: *Esse est percipi)* and we may then say that things do not exist when we do not perceive them, excepted most obviously in the Mind of God, because He perceives everything. Here, I cannot resist to a bit of teasing. Indeed, Berkeley, the most idealist philosopher among the idealists, when he states that being is being perceived, that is to say that what is not perceived is not a being, simultaneously appears as the most positivistic philosopher among the positivists.

For Kant, in deep contrast with Berkeley, a contrast from which I feel authorized to name Kant a realist (because, for him, "things" do exist even if they are not perceived), although not the most realist among the realists, there would not be any deep problem with the unobserved swarming quantum world. It could possibly approach the transcendental Thing in Itself that we are not allowed to watch, unknowable for us. We are now going to see that possible proposed solutions to the measurement problem can indeed be more or less precisely located on a segment, with idealism on the left and realism on the right. And, if the measurement problem is difficult to put in mouth, some of the solutions are difficult to swallow for some people.

In particular, I shall not deeply discuss the Everett relative states or many-worlds interpretation which invokes the splitting of the universe, during the measurement process, between as many branches as the number of possible Eigenvalues [78]. This means in particular that, for Everett, indeterminism is an illusion resulting from the fact that we can only be aware of the particular branch to which we are attached, the restoration of determinism being then achieved if we consider the whole set of branches. This is an approach which is the most extravagant to my mind, although if it was the correct one, then it would make me much ashamed. If this would happen (I believe we shall never know), I would find a relief in the fact that Bell also did not like it [79]. For de Witt [80], the interpretation of Everett is however consistent with a meta-thorem telling that *the mathematical formalism of the quantum theory is capable of yielding its own interpretation*, an efficient way indeed to affirm that there is essentially no problems of interpretation in quantum mechanics, and in particular no measurement problem. Even if I find Everett's approach most repulsive, we have to admit that it contains some most interesting ingredients, in particular the fact that quantum mechanics is viewed as the universal theory, and that, therefore, there should not be any division between a classical universe and a quantum universe. An expedient and efficient way to summarize what we could feel concerning Everett's theory may be found in a quotation of an expert in superstring theory, named Maldacena, who stated that, when he thinks about Everett's theory from the standpoint of quantum mechanical theory, it is the most reasonable vision to believe in, but that, in everyday life, he does not believe to it [81].

If we decide to avoid considering the many-worlds interpretation, the measurement problem in quantum mechanics then arises from the fact that the linearity of Schrödinger's equation and the completeness of the wave function is not compatible, in general, with the fact that a quantum measurement provides a single outcome, i.e. that interference terms in a superposition of states (of the so-called apparatus) are suppressed to lead to a single outcome without any interference. Among the various approaches to be considered, let us note two possibilities discussed just below, namely (i) spontaneous collapse obtained by introducing non-linearities in Schrödinger's equation and (ii) decoherence theory in which interference terms are not suppressed, but diluted in the environment. We shall later on consider the solution provided by Bohmian mechanics.

More on the measurement problem (and on the problems of interpretations) by Jammer [24, 82], Wigner [74], Home and Whitaker [83], Holland [40], d'Espagnat [84, 85], or Bassi and Ghirardi [86], among others. A book by Louis de Broglie, summarizing some objections against quantum mechanics, and discussing the measurement problem in quantum mechanics versus a solution in the framework of the double solution is also to be recommended [64]. See also Bell and his charges against 'measurement', and even against the word 'measurement' [87].

The issue of awareness

Along a line initiated by von Neumann in 1932 [26], it is explained that, eventually, the collapse of the state has to occur in the mind of the observer. This is motivated as follows. The measurement process is to be seen as the result of the interaction between a microscopic object under study and a macroscopic measuring device. For Bohr, the macroscopic measuring device must be treated as a classical system, and the results obtained have to be treated in an objective space and time framework, allowing one in particular to share the results with other scientists, producing an inter-subjective agreement which is a contribution to the objectivity of science (without such an inter-subjective agreement, no objectivity would remain and science would become impossible). It is not quite certain whether Bohr always adhered to this point of view that we may call the point of view of the orthodox Bohr. For, in a book by Omnès [88], the author reports on a Gedanken experiment from Bohr concerning the uncertainty relations, and commented that this experiment implies that macroscopic objects must possess some quantum properties if the theory is to be coherent.

Nevertheless, the point of view of the orthodox Bohr produced some kind of rather peaceful concord between many people. It was in particular accepted by Landau and Lifchitz [56] who furthermore remarked that quantum mechanics is unique, in the sense that it requires its own limit theory (classical mechanics) to be understood. However, in utmost rigor, the point of view of Bohr cannot be taken as the ultimate answer. And this is because the measuring device is built from molecules, atoms, electrons, nucleons, quarks ... that is to say it is a genuine quantum object (even if it is big). It must therefore comply with the rules of quantum mechanics. Therefore, if, before the measurement interaction, the microscopic object to be observed is in a superposition of states, then, after the measurement interaction, this microscopic object plus the macroscopic device must also be in a superposition of states. In the quantum mechanical framework, there is therefore no room for any wave function collapse, a fact which may be viewed as a case of incoherence among the postulates. This inconsistency has also been pointed out by Squires [13]: *The theory tells us about the results of measurements but instruments that are capable of making such measurements cannot obey its laws.* Therefore, since a wave function collapse cannot occur in a physical quantum world, von Neumann concluded that it must occur in the mind.

Some people may find that this is a strange solution, even repugnant, to a strange problem: It looks like too much biased toward the idealistic left side of the idealism-realism segment. We have however to be aware of the fact, well stressed by Descartes and many other philosophers, that all we know is primarily in our mind. The existence of an external world, outside of us, although obvious and immediate it may seem to us, is an inference, already some kind of theoretical construction. Once this is accepted, we have to recognize that physics is making theories on a theory. And also, once it is accepted, we have to face the tremendous mind-body problem which never received a satisfactory enough solution. Would it be that quantum mechanics could, for the first time, drive this problem from philosophy to science? For example, Stapp [55] advocated a psycho-physical theory. In his framework, the physical world, just as described by our laws of physics, would be a structure of tendencies living in the world of mind. More idealistic than Kant, you die.

Unfortunately, von Neumann's solution to the measurement problem generates its own problems, as we may illustrate by telling the story of Wigner's friends. For Wigner indeed [89], the measurement process may be viewed as satisfactory *as long as only a single person's observations have to be considered. If another person undertakes an observation, the outcome of which is later to be communicated to me, one is forced to consider him as an apparatus and ascribe a state vector to him. After his observation, his state will be part of a linear combination of several observational results and this appears to be an unreasonable assumption considering that we always felt, after an observation, to have defined one definite result. There is perhaps no logical or mathematical contradiction here but the conclusion that the state of a friend is a linear combination of several states, indicating different contents of his mind, seems very unnatural. This leads at least me to the opinion that quantum mechanics, in its present form, is not applicable to living systems, whose consciousness is a decisive characteristic.* This argument is still more striking if, instead of only one friend, we consider a chain of friends, with a primary observation made by a first friend in a small laboratory, observed by another friend outside of the laboratory, in an university building, observed by another friend from the garden of the university, observed by another friend ... with Wigner observing from Mars. Furthermore, may be Wigner had more friends than he ever thought. Indeed, Shimony [51], referring to a certain Burgers, mentioned a much friendly, although *of course, ... highly speculative,* possibility: *That the proto-mental properties of elementary particles may function as hidden variables which help to explain the reduction of wave packets.*

Stapp [90] mentioned another difficulty associated with the free will of an human observer as follows. *One immediate apparent difficulty with the inclusion of the observer-scientist in the system they are studying is this: The observer-scientist can apparently invalidate any quantum-theoretical predictions they make about their own behaviors simply by acting contrary to those predictions.* Stapp also raised another closely related problem: *Consider an observer-scientist who is observing the instruments that record his own brain pattern. Suppose his observation of instrument-result A generates a brain pattern that produces an instrument-result B; and suppose his observation of instrument-result B generates a brain pattern that produces an instrument-result A. Then the observation of either state will replace it by the other.* Therefore, said Stapp, *the observer-scientists cannot obtain detailed knowledge about the states of their own brain without altering those states.* Hence, *situations in which the observer-scientists are included in the quantum system they are studying are logically different from those in which they stand outside that system.* This is reminiscent of Bohr because *the fragmented character of contemporary physical theory is an aspect of the exclusion — often mentioned by Bohr — of living systems from the domain of phenomena adequately treated by quantum theory.* In this sense, due to this exclusion, quantum mechanics would not be complete nor satisfactory.

Hence, according to Jauch [91], there is an *evidence that quantum mechanics has left us with an enigma notwithstanding the profound work of Bohr and the "Copenhagen school"*. To remove the skeleton from the closet, more realistic approaches should allow one to get rid of the mind of an observer and still more realistic approaches would get rid of the observer himself.

Spontaneous collapse

A very economical way (may be *too cheap*) to proceed is to amend quantum mechanics. Since the difficulties raised by von Neumann is the result of the linearity of quantum mechanics, associated with the concept of superposition of states, why could we not just jump outside of a too much restricted framework, and introduce non-linear terms in Schrödinger's equation? As I already commented, in particular in the foreword, this is not forbidden (at least, for the time being). We just have to introduce non-linearities in order to solve the difficulties, however adjusted in such a way that we do not take any risk of spoiling empirical predictions which agree with experimental results.

Such a non-linear equation has been proposed by Ghirardi, Rimini and Weber, in short GRW [92]. In their paper, GRW stated that they *present ... an attempt of ... a unified description through the discussion of a dynamical model in which linear superposition of states ... are naturally suppressed.* To summarize, GRW postulated that, most of the time, Schrödinger's equation indeed governs the evolution of the quantum state of a quantum system, but this is not the case any more for a discrete set of instants which are stochastically selected. At such instants, spontaneous reductions of the spreading of the states occur in the configuration space. When the system possesses a small number of degrees of freedom, the rate of these stochastic reductions is small, therefore essentially preserving the predictions of quantum mechanics. Conversely, it is fast for macroscopic systems possessing a huge number of degrees of freedom. In particular, this would defuse Einstein's objection concerning the localization of macroscopic objects. Also, in a measurement process, quantum mechanics would tell us that a needle in a measuring apparatus remains in a superposition of states, something that we never observe. Conversely, in GRW theory, this superposition of states would be destroyed in a very short time, through a spontaneous transition to a well localized state.

Three years after GRW's work was published, Shimony [93] stated that this work, producing a very impressive unification of microdynamics and macrodynamics, looked to him most promising. Later on, Bohm and Hiley [68] summarized this proposal from GRW by saying that they *propose nonlinear, nonlocal modifications of Schrödinger's equation that would cause the wave function actually to collapse. The modifications are so arranged that the collapse process is significant only for large scale systems containing many particles, while for systems containing only a few particles, the results are the same, for all practical purposes* (for all practical purposes: An expression made famous by Bell), *as those of the current linear and local form of Schrödinger's equation.* In 1991, a proposal similar to the one of GRW is provided by Squires [94]. A recent account of the GRW proposal, pertaining to a dynamical reduction program, is given by Ghirardi [95].

GRW may however be criticized on many grounds. In particular, objections to GRW may be found in Shimony [93] or in Albert and Vaidman [96]. One concern, according to Omnès [97], is that people did not succeed in reconciling GRW's approach with relativistic invariance. Another concern [97] is that there exists another approach, called decoherence theory, which does not affect the quantum mechanical framework but is instead developed inside this framework, and which may provide another solution to the Einstein's localization objection, and to the measurement problem. The decoherence process removes quantum interferences with a very small characteristic time, so small actually that it could prevent to observe GRW effects, if they exist at all. This also might mean that GRW spontaneous transitions might be immune to any falsification, in the sense of Popper. More than that, it *must be* immune to falsification since all observations in agreement with experimental facts predicted by quantum mechanics, and verified, must be recovered in any GRW-kind of approach, by construction.

Therefore, any approach of this kind is on the metaphysical side of Popper's demarcation line. Of course, Popper does not claim that metaphysical statements are meaningless and it could be that GRW theory, or better an improved theory of this kind, is correct. But, in the epistemology of Popper, such a theory would not be a scientific theory. This is certainly a weak point. For me however, the most important blow is as follows. The aim of GRW, in a certain sense, was to provide a mechanism to understand the measurement postulates which, in quantum mechanics, have to be accepted as they are, however without any intelligible mechanism to understand them. But, in GRW, there is no mechanism to understand spontaneous localizations which, therefore, also have to be accepted as they are. Therefore, GRW replaced a mystery by another mystery. Nothing deep has been gained. In contrast, decoherence theory is in a much better position.

Decoherence theory

In quantum mechanics, a probability is basically obtained by squaring the modulus of a certain scalar product involving Ψ (Postulate 4), in particular $|\Psi|^2$ provides a density of probability of presence, according to Born's rule. Then, if Ψ is in a superposition of states, the above operation will provide extra-terms which are quantum interference terms (e.g. see Feynman [14] for a pedagogic, and more general, account of such features). If Ψ describes a macroscopic (measuring) apparatus, then the superposition of states prevents, in principle (but not necessarily in practice as we are going to discuss), the apparatus to point to a single outcome.

The basic idea of decoherence theory is that such interference terms are diluted in the environment. Here, von Neumann was right: We cannot get rid of such terms within the framework of a linear quantum mechanics. And therefore, without the postulated action of the mind invoked by von Neumann to explain the measurement problem, we cannot in utmost rigor solve the measurement problem, insofar as we cannot get rid of quantum interferences. Decoherence theory agrees with that, excepted for the postulated action of the mind, but it tells us something more concerning the practice (that is to say for all practical purposes). Indeed, quantum interferences are not removed, but their influence is diluted. This is the way by which classical phenomena emerge from the quantum background. This is also the way allowing a specific and unique outcome to appear in a quantum measurement, from a superposition of states. This may make our mind feeling comfortable by reconciling, to some extent, quantum mechanics and classical physics, and by justifying our macroscopic intuition of the macroscopic world (through a kind of coarse-graining process). This however does not remove the strangeness of the background which remains active: It is diluted, averaged, but still present. For an easy access to decoherence theory and more details, the reader may refer to high quality popularized accounts, such as by d'Espagnat [84, 98], Omnès [15, 88, 97] or by Gell-Mann [20].

In the formulation of decoherence theory, Schrödinger's equation is preserved in all its integrity and the measurement process is viewed as a quantum interaction, i.e. the measuring instrument is treated as a quantum system. Theoretical evaluations and experiments (e.g. [99]) demonstrate that decoherence phenomena occur in a very short time, much shorter in any case that the one associated with GRW's theory. For instance, the inescapable interaction between a piece of dust and the microwave photons of the fossil radiation of the universe is sufficient to force it to a precise localization in space, therefore compensating the natural spreading out behavior generated by Schrödinger's equation. The process is still more efficient for larger macroscopic bodies. This answers Einstein's objection concerning the localization of a moving sphere, the correlated problem of the localization of the Moon, and a similar objection raised by Fermi to Gell-Mann, concerning the question to know why the planet Mars is not diffusively spreading all over its orbit [20]. Examples of values of decoherence times are given in Omnès [88].

Is however decoherence theory the perfect and final solution to the measurement problem? Well, not necessarily. At least this may be (and is) controversial. Because there is no rigorous solution to the measurement problem in the framework of the linear quantum mechanics (just because it is linear), there must be some lack of rigor in decoherence theory. How this lack of rigor manifests itself? The answer is that it manifests itself by the occurrence of vanishing probabilities. The question is now to know how we should interpret these vanishing probabilities, and what we should do with them, without however providing a clear-cut final solution satisfying any one, being content, in this controversial matter, with questions rather than with answers.

Let us denote by τ the (very small) decoherence characteristic time, that one which leads to the disappearance of quantum interferences. In technical terms, such a disappearance may be expressed by saying that the density matrix relaxes very quickly to a diagonal form. The non-diagonal terms basically tend to zero according to an exponential decrease of the form $\exp(-t/\tau)$. With τ very small, this is indeed a very efficient relaxation process. To be more explicit, let us provide some figures. Assume a state made out from the superposition of two states A and B. Due to the interaction with the environment, let us assume a decoherence time equal to 10^{-36}s. This is typical for a piece of dust in a terrestrial atmosphere [88]. After only one femtosecond (10^{-15}s), the value of $\exp(-t/\tau)$ is equal to $\exp(-10^{21})$ which is indeed ridiculously small, so small indeed that we can seemingly set it to zero, straight away. Then, the superposition of states, in practice, does not exist any more. In other words, decoherence theory, pertaining to the framework of the usual quantum mechanics, explains why we apparently observe macroscopic objects as classical objects, although they have to obey quantum laws. But, what it explains is an *appearance*, an *appearance* valid for all of us indeed, but not the "real thing" [98]. For d'Espagnat, this "real thing", the "Real", is "non-physical" in this sense that it is prime with respect to particles and fields, likely prime too with respect to space-time, and does not come under physical laws, although it is willingly conceived as the *source* of the physical laws [98].

Let us now make our discussion of decoherence theory more explicit by using the famous sadistic example of Schrödinger's cat experiment. Let the states A and B denote "cat alive" and "cat dead" respectively. After the superposition of states has disappeared in a very short time, we may for instance be left with only "cat alive", the most sweetest option. Let us note that the interaction between the cat states and the environment inside the box containing the whole experiment is sufficient to drive the decoherence. Therefore, when we open the box, we observe that the cat is indeed alive. With this scenario, the manifestation of the living cat does not suddenly emerge when we open the box, but was already actual inside the box before our opening it. However, more precisely, what we have is still a superposition of states of the Eigenstates "cat alive" with a probability $(1 - \epsilon)$ and "cat dead" with a probability ϵ, with ϵ vanishingly small. It is only by declaring that ϵ, which is vanishingly small, can be set to zero that the Schrödinger's cat paradox is indeed solved. According to Omnès [15], such a reduction from "approximately zero" to "zero" relies on what is known as the "zero Borel axiom". According to this axiom, a probability which is sufficiently small must be considered as being zero, and all associated events must be declared as impossible. But this is exactly where the shoe hurts, where there is a stone.

Indeed, not everyone agrees with the decoherence theory interpretation, as acknowledged by Omnès himself [15], one of the defenders of this approach. Also, Bernard d'Espagnat, following Bell, does not contest the practical validity of the conclusions of the decoherence theory, which is admittedly valid for all practical purposes. But this does not mean that decoherence theory actually solves the measurement problem: In utmost rigor, from the deepest fundamental point of view, a vanishing probability is not a probability equal to zero. To correctly understand what is here going on, let us discuss the concept of vanishing probabilities, with two simple examples which, as we shall see, are from two different natures.

In the first example, let us assume that we have to take, today, a decision which may however be differently motivated, depending whether, to-morrow, a certain event A or another incompatible event B will occur. Let us also assume that we are informed enough to evaluate the probabilities of these events, and that we find that the probability of the occurrence of the event B is vanishingly small. Since we are supposed to make a decision, we shall use the zero Borel axiom and most reasonably bet that the event A will occur with probability one. So we can take a decision, the most reasonable decision, which in practice is without any doubt the best one.

In the second example, we leave the world of everyday experience of the first example, to go back to the quantum world. And we carry out two different versions of the Schrödinger's cat experiment. In the first version, we do not use any more a machinery which can, with a certain probability, deliver a poison, making the cat dead or alive with certain probabilities. We rather use another machinery which may drop or not, with certain probabilities, a piece of sugar in water, so that now our basic Eigenstates are "sugar dissolved" and "sugar not dissolved", assuming for the comfort of this Gedanken experiment that the process of sugar dissolution is instantaneous. Once decoherence has operated, in a very short time, we are for instance left with "sugar dissolved" with a probability $(1 - \epsilon)$, or "sugar not dissolved" with the complementary probability ϵ, or the converse way. Assuming that ϵ may be identified to 0, we are sure, when opening the box, that we shall observe only one of the Eigenstates, and not a superposition of states.

Let us now go on proceeding with a second version of the experiment, which seems logically identical, but that we should be reluctant to interpret in the same easy way. In this second version, we return to the use of Schrödinger's machinery, with poison liberated or not, but we make it more sadistic by replacing the cat by a human being (even if this is legally strictly forbidden). Now, there is something very special with a human being. He can say: *Cogito, ergo sum* (awareness again). But what can be the meaning of "human dead", with a vanishing probability ϵ of saying that he is thinking? Should we state as Bohr did somewhere that quantum mechanics does not apply to living entities [52]. But where is the frontier between inanimate objects and human beings? Or even: Does such a frontier actually exist? In the present case, we do not claim, as with von Neumann, that awareness is necessary to produce the collapse of the wave function. But the above example shows that thinking about awareness (being aware of awareness) may help us to question the limits of our theories.

According to Le Bellac [16], there is a converging agreement according to which the solution proposed by the decoherence theory is satisfactory for all practical purposes, but that it is not relevant (it is a non starter) for what is supposed to be a fundamental theory.

In 1971, Wigner [89] could write that *the probabilistic behavior of the measurement process, as postulated by the standard theory, cannot be explained,* and also: *Clearly, we enter a chain of questions which has no ending and must conclude that the self-contained nature of quantum mechanics is an untenable illusion.* Those who believe that decoherence theory indeed provides a solution to the measurement problem may find such quotations rather old-fashioned. However, those who are not convinced may find that they are still valid and relevant.

Participatory universe

The existence of a quantum of action is the most fundamental, and actually the only one, ingredient which alone is sufficient to generate all the difficulties and strangenesses encountered in quantum mechanics. That the existence of a minimal quantity could produce such devastating consequences was already anticipated by Aristotle. In his treatise on heavens ([7], book I, Chapter 5, entitled: *The Universe is not Infinite*), he wrote that, if we introduced a minimal quantity, it would shake the most important aspects of mathematics. This did not actually happen in the intellectual realm of mathematics, but does happen in the sensible realm of physics.

Indeed, if we ask the question to know what is ultimately the deep reason for the existence of the measurement problem, we may reply, quoting Bohr [100]: *The finite magnitude of the quantum of action prevents together a sharp distinction being made between a phenomenon and the agency by which it is observed.* Effectively, if the quantum of action were equal to zero, or vanishingly small, then the interaction during a measurement could be made vanishingly small too, i.e. could be made without, as some people would say, "disturbing" the observed object (whatever this object is or could be). We would have, as in classical physics, the opportunity of measuring something without interacting with this something (or interacting very weakly, or even interacting in such a way that corrections could be evaluated to adequately modify the net results of the measurements). Bohr, on this topic, also expressed his view as follows [100]: *The discovery of the quantum of action shows up, not only the natural limitation of classical physics, but, by throwing a new light upon the old philosophical problem of the objective existence of phenomena independently of our observations, confronts us with a situation hitherto unknown in natural science. As we have seen, any observation necessitates an interference with the course of the phenomena, of such a nature that it deprives us of the foundation underlying the causal mode of description. The limit, which nature herself has thus imposed upon us, of the possibility of speaking about phenomena as existing objectively finds its expression, as far as we can judge, just in the formulation of quantum mechanics.*

In the words of Rosenfeld [101], an ardent defender of the orthodoxy, we have no more choice in quantum mechanics: The existence of the quantum of action is incompatible with the classical laws, and complementarity therefore appears as the only rational possible interpretation of the situation. And also: ... by making directly intervene the interaction between the studied system and the observation means, it (quantum mechanics) questions the determinism and the objectivity of physical laws ... This intervention of the idea of probability in the theory of quanta, as a pure contingency in the words of Louis de Broglie [62], is a stumbling block for many physicists, including some of the most distinguished ones. Regarding this, one likes to quote a witticism from Einstein "That the good God plays dices, let us accept it: But that He plays dices according to rules, this is something I cannot conceive". But, in quantum mechanics, the very definition of phenomena requires the specification of the circumstances of the observation. Therefore, there is nothing to oppose against the fact that phenomena, essentially depending on the quantum of action, could present an intrinsic statistical character, inherent to their nature and from this irreducible.

Of course, we could possibly imagine that there is a wave function of the universe, something that we implicitly admitted above when we quoted Bell discussing quantum mechanics for the whole world. But this was just a manner to tell the things. One of the main ideas of the Copenhagen interpretation is that we need an external observer, in agreement with the quantum mechanical postulates which describe how the observation process takes place. Indeed, for an observation to be made, we need someone making the observation. So, in utmost rigor, quantum mechanics does not apply to the whole universe if, by definition, there is nothing outside of it. May be we need God to give sense to the concept of a wave function of the universe, but this is certainly driving us too far away. For this reason alone, quantum mechanics cannot be the most fundamental theory. It can only tells us something about things interacting with things in our world. And then, for us, who are embedded into that world, and who, when observing, have to interact with something which is not us, this being done with the limitations imposed by the existence of the quantum of action, this universe must be a participatory universe.

The idea of a participatory universe has been discussed by Patton and Wheeler [102], and Wheeler [32, 67]. Wheeler stated [67] that *the act of measurement has an inescapable effect on the future of the electron. The observer finds himself willy-nilly a participator. In some strange sense this is a participatory universe.* He also stated [32] that *useful as it is under everyday circumstances to say that the world exists "out there" independent of us, that view can no longer be upheld.*

In this conception, we may possibly get rid of the awareness of the observer which is no more the main issue, but we cannot get rid of the observer himself because he pertains to the universe and is irremediably connected to it through any interaction, at least by one unit of the quantum of action. This too holds for an unanimated observer such as an automat recording the results of a quantum experiment. The observer, whatever it is or whoever he is, cannot really stand out of the world like a transcendent God who would let the universe running, possibly still knowing what happens thanks to some kind of transcendent vision, or like an "inhabitant of Sirius". He cannot be isolated from its surrounding but is co-extensively immerged in it. This does not mean by any means that we have to accept a strong version of idealism, that the universe does not exist if it is not perceived. It simply means that, if it is observed, a participative interaction is necessary. The observer in the universe is radically in, and not outside, and this, finally, is nearly some kind of obvious or even tautological statement. Quantum mechanics, from this point of view, makes obvious what should has been obvious from the beginning.

This might be the right place to recall a statement from Eddington [11]. He said that, at the frontiers of the unknown, we discovered strange tracks. We invented exhaustive theories to elucidate their origin. Eventually, we succeeded to reconstitute the creature who left them. And here it is: The tracks are our own tracks.

But also we are here pointing out to a problem which is called vertiginous by Patton and Wheeler [102], a problem of self-reference. We have to think the universe, but we are embedded in it in a participatory way. We would may be prefer to stand out of it, as an external observer, but this is forbidden to us. I do not believe that I am able to have, at the present time, any satisfactory intuition, or any entry, to this issue. It really gives me a vertigo. I also cannot prevent me of believing, due to the word self-reference, that this issue might have some connections with Gödel's completude theorems (to which we shall return later). Therefore, after all, this long-standing measurement problem of quantum mechanics might very well go on standing in front of us (or in relation to us) for a long time. It is fortunate, for the applications of quantum mechanics, that it does not matter too much for all practical purposes.

Chapter 5

Philosophical Aspects of Quantum Mechanics

The themes to be discussed in this chapter are essentially the same than the ones discussed in the previous chapter, or at least strongly correlated with them. However the previous chapter discussed mostly scientific features, possibly with a bit of philosophy, while the present one is going to discuss philosophical issues, possibly with a bit of science. Actually, for a long time, philosophy encompassed all aspects of human questioning. Progressively, when time went on, philosophy and science separated and, today, most scientists believe that the divorce is accomplished. And we might be the children of divorced parents.

Philosophy is however the Great Mother and Science is the Great Son (having became a Great Father). When dealing with hidden variables, we have necessarily to put together again the whole family. Furthermore, the topic of hidden variables is not only a pure scientific topic (if there exists any) but also a philosophical topic. It is impossible to discuss the issue without possessing some necessary amount of basic philosophical knowledge.

The König of Königsberg, transcendental critique

Immanuel Kant (1724–1804), the philosophical König of Königsberg, the intimidating author of what he calls himself a Copernician revolution in philosophy (so intimidating that even some expert students in philosophy might prefer to speak on any other philosopher rather than on Kant), is an inescapable philosopher when dealing with philosophical aspects of quantum mechanics. We cannot pass around him, we have to pass through him. There was a time for being before Kant, and a time for being after Kant. He devoted his life to the study of three questions: What can I know? What should I do? What can I hope? By providing a deep reply to the first question, in his *Kritik der reinen Vernunft* [103], he created the transcendental critique and became acknowledged as the Founding Father of the theory of knowledge (or, at least, as one of the Founding Fathers).

His pertinence for a discussion of quantum mechanics, particularly of its interpretations and epistemological nature, is confirmed by the many references made to him, such as by d'Espagnat [84, 85, 104], Omnès [15] or Heisenberg riding under Kant's banner, and revisiting his philosophy in the context of modern physics as in [53] and also in [10] where chapter 10 is devoted to the discussion of the relationship between quantum mechanics and Kant's philosophy.

The two main ingredients of Kant's philosophy which concern us are (i) the partition of the objects of knowledge between the Thing in Itself (noumenon) and the things as they are, not in themselves, but for us, knowable thanks to the sense channels: The phenomena forming the Phenomenon (ii) the existence of *a priori*'s, that is to say of some kinds of knowledge which are available in advance of experience, and which are independent of experience (I am avoiding to introduce the distinction between the three kinds of propositions put forward by Kant, namely analytical propositions which are necessarily *a priori*, i.e. before experience, synthetic propositions which, according to Hume, are always *a posteriori*, i.e. coming after experience, and the third category, introduced by Kant, concerning the possibility of synthetic propositions *a priori*).

For the first ingredient, let us recall that The Thing in Itself is the origin of our sensations. It is a background from which we cannot know anything (or, if you want to be more cautious, it happens that if we knew something about it, we could not know that we know). It is outside of space and time, and is not conditioned by the law of causality. It is an independent reality, the noumenal, outside the realm of any possible experience.

The underlying quantum world, I mean the one between observations, the one which underlies our sense data and which is at the origin of our sensations, cannot be identified with the Thing in Itself. Indeed, this underlying quantum world is speakable, since we can at least say that it is driven by Schrödinger's equation. In contrast, the Thing in Itself is absolutely unspeakable, in the strongest sense of the word. But the concept of Thing in Itself is nevertheless somehow reminiscent of the quantum hidden mysteries. With our quantum mechanics, we might be much closer to the Thing in Itself than with classical physics.

Still concerning the first ingredient, we now have to discuss the Phenomenon. While the Thing in Itself is a background, the Phenomenon is our foreground. It is what seems to exist, in contrast with the Thing in Itself which is what exists. The Phenomenon is our representation and only such a representation can be the object of science, this science which, by a self-referent skilful trick, produces itself a representation.

The second ingredient concerns the question to know how the Phenomenon is formed from the Thing in Itself. For Kant, all our knowledge begins with sense data, then flows to our understanding, and is shaped by our reason (*nihil est in intellectu quod non prius fuerit in sensu*). This may seem innocuous, but the mentioned shaping is to be understood with a very strong meaning. Things may succeedingly be submitted to our ideas only because our mind (the transcendental self) imposes its views and its framework. Our representation of the foreground depends on the organization of our mind in such a way that, what we know from the things, is only what we are ourselves printing on them. In other words, the order of nature is legislated by the mind. This is the Copernician revolution of Kant: Our mind is not turning around the world, but the world is turning around our mind.

It is worthwhile however to note that the possibility of such a philosophical Copernician revolution was anticipated by an intervention of the honest Sagredo in the famous dialogue of Galileo Galilei on the two systems of the world [34]. Commenting an opinion of a certain author (Scipio Chiramonti), who, as a peripatetician, has been fighting against the astronomical Copernician revolution, Sagredo stated that the intellectual behavior of this author could be understood by assuming that he believed that the nature first built the brain of men and thereafter arranged the things in adequation with the ability of their minds. Sagredo rather believed, he said, that the nature started by making the things according to its own way, and that it thereafter fabricated human reasonings capable to grasp, although painfully, something from its secrets. We are here facing the rather ironical case of someone (Scipio Chiramonti again) being an opponent to the astronomical Copernician revolution of Copernicus, and akin to a defender of the philosophical Copernician revolution of Kant.

A much softened version of the Kantian revolution, as expressed for instance by Heisenberg [10], might better satisfy scientists nowadays. For Heisenberg, the nature was made in such a way that it could be understood or, much better, that our intelligence is made in such a way that it could understand nature. The motivation for such a conviction was that it was the same regulating forces which acted to build the nature in all its forms, and which are at the origin of the structure of our soul, and therefore of our intelligence too. For an orthodox Kantian, this point of view of Heisenberg may hardly look like Kantian. For a modern scientist, it could look like a reasonable re-interpretation of Kant.

According to Kant, the components deeply anchored in our mind, used to shape our experience, are space, time, causality forming what I have already called the classical trio, and which are Kantian *a priori*'s. For Kant, space and time are not things in themselves. They are not empirical objects of our knowledge, abstracted from our experience. On the contrary, they constitute the very *a priori* conditions of the human experience. They are *a priori* forms of our sensibility, the conditions of the existence of the things as phenomena. As for the principle of causality according to which everything which happens is the successor of something which happened previously, *in compliance with a rule*, it is an *a priori* form of our understanding. It is ourselves who introduce order and regularities in what we call the world. We could not find them in the world if they were not already in our mind, and they are in our mind due to its very nature.

Let us remark an important point in the above definition of the principle of causality. The plain succession of events along the time does not suffice to define a causality. It must furthermore be accomplished according to a rule. Therefore, the principle of causality requires time plus regulation. However, Kant does not extend himself sufficiently on the kind of rules to be accepted. The rules could be statistical or indeterministic as in quantum mechanics (such rules are rules indeed). With such an understanding of Kant's definition, we could say that quantum mechanics is both indeterministic and causal. This is certainly something we would not like to say and certainly not something that Kant would have liked to say. The rule invoked by Kant must be deterministic indeed, so that the words deterministic and causal can be identified. But, this being done, the definition given above becomes logically shaky. It might indeed now be rewritten by saying that the principle of causality is the principle according to which everything which happens is the successor of something which happened previously, in compliance with a causal rule. We are then facing a circular definition. To avoid it, a better rewording is necessary. Here is a possible one: Successors are determined by predecessors in an unique way (no alternative) in compliance with a rule. It must also be clear that causality only concerns the necessary connection between events when time goes on, but does not concern the ability of human beings to evaluate in advance how this connection will develop, *even if the rule is known.* An emblematic example is the case of a chaotic process which is both deterministic and unpredictable.

The doctrine of Kant, which tells us that the Phenomenon is only a representation, not the Thing in Itself, and that we may understand the Phenomenon because we give him a structure which is the one of our mind, is called transcendental idealism. Clearly however, the *a priori*'s of Kant deeply conflict with the quantum mechanical Copenhagen interpretation which rejects the classical trio.

Although philosophers usually drastically oppose idealism and realism, I would like however to express my opinion that the transcendental idealism of Kant is actually a realism. It is a realism because it accepts, and even assumes, a deep world outside of us, the Thing in Itself which is the external source of our internal representations. In utmost rigor, excepted may be for Berkeley's conception (although the God of Berkeley could be seen as being the external reality), there is only one consistent idealism, namely solipsism, that only mad people could seriously defend. The fact that Kant idealism is a realism can be made something like obvious if we simplify Kant philosophy, with however the risk of a bit of distortion. A possible summary could be, in plain nowadays words, that (i) the world which is at the origin of our sense data cannot ultimately be known and (ii) the analysis of our sense data depends on the structure and capacities of our brain. Most of the realists, even naive, might agree with such statements. What is forgotten however in this degenerated account, and which will require itself a critique, is the validity of the assertion that space, time, causality are *a priori*'s forced by the way our neuronal net works. Let us however remark that naive realists such as Einstein (this will have to be refined) commit, in Kant's spirit, an usual transcendental mistake: The one to confuse the Thing in Itself and the Phenomenon. Or, in other complementary words, for Einstein, a scientific theory should not only allow us to make efficient predictions, it must also inform us on the structure of the world, not only things for us, but also things independently from us.

It is also interesting to comment on the relationship between realism and the participatory universe. If we define the realism as the doctrine according to which the regularities observed in the phenomena have their origin in a kind of reality which is independent of the *existence* of the observers (and, once again, this implies that Kant is a realist), there is no conflict between realism and the participatory universe of quantum mechanics. Indeed, our definition, emphasizing the word *existence*, tells us nothing on the modes of interaction between the observers, *when they exist*, and the underlying reality. In other words, in a participatory universe, the world exists independently of the observers when they do not exist but, when they exist, they have to participate.

As a last word, I would like to mention that Kant also introduced a solidarity principle according to which all substances, insofar as that they are simultaneous, form an universal community. This looks like an impressive anticipation (or retrodictive reminiscence?) of quantum entanglements, later to be more extensively discussed.

Criticisms of the critique. The Phenomenon

Kant's critique has been criticized, at the first place by Schopenhauer (1788–1860) in his famous book [105]. This book from Schopenhauer starts with a first chapter entitled *The Principle of Sufficient Reason* (another expression for the principle of causality, borrowed from Leibniz), elaborating on its *a priori* value, and also an appendix entitled "Critique of Kant Philosophy" which, as a whole, is rather severe. Later on, the philosophers of the Vienna Circle were also critical against the critique [106], as well as many other philosophers. The preface of Serrus to the *Critique de la Raison Pure* [103] also contains a critique of the critique. I shall myself provide a may be more ingenuous critique, however directed to quantum mechanical concerns.

The first issue to be discussed deals with the dichotomy between the Thing in Itself and the Phenomenon which, obvious as it seems to be, is nevertheless a bit problematic. There is no severe criticism to be done against the Thing in Itself, excepted if you do not believe in it. It is easy to conceive that the Thing in Itself is THE Thing in Itself, that is to say that there is one and only one such Thing. Also, it is not too hard to accept that this ultimate background is outside of space, time, and causality, as advocated by Kant and Schopenhauer. In any case, this concept is clearly enough defined as the unknowledgeable origin of our sensations and representations.

But what is the exact meaning of THE Phenomenon? It does not look like being a single mental object, unified in a single unified mind, excepted possibly if you are a follower of the rather mystical vision of Teilhard de Chardin [107] advocating the existence of a noosphere, that is to say a sphere of human thoughts, a collective consciousness associated with the human kind. Instead, the Phenomenon rather looks like being exploded into many individual phenomena, one phenomenon for each individual, and, furthermore, each of these individual phenomena is not a well defined fixed entity because it is not static, but evolves all along the life in a dynamical way. We could possibly define a phenomenon for each individual but it should be specified at a given time, leading to the concept of an individual phenomenon as a continuum of mental states. But should we limit ourselves to awareness? What about subconscious psychical elements? What about the phenomena to be associated with each dog, cat, butterfly, earthworm ... flower... stone and piece of dust ...?

But, after all, the concept of Phenomenon has been designed by Kant for us, human beings (and more generally for reasonable living entities, anticipating the existence of aliens?), not for cats and dogs. However, even so, it is a very abstract and unnatural concept, even a metaphysical one. The illusory Phenomenon of the human kind is dispatched in libraries and bookshops, journal papers, proceedings ... and expressed in many evanescent or evaporating oral presentations, or posters. It is also of a dynamical nature, justifying the history of sciences, through the growing and accumulating process of the so-called progress of sciences, however regularly disrupted by the death of theories, revolutionary breakthroughs, and the burning of libraries, whether they are accidental or the achievement of vandals.

Now, what about the aim of science? Whether it is merely to achieve correct predictions (to *save the appearances*) or to tell us something of an ontological value, it cannot be to reach the Thing in Itself which, by definition, cannot be reached. It can only deal with the Phenomenon (if we succeed to define it properly). And this gives us indeed the opportunity of forming a decent concept of this concept. Individual phenomena (mine? yours? the one of the neighbor?) cannot, in a strict sense, be attached to the notion of science because these phenomena are highly subjective, and most often deeply conflicting. They must be subsumed in an objective agreement, based on discussions, arguments, experimental checkings, corroborations and falsifications, pointing out to a potential universal agreement if enough time and energy are spent to direct ourselves toward this goal. The Phenomenon may then be viewed as an universal potential representation resulting from an inter-subjective accord. Such an inter-subjective game rules out individual fluctuations and generates the objectivity of science. Although it provides some kind of probing of the unfathomable Thing in Itself, it is a societal feature whose objectivity is the result of the methodological condition that science has to be shared, or potentially shared, by individuals. This was the view of Bohr. And this view led him to the idea that the outcomes of quantum measurements, to be possibly shared by anyone and everyone, should be eventually expressed in a classical framework, with measuring instruments being viewed as classical instruments. A bridge too far, as we shall discuss.

And, at the end of the progress of science, available to human beings, there might be an asymptotic Phenomenon, that we shall never reach but always approach, our only possible ultimate landscape of knowledge. We may call it the World (although, later on, we shall question his very existence, or our ability to approach it closely enough, as we already questioned the Theory of Everything, and as many philosophers would relevantly do).

Criticisms of the critique. The classical trio

Kant outmoded

Another weak point of Kant's philosophy concerns the validity of his *a priori*'s. Indeed, to begin with, our concepts of space and time have considerably evolved since Kant. They do not identify any more with the absolute space and time of Newton, and the very concept of simultaneity, essential to Kant, has been ruined, and relegated to the cemetery of dead ideas. It is well documented that Kant (1724–1804) has been much influenced by or even impregnated with the physics of Newton (1642–1727) so that the classical trio pertains both to Newtonian physics and to Kantian philosophy. Of course, we cannot blame Kant for having ignored the immense advances later made in mathematics, with the differential geometry, the tensor calculus, the Riemann spaces ... and the geometrization of physics achieved by Einstein with general relativity, in which the physical space, instead of being simply a static arena, became dynamically intertwined with time. But, during Kant's lifetime, geometry had already made some steps further away, and it is not too unfair to say that Newton was behind some of his contemporaries. Nevertheless, in any case, the huge developments of the concepts of space and time after Kant shed beams of shadow on Kant's doctrine. Of course, we shall still make use of Newton-Kant space and time for our everyday experiences (in about the same way that we are still going on saying that the Sun rises), but scientists did not have too much difficulties to tame their modern avatars.

On an genealogy of causality

Concerning the concept of causality, and its validity as an *a priori*, I shall not refer to scientific developments after Kant because this would lead us directly to quantum mechanics, this quantum mechanics which is actually in question in this book. Starting from the idea that quantum mechanics rejects strict causality, to take advantage of this idea to *a priori* accuse causality of falsity, and finally to jump from this falsity to a justification of indeterminacy in quantum mechanics, would obviously be somewhat circular, and something that would make any logician roaring. Rather, we may rely on a genealogy of the concept of causality. Building such a genealogy would be a very ambitious plan and, if it was achieved, it could not fit anyway into a section of this book. I shall therefore be content with an unfleshed skeleton.

Presumably, the "prehistoric history" of causality has to be found in the introspection, when human beings, aware of their free will (even if it is an illusion) extended it to all objects of the outside world. There might be some embryonic idea of causality in such an animist conception, because free will produce effects and therefore may be viewed as a cause. But it is certainly not a mechanistic causality and, more precisely, it is not a causality in the sense of Kant because, for him, causality must proceed *in compliance with a rule*. Free will, on the contrary, at least for certain philosophers and theologians, does not work in compliance with a rule: It is the rule, produced by a psychical power. The extraction of mechanistic causality as a set of rules governing the events has been a slow long lasting process, in which the free will has progressively been denied to inanimate objects, to animals, became the privilege of human minds, and eventually, even denied to human minds by materialist doctrines. Such a genealogy demonstrates that the modern concept of causality is far from being an *a priori*. Otherwise, it would have been dominant from the beginning. The principle of causality is already the result of a theoretical elaboration.

Animism can still be observed today in many countries, such as in Japan with its inspiriting Shinto (the Way of the Gods), or in Africa. But it actually gave birth, along many filiations, to various religions and philosophical systems. We may also observe it in the everyday life, around us, by watching children, closer to the primitive mentality than to the physical causation of scientists, and endowing objects with some kinds of willing animal personalities.

The invocation of causes, in the modern sense, can already be found in the ancient times. An author of the school of Hippocrates, four centuries B.C., regarding epilepsy, could write: It looks to me that the disease is no more divine than any others. It has a natural cause as for the other diseases. Men believe it is divine only because they do not understand it. But, if they called divine everything they do not understand, really, divine things would be endless [8]. But this single quotation does not reflect the complexity of the debates in the time of the old Greeks [3, 4]. Questions on causality, and debates on causality, indeed took place, but apparently in a rather disparate controversial way, without however any real attempt to produce any philosophical systematization, in contrast to what will later be achieved by Aristotle.

The perception of the problem, in a psychologically natural filiation with animism, can be identified as being mythical and/or epic with Homer (without forgetting all myths produced by the human mind, from many parts from Asia, or from civilizations prior to the Bible, actually from everywhere, proceeding by images to picture the inexpressible, or to express the unthinkable, a long time before the advent of the babblings of our formal thinking). For Heraclitus, everything was filled with souls and demons. Empedocles expressed himself on the nature of the numerous demons who, wandering here and there, interact permanently with human affairs. For the Pythagoreans, the air is filled with souls, presumably from the heroes and demons. With the atomists (Leucippus and his disciple Democritus), a modern wind is however blowing. For Leucippus, nothing happens fortuitously but everything proceeds from reason and necessity. Similarly, for Democritus, everything which is generated is ruled out by necessity because the cause of the generation of everything is the eddy, named necessity. Also, the list of books and works produced by Democritus contains about ten items exhibiting the word "cause" in their titles. Later on, Aristotle will criticize Democritus for having omitted the final cause, that is to say by explaining everything from the necessity of what we call today the laws of nature (how modern, indeed). It has been reported that Democritus would have say that he would prefer to find a single causal certainty rather than becoming the king of the Persians. According to some sources, the opinion of Democritus could have been a bit less extreme, admitting several kinds of causes viewed as possible. Things could happen due to necessity, destiny, free will or by coincidence. According to a belief from this time, the death, this most extraordinary thing in a human life, that we can anticipate and, may be, not remember after the death (in contrast with birth that we, may be, did not anticipate and cannot remember), cannot be the result of a mechanistic causality. Before, the preoccupation of Thales has been to replace the mythical explanation by a physical explanation, while Anaximander noticed that his contemporaries were still very close to myths and to the primitive chaos.

Plato, in a very modern way and may be explicitly and extensively for the first time, connected the causality and the world of mathematics. In his Timaeus [6], he stated that, in order to reach a partial knowledge of the future, it is necessary first of all to admit that everything which happens has to happen under the effect of a cause. Afterward, he proposed for the sensible world a mechanistic explanation exclusively founded on a mathematical formulation of the constitution of the elementary components and of their transformations. This causal necessity is however subordinated to higher level causes involving intellect and finality. Clearly, we have to recall here our story on the beginning of theoretical physics. Indeed, for Plato, sensible things in the sub-lunar world keep on changing (in contrast with the permanent supra-lunar world), but the changes exhibit a sufficient amount of regularities to allow us to know them and to speak of them. The ultimate explanation of these changes is to be found in mathematics. As we have seen, each element is associated with a perfect solid, and the evolution of the elements (therefore of the perfect solids) is connected with the mathematical structure of the spirit of the world. Furthermore, some amount of animism is still present because the planets and the stars are divinities, that is to say living beings, endowed with a body (essentially made out from fire) and with a soul involving an intellect allowing them to understand what the demiurge is expecting them to do.

Aristotle, the disciple of Plato, after reviewing the work of his predecessors, offered in his physics [33] and his metaphysics [108] a clear systematization of the concept of causality and distinguished four kinds of causes: Material, formal, efficient, and final. He essentially gave an explanatory priority to the final cause (consistently with the fact that he was viewing the universe as a kind of organism), which is rejected from our physics, while the efficient cause, the only one we retain in our physics, the one of Newton and of Kant, was relegated by Aristotle to a subsidiary role, excepted under special circumstances. For instance, as an example of such a special circumstance, an eclipse of the Moon does not have a final cause but an efficient cause which is the interposition of the earth between the Sun and the Moon [108]. Beyond this apparent rational theory of the causality by Aristotle, there are many remnants of animism, may be more however in the mind of Aritotelicians than in the mind of Aristotle himself. The system of the world of Aristotle used 55 spheres which, besides the Prime Mover, or Unmoved Mover, required fifty-five secondary unmoved movers, to make the whole system working. Extrapolating, we arrive to the idea that each body was accompanied by an unmoved mover, so that the universe had to be constantly operated by invisible hands [8].

At the time of Kepler, the idea of blind natural forces, without any aim, which is so obvious to us, was not yet extracted from animistic visions. For Kepler, said Heisenberg [11], the perfect concordance between the organization of the sensible things — oeuvre of God — and the mathematical and intelligible laws — thoughts of God — is the fundamental idea of the *Harmonices mundi*. Platonician and neo-Platonician motivations led Kepler to the belief that the reading of the oeuvre of God — the nature — is nothing else that the knowledge of the relations existing between quantities and geometrical figures.

Even Newton, the spiritual father of Kant (we must say), was not completely liberated from animistic views. For him, an universe controlled by gravitation should have to collapse and, anticipating the theory of chaos, small irregularities in the motion of planets would accumulate, eventually spoiling the magnificent order of the system. Then, using one hemisphere of his brain to betray what the other hemisphere had elaborated, he had to invoke God, to keep stars and planets on their right orbits, to avoid the crunching catastrophe and the deterioration of the world to a cosmic confusion [8]. This is easy to achieve for God because, as Newton believed, the absolute space is *Sensorium Dei* [58]. Also, it is known that Newton stated that, concerning the deep origin of the gravitation, he did not feign any hypothesis (*hypotheses non fingo*). This famous remark can be found in an essay appended to the third edition of the *Principia* in which he wrote: *I have not as yet been able to discover the reason for these properties of gravity from phenomena, and I do not feign hypotheses. For whatever is not deduced from the phenomena must be called a hypothesis; and hypotheses, whether metaphysical or physical, or based on occult qualities, or mechanical, have no place in experimental philosophy.* This looks very positivist. However, in a letter to Bentley, the mask is removed. Newton stops pretending that he does not know the cause of gravitation. For him, gravitation must be caused by an agent permanently acting according to certain rules, whether this agent is material or immaterial [58]. Reading between the lines, Koyré [58] stated that an immaterial cause, as suggested by Newton, could only be a spirit, either some kind of spirit of the nature, or more simply God Himself. Koyré provides us with other quotations from Newton confirming that he had in mind the action of an agent equipped with free will, clever in mechanics and geometry, to explain the orderly functioning of the world.

The monadology of Leibniz, this scientist-philosopher who shares with Newton the merit of the invention of the infinitesimal calculus, can also be viewed as a resurgence of animism. As we see, even at the time when the mechanistic vision of the world developed its rational mathematical basis, the frontier between the natural and the supranatural realms was still more than fuzzy. The matter could still be underlied by animal monadical spirits, and natural laws sustained by the make-shift job of God. And even recently, we had to deal with Gödel, often considered as the most important logician since Aristotle. What should we think of him when we are told that he was believing to the existence of angels, demons, ghosts, and to the Devil [109]? Cassou-Noguès could ask the question to know whether Gödel was mad or simply Leibnizian [109]. In any case, the very existence of Gödel alone is sufficient to feel suspicious with respect to the Kantian conception of causality, and to its *a priori* character. It seems however clear that Gödel considered the causality as the fundamental philosophical concept. But, for him, the causality is independent of temporality. It rather looks like the result of a Leibnizian preestablished harmony between monads, fixed forever when God imagined the world. If we decide to state that Gödel was mad, what about Leibniz? Should we, by definition, state that Kant was right and that any one who does not agree with the validity of the time anchored Kantian conception of causality, like Gödel, is mad, or better, to use a word from Kant himself, not *reasonable*?

Heisenberg also acknowledged that our conception of causality, such as depicted by Kant, is not inborn. In one of his books [11], he stated that the application of the concept of causality is relatively recent. In ancient philosophies, he said, the word *causa* had a meaning much more general than the one corresponding to the today meaning. He added that, referring to Aristotle, the *Scholastic* for example used four forms of "cause". There is the *causa formalis* which would correspond today to the structure or to the conceptual content of a thing; the *causa materialis* that is to say the matter from which a thing is made out; the *causa finalis* which is the aim of a thing and, finally, the *causa efficiens*. The *causa efficiens* only corresponds more or less to what we denote today by using the word of cause.

But a most important philosophical and epistemological issue is to know whether we can really and definitively escape from some kind of animism because, even if we deal with a mechanistic interpretation of nature, in terms of laws of nature, we do not know what is the origin of these laws. There will always be some room for something which is not mechanistic underlying the mechanisms. This may be illustrated by reporting an argument of Salviati, a character taking the place of Galileo Galilei in one of his famous dialogues [34]. Addressing Simplicio, Salviati expressed himself as follows: ... any one knows what is called gravity. But what I am asking you is not the name but the essence of the thing: From this essence, you do not know more than concerning the essence which makes the stars running, excepted the name ... but it is not this which makes us better understanding what is the principle ... which makes the stone moving down...

Concerning this issue, Newton accepted the idea that the motion of planets could be assisted by spirits, or angels, or God, or whatsoever. We have here a point of view which, from a purely logical standpoint, cannot be rejected. If we do not refer to a superior principle, may be immaterial, the origin of the laws of nature is an enigma. If we intend to solve this enigma by accepting such a superior principle, possibly God, we are facing another enigma. Kant would have named such a situation an antinomy. In any case, we are facing an aporia.

On a genealogy of space and time

In the same way that there has been a genealogy of causality contradicting Kant, that we sketched with a brief history, there has been a genealogy of space and time, contradicting Kant too, that we now sketch with another brief history. Let us begin with the space which is associated, according to Kant, with geometry (more specifically with the Euclidean geometry implying the infinity of space).

When we examine the oldest conceptions of the human kind concerning the structure of the system of the world, we are fascinated to observe how much some of them may depart from what seemed *a priori* so natural for Kant, and from what appears *a posteriori* so compulsory to us. For Egyptians, during the night, the Sun is sailing on a golden small boat; and the Earth is an upside-down round bowl [110]. For Thales of Miletus, the Earth was flat, floating on water, the primordial element of the universe ([7], book II, Chapter 13), and the canopy of heaven was a kind of star-spangled ceiling. For Anaximander, the Earth is a motionless cylindrical disk, flat and thick, located at the center of Everything [4, 110]. Also, as mentioned by Aristotle ([7], book IV, Chapter 1), people believed that there is only one hemisphere, the one above us.

These examples form only a microscopic sample among a thousand and one heaps of things to report, to be imported from so many cultures, from western and eastern countries, histories and legends. Hawking [111] reported an anecdote concerning a famous scientist giving a conference on astronomy, and explaining how the Earth is running around the Sun, and how the Sun was running around the Milky Way. At the end of the conference, an old woman stood up and claimed that all this was just a lot of fibs because, actually, the Earth is flat, standing on the back of a giant tortoise. I guess that the old woman might have been impressed by some Polynesian tradition, most akin to some Greek ones. The funny anti-Kantian conception of the old woman should not disturb us too much. After all, we might just claim that she was fairly mad or even stupid. But should we believe that our ancient predecessors were all mad or stupid, that our most distinguished colleagues of these times possessed a brain closer to the one of a monkey than to the one of a Homo Sapiens Sapiens? Or, may be, the Renaissance was the time when a new mutation occurred in the human genome, giving birth to a new race (call it Homo Sapiens Sapiens Sapiens)? Is not it more reasonable to explain the facts (to *save the appearances*) by admitting that the Newtonian and Kantian concepts of space are very far from being *a priori*?

Similarly as for the ancient perceptions of space, we may find ancient perceptions of time which are mismatched with the conception of time of Kant. For instance, the idea of a creation of the world by God with a beginning of time (that some Christian theologians assimilated with the Big Bang, confirming, they believed, the Genesis of the Bible) and an end of time (the apocalyptic Armageddon and the Last Judgment, or more recently the parousia of Teilhard de Chardin), corresponds to the vision of an universe closed in time, in the same way that it was closed in space. The date of creation could even been evaluated to a few thousands years ago by genealogists having studied the Bible (a belief shared by Descartes [112], still common at this time as can been seen from Spinoza [113], and still believed by some people nowadays) while, for the first Christians, and even some modern prophets, the date of destruction is to-morrow. There is, behind such perceptions, an amazing narrowness of horizon which astounds any modern scientist, surely a deep insuperable chasm between the soul of before and the soul of now. The essential reason why the narrowness of time experienced by some cultures is in conflict with the transcendental philosophy is that, for Kant, the mathematical basis of the time is arithmetic, or, in other words: Arithmetic, this synthetic *a priori*, is related with the pure intuition of time. And therefore, time should be infinite, without beginning nor end, as is the set of natural numbers, or better the one of real numbers.

As we know, ancient conceptions of space and time culminated in the prodigious System of the World described by Aristotle in his book *On the Heavens* [7]. Aristotle's system distinguishes between the sub-lunar world and the supra-lunar or heavenly world. The Earth, motionless, located at the center of the Universe, is the realm of changes, generation and corruption, produced by various combinations of the four elements. Simple motions are rectilinear from up to down for heavy bodies or from down to up for light bodies. In contrast, the heavenly bodies are perfect realities, not submitted to generation nor corruption, that is to say imperishable, and they must therefore move in a perfect way, that is to say following a circular motion, the only one which can last forever (for Aristotle, motions along straight lines cannot last forever because the world is closed). The matter of the celestial bodies is named the fifth element according to commentators, or named ether according to a tradition (not according to Aristotle himself who viewed it as the first body due to its ontological prior significance [114]). In the celestial realm, there are spheres supporting the circular motions of the Sun, of the planets and of the stars. The outermost sphere of the universe is the one in which the fixed stars are set and inserted (like precious jewels). It is animated by a regular motion from east to west. Planetary spheres exhibit more complicated motions and, to *save the appearances*, a rather complicated system of fifty-five interacting spheres is required, plus the outermost sphere of fixed stars [114]. After this extreme sphere, there is nothing, no space for things to be located and also no vacuum!

One knows of the famous German engraving, from the sixteenth century, exhibiting an inquisitive traveller who reached the end of the world, on a flat Earth, piercing the base of the vault, and discovering the complicated mechanism driving the motion of the celestial bodies. Such a medieval picture conveys some kind of feeling of what was the old world, although Aristotle would have been horrified by such a misleading representation: The Earth is not flat, but spherical, and it is a non-sense to believe that we can pierce the vault. However, on the way to the indefinite world of nowadays, the problem of the vault as an ultimate separation between nothing and something has been a crucial issue. It could take the form of a simple classical question to know what would happen if someone pushed his head through the vault. We might argue that, on one hand, this should be easy since there is no resistance to expect from nothingness but, on the other hand, we might also argue that there is no room in nothingness to set the head [58]. For Giordano Bruno discussing the issue, the Aristotelian conception of a closed space is not only erroneous, but also absurd [58], although, as explained by Koyré, the details of the arguments used by Bruno to contradict Aristotle are certainly flawed. For Koyré, Bruno did not understand Aristotle. Very likely, we are here facing an example of what Kuhn would have called an incommensurability between two visions [115]. The visions of Aristotle and Bruno imply two different structures of the mind which made any communication between them most uneasy.

As a result, we feel uneasy in Aristotle's world and Aristotle would feel uncomfortable in ours.

The Aristotelian system of the world, non-Newtonian and non-Kantian, has been accepted, studied, and "understood" for centuries, something which would be impossible to grasp if we believed that Kant *a priori*'s were deeply anchored in the very structure of the human mind. A famous and beautiful book by Koyré [58] exposed the history which allowed the human mind to pass from the closed world of Aristotle to the opened world of today, without forgetting another beautiful book by Koestler [8]. Most individuals of the Renaissance (including scholars and obviously theologians), frightened by the opening of cosmic doors and windows, rejected from their predominant location at the center of the Universe (just a cosy corner!) built for them by God, most understandably have felt a hopeless angst when they had to become orphan of the comforting world depicted in De Caelo [7]. As stated by Pellegrin [114], the mental resistance to accept the new world must have been huge and desperate, due to the fact that the finite world of Aristotle, familiar and stable, was really a human scaled world. The intensity of this resistance can hardly be imagined nowadays, but only approached by those who would achieve the effort to deeply feel, with bones and nerves, the consistency of Aristotelian conceptions. The impregnation of the human mind by the old world, and the reluctance to welcome the new world, is vividly illustrated by the character of Simplicio, a convinced but honest peripatetician, in the famous dialogue of Galileo Galilei on the two systems of the world, namely the Ptolemaic one inherited from Aristotle and the Copernician one [34]. Such resistances and reluctances do not plead in favour of the *a priori* transcendental nature of space and time.

Many other, however less dramatic, examples may be found in the history of sciences. They all demonstrate that, along the history of our understanding, the structure of the human mind changes too. Not only do we learn and understand new facts, but we also modify our understanding of the word understanding (see the end of chapter 10 in [10]).

Popper's criticism of the critique

A very pertinent criticism of the critique, somehow summarizing the above discussion, is available from the epistemology of Popper. As he stated [61], to pretend that the immediate or direct character of a belief establishes its truth, this is the fundamental mistake of the idealism. For him, Kant was in error because he did not realize that the imposition of our mind onto the world rarely succeeds, that we make trials and that we mislead ourselves again and again, and the result — our knowledge of the world — is indebted to the resistance of the reality as well as to our ideas which happen ones after the others. In particular, he noticed, if we say that non-Euclidean geometries cause the ruin of the Kantian theory of space, then we necessarily have also to say that relativity causes the ruin of his theory of time. Indeed, Kant explicitly asserts that there is only one time, an assertion for which the validity of the intuitive idea of simultaneity is decisive (in irreducible conflict with Einstein teachings). According to Margenau [116], the time for Newton *was a unique process of flow, independent of the circumstances of the observer. This empirical uniqueness was epitomized in Kant's system by a rationalization which attached transcendental necessity to the uniqueness of time, thus lifting it above empirical examination. Relativity shattered this isolation and again made time a matter for experimental inquiry.*

In contrast, in 1935, a philosopher, Grete Hermann, deeply convinced of the definitive validity of Kant's critique, tried to demonstrate that quantum mechanics does not require us to abandon or even to modify the Kantian conception of the category of causality [117], something which was difficult to ratify at this time, and still more difficult to ratify nowadays. The precious analysis of Léna Soler in Ref. [117] indeed confirms that the attempt of Hermann to reconcile quantum mechanics and Kant's critique is not pertinent. The point of view of Hermann is also discussed by Heisenberg [10]. It is actually a logical and systematic commitment for all Kantians and neo-Kantians that, eventually, any theory should end up with a consistent deterministic account. For them, quantum mechanics is therefore not complete in the sense that it is not yet deterministic (but should become deterministic, soon or later).

It is important however to remark, for later use, that the conception of Popper, as stated above, tells us nothing on what we have to do in the case when several theories agree simultaneously with experiments, a question which we will have later to discuss. This is an opportunity to state that the epistemology of Popper is not perfect, and should not be viewed as the ultimate epistemology, the "true" epistemology. Many criticisms have indeed been addressed to Popper. See in particular Harré [118] who stated that *to define the terminus of enquiry in terms of truth is an act of faith*. However, an heritage of Popper is a convenient language that I have already used and that I shall use again, in the same way that we may conveniently use common sense for everyday life, even if we are ready to give it up in the laboratory.

Therefore, we may conclude, after this brief but sufficient survey, that we should not be founded to make our understanding of the World founded on Kant's *a priori*'s. They were (unduly) too much dependent on the physics and most mathematics of his time. They are dated and not out of date, valid forever. For Spengler [119], Kant affirms the validity of his principles as valid for all human beings, and for all times. But, the categories he invented pertain only to some western minds. The structures of the mind of Aristotle, of Russians, of Chinese, of Greeks ... are different, although possessing their particular consistency. The philosophy, from Bacon to Kant, he concluded, is merely a curiosity. Therefore, if we rely only on this epistemological story, we should not cry if we are forced by quantum mechanics to give up Kantian categories. But, are we forced to give them up? Well, this is one of the main subjects of this book.

Neuronal flexibility

I am not going to make the mistake of saying that mind is located in brain. We do not know what is mind, and we do not neither know whether it is located somewhere, or even if it makes sense of speaking of its location. The famous scholastic question to know how many angels could sit on the point of a needle, even if it makes us today arrogantly laughing, was not a stupid question but a rather deep one. It was prefiguring the classical Descartes opposition between Mind having thought as the distinctive attribute, and Matter whose distinctive attribute is spatial extension. Let us however accept to say that the functioning of the brain is strongly correlated with our mental states, and that its neuronal structure (incredible power and inescapable limits) has something to do with the quality of the theories we are able to produce.

It is then reasonable to turn ourselves to neurology sciences whose current developments are much impressive. And what neurologists can tell us today is that the brain is an organ exhibiting an incredible flexibility, both neuronal and presumably glial. Kant *a priori's* would possibly to be found in some fundamental, inalienable, structures of our brain. But we are very far to be able to identify such structures, and their mental counterparts, if even they exist. If they exist, they might find their origin in the long lasted evolution of life all along several billions of years, whose most intimate details are lost forever. But we also cannot dismiss the possibility that the World, connected in some way with the Thing in Itself, could contain transcendental information conveyed to the brain. In other words, the World could be connected with Plato ideals, or receive the fingerprints of God. This point of view would certainly have pleased Descartes.

We are not in the position to be empirically able to decide between the different undecidable possibilities just discussed above. There could be *a priori's*, but we do not know enough to point them out. This mystery would require us to be able to evaluate the consequences of the flexibility of the brain. As an extreme example, I remember of a quantum mechanist expressing the idea that all problems about quantum mechanics will simply evaporate after a few more generations (after a cure of desintoxication?). I do not believe it. And, if this were going to happen, it would be the end of human questioning. Robot or super-human? Would this be a renunciation or a progress? But it must be admitted that this quantum mechanist might be right: After all, classical mechanics too is counter-intuitive and if you do not feel so any more, it is because we have been well trained to it. In the same mood, Charpak told us that we would show how the mathematical form of the laws allow one to surpass all apparent contradictions thanks to a superior harmony and why the apparent paradoxes of the theory are not in it, *but in us* [17]. Another way to state this point of view is to claim that quantum mechanics is firmly resistant to our prejudices.

I remember having learnt classical mechanics, and, after, relativistic mechanics, and, after, having taught relativity for many years and, after, having to teach classical physics, in particular fluid mechanics. And I remember how my intuition, formed by many years of friendly habituation to relativity, was reluctant to go back to the Newtonian framework. When hiking in mountains, I have now some kind of immediate 4D-perception that the summits over there are far away in space and in time. I never recovered my previous naive freshness of what is a Newton-Kant landscape. The fact that the neuronal flexibility is indeed an established property of the brain (and therefore, somehow, of the mind, at least of the phenomenal self if not of the transcendental self) opens the way to an evolutionary Kantism, in which categories could change, in the same way that they could be different for other unknown reasonable aliens, and in the same way that they are different for cats and dogs.

Neuronal flexibility (and stability) of our brains is associated with the notable flexibility (and stability) of our theories. We have here to remind the story, already discussed, concerning the two predominant systems of the world [8, 22, 34, 58]. The system of the world of Aristotle with its heavenly finiteness and incorruptibility, its teleological and its organism-like functioning, has been convincing enough, something like obvious to those who were educated in its arms, and stable, for two thousands years. After a painful transition, we became familiar with the new mechanistic system, exhibiting heavenly indefiniteness and a host of violent and cataclysmic events (novae, supernovae, collisions of stars, black holes, and galaxies, and the most violent of them: The Big Bang). We have been flexible enough to accept the new world and more than that to look at it as an obvious solution to the problem of the structure of the universe. Not only did we dramatically modify our fundamental conceptions, but we also succeeded to revolutionize the very structures of our minds. Such a possibility, Kant, not aware of modern physics, but expectedly aware of the theories of the past, should have extensively commented, and taken into account.

In contrast, most of the scientists contemplating the finite world would nowadays do it with some amount of condescension, possibly considering it as a mere kind of superstition. In the same way that we have been able to switch from the system of Aristotle to the one of Newton and, after, to the one of Einstein, shall we eventually switch from the classical systems of Newton and Einstein to the quantum system of Bohr, Heisenberg, Dirac and many others? And stop raising "metaphysical" issues? Shall we eventually just accept quantum indeterminism and quantum holism with serenity?

In the words of Van Fraassen [21], *need our physics be so peculiar? What do the phenomena demand? The early quantum theory startlingly depicted matter as not continuous, and energy transmitted only in quanta. But more curious features would appear: Correlations between separate, distant events, looking almost like telepathy in nature. Although the quantum theory clearly developed in response to experimental results, the suspicion could not help but arise that physics could have succeeded in a more traditional format ... cultural pressures perhaps inclined physicists to greater sympathy than their predecessors had with indeterminism and holism.*

However, I do not believe that quantum mechanical problems will evaporate just because, due to the flexibility of our brain, in particular regarding its response to cultural pressures, they would become no-problems, that the students of the student of our students would simply could not see any more what the problems were, that these problems would become dead problems and no more living problems, in the same way that it is now difficult for us to relive the scandal produced and deeply felt by the Pythagorean disciples when they heard about the existence of the square root of 2, with many other similar examples to be easily found in the history of sciences.

To summarize, we are left with several questions such as the following ones. Can we understand quantum mechanics, if we have to abandon the classical trio? Is our brain (our mind) sufficiently flexible to eventually allow us to understand quantum mechanics as it is, to accept it right away, without any feeling any more of strangeness? Do we really need to surrender or should we think more about it? Are we so close to the Thing in Itself that his bright light makes us blind and stammering? And is mathematics, this incredibly efficient tool which allows us to get rid, at least in part, of our everyday language, and to pass through ghostly appearances, the only machete at our disposal to cut a path into the jungle which protects a forbidden world behind hidden worlds?

Scientists-Philosophers

We are now equipped enough to survey the philosophical attitude of some scientists with respect to quantum mechanics, using Kant's critique as a reference to comment them.

We already discussed the vision of Einstein. To recall it, here is another quotation (from [120]): *If one asks what, irrespective of quantum mechanics, is characteristic of the world of ideas of physics, one is first of all struck by the following: The concepts of physics relate to a real outside world ... An essential aspect of this arrangement of things in physics in that they lay claim, at a certain time, to an existence independent of one another, provided these objects are situated in different parts of space.* This illustrates the insistence of Einstein on the existence of a real world made out from separated objects. The conception of objects, existing in different parts of space, which is in conflict with the non-separability exhibited by the quantum world, is also a conception of space. Space is viewed as an arena in which objects are located, even if, in general relativity, the objects themselves structure the geometry of the space (more exactly of the space-time). More generally and precisely, events take place in the space-time, according to the principle of causality, and this provides a description of the real world. Kant's category of causality is still living and alert, but space and time Kant's categories are dead.

I am not aware of Einstein referring to the Thing in Itself. But, if it had done it, my guess is that Einstein would have pretended that the Thing in Itself is, at least in part, knowledgeable. It would not have made the distinction between Thing in Itself and Phenomenon. Einstein however, regarding the categories of Kant, wrote that his theoretical attitude *is distinct from that of Kant only by the fact that we do not conceive of the "categories" as unalterable (conditioned by the nature of the understanding) but as (in the logical sense) free conventions. They appear to be a priori only insofar as thinking without the positing of categories and concepts in general would be as impossible as breathing into a vacuum* [121]. Another issue for Einstein is that of the strong objectivity insofar as it requires the possibility of a passive observer: *Physics is an attempt conceptually to grasp reality ... independently of its being observed. In this sense one speaks of "physical reality"*. For a complementary analysis of Einstein's conception of reality, see Margenau [116], and for the relationship between Einstein and the logical positivism, see Frank [122]. More generally, see Schilpp [123], in particular a paper by Born [124] in which we may find the following quotation from a letter from Einstein: *In our scientific expectation we have grown antipodes. You believe in God playing dice and I in perfect laws in the world of things existing as real objects, which I try to grasp in a wildly speculative way.*

As Kant, Einstein is a realist but his realism is very far remote from Kant transcendental idealism. Borrowing an established terminology, we might say that Einstein is a naive realist, at least some of his statements (but not all of them) express a naive realism. Einstein did not give his adhesion to the club of the defenders of the classical trio (space, time, causality) that he reduced to a classical duo (space-time, causality), but he adhered to a strong version of realism. Independently of the dramatic break-through he achieved, he is a son of Newton, at least of the one who built the first coherent mechanistic view of the universe viewed as a causal machinery. We may also refer to Einstein to define a strong version of objectivity: Things, in the outside real world, evolve according to the principle of causality, independently of us, human beings. This is obviously in deep contrast with the quantum mechanical vision of a participatory universe, in which human beings and the universe are entangled together.

Actually, Einstein is in the filiation of Descartes rather than in the one of Kant. The philosophy of Kant is in part a reaction against Descartes, more generally against dogmatic metaphysics and, from this standpoint, we may view Einstein as a pre-Kantian in the same sense than Descartes is a pre-Kantian, that is to say they arrive "before" Kant. We know that Descartes tried to reconstruct everything with the power of the reason as a tool. For him, the faithfulness of reason was secured by God who, because of His infinite goodness preventing Him to mislead His creatures, is the guarantor of the possibility of knowledge. The God of Einstein was not the God of Descartes, but rather the one of Spinoza. But Einstein too was confident, and self-confident, with respect to the power of mind. For Descartes, and as far as I can see, for Einstein too, there were not such things like *a priori*'s (in the sense of Kant). Even today, I believe that it is not exaggerated to say that most scientists are, or have been, instinctively, naive realists.

For Bohm [60], there is a focus on fields and particles which are viewed as the only entities that the mind may conceive, some kinds of Bohm's *a priori*'s. This statement refers indeed to the *widespread impression that the human brain is, broadly speaking, able to conceive only two kinds of things, namely fields and particles*, but we are not exactly dealing with *a priori*'s in the sense of Kant because Bohm's *a priori*'s are under the influence of the empirical world: *We can only conceive of what we meet in everyday experience or at most experience with things that are in the domain of classical physics where, as is well known, all phenomena fall into one or the other of these two classes.* With the reference to the capacity of brain (say the mind) we are however not too far away from Kant. Also, obviously, fields and particles (in the mind of Bohm) have to evolve in space and time, according to causality. Nevertheless, for Bohm, the epistemological problems of the theory of knowledge are not really relevant in classical physics, for, according to Bohm and Hiley [68], *in classical physics there was never a serious problem either about the ontology, or about the epistemology. With regard to the ontology, we assumed the existence of particles and fields which were taken to be essentially independent of the human observer. The epistemology was then almost self-evident because the observing apparatus was supposed to obey the same objective laws as the observed system, so that the measurement process could be understood as a special case of the general laws applying to the entire universe.* As a result, classical physics does not have to worry with the distinction between Thing in Itself and Phenomenon or World. The Thing in Itself, as for Einstein, is just there, right outside, and we can grasp it in an observer-independent objective way. It will be the commitment of Bohm to drive back quantum mechanics to such an objective framework. Therefore, not surprisingly, his interpretation of quantum mechanics, he will call it the ontological interpretation.

The desire of some scientists-philosophers to return to an objective ontological interpretation has been shared by many, and put forward simultaneously with the development of the less objective Copenhagen interpretation. Schrödinger, with the same mood as Einstein, expressed himself in his memories by saying (from [125], or see Chevalley's introduction in [52]) that *it has even been doubted whether what goes on in the atom could ever be described within the scheme of space and time. From the philosophical standpoint, I would consider a conclusive decision in this sense as equivalent to a complete surrender. For we cannot really alter our manner of thinking in space and time, and what we cannot comprehend within it we cannot understand at all. There are such things, but I do not believe that atomic structure is one of them.* There is no explicit reference to Kant here, but Kant's spirit is blowing on the waters. We shall have many opportunities to mention another opponent to the Copenhagen interpretation, sharing the same desire to restore the classical trio, namely Louis de Broglie.

According to Omnès [15], the lack of objectivity of quantum mechanics, such as expressed above and disliked by Einstein, Bohm, Schrödinger, Louis de Broglie, and others such as Vigier, may exhibit its most repelling aspect in the Copenhagen complementary principle because it changes the obvious clearness that we instinctively associate with reality to an irreducible ambiguity.

The veiled reality

There exists a most appealing philosophical interpretation, due to d'Espagnat, which succeeds to explain quantum mechanics in a modified Kantian framework. D'Espagnat's conception is called the veiled reality (e.g. [69, 84, 85, 98, 126]). I am going to dare to give my interpretation of d'Espagnat's interpretation, with the hope that my translation is not a treachery (*traduttore e traditore*). Going from Kant's terminology to d'Espagnat's one, let us translate Thing in Itself to Independent Reality (in the sense of mind-independent reality), and Phenomenon by Empirical Reality. We may then define three different conceptions concerning the Independent Reality (i) It can be grasped by human beings (Descartes, Einstein) or (ii) it is fully unknowledgeable (Kant) or (iii) it is not totally unknowledgeable, it is not totally accessible, but it is veiled (d'Espagnat).

Einstein's conception defines the strongest objectivity that we may envisage and to which he devoted his life, in an attempt to capture *this huge world, which exists independently of us human beings and which stands before us like a great, eternal riddle* [127]. Because quantum mechanics exhibits an intrinsic indeterminacy, and because it works outside of space and time, that is to say as a whole because quantum mechanics is not compatible with the classical trio (space, time, causality) nor with the reduced classical duo (space-time, causality), we have to state that it is not objective in the strongest sense, the sense present in the mind of Einstein. Also, it does not comply with Kant's *a priori*'s. The acme of the conflict culminates with the heart-breaking measurement problem.

For Kant, there would not be any kind of objectivity associated with the Independent Reality itself, just because we cannot say anything about it. But we might define an objectivity associated with the Empirical Reality, formed from an inter-subjective agreement between us, human beings, who are studying it. This was also basically the conception of Bohr. This kind of objectivity, we might call it empirical objectivity or, may be better, phenomenal objectivity.

On one hand, we have to oppose Kant, with his classical trio of *a priori*'s, and Bohr, with his interpretation of quantum mechanics working outside of the classical trio, or, better said, accepting causality and space/time only in a complementary way. But, on the other hand, both Kant and Bohr refer to the necessity of inter-subjective communication, that is to say to a common structure of human minds. Harré [118] already perceived this analogy between Kant and Bohr: *It seems to me that there is a genuine parallel between the Kantian ontology of phenomena and that proposed by Bohr. For both, phenomena are created by the application of something humanly created to something which, as far as human thought could go, must be taken to be undifferentiated. All we can say of it is that it has certain dispositions to permit shaping in this or that way, as apparatus would permit. New concepts realized in new equipments might reveal new dispositions of the glub* (a word used by Harré to denote the ur-stuff of the world). *But the concepts of physics, as a descriptive science, are constrained to those by which observable, manipulable apparatus and its reactions can be described. Immediately certain epistemological conclusions follow. Since the apparatus must be visible, tangible, and so on, and since the phenomena are shaped up by it, the concepts of classical physics must provide the only recourse for unambiguous communication between people about phenomena. This is Bohr's correspondence principle. The route to it is exactly parallel to the Kantian transcendental deduction of the categories. All we can know of the glub are its affordances, dispositions of which only the consequent is well defined, and the compossibility conditions of the joint display of phenomenal categories in apparatus. This corresponds, perhaps less happily than the transcendental deduction, to the antinomies. This is Bohr's complementarity principle.*

Furthermore, phenomenal objectivity is also close to the definition of the word objectivity as viewed by Popper [25]. For him, the objectivity of scientific statements comes from the fact that they can be inter-subjectively submitted to tests and therefore, in this language, quantum mechanics is objective. In the epistemology of Popper, this definition actually defines a rule of methodology, and therefore the scientific realm, in contrast with the metaphysical realm with, in between, a demarcation line. Strong objectivity tells more. It requires the independence of the observer and of the observed, and that the observed causally evolves in space and time (or space-time). Strong objectivity can be tested by experiments and is therefore objective in the language of Popper, and quantum mechanics corroborates the failure of this kind of objectivity. However, if we hold fast, as Einstein did, on the extra-conditions for strong objectivity, if we view them as compulsory rational demands that each satisfactory physical theory should satisfy, then they are of a metaphysical nature. Again, this is in agreement with Popper who did not accept nor reject the principle of causality. He is simply content to exclude it from the realm of science insofar as it is a metaphysical principle.

With d'Espagnat, the Independent Reality is veiled but we can grasp something of it. Admittedly, we do not know whether what we have grasped is really a fish from the most deepest Independent Reality, or if it has been swimming not too far from the surface, but the frontier between Independent Reality and Empirical Reality is not so clear-cut as Kant would have believed. In particular, non-locality in quantum phenomena (to be viewed as an empirical matter of fact, independently of the future fate of theoretical quantum mechanics) is something so strong and strange that, d'Espagnat believes, it is not only an element of the Empirical Reality but also of the Independent Reality. This would clearly agree with the Kantian conception that the Independent Reality is outside of space and time. Obviously, the veiled reality is furthermore compatible with the idea of a participatory universe, made inescapable by the existence of the quantum of action.

The kind of objectivity defined by d'Espagnat's conception is called by him a weak objectivity. According to this author [85], what was the most disgusting feature in quantum mechanics for Einstein was not so much its inherent indeterminacy, but namely its weak objectivity. This is not difficult to imagine if we are aware of the desire of Einstein to hold fast on strong objectivity.

But, as a general assessment, what is the most comfortable with the veiled reality is that it is a kind of epistemological agnosticism. It looks very reasonable and does not take any risk. It does not say nothing, and does not say too much. Furthermore, it provides a possibility of synthesis between Kant (no fish from the Thing in Itself) and Einstein (all fishes). Or, better said, it is like a rope, a continuum of more or less skeptical attitudes, with Kant and Einstein at the tips. But this may be going too far from the meaning of d'Espagnat for whom the veiled reality is a conception in which the notion of an Independent Reality is considered as remote and even as nearly unknowledgeable [126]. I might have here committed a treachery, or more indulgently, proposed a modified veiled reality version. See Bitbol [128] for complementary discussions.

Spinoza still alive and Aristotle not dead

Although I liked to discuss the veiled reality in the continuation of my survey anchored on Kant, d'Espagnat liked to discuss it in reference to Spinoza. For him [84, 98, 126] the notion of Independent Reality is close to Spinoza's concept of *natura naturans* (denoting an alive God, identified with nature in an active sense) while the notion of Empirical Reality is close to Spinoza's concept of *natura naturata* (denoting a passive God with nature already created) [129, 130].

The use of the works of philosophers to make comments may however be somewhat flexible. Indeed, while d'Espagnat referred to Spinoza in connection with the veiled reality ant its weak objectivity, it is well known that Einstein also referred to Spinoza, but in connection with naive realism and strong objectivity so that, as often mentioned, the God of Einstein is the God of Spinoza. As a relevant quotation, Wheeler [67] discussing Einstein's views could state: *Spinoza's influence on his thinking about cosmology Einstein could shake off, but not Spinoza's determinism outlook. Proposition XXXIX in the Ethics of Spinoza states "Nothing in the universe is contingent, but all things are conditioned to exist and operate in a particular manner by the necessity of divine nature". Einstein accepted determinism in his mind, his heart, his very bones. What other explanation is there for his later-life position against quantum indeterminacy than this "set" he had received from Spinoza?* And Cushing [131] observed: *Einstein's general position on foundational questions in physics ... can be seen as a search for the God of Spinoza.*

Einstein believed that God manifests himself in the rational structure of the external physical world, which the physicist tries to capture in a causal space-time theory. Einstein's basic Weltanschauung was that of a rational, causal world that could be comprehended in terms of an objective reality. And we have this well known content of a telegram of Einstein to a New York rabbi asking him whether he believed in God: *I believe in a God who reveals himself in the harmony of all being, in the God of Spinoza* [2], or with another version: *I believe in Spinoza's work, who reveals himself in the harmony of all being, not in a God who concerns himself with the fate and actions of men* [132].

Nevertheless, although such many references to Spinoza do exist, this philosopher is much less useful to discuss quantum mechanical features than was Kant. But there is also another long venerated, much influential, philosopher of the antiquity, a genius of science, named Aristotle. We already met him in connection with different issues, such as the criticism of the critique, and an ingredient of the classical trio, namely causality, and we shall have many other opportunities to refer to him. There is no doubt that Aristotle's vision of the world is outmoded, and that his influence, in some sense, has been possibly negative. Some commentators could say that he blocked the development of the human knowledge for so many centuries, and that he participated objectively to what is sometimes called, in my opinion unfairly, the medieval obscurantism. But I prefer to state that he pushed the spirit of the old Greeks to some kind of logical conclusion and that this had to be done. Having done this, something that I feel like a compulsory passage, he provided us with an extraordinary epistemological laboratory that we would have missed if it did not exist. Now, a genius cannot be completely wrong and, notwithstanding his mistakes, insufficiencies and shortcomings, he has been a master of thought and is still an inspiriting matter to think. In this subsection, I would like to insist on the fact that Aristotle is not only a dead old beard man from olden days, but that it is still alive: Quantum mechanists may like to refer to him.

For instance, with two of the causes listed by Aristotle (the material cause and the formal cause), we may associate two substantives (matter and form respectively). Forgetting many controversial issues which should be reserved to philosophical experts, we may say that the matter is the substrate (for instance the bronze of a statue) and that the form informs the matter (for instance makes a statue out from the bronze). We shall later see, when appropriate, that such a terminology may be used to discuss quantum teleportation.

Another most important terminology from Aristotle concerns the concepts of potentiality (or potency) and actuality. For Aristotle, the potentiality is a real way of being. But this way of being refers to a virtual reality that possesses the capacity to become actual, that is to say to be in actuality, the second way of being [108]. As we already mentioned, Heisenberg [53] used the same terminology with respect to the measurement process in quantum mechanics. This was a fortunate borrowing, providing a convenient language, psychologically striking, to describe quantum measurements, much used and praiseworthy for some thinkers, although disproved by others. Also, by referring to an ancient era, it provides some kind of classical respectability to a non-classical quantum realm. Bohm also, in his 1951-book [41] recurrently liked to use the word "potentiality".

In this language, before a quantum measurement, a Ψ provides a catalogue of potentialities, i.e. a list of all possible outcomes of the measurement, say a list of a_n's in the case of a discrete spectrum (with associated probabilities). This language allows us also to say, for instance, that a quantum object is potentially present in the whole wave function (this is the case of a continuous spectrum). The location, before the measurement, is not simply something which is unknown, but indeed it does not exist, or better said, it does not exist in actuality. It only exists as a potentiality for the quantum object to manifest itself here or there. Once a measurement is made, one of the a_n's, or a definite location, has become actual. It entered the stage of actual things, a state of affair which cannot be modified any more in the future. The measurement process is then to be viewed as a transition from potentiality to actuality. The actuality is fabricated by the measurement process and the transition from potentiality to actuality is so violent that it is indeed a jump, a quantum jump, from indefiniteness to definiteness, a real collapse. Furthermore, what is potential can be modified, like in delayed-choice experiments. What is actual is actual forever.

This point of view, seemingly introduced by Heisenberg for the first time, approaches something like an ontology, or can even be viewed as an ontology. In this ontology, our actual foreground emerges from a background full of and filled with potentialities. This vision is not only in agreement with the non-relativistic quantum mechanics, but also remains valid when relativistic effects are implemented, such as the emergence of particles from the vacuum. For d'Espagnat [84], Heisenberg produced an approach ... which provides a synthesis between what is effective and what is virtual. For Bitbol [133], atoms or elementary particles ... form a world of potentialities or possibilities rather than a world of things or facts. I shall add, in other words, with a Kant-d'Espagnat flavour, that when we do not observe it or do not interact with it, the world stands outside there as the Thing in Itself, or the Independent Reality, a world of possible things and facts not perceived. But, when we observe it, which requires a participative interaction due to the existence of the quantum of action, then it becomes a veiled glint in actuality for us; no more: Being is perceive (*esse est percipi*), or being perceived, but: Being for us is being in actuality.

Due to its suggestive power, the revival of the potentiality/actuality duality initiated by Heisenberg in quantum mechanics has become of rather common use by scientists-philosophers, most often without any reference any more to Aristotle nor to Heisenberg, just as it was a natural way of thinking. But Heisenberg did not simply borrow the idea to Aristotle: It was indeed a borrowing with modifications and added value. For Aristotle, the transition from potentiality to actuality is an easily understandable process. For instance a man sitting has the potentiality of standing up. If the man, after sitting stands up, then "standing up" has became actual. There is nothing mysterious about that. In contrast, the transition from potentiality to actuality in quantum mechanics is still obscure. Actually, nothing is really said to make us understand how this transition occurs. This is again the measurement problem which is left intact and unsolved. It has just been given a fancy dress under misleading linguistic clothes.

It is compulsory to mention here another reference to Aristotle, often acknowledged as the founder of logic. For him, a proposition is True (receiving the value T) or False (receiving the value F). This generates a two-valued logic, or Aristotelian logic, based on the values T and F. It is nowadays known as the *Tertium non datur* logic (third is not given, or there is no third option). Sure, a proposition cannot be both true and false but, in quantum mechanics, it has been argued that a proposition could be neither true nor false. It could be Indeterminate (receiving the value I). For instance, before a measurement, the value of the spin component of a spin-$(1/2)$ particle along a given axis is neither $(1/2)$ nor $(-1/2)$. Therefore, the following proposition is Indeterminate: Before a measurement, the spin component ... possesses a value x. It is indeterminate insofar as x itself is indeterminate. But we could play with the words and say that "x indeterminate" does not mean that x does not possess a value, but simply that this value is indeterminate. Therefore, in utmost rigor, the logical value M (for Meaningless) is better to choose, due to the following meaningless proposition: Before a measurement, the spin component ... possesses a physical value. This opens the way to the use of three-valued logics, however somewhat controversial, in quantum mechanics ([134, 135] and several papers in [136]). There exist logics which are more than three-valued, and even infinite-valued logics [137]. Deviationist logics are also discussed by Quine (chapter VI in [138]), in particular in connection with quantum mechanics and Heisenberg uncertainty relations.

More on positivism

When quantum mechanics developed itself to a mature theory, say from Max Planck (1900) at the very beginning of the twentieth century, to Schrödinger, Heisenberg, Born, Jordan ... in the late twenties of the same century, and to Dirac, von Neumann in the early thirties, the philosophical Zeitgeist was positivistically oriented.

It has been sometimes suggested that the Founding Fathers have been in spiritual contact with the neo-positivists of the Vienna Circle, but as far as I know this is doubtful. I do not know any evidence (may be my lack of knowledge) of such an interaction. As a matter of fact, let us recall that Carnap, one of the most active participants to the Circle, arrived in Vienna in 1926, the year when Schrödinger published his equation, at a time when already most of the foundations of quantum mechanics were embedded in a solid ground. I indeed found an evidence of an exchange of ideas between Heisenberg and a representative of Vienna Circle (Chapter 11 of [10]) but the date of this exchange is uncertain (I can only claim that it did not occur after 1933) and, in any case, there is no evidence, at least known to me, that such exchanges might have influenced Heisenberg. But, even without any influence from the Vienna Circle, such philosophical doctrines known as positivism, empiricism ... were already well developed, all along a venerable tradition with Locke, Hume ... Comte... and Mach.

In particular, we know the influence of Mach on Einstein. Special relativity, with its compulsory uses of rulers to make length measurements, and of clocks to make duration measurements, its careful elimination of the hopeless simultaneity, and its rejection of the unobservable and of the superfluous ether, looks like of a strict positivist allegiance. But for some commentators, Einstein has never been really a complete positivist. This may be testified by his work on Brownian motion where his aim was to guarantee the existence of atoms, which was vigorously dismissed by the positivism of Mach. Indeed, Einstein stated that the validity of the law of Brownian motion made a lot *to convince the sceptics, who were quite numerous at that time (Ostwald, Mach) of the reality of atoms*, adding: *The antipathy of these scholars towards atomic theory can indubitably be traced back to their positivistic philosophical attitude* [127]. The importance of such works from Einstein to convince physicists of the reality of atoms and molecules is also acknowledged by Born [124]. Furthermore, although I previously pinned Einstein on the blackboard as a naive realist on the basis of some quotations, it could be viewed as an idealist oriented mind if we use other quotations, in particular when he asserts that concepts and theories are free inventions of the human spirit. A balanced report of Einstein's epistemological position is available from the fifth chapter of a book by Heisenberg [10], to which the interested reader might like to refer, in which Einstein discussed Mach and relativity, and claimed that he does not want to be the defender of a naive realism. Eventually, it is not surprising that Einstein could have been named an epistemological opportunist, a reasonable attitude for a scientist indeed.

From this point of view, the case of Heisenberg, indeed a defender of quantum mechanics (in contrast with Einstein), is more representative. As we know, he built his approach by refuting the use of any unobservable quantity. On this point, Einstein opposed Heisenberg in a private discussion. But (he said to Heisenberg), for sure you do not seriously believe that only observable quantities can be included in a physical theory ... And, he insisted: Concerning principles, it is completely erroneous to intend to base a theory uniquely on observable quantities [10].

Heisenberg's scientific work may indeed be viewed as the result of an epistemological program. The mood prevailing at this time is well exemplified by the following statement from Heisenberg, in 1927: *As the statistical character of quantum theory is so closely linked to the inexactness of all perceptions, one might be led to the presumption that behind the perceived world there still hides a "real" world in which causality holds. But such speculations seem to us, to say it explicitly, fruitless and senseless. Physics ought to describe only the observations. One can express the true state of affairs better in this way: Because all experiments are subject to the laws of quantum mechanics ... it follows that quantum mechanics establishes the final failure of causality.* This statement may be understood as an attack against hidden-variables theories. But the whole doctrine of positivism constitutes such an attack.

To be fair however, and complete enough, it is necessary to report that we possess evidences that eventually Heisenberg departed from, or took a bit of distance with, positivism. Such evidences are available from the chapter 17 of one of his books [10], to which the reader might refer for details. I shall be content with extracts. In particular, Heisenberg stated that words like 'intention' or 'mission' come from the human sphere; concerning the nature, we could at best use them as metaphors. But our old parallel between the astronomy of Ptolemy and the modern theory, favoured since Newton, for the description of planetary movements can possibly here still help us. If we adopt as a criterion of truth the faculty of calculating in advance, then the astronomy of Ptolemy was not more erroneous than the one of Newton. But if we compare nowadays Newton and Ptolemy, we nevertheless have the feeling that Newton formulated the orbits of celestial bodies, thanks to his equations of motion, in a way more accurate and more complete than Ptolemy; that he had in some way described the design by which nature is built... What can I say more concerning my scientific conception of truth? To this, Pauli replied: Well, positivists would say that you are playing with muddled chattings; and obviously they would say with pride that such a thing could not happen to them. But where can we find more truth, in confusion or in clarity? At this point Pauli quoted Niels Bohr quoting Schiller saying that in the abyss stands truth. But, Pauli added: Is there any abyss, and is there any truth? ... Thinking about this, Heisenberg came to the conclusion that for a positivist, the world is divided between what can be clearly stated, and on what we should remain silent... But, deeply, he stated, this is the most absurd philosophy. Because about nothing can be clearly expressed. If we eliminate everything which is not clear, there will likely only remain tautologies completely without any interest. That is enough for my quotation from Heisenberg.

Basically, positivism may be said to rely on the assertion that every thing which is not observable does not exist or, at least, if this statement is found too strong for some people, that it should not appear in a physical theory. The most convinced proponents of positivism might also claim that unobservable entities are of a metaphysical nature, a judgment which, in the current mood of the dissolution of metaphysics, is to be viewed as highly pejorative, and dishonor them as ingredients for any decent scientific commitment. A corollary of positivism is operationalism according to which scientific concepts must be defined from the concrete operations by which they are measured or applied. Hidden-variables theories do not satisfy such constraints.

We must inescapably claim that positivism has been a source of progress for science (e.g. the young Einstein with relativity). But other progresses in science have been the consequence of non-positivistic attitudes (e.g. the hypothesis of atoms). I am not going to say that anything goes, in the somewhat extreme meaning of this expression endorsed by Feyerabend [139], but I am inclined to recommend to scientists some kind of epistemological agnosticism, or even opportunism.

As a matter of fact, a splendid history much relevant to positivism is the history of the debate between geocentrism and heliocentrism as reported for instance by Koestler [8]. I shall better discuss the issue when we shall examine the Duhem-Quine under-determination thesis (to avoid any confusion, see an important remark on terminology in the corresponding section, near the end of the book), but, for the time being, I shall be content with a relevant quotation: "What makes the pre-Keplerian astronomy close to modern physics is that, both of them have been developed as close systems in a relative isolation, manipulating a set of symbols according to certain rules. Both of them succeeded. Modern physics provided us with nuclear energy, the astronomy of Ptolemy provided predictions whose accuracy astonished Tycho Brahe. Medieval astronomers manipulated their symbols for epicycles as does today physics with the wave Ψ of Schrödinger or the matrices of Dirac. And this worked, although they knew nothing concerning gravitation, or elliptical orbits, believed to the dogma of the circular motion, and completely ignored why it worked". It is certainly the right time to recall here the famous sentence, and the title of a book, by the mathematician René Thom, telling us than predicting is not explaining [140]. More than that, Harré [118] provided examples of ridiculous consequences which could be drawn from a blind acceptance of the most dogmatic positivism we could imagine: *... So, the back of the moon did not exist before the Apollo fly-pasts, nor did bacteria before the invention of the microscope. Henry VIII is now thought to have died of* *syphilis. What can we make of the statement that Henry VIII and Al Capone died of the same disease? It must be nonsense, since for us Renaissance syphilis is an unobservable.*

The following opinion of Bohm [30], a defender of hidden variables, clearly illustrates some limits of positivism: *We can in no way deduce, either from the experimental data of physics, or from its mathematical formulation, that it will necessarily remain forever impossible for us to observe entities whose existence cannot now be observed. Now, there is no reason why an extraphysical general principle is necessarily to be avoided, since such principles could conceivably serve as useful working hypotheses ... For the history of scientific research is full of examples in which it was very fruitful indeed to assume that certain objects or elements might be real, long before any procedure were known which could permit them to be observed directly.* Obviously, a very good example illustrating this quotation of Bohm concerns, once again, the history of atoms, objects which were first postulated, since at least Democritus and Leucippus, forgotten for a long time, born again with modern chemistry, denied by Mach and other positivistic oriented individuals, before being accepted in face of experimental evidences. But, before the experimental evidences of the existence of atoms, they were rejected by positivists. Our old Greeks already said something on this issue [3, 4]. They said that atoms are visible only by the reason, because the reason alone is able to postulate the existence of invisible things.

We may indeed conceive that something unobservable might be vital to produce a coherent theory. We should be allowed to introduce such entities if they are useful, fruitful, if they help us to produce clear pictures to understand phenomena, not only to predict them, or to avoid paradoxes. We should be allowed also to introduce extra-physical principles if they possess the same virtues. We have learnt something concerning the harmful consequences of an excess of metaphysics. We should have also learnt something concerning the harmful consequences of a total rejection of metaphysics. Experiments are not to be forgotten, this is a methodological rule. But the agreement with experiments is the crowning of a theory. It should not necessarily lies at its foundations.

The mood of the previous paragraph is in agreement with Feynman [14] when he says that there is a thing on which people have much insisted after the development of quantum mechanics, namely the idea that we should not speak of things that we cannot measure. However, he adds that it is not true that we can make the science progressing by using only concepts which are directly born from experience. We should then ask ourselves whether positivism would not be a renunciation, or a too cautious attitude aiming to avoid taking any chance. If erected as a demand locking us up in a prison camp, then positivism is quite alike a metaphysical dogmatism. In particular, Harré [118] discussed *the rise to dominance in the academic community of a corrupting and deeply immoral doctrine, logical positivism.*

A moderate appraisal of what is positivism may be found in Nagel [141] that I would like to quote *in extenso*: *An influential school of philosophy thought centered in Vienna, which actually called itself "logical positivism" or "logical empiricism" was strongly sympathetic to Mach's phenomenalism. The school agreed fully with Mach's insistence that all concepts employed in science must be defined in terms of the procedures to be followed in applying them to concrete subject matters and that the procedures themselves are to be defined in terms of immediately apprehended sensory qualities. This strong requirement for admissible theory constructions was believed by members of the school to be completely satisfied by the theory of relativity ... The view that all concepts entering into scientific statements must be operationally defined can be called operationalism ... The introduction of operational standards of definitions and analysis undoubtedly contributed to the clarification of concepts in many domains of inquiry. On the other hand, the puritanitical rejection of ideas that sometimes followed by the adoption of operationalism, because no operational definitions could be given for them (for example, for the notion of photons), but which nonetheless may play important roles in inquiry, became an obstacle in some branches of science to the construction of imaginative and promising theories ... Phenomenalism is at best a program of analysis, a program that in the opinion of a majority of students is most unpromising.*

It is impossible to discuss positivism, a doctrine which insists so much on the significance of experiments in the formation of knowledge, without referring to Popper [25, 61]. He criticized what is now called the bucket theory of the mind and of the knowledge, according to which all our knowledge would come from experiments, like pouring water in a bucket, only a passive process of information. But experiments are used for corroborations and falsifications. As a whole, the process is an active and creative one, and also a rational process, with conjectures and refutations, producing what can be called an objective knowledge [61]. This conception, to which I adhere, could be viewed, depending on your mood, as a neo-neo-positivism, softening its rigidity, or as a refutation of it, destroying its rigidity. But it agrees fairly well with the view of Einstein when he said: *A theory can be tested by experience, but there is no way from experience to the setting up of a theory. Equations of such complexity as are the equations of the gravitational field can be found only through the discovery of a logically simple mathematical condition which determines the equations completely or (at least) almost completely* [127]. Popper would agree. For him [25], in the history of science, it is always the theory, not experiments, always the idea, and not the observation, which open the way to a new knowledge (although experiments prevent us to run into blind-alleys).

Ironically enough, we have a very good example (one example is enough if it is good) of a feature contradicting a strict positivism in quantum mechanics itself. Indeed, the wave function Ψ, or more generally the state vector $|\Psi\rangle$, the most fundamental entity of the quantum mechanical formalism, is strictly unobservable (in exactly the same way, and this will become much significant at the end of the book, as the action S in Hamilton–Jacobi's formulation of classical mechanics). It is unobservable in the common sense meaning of the word, and it is also unobservable insofar as it is not an observable of the quantum mechanical formalism, in the quantum mechanical sense of the word. This is consistent with the fact that an observation requires a measurement and, therefore, between observations, the unobservable Ψ evolves in an unobservable way (according to Schrödinger's equation). In connection with this important remark, Bell [142] could state that a so-called hidden quantity *rather than* Ψ *is historically called a hidden variable is a piece of historical silliness*. The formalism of quantum mechanics, usually presented as a positivist exercise, therefore relies heavily on an entity necessary for its consistence, but which is not directly emerging from the experience. Furthermore, not only Ψ is unobservable but the structure of quantum mechanics implies that, at least in this framework, it should remain unobservable forever. All this is reminiscent indeed of what our old Greeks, quoted above, were saying of the atoms: The wave function escapes our immediate observation, either by an apparatus, or by our sense, and can only be grasped by us using our reason and our spiritual ability of thinking. Hence, is quantum mechanics metaphysical?

The previous question has a meaning in some languages, but we are allowed to use other languages and points of view. For Harré [118], we may distinguish between three realms of theories. In particular, Realm 3 theories use *beings which are beyond all possible human experience and yet which are the typical referents of physical theory*. Therefore, they refer *to things which, if real, could not become phenomena for human observers, however well equipped with devices to amplify and extend the senses*. This state of affair, depending on the case under study, might be provisional (the status of a Realm 3 theory may change as the result of technological advances) or forever. Realm 3 theories therefore contain ingredients that a positivist scientist would like to consider as metaphysical. An example of entities pertaining to the third realm, pointed out by Harré himself, concerns quantum states. Therefore, quantum mechanics is a metaphysical theory, insofar as it pertains to the third realm.

Bohr–Einstein debate

Everything, or nearly everything, we have discussed in this chapter and in the previous one, concerning philosophical and epistemological issues, and problems of interpretation, may be somehow crystallized in the Bohr–Einstein debate between two kinds of extreme views of the quantum mechanical world. This debate already very active during the Solvay Congress of 1927 [46, 47] reached a kind of summit with the Einstein–Podolsky–Roosen (EPR) paradox in 1935 [143] and Bohr's reply [144, 145], and generated time forward propagating waves. This section will not discuss this paradox that is better reserved for a later time, and for a more appropriate place. But, fortunately as far as the organization of this book is concerned, we do not need to discuss it to understand the essence of the Bohr–Einstein debate.

The Solvay Congress of 1927 may be viewed as the place where the position of Bohr, more generally the Copenhagen interpretation, received its birth certificate and, more than that, where Bohr established its dominance over more classically oriented opponents, from this time on defeated, such as Einstein, Schrödinger, or Louis de Broglie. A dramatic part of the dialogue is when Einstein opposed himself to Bohr with a Gedanken experiment, aiming to demonstrate a violation of the Heisenberg uncertainty principle (which is not a principle, since it can be derived from the postulates of quantum mechanics, and does not concern any uncertainty, since quantum objects do not possess properties before being measured), that, soon after, Bohr demolished by using the very general relativity created by Einstein [46]. The prevalence of Bohr's point of view which followed amply explains why the Copenhagen interpretation is so often known as the orthodox interpretation, with scientists illuminated by the lights of the right church, and others rejected in the darkness of appalling heresies. May be better said in a softer way, the proponents of alternative interpretations, facing the many successes of quantum theory which created a very optimistic mood and atmosphere, became soon after increasingly old-fashioned. As stated by Squires [13], *the elegance, simplicity and economy of quantum theory contrasted sharply with the contrived nature of a hidden-variables theory which gave no new predictions in return for its increased complexity; the whole hidden variables enterprise was easily dismissed as arising from a desire, in the minds of those too conservative to accept change, to return to the determinism of classical physics; ... mumbo-jumbo of the Copenhagen interpretation;*

this interpretation, due mainly to Bohr, acquired the status of a dogma; it appeared to say that certain questions were not allowed, so, dutifully, few people asked them.

According to Heisenberg [146]: *What was born ... in 1927 was not only an unambiguous prescription for the interpretation of experiments, but also a language in which one spokes about Nature on the atomic scale, and insofar a part of philosophy. Indeed, the way in which Bohr had thought about atomic phenomena since 1912 had always been something intermediate between physics and philosophy, and he ... formulated the new interpretation of quantum theory in the philosophical language to which he had become accustomed by 15 years acquaintance with atoms, and which seemed best suited to the problems involved. This was not, however, the language of one of the traditional philosophies, positivism, materialism, or idealism; it was different in content, although it included elements from all these systems of thought* (By the way, this book from Heisenberg also provides an early review of hidden-variables theories).

The salient features of Bohr's view may be summarized as follows. A quantum object has no property at all when it is not observed. The concept of property is meaningless if there is no observation (measurement). Then, quantum mechanics does not deal with the properties of micro-objects, in themselves, but with relationships involving large-scale phenomena, resulting from some kind of interaction between a quantum object and a macroscopic measuring apparatus, viewed as a classical device. The phenomena resulting from such interactions form indivisible wholes. It would be erroneous, and even meaningless, to attempt to analyze them as made out from different separated parts, even conceptually. The theory is not viewed at something aiming to describe nature itself (although the very existence of this nature is not denied), but merely as a predictive tool for the possible outcomes of measurements, and associated probabilities. Also, because measurement transitions result from discontinuous changes of states, the famous quantum jumps so disliked by Schrödinger, they cannot be represented in the framework of space and time. Another issue, already mentioned but on which we have to insist, is that, for Bohr, any ultimate description of measurement processes has to been made by using classical concepts, a pre-quantum language which is however required to allow inter-subjective agreement (objectivity in the sense of Bohr). This was also the point of view of Landau, already mentioned, stating that the quantum mechanical formulation requires the use of its own classical limit, a kind of feature unprecedented in science [56]. It agrees with Heisenberg's statement according to which ... *the measuring device is necessarily described in terms of classical physics,* or with an analysis from d'Espagnat [85], stating that the primacy of the classical language is underlying the thinking of Bohr. A may be last issue is that Bohr was viewing Ψ as characterizing individual systems, in contrast with Einstein who favoured an ensemble interpretation in which the wave function was actually dealing with an ensemble of systems [24]. The technologies available today allowing us to perform investigations on individual systems lead to a support of the Bohr's point of view, at least in that respect.

The vision of Bohr which defines a kind of weak objectivity (even if in a sense slightly different from the one of d'Espagnat) is obviously in deep contrast with the strong objectivity point of view of Einstein. Against Einstein's objectivity, we may again convoke Heisenberg [146]: *The criticism of Copenhagen interpretation of the quantum theory rests quite generally on the anxiety that, with this interpretation, the concept of "objective reality" which forms the basis of classical physics might be driven out of physics ... this anxiety is groundless, since the "actual" plays the same decisive part in quantum theory as it does in classical physics. The Copenhagen interpretation is indeed based upon the existence of processes which can be simply described in terms of space and time, i.e. in terms of classical concepts, and which thus compose our "reality" in the proper sense. If we attempt to penetrate behind this reality into the details of atomic events, the contours of this "objectively real" world would dissolve not in the mist of a new and yet unclear idea of reality, but in the transparent clarity of a mathematics whose laws concern the possible and not the actual. It is of course not by chance that "objective reality" is limited to the realm of what Man can describe simply in terms of space and time. At this point we realize the simple fact that natural science is not Nature itself but a part of the relation between Man and Nature, and therefore is dependent on Man. The idealistic argument that certain ideas are a priori's ideas, i.e. in particular come before all natural science, is here correct. The ontology of materialism rested upon the illusion that the kind of existence, the direct "actuality" of the world around us, can be extrapolated into the atomic range. This extrapolation, however, is impossible.*

D'Espagnat [85] imagined that Einstein's reluctance against Bohr–Heisenberg interpretation could have been expressed by the following and most emblematic question: What, in the theory, corresponds or may correspond to a physical reality? Get out of fuzziness and tell me. Bohr's reply could have been (and actually has been in his rebuttal to EPR-paper) a mere and deep rejection of the very pertinence of the question. Bohr indeed did not interpret quantum physics as describing the reality here outside, but as expressing a human knowledge which has to be transferable to other human beings.

In this section, I did not discuss too much the point of view of Einstein that we, by now, know rather well. Then, it could be fair to give him the opportunity of one of the last words, which contains in my opinion both a bit of sadness and of humour, may be worst. According to Pais [147], *Einstein was neither saintly nor humourless in defending his solitary position on the quantum theory. On one occasion, it could be acerbic. At one time he said that Bohr thought very clearly, wrote obscurely, and thought of himself as a prophet. Another time, he referred to Bohr as a mystic.* This is rather in resonance with Louis de Broglie saying [62], regarding Bohr, that he was one of the greatest scientists of his time, but a bit like a Rembrandt of contemporary physics, because he sometimes exhibits a certain inclination for the "chiaroscuro". Such appraisals are however somewhat objectively resonant with the attitude of Bohr when he stated that everything he said is not to be understood as an affirmation but as a question [148].

It is likely correct to believe that Einstein was indeed disgusted by the Copenhagen interpretation, he and in fact many others. So many reluctances that I expressed in many occasions up to now are far enough sufficient to have motivated the search of alternative interpretations, namely the search for hidden-variables theories. The opposition of Einstein to the orthodoxy has been regretted by many, for instance by Pauli writing that *the writer* (Pauli) *belongs to those physicists who believe that the new epistemological situation underlying quantum mechanics is satisfactory, both from the standpoint of physics and from the broader standpoint of the conditions of human knowledge in general. He regrets that Einstein seems to have a different opinion on this situation* [149]. But Einstein persisted and signed, in particular in 1953 [150].

Our previous expositions aimed to a good understanding of the underlying motivations for a search of hidden variables. We are now equipped enough to effectively enter into the hidden worlds of the quantum world. More on the Bohr-Einstein debate can be gained from [123], in particular from [151], and also from [13, 24, 52, 100, 147], among many others.

Chapter 6

A Brief History of Causal Theories

Terminology

What do we mean by causal theories? We first mean theories which are alternatives to quantum mechanics, in which the intrinsic character of the measurement indeterminacy, likely one of the most striking features of quantum mechanics but may be not the most striking one, is removed. Such a removal is carried out by the use of hidden variables. Therefore causal theories are causal in the sense that they are deterministic. However, following a kind of tradition inaugurated by Louis de Broglie, we shall reserve the name of causal theories to two specific theories, namely the double solution of Louis de Broglie, and the pilot wave theory initiated by Louis de Broglie again and, later on, developed further by Bohm. With this complementary precision, a deterministic theory is not necessarily a causal one. Therefore, from now on, in opposition with what has been done up to now in this book, the word "causal" and "deterministic" should not be any more taken as synonymous words. I hope that this subtlety of language, introduced to conform to Louis de Broglie's vocabulary, will not disturb the reader.

We may also hesitate between using the expressions "hidden-variables theory" or "hidden-variables interpretation". Causal theories do not intend to invalidate the formulation of quantum mechanics which is supposed to provide correct predictions, that is to say predictions in agreement with experimental results (this is indeed the case up to now). Causal theories are essentially constructed under the constraint that they should produce the same predictions as quantum mechanics (this will be softened when appropriate).

If quantum mechanics were, one day or another day, to be falsified by experiments, then the causal theories, because they produce the same predictions as quantum mechanics, would be simultaneously falsified too. However, they aim to the production of a picture of the world which will be much different (deterministic picture, in particular) than the one provided by quantum mechanics (with its intrinsic indeterminacy). They therefore provide a different interpretation of the world, and an interpretation which is different from the Copenhagen interpretation, so that the word "interpretation" is legitimate. But because they allow one to make predictions for experiments, they also really deserve the word "theory". We should not fight too much on such points of terminology, and I shall use one word or the other (interpretation or theory), not indifferently however, but depending on the kind of emphasis I want to convey. The difference between "theory" and "interpretation" in this context will be similar to the one I used for "quantum mechanics" and "Copenhagen interpretation". However, insofar as interpretations are different with quantum mechanics and causal theories, we should also consider quantum mechanics and causal theories as different theories, even if they agree on the predictions of outcomes of experiments. Sometimes, I may also possible use the word "model".

There also exist hidden-variables theories which are not deterministic but stochastic. Stochasticity should however not be confused with indeterminacy. For instance, in the deterministic classical physics, we may use the word "stochasticity" to characterize the theory of Brownian motion although, in the classical framework, we do not question the fact that, ultimately, a Brownian motion is the result of a deterministic process.

Sources

Before digging the field of causal theories, it is convenient to have in mind a historical background of them, in particular to understand how they developed against the prevalent formalism and interpretation. This is a fairly simple exercise because we possess many invaluable testimonies of this history, in particular from the two main protagonists of the business, namely Louis de Broglie and Bohm, e.g. [29, 37, 60, 62–65, 68, 152, 153], for a selection. Complementary historical analysis and reports are also available from Vigier [50], Jammer [24, 82], Pinch [154], Wheeler and Zurek [100], Holland [40], or Cushing [131], among others. The latter references may possibly provide a more general overview, not only on causal theories, but on various hidden-variables theories. Such a more general point of view is also discussed by Freistadt [155], Belinfante [1], Pipkin [156] or d'Espagnat [85].

Louis de Broglie dissatisfied

Virtually all reasons for disliking quantum mechanics and the Copenhagen interpretation, discussed in previous chapters, provided strong motives for Louis de Broglie to commit himself to the development of alternative theories. Indeed, he certainly has been the most severe and most eloquent opponent to the orthodoxy, not only in the deepness of his soul but also publicly, and for most of his life, excepted for a resigned interruption that I shall tell. I already provided, in several places, several quotations from Louis de Broglie where he expressed his criticisms. He took many opportunities to claim his displeasure, and even his reluctance, in particular with respect to the following items: The impossibility to represent the motion of a quantum object in ordinary space and time and, in contrast, the use of an abstract configuration space; more generally, the impossibility of reconciling quantum mechanics with the clear ideas of classical physics; or the fictitious and subjective character of the wave function (its probabilistic interpretation, or his lack of objectivity when it is interpreted as representing only a state of knowledge). Indeed, for Louis de Broglie whose mind was educated with classical physics, and therefore immerged in the classical trio, the ideas of the new physics, out of space, out of time, and out of causality, generated a head-on conceptual collision.

Also, Louis de Broglie complained about what has been called elsewhere the complacency to quantum mechanics, for instance in [65] where he expressed his increasing conviction that the reasonings underlying the present interpretation of quantum mechanics are not as decisive as they look, and contain many small cracks. He afterward also expressed his feeling that most of the present theoreticians, capitulating in face of exaggerated abstract tendencies, renounced too easily to the building of an intelligible picture of quantum physical phenomena.

But, very likely, the most important clash in Louis de Broglie's mind arose from the conflict between the concepts of particles and of waves. This may be inferred from the very writings of Louis de Broglie, in particular when he stated [62] that the duality between waves and particles constituted the big tragedy of contemporary physics. Waves or particles, this is something, concerning the nature of light, which already pertained to the history of classical physics. Regarding this issue, Huyghens, a defender of the undulatory nature of light, had won against Newton who defended a corpuscular point of view, in a fairly famous Huyghens–Newton controversy. The problem of the nature of light was then eventually solved to the advantage of Huyghens, but only till Einstein made it vividly alive again when he introduced light quanta, a resurrection of the corpuscular point of view, not only in agreement with the experimental features of the photoelectric effect, but also afterward with the experimental features of the Compton effect. But the problem was not only alive again, it was deeply different. Instead of having to choose between wave or particle, it was necessary to admit that light was wave and particle, at the same time, in a dual way that had to be explained. Next, concerning matter, there has been no controversy at all, at least for a long time. Matter was made out from particles, but only until Louis de Broglie, in his celebrated thesis [157], advanced the hypothesis that matter particles should manifest an associated wave aspect, an assumption soon to be confirmed by Davisson and Germer experiments. As a matter of fact, this hypothesis had been seminal for the work of Schrödinger and for the advent of his equation, two years later. Light and matter were then unified, at least from the point of view of what has been called for a long time wave-particle duality (a terminology no longer to be recommended nowadays).

Quantum mechanics succeeded in a magnificent way to dissolve the duality by a synthesis. Quantum objects are not classical waves and they are not classical particles: Not wave *nor* particle, but something else. For Louis de Broglie however, as we have seen, the price to pay was too heavy. In building hidden-variables theories, he made the choice of another option: Wave *and* particle [24]. The duality became a complementarity, not however in the sense of Bohr.

Louis de Broglie started

As soon as Louis de Broglie conceived his first ideas on wave mechanics, there was no doubt for him that it was necessary to find a new theory, synthesizing the concepts of waves and particles, and retaining as far as possible their traditional aspects. Concerning particles, although quantum mechanics rejects the very concept of trajectory, Louis de Broglie wanted to preserve it. For him, the two points of view (rejecting and preserving) were not deeply conflicting because [63], from the fact that quantum measuring processes do not allow us to simultaneously assign a position and a state of motion to a corpuscle (a word he liked, and that I shall use in this section, in his honour), it does not necessarily follow that, *in reality*, the corpuscle does not have any position nor velocity. But, concerning waves, they also had to be preserved. In particular, Louis de Broglie was impressed by the fact that the theory of atoms introduced integers to deal with the stationary motion of electrons while, in classical physics, such occurrences of integers only arose in interference phenomena, and in oscillatory normal modes. Therefore, for Louis de Broglie, corpuscles and waves were to become, in some way, two aspects of the same thing. As he expressed during the Solvay Congress of 1927 [46], the existence of elementary corpuscles for matter and radiation being admitted as a matter of fact, these corpuscles are supposed to be endowed with a periodicity. In such a vision, a matter point is no more conceived as a static entity only concerning a tiny part of space, but as the center of a periodic phenomenon spreading around it. This idea, that a periodic phenomenon must be associated with a corpuscle, has been an obsession to which Louis de Broglie dedicated most of his life.

The way to achieve the synthesis was to consider the corpuscle as a kind of permanent singularity embedded in an extended wavy phenomenon. This has led him to a theory known as the double solution. The name of the theory comes from the idea (about 1925–1927) that the equation of motion of wave mechanics (Schrödinger's equation of course) is satisfied simultaneously by two coupled equations. One solution, denoted u, represents a singularity, describing really the corpuscle associated with the wave while the other solution, with a continuously variable amplitude, describes only the statistical behavior of the motion of a cloud of particles. The first singular solution exhibits very high values of amplitudes, even infinite in an early version, around a moving point of space (Note: In this chapter, we start by commenting on particles; hence these particles chronologically correspond to the first solution, but, in the next chapter, particles are associated with the second solution). We are then facing a local moving singularity which can be identified with a corpuscle. For Louis de Broglie, although the motion of this corpuscle should admittedly satisfy new dynamical laws, it should develop itself along a trajectory, with a well defined velocity at each point of it. The second solution (in this chapter) formally identifies with the Ψ of quantum mechanics, but with a different interpretation. For Louis de Broglie, the Ψ of the double solution, as the Ψ of quantum mechanics, does not exactly describe the physical reality. It leads indeed to a statistical theory, representing some kind of average, when the actual trajectories followed by the corpuscles are totally unknown, with the exhibition of two layers, the second layer, the deepest one, being of a classical nature and therefore deterministic. Then, at the level of the first layer, Ψ can be viewed as fictitious, statistical, subjective ... without however any epistemological damage.

Also, an important ingredient of this theory is that the phases of Ψ and u have to be equal. This rule, named concordance of phases, ensures the tight solidary coupling between the real wave embedding the corpuscle and the continuous wave of wave mechanics. The common phase therefore possesses a deep meaning so that Ψ

was given by Louis de Broglie the name of a phase wave. The quantum mechanical indeterminacy then results from our lack of knowledge concerning the initial conditions required to build the trajectories of the corpuscles in the cloud. Therefore determinism is restored in a classical way, with indeterminism as the consequence of a lack of knowledge, instead of being of an intrinsic nature. We obtain a hidden-variables theory with hidden variables associated with quantities describing the hidden trajectories. This is a brief summary of the first causal theory of Louis de Broglie.

To gain a better understanding of the theory of the double solution, and more important of where it may come from, let us return to the Hamilton–Jacobi's formulation of classical mechanics. We already mentioned the significance of this formulation for the purpose of this book. We are going indeed to check this significance here, and we shall have other opportunities for further returns. Let us recall, as sketched in Fig. 1(b), that, in Hamilton–Jacobi's formulation of classical mechanics, we are facing trajectories which are orthogonal to iso-value surfaces of the action S. Actually, we may consider a cloud of particles, generating a compact set of trajectories, embedded in and orthogonal to a field phenomenon extended in space. This is a classical structure that Louis de Broglie attempted to generalize to a quantum mechanical structure. The motivation for this attempt was that Louis de Broglie believed that it was wrong to abandon the notion of trajectories, followed by classical particles. The Hamilton–Jacobi structure then provided a natural already ready framework for an efficient generalization because, now, in quantum mechanics as in classical mechanics, we have trajectories embedded in a field. In Fig. 1(b), the trajectories are observables and the field S may be considered as hidden. The corresponding structure for quantum mechanics is given in Fig. 1(c). Quantum trajectories are now orthogonal to a field formed by the phase of Ψ. We shall demonstrate, when appropriate, that, in the classical limit, this phase of Ψ identifies with the field S (within a multiplicative constant). But, in contrast with the structure of Fig. 1(b), the trajectories of Fig. 1(c) are the hidden quantities. Furthermore, both S in Fig. 1(b) and the phase of Ψ in Fig. 1(c) are unobservable. A further analysis of these structures, in the (speculative) last chapter of this book, will eventually lead me to the rejection of causal theories. Anticipating, even if it might be surprising for the reader to hear of this now, let us state that Louis de Broglie insisted to maintain the concept of trajectories, while, conversely, I shall argue that this concept must be rejected, even in classical mechanics (more precisely: In the experimental facts of classical mechanics, and in any correct theory which would try to explain these facts). This is reflected in Fig. 2 although the time has not come to comment about it.

Louis de Broglie defeated

As we already discussed, Brussels in 1927 has been the very place where occurred a living confrontation between eminent experts of the time, the very place where the Copenhagen interpretation has become the dominating one, and the very place where the Bohr-Einstein debate became so acute [24, 46, 47]. It also has been a turning point for the history of the newly born causal theories, the very place where Louis de Broglie has been defeated. This happened as follows.

In the spring of 1927, Lorentz asked Louis de Broglie to prepare a report on wave mechanics for the forthcoming Solvay Congress to be held in October. Louis de Broglie, at this time, was fighting with the many mathematical difficulties he encountered when trying to make his double solution satisfactory enough. He then found too hazardous to expose his not yet ready theory, and decided to fall back on a simpler half-hearted point of view, named the pilot wave theory, the second causal theory of Louis de Broglie, that he himself called a degenerated version of the double solution.

In the pilot wave theory, Louis de Broglie held fast on the existence of corpuscles and also admitted the validity of the wave equation of wave mechanics, even if only of a statistical nature. He then placed the corpuscle inside the wave and assumed that its motion is driven by the propagation of the wave. The driving process is described by a guidance formula which made trajectories orthogonal to the isophases of Ψ (again our now familiar Hamilton-Jacobi's picture).

Therefore, the corpuscle is viewed as an entity which is piloted by the wave Ψ, hence the name given to the theory. Something is lost however with respect to the double solution. The corpuscle is guided by the wave but it is no more incorporated in a tight way inside the wave. Also, the wave-particle duality is acknowledged as a matter of fact, but there is no more any understanding of its nature. However, the intuitively appealing notion of a localized particle with a deterministic motion along its trajectory is preserved [158].

During the Solvay Congress, discussions were sharp: Bohr, Heisenberg, Born defended the Copenhagen probabilistic interpretation. In particular, Bohr re-expressed his idea of complementarity (presented previously in a Como lecture), the fact that underlying quantum events are outside of space and time, and that the space/time description and the causal description are complementary [159]. Schrödinger attempted to put forward its interpretation of a corpuscle viewed as a wave packet, i.e. he did not believe to the existence of corpuscles as basic entities. He expressed his displeasure in face of what would become the orthodox interpretation. For him, it seems that there is an indication that, when our knowledge has progressed enough, everything will eventually become understandable again in a three-dimensional space. But he did not support Louis de Broglie [160]. Lorentz rejected the probabilistic approach and insisted on the fact that physics should rely on clear notions in the framework of space and time. Pauli, attacking the pilot wave, stated that he did not believe that the representation of Louis de Broglie could be developed in a satisfactory way. Einstein, deeply hostile to a genuine probabilistic interpretation, vigorously attacked it, against Bohr ... and did not win. But more important and most disappointing for Louis de Broglie is that Einstein, although in principle in favour of a deterministic interpretation, did not support him. For details, we may return to the printed discussion following the presentation of Louis de Broglie [158] which is not very enthusiastic, and also to the final discussion following the bulk of the Congress [46]. In particular, the final discussion contains Einstein's objection to quantum mechanics, with his hemispheric screen already discussed, and Pauli's objection to the pilot wave, with his rotator and inelastic scattering, to be discussed later.

It may be interesting to try to understand the position of Einstein regarding Louis de Broglie's proposal. A possible explanation is the fact that Einstein himself already had thought of something similar to the pilot wave approach, and eventually rejected it. He spoke of ghostly waves (Gespensterfelder) guiding the photons [52]. According to Wigner [74], *Einstein ... believed in the concentration of the energy in quanta and that these quanta have structures similar to particles. However, their motion is governed by what he called Führungsfeld — that is to say "guiding field" — and this obeys the equations of electrodynamics. In this way, the existence of interference phenomena could be reconciled with the concentration of energy in very small — perhaps infinitesimally small- volumes. However, the picture could not be reconciled with the conservation laws for energy and momentum, and Einstein firmly believe in these. As a result Einstein never published the Führungfeld idea.* In another reference [161], Wigner provided a complementary view to the above one by stating that the Führungsfeld *obeys the field equation for light, that is Maxwell's equation. However, the fields only serve to guide the light quanta or particles, they move into the regions where the intensity of the field is high. Yet Einstein, though in a way, he was fond of it, never published it. He realized that it is in conflict with the conservation principles: At a collision of a light quantum and of an electron for instance, both would follow a guiding field. But the guiding fields give only the probabilities of the directions in which the two components, the light quantum and the electron, will proceed. Since they follow their directions independently, it may happen that in one collision the light quantum is strongly deflected, the electron very little. In another collision, it may be the other way around. Hence the momentum and the energy conservation laws would be obeyed only statistically — that is, on the average. This Einstein could not accept and hence never took his idea of the guiding field quite seriously.* This hidden-variables theory of Einstein, although unpublished, was the subject of an oral presentation, made by Einstein, on 5th may 1927 (a few months before the Solvay Congress), at a meeting of the

Prussian Academy of Sciences in Berlin [162]. Such circumstances certainly shed some light on the behavior of Einstein, with respect to Louis de Broglie's proposal, at the Solvay Congress of 1927.

As a whole, besides the laconic indifference of Einstein, the pilot wave theory of Louis de Broglie was severely criticized, possibly in a much more acerbic way than the one we could draw by simply studying the accounts of the Congress [46]. You know, during a congress, there is not only a conference room for public debates but also coffee breaks, corridors, and all sorts of informal gossips, particularly if, as was the case, all participants share the same hotel. According to a testimony from Heisenberg [10], the most animated discussions occurred during meals at the hotel. This is the place where Einstein put forward, several times, his famous objection that God does no play dices, to which Bohr could only reply that it is not to us to prescribe how God should rule the world [10]. It might however be significant that the name of Louis de Broglie is not mentioned at all by Heisenberg in this book [10] where he reported so many details on what has been going on during the Solvay Congress of 1927. This was however not a complete boycott of Louis de Broglie. Indeed, Heisenberg let him express extensively in one of his book [11], although not on the issue of hidden variables.

A dramatic feature is that Pauli made an objection that Louis de Broglie could not successively strike down. This objection (to which we shall return later in a more appropriate place, with a rebuttal by Bohm) was that the pilot wave theory (or model) could not be applied in a coherent way to the case of a two-body scattering process, specifically to the Fermi quantized rotator. Louis de Broglie had become a lonesome warrior.

Louis de Broglie depressed

Returning from Brussels to Paris, thinking more about the pilot wave and on what happened during the Solvay Congress (no support from Einstein, at least publicly, Pauli's objection, generally little support to him even from the parts of Schrödinger or Lorentz ...), Louis de Broglie felt poignantly alone (Bohm's testimony in [154]). It is not a big step forward to believe that he must have felt depressed too.

He thought that Pauli's objection could possibly be refuted, but more important, he found new objections by himself (such objections are discussed in [37]). The most significant one is that, in the pilot wave model, the objective corpuscle was guided by a wave Ψ which, insofar as it was reflecting our state of knowledge, was of a subjective nature. How something subjective could possibly guide something objective? Another objection, again associated with the status of Ψ, concerns the many-body problem because, for this case, it is even more difficult to consider the pilot wave theory as giving a genuine physical picture of phenomena, due to the fictitious and abstract character of the propagation of a wave in the configuration space.

Also, there was little comfort with the double solution. It did not suffer from the above objections but was buried in mathematical difficulties that, actually, Louis de Broglie never succeeded to solve. For Louis de Broglie however, if the causal theories he started to develop were one day or another day able to grow again from their ashes, it would be from the double solution approach. Such a phoenix process would certainly be impossible from the pilot wave model (he erroneously thought), because this degenerated version was absolutely indefensible.

Facing such a situation, Louis de Broglie felt so discouraged that he gave up the causal theories, only a few years after he introduced them. He then capitulated, defeated by the probabilistic interpretation, and even accepted to teach it, although, I guess, reluctantly. Later on, Born will take notice of the acceptance by Louis de Broglie of his defeat, and of his acknowledgment that the probabilistic interpretation (the one of Born) is the only adequate one [163]. In 1932, five years after Brussels, 1927, von Neumann, in his celebrated book [26], produced a proof of the impossibility of hidden variables which, seemingly, closed the issue forever. Apparently, this was a final knock.

Bohm Bang

Ironically enough, it was not the double solution but, in contradiction with Louis de Broglie's expectation, the pilot wave theory which made the legendary Arabian bird born anew from ashes. In 1952, Bohm published two papers on a pilot wave approach [29, 30] in which he pointed out the fallacy of von Neumann's proof of impossibility, provided a rebuttal to Pauli's objection and proposed a consistent solution to the measurement problem, among other issues. From an epistemological and also methodological point of view, one the most interesting features of Bohm's approach is the way he dealt with one of Louis de Broglie objections, namely the one concerning the subjectivity of Ψ and the damning question to know how such a subjective entity could pilot an objective corpuscle. For Bohm, the answer is incredibly simple. By a kind of *fiat*, he just decreed from the beginning that Ψ is an objective wave. This is however made consistent by the fact that the solution to the measurement problem given by Bohm is an objective solution (wait to see how).

Bohm and de Broglie

I shall not investigate deeply whether Bohm effectively was aware or not of Louis de Broglie's work when he started to develop his theory, although we have the following testimony from Bohm [60]: *Then, in 1951 ... the author* (that is to say Bohm) *began to seek such a model, and indeed shortly thereafter, he found a simple causal explanation of the quantum mechanics which, as he later learned, had already been proposed by de Broglie in 1927.* But, in any case, it is well attested that he was aware of it in 1951 because, one year before having his seminal papers published in 1952 in the Physical Review, he sent preprints to Louis de Broglie. According to the latter, Bohm fully picked up the pilot wave theory as it was exposed at the Solvay Congress of 1927, including certain developments such as the introduction of a quantum potential. As a matter of fact, he also acknowledged that Bohm was not aware of this work produced 25 years before, when he engaged himself on causal theories. He also acknowledged that Bohm achieved a few more steps, in particular regarding Pauli's objection and the measurement problem [62]. The official reaction of Louis de Broglie has then been to publish a C.R.A.S. to recall his indisputable priority in the topic, and also the difficulties which made it repudiating the pilot wave after 1927 [153]. Objections against the pilot wave are also summarized by Louis de Broglie in Refs. [37] and [62].

According to a testimony of Bohm to Pinch [154], after having received the 1951 preprints, Louis de Broglie *wrote back saying that he had already done this sort of thing.* Upon looking at the preprints, Pauli manifested a similar reaction, saying that *it was old stuff, dealt with long ago.* See Pauli [164] reporting on de Broglie-Bohm story. But Louis de Broglie, in his reply to Bohm, also explicitly discussed the objections against the pilot wave that made him give it up, and Bohm found that he could answer them. The rebuttals are included in the Appendix B of the second 1952 paper of Bohm [30], saying: *All the objections of de Broglie and Pauli could have been met if only de Broglie had carried his ideas to their logical conclusion.* Later on, in many places, Bohm referred to Louis de Broglie as the primary father of the filiation. For instance, in Bohm and Hiley [68], he wrote: *Thus de Broglie proposed very early what is, in essence, the germ of our approach,* and he considers his theory as *"an extension"* of de Broglie which answered all the objections. But, as a whole, the primary reception of Bohm's ideas by Louis de Broglie was rather negative. In the words of Belinfante [1], *when Bohm in his 1951 theory revived de Broglie's 1927 theory of the pilot wave without attaching to it a theory of the double solution, de Broglie objected against this by arguments that might have been presented by a pupil of the Copenhagen school.*

Louis de Broglie resuscitated

Although Louis de Broglie's reactions to Bohm's work were not, at the beginning, very enthusiastic (to say the least), he changed his mind rather quickly, likely in part due to the relevant answers of Bohm to his objections. According to Pinch [154], *these answers were in part enough to convince de Broglie who then took up his original interpretation again.* In the words of Cushing [131], *David Bohm's 1952 paper was to have a profound reconversion effect on de Broglie.* This was a new start for Louis de Broglie who felt encouraged to take up his ideas again, and to develop them in various ways, with however his previous recurrent belief that the double solution was on the right track. As testified in 1956 [63], Louis de Broglie has been wondering since 1951 (the year when he received Bohm's preprints) whether his first idea, the double solution, after all, was not the correct one. In fact, already in 1952 [152], he published a work in which he examined how the objective and causal theory of the double solution could be taken up again.

Improvements and modifications are then introduced by Louis de Broglie. May be the most innovative one concerns a new hypothesis telling that the equation of propagation of the singularity wave u should be nonlinear, in contrast with the one for Ψ which remains linear, although both equations are nearly everywhere identical. Such a late focus on non-linear equations was motivated by some developments due to Einstein, in general relativity and unified field theory, in which besides the gravitational field there also exist superimposed gravity singularities having to satisfy non-linear equations. We are here facing a strong similarity between the original ideas of Louis de Broglie's double solution and latest researches of Einstein, a similarity pointed out by Vigier, and that Louis de Broglie tentatively exploited [50, 60, 63].

Madelung

In the margin of the main stream represented by Louis de Broglie and by Bohm to build hidden-variables interpretations of quantum mechanics, and besides the aborted attempt of Einstein, it is fair and compulsory to mention, although only briefly, another attempt due to Madelung. This latter author published in 1926 a hydrodynamical interpretation of quantum mechanics [165]. The most striking point is that Madelung formulation (the equations and the mathematics) are essentially the same than in the pilot wave theory, in particular the one of Bohm in 1952, but with a very different interpretation. However, in the words of Cushing who provided an exposition of Madelung's approach [131], it *did not carry conviction.* Of course, Madelung was much less famous than Louis de Broglie and did not had the opportunity to expose his position in Brussels, 1927. But, although there was not any real and deep debate on his position, he is mentioned in several places by Louis de Broglie and Bohm. In particular, Bohm [60] acknowledged that the first steps towards an alternative interpretation of the quantum theory were taken ... by de Broglie and Madelung. See also Jammer [24] discussing Madelung, stating that he proposed *the earliest "hydrodynamic interpretation"*, and that *so far Bohm's analysis resembles Madelung's hydrodynamical model, de Broglie's theory of pilot waves, and similar interpretations.* Do not forget to make a detour by another work from Cushing discussing also Madelung's approach, and more generally the de Broglie–Bohm story [166]. Later on, Madelung's model will be extended by Takabayasi [167, 168] and Schönberg [169, 170]. While Madelung only considered fluids with potential flows, Takabayasi discussed the existence of complicated fluctuations about a motion of constant velocity, and Schönberg considered a turbulent medium. The relationship between quantum mechanics and fluid dynamics is also discussed by Bass [171].

Chapter 7

The Double Solution

General warnings

Although Louis de Broglie eventually did not succeed in building a satisfactory double solution theory, in particular due to the overwhelming mathematical difficulties he encountered, it is useful or even compulsory to provide in this book a fairly extensive discussion of it, and to keep it in mind. This is intended to have a pedagogical virtue by making us more familiar with the logic underlying the development of causal theories. Let us note also that the lack of success in building a theory should not necessarily be taken as a final motivation to definitely reject it. Someone, more clever, more imaginative, or simply more lucky, could possibly succeed where predecessors have failed. True, I shall eventually propose to dismiss all causal theories (but there will be a loophole). This will be however another story and, even dismissed, a theory might be worth to be studied in itself, not only for a better understanding of the problems raised, but also because it could contain some pieces of inspiration for future research.

Most expositions by Louis de Broglie are carried out in the framework of special relativity, and often by referring to the many-body problem. However, to deeply understand the double solution, it is sufficient to work in the classical limit, and most often with the one-body problem. By doing so, I shall simplify the mathematics without damaging the physical contents I intend to convey. Very often, Louis de Broglie himself specifies his equations to what he called the Newtonian approximation, which is obtained by taking two limits, namely $v/c \to 0$ and $\hbar \to 0$ (obvious notations).

Besides pedagogical purposes, I have another reason to restrict myself to a non relativistic framework. Indeed, my final objections against causal theories will be developed in the classical framework, which will provide a sufficient framework for this, but can be made immediately propagated to the relativistic framework. Let us note however that the relativistic framework used by Louis de Broglie is a small energy framework, in particular in which the number of particles is conserved. Obviously, annihilation and creation processes would make the concept of objective particles more problematic. My account of the double solution relies on several works by Louis de Broglie, essentially Refs. [37, 62–65, 172]. The original reference is however dated 1927 [173]. A more recent account of the theory is due to Andrade e Silva [174].

Hamilton–Jacobi made statistical

The formal anchoring of the double solution (and also of the pilot wave) is again this beautiful, and much fruitful, Hamilton–Jacobi's formulation of classical mechanics. Let us then return to Fig. 1(b). Instead of considering only one trajectory perpendicular to the surfaces $S = C$, as depicted in the figure, let us consider all available trajectories. By the way, these trajectories (in a mechanical language) are akin to rays (in a geometrical optics language), associated with the propagation of waves, with the surfaces $S = C$ being viewed as wave fronts.

Trajectories in our cloud of trajectories are distributed according to a certain spatial density ρ, certainly related to S, although the exact relationship between ρ and S is not required in this section. The value of $S = S(x_j, t)$ then allows one to evaluate the density of probability for a particle to be at location x_j at time t. The concept of probability in this picture is a classical concept, in contrast with the concept of probability in quantum mechanics. It is just the consequence of our lack of knowledge concerning initial conditions for the particles. This is due to the fact that we are considering a cloud of particles and that we did not accordingly specify any initial condition. Indeed, if initial conditions were specified, we would not be dealing with a cloud of particles any more, but with a single trajectory, and we would know the location of the particle at a given time along the trajectory. Instead, by considering a cloud of particles, we have jumped from an individual description to an ensemble description, requiring the use of statistical concepts.

Let us now assume that we make some kind of measurement and observe a particle at location $x_{j,0}$ at time t_0, with a velocity $v_{j,0}$. Then, from the equations of motion of classical mechanics, and knowing the initial conditions that we have just measured, we know which is the trajectory of the particle. And we also know that all other trajectories have to be dismissed. Our act of observation therefore produced a dramatic modification of the way we used to describe the physical situation under study. Before the measurement, we had a cloud of particles with a density of probability of presence different from zero in a significant region of space. After the measurement, we have registered a collapse of this density which becomes zero almost everywhere excepted for the actual trajectory revealed by the measurement.

The S-collapse so obtained is however of a very classical nature indeed. Before the measurement, the density of probability of presence was diffuse in space as a result of our lack of knowledge and, from this point of view, it is admittedly subjective. As soon as, through an observation, we know which trajectory is actually followed by the particle, the subjective density of probability of presence loses its meaning, and has to be redefined by incorporating the result of the observation.

The analogy with quantum mechanical features should then be obvious. In the quantum mechanical framework, we are also facing a subjective field, here the wave Ψ, which by an act of observation collapses to another wave Ψ which has to incorporate the outcome of the measurement. Quantum mechanics however provides an interpretation of this collapse by invoking an intrinsic (ontological) indeterminacy while, in contrast, there is no such indeterminacy in the S-collapse. Therefore, the analogy is not perfect. But would not it be possible to make it perfect?

Let us start from classical mechanics. If we only knew the field S, and possessed a classical mechanical formalism relying only on S, we could possibly turn classical mechanics to a probabilistic theory in which hidden variables would be associated with hidden unknown (deterministic and objective) trajectories. Such a project or prospect however does not look very appealing, just because classical trajectories are not supposed to be hidden: They lie on the foreground of classical mechanics, not in a background. But, conversely, let us now start from quantum mechanics. We then have the obvious idea that we could possibly restore determinism in a quantum mechanical framework by invoking the existence of (deterministic and objective) hidden trajectories. Such trajectories would then form a background, with Ψ on the foreground. And the analogy between S-collapse and Ψ-collapse would become nearly perfect.

With such considerations, we now possess a qualitative entry key to enter the details of causal theories, in particular of the double solution. We are now going to evolve from qualitative arguments to more formal and quantitative ones, concentrating ourselves on matter waves (we shall not need, in this book, to discuss light quanta, although they are also discussed by Louis de Broglie).

The first solution

PEDESTRIANS ($\sim\sim\sim\sim$) may skip this section, or be content reading qualitative comments without referring to equations. The previous chapter on the brief history of causal theories might be sufficient for PEDESTRIANS ($\sim\sim\sim\sim$) to have a fair enough understanding of what is going on. Otherwise, the next section is a summary for PEDESTRIANS ($\sim\sim\sim\sim$). In this chapter, we furthermore conveniently consider that the first solution concerns Ψ (the name of first solution and second solution can be interchanged without any damage).

Louis de Broglie admitted the existence of Ψ and the validity of its equation of motion, namely Schrödinger's equation. We use the

$|\mathbf{r}\rangle$- representation, with the state vector $|\Psi\rangle$ replaced by the wave function Ψ, and rewrite Eq. (3.4) as:

$$i\hbar\frac{\partial\Psi}{\partial t} = -\frac{\hbar^2}{2m}\frac{\partial^2\Psi}{\partial x_j^2} + V\Psi \tag{7.1}$$

in which the Hamiltonian has been written as the sum of the kinetic energy operator and of the potential energy operator, with usual notations. The wave function Ψ can always be given the following form:

$$\Psi(x_j, t) = a(x_j, t)\exp\frac{i}{\hbar}\varphi(x_j, t) \tag{7.2}$$

in which a denotes a real amplitude and φ, a real function, which will conveniently be called the phase. Inserting Eq. (7.2) into Eq. (7.1), we then obtain two equations reading as:

$$\frac{\partial a}{\partial t} + \frac{1}{m}\frac{\partial a}{\partial x_j}\frac{\partial\varphi}{\partial x_j} + \frac{1}{2m}a\frac{\partial^2\varphi}{\partial x_j^2} = 0 \tag{7.3}$$

$$-\frac{\partial\varphi}{\partial t} = \frac{1}{2m}\left(\frac{\partial\varphi}{\partial x_j}\right)^2 + V - \frac{\hbar^2}{2ma}\frac{\partial^2 a}{\partial x_j^2} \tag{7.4}$$

In the classical limit ($\hbar \to 0$), Eq. (7.4) reduces to:

$$-\frac{\partial\varphi}{\partial t} = \frac{1}{2m}\left(\frac{\partial\varphi}{\partial x_j}\right)^2 + V \tag{7.5}$$

which may be identified with Hamilton–Jacobi Eq. (2.1). Therefore, in the classical limit, the phase $\varphi(x_j, t)$ of the wave function $\Psi(x_j, t)$ may be identified with the action $S(x_j, t)$ of the Hamilton–Jacobi's formulation of classical mechanics. In more rigorous terms, let us note however that the genuine phase of Ψ is $\varphi/\hbar$. Therefore, this genuine phase identifies with S, within a multiplicative constant, a result that I promised to establish in the previous chapter. This promise is now satisfied.

We then see that Schrödinger's equation may be viewed as the result of a deformation of Hamilton–Jacobi's formulation providing a generalization to quantum mechanics. Indeed, starting from Hamilton–Jacobi's equation, and using a principle called the elevation principle, we may obtain by deformation a set of generalized Schrödinger's equations, possibly including nonlinear terms, with Schrödinger's equation *stricto sensu* being the simplest equation among the set. We shall return to this issue later, in Chapter 16. As far as Eq. (7.3) is concerned, I shall not discuss it under its present form. Let us wait a bit before dealing with it.

The first solution for PEDESTRIANS ($\sim\sim\sim\sim$)

Louis de Broglie admitted the existence of Ψ and the validity of Schrödinger's equation which determines the evolution of Ψ with respect to time (this is what is called first solution in this chapter). As usual, Ψ can be written as $a\exp(i\varphi/\hbar)$, in which a is a real amplitude and $\varphi/\hbar$ is the phase. In the classical limit, i.e. when $\hbar \longrightarrow 0$, it may then been shown that φ may be identified with the action S of the Hamilton-Jacobi's formulation of classical mechanics.

The second solution

PEDESTRIANS ($\sim\sim\sim\sim$): Same recommendations than for the first solution. Otherwise, the next section is for you.

While the first solution deals with waves of the form of Eq. (7.2) satisfying the time evolution Eq. (7.1), the second solution deals with hidden particles. In the Hamilton–Jacobi's formulation, the velocity of particles is given by Eq. (2.7), with trajectories orthogonal to the surfaces $S = C$, or $S_0 = C_0$, as depicted in Fig. 1(b).

In the quantum mechanical formulation, we similarly introduce particles with velocities again given by Eq. (2.7), and S changed to φ (compare Eqs. (2.1) and (7.5)), leading to:

$$v_j = \frac{1}{m}\frac{\partial \varphi}{\partial x_j} \tag{7.6}$$

Therefore, in the same way that classical velocities are proportional to the gradients of S (Fig. 1(b)), quantum velocities are proportional to the gradients of φ, that is to say orthogonal to isophase surfaces, as depicted in Fig. 1(c). This may be viewed as an hypothesis, resulting from an extrapolation to quantum mechanics, beyond the domain of validity of classical mechanics (or in another language, beyond the domain of validity of geometrical optics). The validity of such an extrapolation would be consistent with the fact that the hidden trajectories are considered as real trajectories of objectively existing entities. But we could take another point of view, although not in the spirit of causal theories such as born in the minds of Louis de Broglie and Bohm. According to this second point of view, we could consider Eq. (7.6) as a relation of definition of a certain velocity, absolutely licit whatsoever, as is always a logically consistent definition. But we are not required to assert any more that such a definition would concern any real objective particle. It would be enough if it concerned fictitious particles, generated by playing with formulae, without any ontological significance. But, with such a definition, we are playing with formulae, and not making physics any more in the sense of the causal theories.

In the objective causal framework, Eq. (7.6) (which is then to be taken seriously) is called the guidance formula. This terminology reflects the fact, exhibited by the equation, that the wave Ψ, or more properly said its phase φ, guides (or pilots) the particles.

Now, using Eq. (7.6), Eq. (7.3) becomes:

$$\frac{\partial a}{\partial t} + v_j \frac{\partial a}{\partial x_j} = 0 \tag{7.7}$$

in which we have used the continuity equation for an incompressible fluid, namely:

$$\frac{\partial v_j}{\partial x_j} = 0 \tag{7.8}$$

Invoking the above continuity equation is consistent with the idea that particles should be some kinds of rigid, non-interpenetrating, objects. Under such circumstances, we may manipulate further Eq. (7.7) to obtain:

$$\frac{\partial a^2}{\partial t} + \frac{\partial \left(a^2 v_j\right)}{\partial x_j} = 0 \tag{7.9}$$

which is another continuity equation. To interpret it, let us recall that the density of probability of presence of the particle is given by, in quantum mechanics:

$$\rho = |\Psi|^2 = a^2 \tag{7.10}$$

in which Ψ is assumed to be correctly normed. The continuity equation 7.9 then leads to still another continuity equation reading as:

$$\frac{\partial \rho}{\partial t} + \frac{\partial(\rho v_j)}{\partial x_j} = 0 \tag{7.11}$$

This continuity equation is well known in quantum mechanics. See for instance Blotkhintsev [38] for a discussion. But, because ρ is defined as the density of probability of presence of a particle, this relation is still of a statistical nature.

To made a genuine deterministic theory, Louis de Broglie introduced what he called a second wave function (the second solution), denoted u, to be associated with the objective particle, i.e. describing the physical reality, and reading as:

$$u(x_j, t) = f(x_j, t) \exp \frac{i}{\hbar} \varphi'(x_j, t) \tag{7.12}$$

to be compared with Eq. (7.2). The continuous amplitude $a(x_j, t)$ has been replaced by another amplitude $f(x_j, t)$ which however must describe a singularity representing the particle. The association of the singularity with the particle allows one to simultaneously speak of the location and of the velocity of the particle, in a clear way. Also the phase $\varphi(x_j, t)$ has been replaced by another phase $\varphi'(x_j, t)$. To better tie up the two solutions, that is to say the statistical wave Ψ and the objective wave u, Louis de Broglie then set down a postulate of concordance of phases, according to which:

$$\varphi(x_j, t) = \varphi'(x_j, t) \tag{7.13}$$

This rule of concordance is already announced in the thesis of Louis de Broglie [157] where such a concordance is observed, when time goes on, between a periodical phenomenon associated with the corpuscle and an accompanying wave.

Therefore, with this scheme, Louis de Broglie proposed a causal interpretation in which, according to his original intuition, the particle is embedded in an extended wavy phenomenon. Furthermore, the equation of propagation governing u is supposed to be the same than the one governing Ψ, i.e. both u and Ψ satisfy Schrödinger's equation. The main mathematical difficulties involved in the theory concern the explicit exhibition of singular solutions of the kind u, something that Louis de Broglie never succeeded to achieve properly.

The second solution for PEDESTRIANS ($\sim\sim\sim\sim$)

The second solution (in this chapter) deals with hidden variables, namely with the existence of particles (corpuscles) following true trajectories, on which, at any time, they have a true location. Furthermore, on these trajectories, particles possess a true velocity proportional to the gradient of the phase of the first solution Ψ, according to an equation named the guidance formula. This terminology reflects the fact that the wave Ψ, or more properly said its phase φ, guides (or pilots) the particles. A particle is represented by a second solution denoted as u, describing a singularity. Therefore, with this scheme we have sketched here, Louis de Broglie proposed a causal interpretation in which, according to his original intuition, the particle is embedded in an extended wavy phenomenon. Furthermore, the equation of propagation governing u is supposed to be the same than the one governing Ψ, i.e. both u and Ψ satisfy Schrödinger's equation. The main mathematical difficulties involved in the theory concern the explicit exhibition of singular solutions of the kind u, something that Louis de Broglie never succeeded to achieve properly.

Non-locality and double solution

The formulation may be generalized to the case of N non-identical particles. In such a case, that I however simplify to the case $N = 2$, there will be coordinates $x_j^{(1)}$ for the first singularity and coordinates $x_j^{(2)}$ for the second singularity. This allows one to define a classical configuration space formed from the objective couples of coordinates $(x_j^{(1)}, x_j^{(2)})$, not to be confused with the corresponding abstract configuration space of quantum mechanics. Such an extended framework is more adequate to discuss the fact that the double solution theory is non-local and contextualist, two most important features that I shall clarify later using more appropriate opportunities, in particular regarding Bohm's pilot wave and quantum mechanical features.

Let us nevertheless introduce these features in a qualitative way. Regarding them, let us remark that the motion of one singularity (to start with a 1-body problem), being linked to the evolution of the wavy phenomenon in which it is embedded, depends on all circumstances encountered by this associated wavy phenomenon during its propagation. Therefore, the motion of the particle is actually not completely classical but is influenced by all obstacles encountered by the associated wave, thereby providing an explanation for interferences and diffraction experiments, such as in the two-slit experiment [65]. Furthermore, going now to the extended framework, the motion of the two particles in a 2-body problem (with an obvious generalization for the N-body problem) are no more independent since they both depend on the same wave Ψ. This is in essence what is meant when we say that the double solution is non-local and contextualist, two notions which at the present time seem identical but, as we shall later see, are actually different in a rather subtle way.

More vividly, as stated by Louis de Broglie, each singular region should be considered as a center of forces influencing the propagation of the wavy phenomena associated with the other particles and therefore influencing the motion of these other particles. This should result in a complicated motion of the set of singular corpuscles/regions. This idea will be developed much further by Bohm.

Non-locality is also put forward when Louis de Broglie discusses the pressure exerted on a wall by an incoming flux of particles [62]. As he said, from the point of view of wave mechanics, there exists in the vicinity of the wall a state of interferences produced by the superposition of incident and reflected waves. The use of the formula giving the velocity of the particles (the guidance formula) demonstrates that the particles do not hit the wall. Then, how could they possibly exert a pressure? This can only be the result of stresses present in the interference region. For Louis de Broglie, such stresses are internal stresses spreading in the surrounding space, corresponding to the fact that the particle is not only a point entity, but also the center of an extended phenomenon.

Also, the fact that the configuration space of the double solution is a classical one is obviously important in the mind of Louis de Broglie. It allows one to think of a system of N particles as being made of N waves propagating in the "real" Newtonian space, rather than as being made of only one wave propagating in the abstract quantum mechanical configuration space.

To avoid too much redundancy and too many repetitions, I shall not discuss here the most important concept of quantum potential, to be more appropriately introduced when we shall examine the details of the pilot wave theory. Let us simply say that this quantum potential is a non-local potential associated with a quantum field which would exist even in the absence of any classical field, such as the electromagnetic or the gravitational fields. And again for a similar reason, I am postponing the discussion of objections to the double solution which would be shared by the pilot wave, and therefore addressed to all causal theories.

Blindly forging ahead?

After having given up his causal theories for a long time, Louis de Broglie started to ride the wild horse again, impulsed by Bohm's coming, with a small school of collaborators, mainly including Vigier (that he appreciated enough to make him participating, with a paper, in [62]). While Bohm further developed his pilot wave, Louis de Broglie however focused on his cherished double solution and attempted to complete it, in various ways.

One of the modifications he introduced concerns the postulate of concordance of phases that he eventually found too severe, and could not be mathematically justified. Without relaxing it totally, he made it more flexible by being content to simply assume that φ and φ' coincide at each time in the vicinity of the singular region constituting the particle, i.e. not for all x'_js at any time as demanded by Eq. (7.13), see [62]. Under this weaker form, Louis de Broglie expressed the hope that the postulate could more easily find a mathematical justification.

Next, concerning the particle singularity, the most difficult task was to demonstrate that, given a specific situation (the form of Ψ, its evolution equation, and boundary conditions), there also exist u-solutions describing a moving singularity. In the earlier version of the double solution, the second solution u had to satisfy the same equation as the first solution Ψ, that is to say a linear equation, already leading to formidable mathematical problems. But, later on, Louis de Broglie came to the idea that the equation for u should be non-linear, becoming only approximately linear and identical to Schrödinger's equation far enough from the singularity. As already said, this was inspired by Einstein's works and his bunched fields [65], and by the influence of Vigier who had been working for a while, independently of Louis de Broglie, at the Institute Henri Poincaré, digging the double solution and space-time singularities ([153, 175], and see furthermore [176]). But it was also the consequence of a theorem due to Sommerfeld, which ruled out the possibility to introduce singularities, in the genuine mathematical sense of the word, in a linear framework [62]. Nonlinearities would also have the virtue of preventing the spreading of the waves u, as in soliton theory (known to Louis de Broglie), so that particles would preserve their individuality [63]. Also, as we already discussed, there is no rigorous solution to the measurement problem in a linear framework, a fact acknowledged by Louis de Broglie when he discussed quantum mechanical transitions. For him, these transitions, rather than being indivisible unanalyzable quantum jumps, could be transitory continuous processes of a very short duration, where non-linear phenomena would play a significant role, even outside of the singular regions [65]. This would modify our vision of stationary states in quantum mechanics. Indeed, one is able to compute the (unobservable) stationary states but they have no real interest because nothing happens in such states. What is interesting (and can be observed) are the transitions between stationary states, but such transitions are not computable: They are just jumps. Louis de Broglie's proposal concerning non-linearities was for him an opportunity to modify the status of this problem. Needless to say, this proposal would also open an opportunity to solve the measurement problem as commented by Louis de Broglie [65]. Clearly however, the introduction of nonlinearities cannot simplify the mathematical difficulties involved in the theory, but can only make them still more formidable. Nowadays, researchers working for instance in the field of non-linear dynamics and chaos theory are well aware of the fact. Therefore, there is no wonder if [155] *the double solution presents stupendous mathematical difficulties which de Broglie was unable to overcome.* Indeed, as noted by Belinfante [1], to the best of his knowledge (and to mine), any specific (translate : Useful) solution for u was never published.

Also, in the earlier version of the double solution, the singularity was viewed as a mathematical singularity *stricto sensu*, that is to say becoming infinite at the location of the particle (as a certain negative power of distances). But, later on, Louis de Broglie resuscitated came to the conviction that it should rather be a small moving singular region such that, inside this region, the values of u would become large enough (however no more infinite) to imply the breaking down of the linear approximation. Louis de Broglie suggested that the size of this region could be of the order of 10^{-13} cm, a critical characteristic length also put forward by Bohm, a guess certainly inspired by the development of the physics of elementary particles (recall that 10^{-13} cm, i.e. a fermi, is a characteristic nuclear size). This idea generated the hope that quantum mechanics could eventually break down at such length scales for the study of nuclear phenomena or in high-energy processes [63], and that causal theories could then possibly find some empirical support. This did not happen.

Louis de Broglie also discussed the N-body problem for the case of identical (indiscernible) particles, bosons and fermions. This could be incorporated in the double solution if we could demonstrate that, in the case of fermions, the u-waves could only contain one singularity, providing a causal understanding of the Pauli's exclusion principle, in contrast with the case of bosons where the u-waves could support many singularities [62, 63]. Louis de Broglie went on further with many other speculative ideas such as concerning the implementation of Dirac's equation in the double solution [62], the possibility (examined in collaboration with Vigier) that the space-time metric tensor in general relativity could depend at each point of the value of the wave u [63], the concomitant hope that the theory of u-waves could serve for a synthesis between general relativity and the theory of quanta (this again did not happen), the generalization of the double solution to the case of particles with spins (dealing with spinors, say several wave functions instead of a single one), the idea that there could exist wave packets of the u-kind without any singular region, the existence of a sub-quantum mechanical level (the vacuum?) which would introduce a random element in the theory, and would permit to complete the dynamical picture of particles by a thermodynamical picture, what Louis de Broglie called a hidden thermodynamics of particles (an idea to be much developed by Bohm)... After all these attempts, was Louis de Broglie still believing in himself? May be not if we follow Bonk [177] stating: *Another indication of de Broglie's failure to entrust full explanatory power to his approach is his statement that energy and momentum are not conserved which is true only if the quantum potential is not properly taken into account. But why is that? A possible explanation is that he simply did not believe there is anything corresponding to the term in nature, or that there will ever be independent experimental indication of such forces.* But certainly yes, if we follow Louis de Broglie himself who, in a work dated 1970 [178] which can be viewed as a last will and testament (although actually Louis de Broglie died in 1987), summarized his ideas and concluded : *Today, I am convinced that the conceptions developed in the present article, when suitably developed and corrected at certain points, may in the future provide a real physical interpretation of present quantum mechanics.*

Hence, after all, and after a whole life, Louis de Broglie seemingly went on believing that he was on the right track. Up to now, this belief has been deceived and the double solution remains an unachieved program of research. Will someone take again the burden on his shoulders or will the double solution remain an unique and solitary creation? Should we conclude that Louis de Broglie was really blindly forging ahead, desesperately running into blind-alleys, or could we imagine that someone will find a way out from the marshes?

Chapter 8

Pilot Wave

It is easy to pass from the double solution exposed in the previous chapter to the pilot wave to be discussed in the present one. Essentially, what we have to do is to forget anything concerning the second solution, that is to say concerning the u-waves and their singularities, and to preserve the first solution Ψ and the guidance formula driving the underlying hidden objective particles. Then, these particles are piloted by the wave Ψ, hence the name of the theory. We are still left with two versions, the pilot wave theory of Louis de Broglie, and the pilot wave theory of Bohm. They essentially differ by the epistemological status of Ψ, subjective for Louis de Broglie and objective for Bohm, and by the fact that Bohm developed the theory much further, up to its logical conclusion, as he wrote. In particular, objections which have been addressed to the version by Louis de Broglie have been successfully enough answered by Bohm. Therefore, to avoid any inconvenient redundancy on this matter, we shall now on discuss Bohm's version, and the terminology "pilot wave", excepted possibly when specified otherwise, will refer to this version.

Early Bohm

As for any creative people, we must distinguish several Bohm's. In particular, there has been a pre-early Bohm, the one who wrote a famous book on quantum mechanics [41] in which he was a Zealot of the Copenhagen interpretation, and opposed himself to hidden variables. Indeed, he wrote [41]: ... *the quantum theory is inconsistent with the assumption of ... hidden variables ... In other words, no completely deterministic mechanism that could explain correctly the observed wave-particle duality of the properties of matter is even conceivable ... Until we find some real evidence for a breakdown of the general type of quantum description now in use, it seems, therefore, almost certainly of no use to search for hidden variables.* Bohm even devoted a whole section of his book to a *proof that quantum theory is inconsistent with hidden variables.* Concerning the EPR- paradox to be studied later in this book, he also wrote: *We can now use some of the results of the analysis of the paradox EPR to help prove that quantum theory is inconsistent with the assumption of hidden causal variables ... We conclude then that no theory of mechanically determined hidden variables can lead to all of the results of the quantum theory.*

Nevertheless, reading between the lines, we can sometimes observe the next coming Bohm, a renegade to himself, showing the tip of his nose. This next coming Bohm is the one I call the early Bohm who revealed himself in his two seminal papers dated 1952, exposing his pilot wave version [29, 30]. There will be other Bohm's later, essentially concerning Bohm revisiting Bohm. The early pilot wave of the early Bohm (simply named pilot wave) is also discussed in many places, e.g. by Belinfante [1], Wigner [74], Holland [40] and by Cushing [131, 166].

Let us now go on following the early Bohm [29]. We still use Schrödinger's equation (7.1) for the wave Ψ. However, Ψ is now viewed as an objective field from the very beginning, exactly such as an electromagnetic field satisfying Maxwell's equations. Such a decree will appear to be consistent with the further developments of the pilot wave theory, in particular due to the solution it will provide for the measurement problem. But, for the time being, why not? We do not know from first principles why there should be an electromagnetic field, nor do we know *what is* an electromagnetic field. Therefore, why should we not adopt the same complacency for Ψ than the one we have for **E** and **H**, or for their unification in an electromagnetic tensor H_{ij}? In the words of Bohm: *In the last analysis, there is, of course, no reason why a particle should not be* *acted on by a Ψ-field, a gravitational field, a set of meson fields, and perhaps by still other fields that have not yet been discovered.*

At this point, PEDESTRIANS ($\sim\sim\sim\sim$) might smoothly jump to the next section.

Also, we still write the expression for Ψ under the form given by Eq. (7.2). But, we shall make a slight change of notations, from Louis de Broglie's notations to Bohm's notations, by replacing a by R, and φ by S. Then, we may again insert Eq. (7.2) into Eq. (7.1), and obtain Eqs. (7.3) and (7.4) that we now better rewrite, following Bohm, under the form

$$\frac{\partial R}{\partial t} = -\frac{1}{2m}\left(R\frac{\partial^2 S}{\partial x_j^2} + 2\frac{\partial R}{\partial x_j}\frac{\partial S}{\partial x_j}\right) \tag{8.1}$$

$$-\frac{\partial S}{\partial t} = \frac{1}{2m}\left(\frac{\partial S}{\partial x_j}\right)^2 + V - \frac{\hbar^2}{2m}\frac{1}{R}\frac{\partial^2 R}{\partial x_j^2} \tag{8.2}$$

We now introduce the density of probability of presence, as in Eq. (7.10), with a slight change of notation, according to

$$P(x) = |\Psi|^2 = R^2 \tag{8.3}$$

Eqs. (8.1) and (8.2) can then be rewritten as

$$\frac{\partial P}{\partial t} + \frac{\partial}{\partial x_j}\left(\frac{P}{m}\frac{\partial S}{\partial x_j}\right) = 0 \tag{8.4}$$

$$-\frac{\partial S}{\partial t} = \frac{1}{2m}\left(\frac{\partial S}{\partial x_j}\right)^2 + V - \frac{\hbar^2}{4m}\left[\frac{1}{P}\frac{\partial^2 P}{\partial x_j^2} - \frac{1}{2P^2}\left(\frac{\partial P}{\partial x_j}\right)^2\right] \tag{8.5}$$

Again, in the same way we used to discuss Eq. (7.4), we may take the classical limit $\hbar \to 0$ of Eq. (8.5), yielding

$$-\frac{\partial S}{\partial t} = \frac{1}{2m}\left(\frac{\partial S}{\partial x_j}\right)^2 + V \tag{8.6}$$

which is exactly Eq. (2.1) of the Hamilton-Jacobi's formulation of classical mechanics. We also assume Eq. (8.7), for similar reasons and with similar comments, now reading as

$$v_j = \frac{1}{m}\frac{\partial S}{\partial x_j} \tag{8.7}$$

which is again the guidance formula, extending the classical picture of Fig. 1(b) to the quantum picture of Fig. 1(c). Now, from Eq. (8.4), we can derive a continuity equation reading as

$$\frac{\partial P}{\partial t} + \frac{\partial}{\partial x_j}(Pv_j) = 0 \tag{8.8}$$

to be compared with Eq. (7.11). Again, this equation allows one to regard P as a density of probability of presence for particles in an ensemble of trajectories. And, according to Eq. (8.7), these trajectories are orthogonal to iso-value surfaces of S.

Going back to Eq. (8.5), we see that the motion of the particle can be viewed as depending on two potentials: The classical potential V and another extra-potential U, called the quantum potential, reading as

$$U = -\frac{\hbar^2}{4m}\left[\frac{1}{P}\frac{\partial^2 P}{\partial x_j^2} - \frac{1}{2P^2}\left(\frac{\partial P}{\partial x_j}\right)^2\right] = -\frac{\hbar^2}{2m}\frac{1}{R}\frac{\partial^2 R}{\partial x_j^2} \tag{8.9}$$

The equation of motion of the particle then takes the form of a generalized Newton's law according to

$$m\frac{d^2 x_j}{dt^2} = -\frac{\partial}{\partial x_j}\left(V - \frac{\hbar^2}{2mR}\frac{\partial^2 R}{\partial x_j^2}\right) \tag{8.10}$$

Now, a most important fact is that the pilot wave can be made empirically equivalent to quantum mechanics. Following Bohm (see also Holland [40] for a similar discussion), *all the results of the usual interpretation are obtained from our interpretation if we make the following three special assumptions which are mutually consistent:*

(1) *That the Ψ-field satisfies Schrödinger's equation.*
(2) *That the particle momentum is restricted to $p_j = \partial S/\partial x_j$.*
(3) *That we do not predict or control the precise location of the particle, but have, in practice, a statistical ensemble with probability density $P = |\Psi|^2$. The use of statistics is, however, not*

inherent in the conceptual structure, but merely a consequence of our ignorance of the precise initial conditions of the particle.

The formulation we have exposed here is indeed very close to the one of Louis de Broglie (see Louis de Broglie's first solution in the previous chapter), and also close to the one of Madelung [165], but the interpretation given by Bohm, as well as the extension of the consequences of the formalism, make the difference. Later on, Bohm and Hiley [179] will comment on Madelung as follows: *Clearly, the quantum potential interpretation is to be distinguished from a hydrodynamic model (such as that of Madelung). For, in this model, the particle is simply pushed mechanically by the fluid. The notion of a particle responding actively to information in the Ψ field is indeed far more subtle and dynamic than any others that have hitherto been supposed to be fundamental in physics. This is especially evident ... in the many-body system, for which simple models such as the hydrodynamical one, can no longer be applied at all.* This quotation however anticipates a later interpretation, by Bohm (and Hiley), of the quantum potential as a kind of active information potential [179, 180]. For Bohm versus Madelung, see also Jammer [24].

Early Bohm for PEDESTRIANS ($\sim\sim\sim\sim$)

The pilot wave Ψ is assumed to satisfy Schrödinger's equation. However, Ψ is now viewed as an objective field from the very beginning, exactly such as an electromagnetic field satisfying Maxwell's equations. A guidance formula, such as already introduced above, is again put forward telling us how particles are piloted by Ψ. It is then shown that particles experience the effect of two potentials: The classical potential V and another extra-potential U, called the quantum potential, and that the equation of motion of the particles takes the form of a generalized Newton's law.

Furthermore, a most important fact is that the pilot wave theory is designed in such a way that it exactly reproduces the predictions of quantum mechanics. Hence, pilot wave theory and quantum mechanics, although they direct us to two different interpretations, and ontologies of the world, are empirically equivalent.

Successes of pilot wave

The pilot wave is successful in many respects, making it something like unforgettable. It first succeeds in reaching its goal, that is to say to provide a causal objective description of quantum phenomena so that, in the words of Bohm, it is an ontological interpretation of quantum mechanics. It is non-local and contextualist (as is the double solution). This must be viewed as a success because, as we shall see, any acceptable hidden-variables theory must be non-local and contextualist. And last, but far from being the least, it succeeds to provide a satisfactory solution to the measurement problem.

Ontology

With Bohm's theory, we have particles and one piloting wave, but both the wave and the particles are objective, the wave because of a consistent decree, the particles because they are hidden entities viewed as existing really, objectively piloted by the classical potential and by the quantum potential, following deterministic trajectories with well defined values of positions and velocities. Both of them (particles, and the pilot wave) obey deterministic equations of motion. This is also very appealing to the intuition since waves (or fields) and particles are the two kinds of entities populating the classical world.

The description thus proposed in Bohm's interpretation is also independent of the act of observation. The act of observation may possibly "disturb" the particles under study, because it is an interaction process still submitted to the indivisibility of the quantum of action, but it does not create the properties of the particles which possess objective properties before the measurement. The measurement does not create the properties, it just reveals the observed properties. This occurs through a process that can be analyzed in a detailed way (as we shall soon do), something forbidden by the orthodoxy. Things happen objectively in the quantum world, independently on whether there are observers or not. And the entities evolving in the quantum world may be called beables (be-ables in contrast with observ-ables), to use a world introduced by Bell

[79, 181–183]. Indeed beables are defined as entities which possess a reality that is independent of being observed or known in any other way. Among the set of beables, Bell, as we shall discover, has particularly be interested with local beables, the ones (unlike, for example, the total energy) which can be assigned to some bounded space-time regions [100]. Therefore, as stated by Bub and Clifton [184], *the causal interpretation ... is a reformulation of quantum mechanics in terms of beables.*

Bohm's pilot wave and quantum mechanics are empirically equivalent: They lead strictly to the same predictions for measurement outcomes (values and probabilities). This happens by construction and acknowledges the fact that quantum theory is in agreement with experimental results. In particular, in Bohm's pilot wave, we still have $P = |\Psi|^2$, which connects predictions of the pilot wave and predictions from quantum mechanics. But, and this is a main point concerning the issue of interpretations, the meaning of P in the two frameworks is very different. In quantum mechanics, the density of probability of presence concerns a probability of observing (finding) the particle here or there while, in the pilot wave, it concerns the probability for the particle for be-ing here or there. It is really in pilot wave a density of probability for being present, not for being found.

This gives to Ψ a double interpretation [180], *first as a function from which the quantum potential could be derived and, secondly, as a function from which probabilities could be derived.* Furthermore [68], *the observables ... do not have a fundamental significance in our theory but rather are treated as statistical functions of the beables that are involved in what is currently called a measurement.*

We may then agree, in some sense, with Bohm when he calls his interpretation an ontological interpretation. Indeed, with Bohm, it is possible to describe and analyze the quantum world with a model which is intuitively clear and precise. This model rests on a reality which does exist independently of the act of observation, nevertheless reproducing all empirical predictions of the usual Copenhagen interpretation. It allows us to recover the strong objectivity associated with the classical trio (space, time, causality). Einstein was not enthusiastic with the pilot wave version of Louis de Broglie. In the same way, he has not been enthusiastic with the version of Bohm. Nevertheless, the pilot wave theory of Bohm achieved what, according to Einstein, should be the programmatic aim of physics.

But, from the point of view of the Kantian theory of knowledge, the word "ontology" as used by Bohm is a misnomer. It can only be understood if we consider Bohm as a strong realist, searching for a strong objectivity, as was Einstein himself. The so-called ontological interpretation indeed can only be ontological if we forget Kant's teaching, that is to say if we forget to distinguish between the Thing in Itself and the Phenomenon. Kant, from his philosopher Paradise (if such a Paradise for dead philosophers does exist), is likely thinking that the ontology of Bohm is naive. Or, to state it differently, the ontology of Bohm is not an upper-case Ontology of the Thing in Itself but a lowercase ontology of the Phenomenon, or of the World.

Also, from the point of view of the theory of knowledge again, we must once more point out to the fact that the predictions of measurement outcomes by the pilot wave are the same than those of quantum mechanics, so that it cannot be falsified, or, better said, it cannot be experimentally discriminated against quantum mechanics. Any falsification would work simultaneously both for the pilot wave and for the quantum mechanics. Therefore, according to Heisenberg [146], *we are ... not concerned with counter-proposals to the Copenhagen interpretation, but with its exact repetition in a different language.* From a certain positivist point of view, only concerned with measurement outcomes, this might be a last word. But different languages mean different ways of thinking, and therefore different ways of viewing the world. For instance, I remember my surprise when I realized that I could not properly translate *cogito, ergo sum* in Japanese: Two different languages, and two different ways of viewing one of the most important mysteries we are facing to, namely what "I" is. If we do not renounce to the idea that science may possibly tell us something concerning the reality, even in a veiled fashion, then we have to address the question to know which language is the best, a very sensitive issue that we shall have to examine.

Non-locality and pilot wave

In the previous subsection, we saw that the pilot wave provides an attractive reconciliation between quantum mechanics and classical ideas. Then, it might seem up to now that the ontology of Bohm is the one of classical mechanics, in all aspects. Such is however not the case. Indeed, the pilot wave theory introduces highly non classical features, namely non-locality and contextuality, as did the double solution. This should not be too much surprising. It might be guessed from the facts that the pilot wave and the quantum mechanics are empirically equivalent, and that quantum mechanics is non-local and contextualist (as we shall later deepen). Also, we shall demonstrate that any acceptable hidden-variables theory must be non-local and contextualist and, therefore, if this was not the case for the pilot wave, it would have to be rejected straight away. We are now going to develop this issue in a quantitative way. PEDESTRIANS ($\sim\sim\sim\sim$) may skip this quantitative discussion, involving equations, and land on the end of the present subsection, just after the last equation of it.

Our discussion of non classical features, regarding the double solution, was essentially qualitative and the arguments we used in this double solution context still hold for the present pilot wave context. However, we are now going to provide a more careful discussion, relying on the formalism of the pilot wave. For doing this, it is most convenient to deal with a many-body problem. Let us start with $N = 2$. Hence, we now have an objective wave function of the form $\Psi(x_j^{(1)}, x_j^{(2)})$, with $x_j^{(1)}$ and $x_j^{(2)}$ being objective Cartesian coordinates associated with particles 1 and 2 respectively. Instead of Schrödinger's equation 7.1 for the one-body problem, we now deal with a Schrödinger's equation for a two-body problem, reading as

$$i\hbar\frac{\partial\Psi}{\partial t} = -\frac{\hbar^2}{2m}\left[\left(\frac{\partial^2\Psi}{\partial x_j^2}\right)_{(1)} + \left(\frac{\partial^2\Psi}{\partial x_j^2}\right)_{(2)}\right] + V\Psi \qquad (8.11)$$

in which the subscripts (1) and (2) attached to the Laplacians are associated with particles 1 and 2 respectively. The wave function is next written in a way generalizing Eq. (7.2) according to

$$\Psi(x_j^{(1)}, x_j^{(2)}) = R(x_j^{(1)}, x_j^{(2)}) \exp \frac{i}{\hbar} S(x_j^{(1)}, x_j^{(2)}) \tag{8.12}$$

and, always assuming $P = R^2$, we obtain the generalizations of Eqs. (8.4) and (8.5), see also Eq. (8.9), taking the form

$$\frac{\partial P}{\partial t} + \frac{1}{m} \left[\left(\frac{\partial}{\partial x_j} \right)_{(1)} P \left(\frac{\partial S}{\partial x_j} \right)_{(1)} + \left(\frac{\partial}{\partial x_j} \right)_{(2)} P \left(\frac{\partial S}{\partial x_j} \right)_{(2)} \right] = 0 \tag{8.13}$$

$$-\frac{\partial S}{\partial t} = \frac{1}{2m} \left[\left(\frac{\partial S}{\partial x_j^{(1)}} \right)^2 + \left(\frac{\partial S}{\partial x_j^{(2)}} \right)^2 \right] + V(x_j^{(1)}, x_j^{(2)})$$

$$-\frac{\hbar^2}{2mR} \left[\left(\frac{\partial^2 R}{\partial x_j^2} \right)_{(1)} + \left(\frac{\partial^2 R}{\partial x_j^2} \right)_{(2)} \right] \tag{8.14}$$

These equations describe the motion of the two particles in a six-dimensional configuration space which therefore does not identify with the three-dimensional real space, but which is however an objective classical configuration space, with actual trajectories, not an abstract and fictitious configuration space as discussed by Louis de Broglie. Heisenberg [146] asked whether the configuration space of Bohm is a real space. We see that, in a strong sense, the answer should be positive.

Now, each particle possesses a velocity given by a guidance formula, simply generalizing Eq. (8.7):

$$v_{j(i)} = \frac{1}{m} \left(\frac{\partial S}{\partial x_j} \right)_{(i)}, \quad i = 1, 2 \tag{8.15}$$

In the classical mechanics limit ($\hbar \to 0$), Eq. (8.14) reduces to the two-body Hamilton–Jacobi's equation. But, following Bohm's interpretation, this Eq. (8.14) also exhibits a two-body quantum potential reading as

$$U = \frac{-\hbar^2}{2mR} \left[\left(\frac{\partial^2 R}{\partial x_j^2} \right)_{(1)} + \left(\frac{\partial^2 R}{\partial x_j^2} \right)_{(2)} \right] \tag{8.16}$$

The generalization to the N-body problem is straightforward and, in this case, the quantum potential obviously reads as

$$U = \frac{-\hbar^2}{2mR} \sum_{n=1}^{N} \left(\frac{\partial^2 R}{\partial x_j^2} \right)_{(n)} \tag{8.17}$$

in which now

$$R = R(x_j^{(1)}, x_j^{(2)}, \ldots, x_j^{(N)}) \tag{8.18}$$

PEDESTRIANS ($\sim\sim\sim\sim$), land on here

The structure of the quantum potential U, depending on R, i.e. on all coordinates of all particles, then shows that, as stated by Bohm, *the force between any two particles may depend significantly on the location of every other particle in the system.* This is the non-local character of the pilot wave.

We might argue that, may be, we should not be surprised with such a feature. After all, the situation looks quite similar to the one in classical mechanics, if only we examine the structure of the classical potential $V(x_j^{(1)}, x_j^{(2)}, \ldots, x_j^{(N)})$. Physically, we indeed know that the motion of the earth depends on the location of far remote objects, just like the Moon, the Sun, all the other planets, and stars, and galaxies... And, in Newtonian mechanics, the gravitational field acts instantaneously through space, i.e. Newtonian mechanics is highly non-local too, a kind of outrageous feature for Newton contemporaries and successors, much uncomfortable for Newton himself. Nowadays, thanks to relativist advances, there is no trouble any more. We know that interactions do not act instantaneously but propagate at a speed which cannot exceed the speed of light and which, in the case of gravitational fields, is exactly equal to the speed of light. Non-locality is therefore a common feature of non-relativistic theories, an artefact due to a too poor level of approximation.

Hence, since we are dealing, and Bohm was dealing, with a non-relativistic quantum theory, why should we bother with a theoretical artefact? There should be anything mysterious with it. Just go to a relativistic extension of quantum mechanics, and the artefact will disappear. But it does not! As we shall extensively discuss, non-locality is also an experimental feature of quantum mechanical phenomena. Whether this produces a head-on collision between quantum mechanics (Copenhagen interpretation, pilot wave, or empirical data) and relativity is an issue that we shall have to investigate. Here, remember that Einstein, in his effort to build a hidden-variables theory, also met this inconvenient non-locality property. He did not like it, and gave up, leaving his work unpublished. Otherwise, he might have been the real acknowledged father of causal theories.

Later on, in his 1993-book (published one year after his death), with the whole Chapter 7 devoted to non-locality, Bohm (together with Hiley) [68] commented on it and on its relationship with relativity: *The guidance relation implies that the particles are guided in a correlated way ... the particles can be strongly coupled at long distances. Their interaction can therefore be described as nonlocal ... in a relativistic context, the nonlocality is still present but it does not introduce any inconsistency into the theory, e.g. it does not imply that we can use the quantum potential to transmit a signal faster than light ... the relationship between parts of a system described above implies a new kind of wholeness of the entire system going beyond anything that can be specified solely in terms of the actual spatial relationships of all the particles ... this means that the guidance condition and the quantum potential implied by it can, under certain conditions, have the novel quality of being able to organize the activity of an entire set of particles, in a way that depends on a pool of information common to the whole set.*

For Aharonov *et al.* [185, 186], see also Drezet [187], non-locality features *challenge any realistic interpretation of Bohm trajectories*. The exact meaning of this sentence may depend on the sense we give to the word "realistic". Keeping on with a terminology developed in previous chapters, we may still assess that the pilot wave is causal, objective, complies with the classical trio (space, time, causality), and ... is a realist theory in a strong sense. However, it should be now clear that non-locality features indeed drive us rather far away from the classical picture of classical mechanics. The original Louis de Broglie's idea to return to a clear and precise classical description of the world cannot be achieved. There is definitively something rotten in the quantum kingdom.

The non-locality feature of the pilot wave (and also of the double solution) had far-reaching consequences for subsequent developments of quantum mechanics. Bell [188] refers to it, and asks himself whether he should be or not a generic feature for any acceptable hidden-variables theory. This is the kind of question which led him to the discovery of his famous inequalities [189], to the assessment of the impossibility of local hidden-variables theories, bringing the issue of hidden variables in the realm of experimental physics (Chapter 12). And the experimental results will confirm that wholeness is a property of the world. Our participatory universe is genuinely and deeply holistic.

Concerning the issue of contextuality, let us again accept to confuse it with the one of non-locality for a bit of time. As stated by Bitbol [133], Bohm's theory is explicitly presented as being contextualist. The form of the guiding wave effectively depends on the configuration of the complete set-up (if not of the complete environment), and the motion of each corpuscle would then be modified by any minor modification of the set-up, and even of the environment. This is contextuality, at least before we shall refine the concept.

Solution to the measurement problem

The most impressive success of Bohm's theory is likely to be that it provides an objective, causal, solution to the long-standing measurement problem. In other words, to refer to a much popularized Gedanken experiment, it solves the Schrödinger's cat paradox (see Bohm and Hiley [68] for a specific discussion of this paradox in a causal framework). In the words of Cushing [131], *one of the most beautiful aspects of Bohm's paper is his treatment of the measurement problem (which becomes a nonproblem)*. The deep reason why it is so is that the measurement process is viewed as a dynamical interaction problem, that is to say a many-body problem which can therefore be described in the many-body framework of the pilot wave formalism, previously sketched. As a result, there is actually no collapse of the wave function and the whole mystery simply evaporates.

In any case, a completely satisfactory quantum mechanical theory, whether it is quantum mechanics (as it is, or completed, or amended) or an alternative hidden-variables theory, should definitively solve the measurement problem. Some people believe that actually quantum mechanics did it, for instance proponents and defenders of decoherence approaches, but the issue remains controversial. Most seemingly, Bohm succeeded. Therefore, with the pilot wave, we possess a theory which is empirically adequate, and which is able to analyze the measurement process in itself, far from the Copenhagen interpretation where this process must remain in the limbo of unspeakable breathes.

This does not however yet prove, by any means, the superiority of the pilot wave with respect to quantum mechanics, or its validity. A satisfactory theory should not be partially true and partially false (whatever the epistemological meanings we give to these words), it should be a coherent and consistent structure to be accepted or rejected as a whole, even if only in a restricted domain of validity. One region of a theory may be partially correct but, if it is in a hypothetico-deductive way contaminated by other insane regions, the whole edifice is spoiled and has to be destroyed. Whether there is some amount of falsity in Bohm's theory, sufficient to reject it, is something we do not consider at the present time. In any case, the solution proposed by Bohm to the measurement problem is an achievement. Conversely, the failure of quantum mechanics regarding this problem, at least in the mind of some experts, is worrying, but does not yet definitively invalidate it. We cannot exclude the possibility that a solution will be eventually found, in the framework of the present theory, or better in a slightly modified framework, with only very minor alterations, without forgetting that, for some people, decoherence theory achieves the aim.

Many papers and books have discussed Bohm's approach to the measurement problem, e.g. by Louis de Broglie [63], Belinfante [1], Wigner [74], Bohm and Hiley [68, 179, 190], Finkelstein [191] or Bub and Clifton [184]. We shall here however rely on the original exposition by Bohm, dated 1952 [30], with however a slight change of notations and minor corrections.

The measurement process is an interaction process, taking place between a quantum object and a certain macroscopic apparatus. In contrast with the original view of Bohr, propagated by other people like Landau, the macroscopic measuring apparatus is not considered as a classical system. It is exactly taken for what it is, made out from atoms, electrons, nucleons ..., that is to say for another genuine quantum object, although involving a huge number of degrees of freedom. Therefore, a measurement process is just a special case of a many-body quantum process. This is sufficient to understand that it will be causal and objective, although non-local and contextualist.

This being said, PEDESTRIANS ($\sim\sim\sim\sim$) may take off, and land on later in the present subsection, a few lines after the last equation of it.

In agreement with quantum mechanics, for a quantum object under study represented by a state vector $|\Psi\rangle$, the apparatus after the interaction will return one of the Eigenvalues a_n of the spectrum associated with the observable being measured (for convenience, we assume a discrete non-degenerate spectrum), with probabilities depending on the state. Each value a_n may be seen as characterizing a quantum state of the measuring device (say, one position among the different possible positions of a pointer). Therefore, the interaction introduces correlations between the state of the quantum system to be investigated and the measuring device used to make the investigation.

Let the observable be denoted by Q, and let x_j denote the (objective) coordinates of the quantum object under study. The measuring apparatus has to be described by a huge number of coordinates but it is sufficient to consider only one relevant coordinate, say the position of a needle, that we shall simply denote by y. Following Bohm, we shall now satisfy ourselves in considering only impulsive measurements, that is to say measurements involving a strong interaction between the system observed and the apparatus. As a consequence, the interaction lasts for a very short time. Therefore, evolutions of the system and of the apparatus, which would have naturally occurred in the absence of the interaction, can be neglected. Then, during the measurement process, when the interaction takes place, the Hamiltonian driving the interaction is an interaction Hamiltonian that we shall denote H_I. Because H_I has to couple the system and the apparatus, it must depend on the observable Q, but also on operators involving y. For the sake of convenience, and an easier understanding, let us follow Bohm again and introduce a specific example

$$H_I = -aQP_y \tag{8.19}$$

in which a is a constant, and P_y the momentum operator conjugate to y.

The whole process has to be guided by an objective wave function, or pilot wave, $\Psi(x_j, y, t)$ which depends both on the system coordinates x_j and on the apparatus coordinate y. This wave function satisfies Schrödinger's equation in a four-dimensional configuration space, according to, during the interaction process

$$i\hbar\frac{\partial\Psi}{\partial t} = H_I\Psi = -aQP_y\Psi = ia\hbar Q\frac{\partial\Psi}{\partial y} = bQ\frac{\partial\Psi}{\partial y} \tag{8.20}$$

in which we have used the representation $-i\hbar\partial/\partial y$ of the momentum operator P_y, and set $b = ia\hbar$.

The wave function can be expanded over the complete set $\Psi_q(x_j)$ of Eigenfunctions of the operator Q, with q denoting the Eigenvalues of the operator Q, according to

$$\Psi(x_j, y, t) = \sum_q \Psi_q(x_j)f_q(y, t) \tag{8.21}$$

in which $f_q(y, t)$ represents the expansion coefficients which must clearly depend on y. Inserting Eq. (8.21) into Eq. (8.20), Schrödinger's equation yields a series of evolution equations for the expansion coefficients, reading as

$$i\hbar\frac{\partial f_q(y, t)}{\partial t} = bq\frac{\partial f_q(y, t)}{\partial y} \tag{8.22}$$

in which we have used the Eigen-equation

$$Q\Psi_q(x_j) = q\Psi_q(x_j) \tag{8.23}$$

We now search for the solution to Eq. (8.22) under the form

$$f_q(y, t) = f_q^0(y + A(q)t) \tag{8.24}$$

showing that the initial value of $f_q(y,t)$ is $f_q^0(y)$. Inserting Eq. (8.24) into Eq. (8.22), we obtain

$$i\hbar A(q)\left(f_q^0\right)' = bq\left(f_q^0\right)' \qquad (8.25)$$

Hence,

$$A(q) = -i\frac{bq}{\hbar} \qquad (8.26)$$

With the solution under the form of Eq. 8.24, we may rewrite Eq. (8.21) as

$$\Psi(x_j, y, t) = \sum_q \Psi_q(x_j)f_q^0(y + A(q)t) \qquad (8.27)$$

We assume that, at the beginning of the interaction process, the system and the apparatus are independent, i.e. they are not entangled. This means that the initial wave function is factorizable, say at time $t = 0$, under the form

$$\Psi_0(x_j, y) = \Psi_0(x_j)g_0(y) = g_0(y)\sum_q c_q\Psi_q(x_j) \qquad (8.28)$$

in which we expanded $\Psi_0(x_j)$ over the complete set $\{\Psi_q(x_j)\}$, the c_q's are expansion coefficients, and $g_0(y)$ is the initial wave function associated with the apparatus coordinate y. The wave function $g_0(y)$ is a wave packet that, without any loss of generality, we take centered at $y = 0$, with a width Δy. Comparing Eqs.(8.21) and (8.28), we deduce

$$f_q(y, 0) = f_q^0(y) = c_q g_0(y) \qquad (8.29)$$

so that, using now Eqs. (8.24) and (8.29), Eq. (8.21) becomes

$$\Psi(x_j, y, t) = \sum_q c_q\Psi_q(x_j)g_0(y + A(q)t) \qquad (8.30)$$

This equation shows that the interaction has generated a correlation between the apparatus coordinate y and the Eigenvalues q's, due to the fact that A depends on q. Such a correlation did not appear in the initial wave function of Eq. (8.28) because $g_0(y)$ could be factorized out. As a result, Bohm then argued that the wave function of Eq. (8.30) becomes very complicated. Let us write it, as now usual for us, under the form

$$\Psi(x_j, y, t) = R(x_j, y, t) \exp \frac{i}{\hbar} S(x_j, y, t) \tag{8.31}$$

We may then conclude that R and S become very complicated too, undergoing very rapid oscillations with respect to both space and time. This must also be true for the quantum potential which, in the present case, reads as

$$U = \frac{-\hbar^2}{2mR} \left(\frac{\partial^2 R}{\partial x_j^2} + \frac{\partial^2 R}{\partial y^2} \right) \tag{8.32}$$

This equation shows that the shape of U will become most particularly complicated when R is small, that is to say when the density of probability of being present is small. The particle and the apparatus, whose momenta are given by gradients of S, will also undergo *violent and extremely complicated fluctuations*. Afterward, Bohm demonstrates a result which is intuitively rather obvious: The packets $g_0(y + A(q)t)$, corresponding to different values of q, will cease to overlap in space, if the product of the strength of interaction (embodied in A) by the duration Δt of the interaction is large enough. If this does not occur, then the measurement is not completed.

PEDESTRIANS ($\sim\sim\sim\sim$), land on here

So, when the measurement is completed, we are dealing with a family of non-overlapping packets in y-space, corresponding to the different values of q, which are separated enough to receive a classical description. Furthermore, the coordinate y has to enter a wave packet, even if the motions are too complicated to allow us, in practice, to predict in which one this will happen. Once this is done, it cannot enter any more into the intermediate spaces between the packets, since the density of probability there is essentially zero. Hence, the wave packet containing y determines the outcome of the measurement. The other wave packets conversely become empty and, from now on, can be ignored. In the words of Bohm and Hiley [190]), *the remaining empty wave-packets will not be effective*, and: *It follows that the result is the same as if there had been a collapse of the wave function to a state corresponding to the result of the actual measurement.*

More discussions on empty wave packets are available from Rietdijk [192], Hardy [193] or Bläsi and Hardy [194]. Interestingly enough, empty waves was already introduced in Bohm's book dedicated to the usual quantum mechanics [41], when discussing tracks in cloud chambers, before his public commitment to hidden variables, just a bit later. In contrast, Schmidt [195] discussed the fact that, relying on the existence of empty waves, one can discriminate the causal from the Copenhagen interpretation of quantum mechanics. However, he postulated that empty waves are active, in conflict with Bohm's assumption that empty waves, which do not embed particles any more, are passive. See also Selleri [196].

A spectacular effect of Bohm's theory of measurement is that it allows one to understand in an objective way the famous two-slit experiment, a paragon of quantum mechanics, in particular to tell us through which slit a particular particle reaches the screen, without however destroying the interference pattern. Nevertheless, this does not contradict quantum mechanics because, although a slit is "chosen", the chosen slit pertains to the hidden world. Also, because Bohm's interpretation agrees with the predictions of quantum mechanics, it does not conflict with Heisenberg's uncertainty relations. We simply have to distinguish between the values observed in the act of measurement, and the true hidden values. As stated by Mermin [197], *the indeterminacy discovered by physical measurements of subatomic phenomena simply tells us that we cannot know the definite position and velocity of an electron at one instant of time. It does not tell us that the electron, at any instant of time, does not have a definite position and velocity.* That is to say, *if two observables fail to commute, then the uncertainty principle does not prohibit both from having definite values in an individual system.*

As stated by Bohm [30], *hidden variables determine the precise results of each individual measurement process. In practice, however, in measurements that we know how to carry out, the observing apparatus disturbs the observed system in an unpredictable and uncontrollable way, so that the uncertainty principle is obtained as a practical limitation on the possible precision of measurements. This limitation is not, however, inherent in the conceptual structure of our interpretation.* The word "disturbance" used in this quotation has been and still is often used when discussing the usual quantum mechanics, for instance regarding the famous Heisenberg microscope. But, in the quantum mechanical framework, where the measurement process is unanalyzable according to Bohr, or has to be treated by the wave function collapse postulate according to von Neumann, this word is like an anachronistic oxymoron (hence my reluctance to use the word "disturbance" in this book). Conversely, in the pilot wave theory, the concept of disturbance is an objective one, and the word is to be taken seriously.

A qualitative summary of what we have just learnt on the measurement process in the pilot wave theory, due to Bohm and Hiley [68], is worth to be presented to end this subsection. It goes on as follows: *... the essential new features of quantum measurement is that there is a mutual and irreducible participation of the measuring instrument and the observed object in each other. As a result, any attempt to discuss this process as measuring "a property of the observed object alone" will not be consistent with our interpretation ... In treating measurements ontologically we shall for the sake of convenience divide the overall process into two stages. In the first stage the measuring apparatus and the observed system interact in such a way that the wave-function of the combined system breaks into a sum of non-overlapping packets, each corresponding to a possible distinct result of the measurements. In the second stage this distinction is magnified by some detecting device, so that it is directly observable at the large scale level ... it is clear that measurement has in effect been treated as a particular case of quantum transition even though one of the systems concerned, i.e. the apparatus, is considered to be macroscopic ... We emphasize however that in our treatment there is no actual collapse; there is merely a process in which the information represented by the unoccupied packets effectively loses all potential of activity ... It is clear then that we are not measuring a state that has already been in existence. Rather the apparatus and the* *observed system have participated in each other, and in this process they have been deeply affected each other ... While the position of the particle considered as an abstract concept is an intrinsic property, the other properties are in general still context dependent.*

Another connected issue is the one of delayed-choice experiments that we already discussed. Let us recall that [67], *these are designed to show that according to the quantum theory the choice to measure one or another pair of complementary variables at a given time can apparently affect the physical state of things for considerable periods of time before such a decision is made.* This may look like a paradox but, if we simply state that a delayed-choice experiment is a measurement, then the pilot wave theory must be able to describe it in a causal, objective, way. Such is indeed the case as discussed by Bohm [68] who demonstrated that his theory removes any paradox. The reader may also refer to Bell [198] who sympathetically discussed the application of the de Broglie-Bohm version of quantum mechanics to delayed-choice experiments.

The solution to the measurement problem has also implications concerning the non-local character of the pilot wave theory. Indeed, any measurement of the location of a quantum object must still provoke a collapse of the wave function, even if this collapse is now the result of an objective deterministic process. This objective collapse by itself is already a non-local process. But, more importantly, the shape of Ψ is dramatically modified, and therefore accordingly produces a dramatic modification of the quantum potential. Therefore, a location measurement in a certain region of space instantaneously modifies the quantum forces exerted on all other quantum objects, even far remote.

Objections to pilot wave and answers

We already had several opportunities to discuss objections raised against the pilot wave, in particular those raised by Louis de Broglie himself, the most striking one being linked to the subjectivity of Ψ, annihilated by Bohm when he just decreed that Ψ was actually objective, consistently with his analysis of the measurement problem. In this section, we shall focus on other objections, possibly already mentioned, which could actually be addressed to both versions of the pilot wave (the one of Louis de Broglie and the one of Bohm), and also possibly to the double solution, and which have been put forward by eminent experts. In all cases, Bohm has been able to provide convincing rebuttals which then should be considered as successes of pilot wave, and therefore implemented in the previous section. However, I preferred to emphasize them in the present separated section which can then be considered as a kind of codicil supplementing the previous one. Specifically, these objections are due to Pauli (three objections from him), Perrin and Einstein.

First Pauli's objection. Inelastic scattering

Although this subsection contains two equations, PEDESTRIANS ($\sim\sim\sim\sim$) might be able to decently deal with it. Otherwise, there is no huge damage.

This objection from Pauli was raised against Louis de Broglie during the Solvay Congress of 1927 and, as we have seen, produced a depressing influence on Louis de Broglie. As stated by Cushing [131], *Wolfgang Pauli criticized de Broglie's theory on the basis of the example of the inelastic scattering of a plane wave by a rigid rotator. Although de Broglie felt he understood the general outlines of a suitable response to Pauli's objection, his rebuttal in fact appeared ad hoc and was not convincing.* Various discussions and comments on this issue are also available from Louis de Broglie [62, 63], Jammer [24], Dewdney and Hiley [199] or Holland [40].

We again follow Bohm [30]. While Pauli's objection originally concerned a problem of inelastic scattering by a rigid rotator, Bohm rather considered a problem of inelastic scattering of a particle by a hydrogen atom, a problem which is actually conceptually equivalent to the one of Pauli, but that Bohm already treated previously in the first of his 1952-papers [29]. Let us then consider an initial factorized wave function of the form

$$\Psi = \Psi_0(x_j) \exp \frac{i}{\hbar} p_{0,j} y_j \qquad (8.33)$$

representing the initial combined system, in a stationary state, made out from the wave function $\Psi_0(x_j)$ of the hydrogen atom in its ground state, and from the oscillating plane wave function of the particle with momentum $p_{0,j}$. Once the interaction between the particle and the atom is completed, the wave function can be written as (compare with Eq. (8.21) for instance)

$$\Psi = \sum_n f_n(y_j)\Psi_n(x_j) \tag{8.34}$$

in which $\Psi_n(x_j)$ is the wave function for the nth excited state of the atom (including the ground state) and $f_n(y_j)$ is the associated expansion coefficient.

Up to now, what we have written down is very similar to our treatment of the measurement problem. This should not be surprising. We are indeed dealing with an interaction problem, albeit not an interaction measurement problem (there is no macroscopic measuring device involved). Then, similarly as before, we may conclude that momenta will again undergo *violent and extremely complicated fluctuations*, due to the quantum potential which relates the motion of the particle and of the atom *in a way that depends strongly on the position of each particle*. Therefore, we cannot clearly see, due to this coupling, how the scattered particle and the atom will eventually reach definite energy values, in complete contradiction with experiments. Hence, according to Pauli, the pilot wave model (as expounded by Louis de Broglie in 1927) is untenable. As we know, such an argument made a devastating impression to Louis de Broglie who, soon after, gave up.

To answer the objection, Bohm argued that the whole argument depends much on the fact that we have assumed an incident plane wave to represent the incoming particle. But [30] *the use of an incident plane wave of infinite extent is an excessive abstraction, not realizable in practice.* Indeed, in practice, both the incident and outgoing parts of the Ψ-field have to be bounded packets. These outgoing parts are obtained from the asymptotic forms of the $f_n(y_j)$'s. Then, similarly as in the pilot wave interpretation of the measurement problem, the outgoing packets, corresponding to different values of n, will eventually become separated (such a separation does not occur if the packets are not bounded). The scattered particle has to enter one of these packets, although in an unpredictable way, letting the other packets empty. Once entered into one packet, the particle will remain embedded in it forever. Hence, both the atom and the particle end with well defined energy values, associated with the selected value of n.

We might object that the argument is fairly qualitative, and that a better argument should be more quantitative, and would then depend on the size of the incident packet. Indeed, if we begin by thinking of a well bounded packet and let its size growing, we shall eventually return to the situation of an incident plane wave, considered by Pauli, meeting again his objection, without however any clear-cut frontier between the case when the objection is valid and the case when it is rebutted. What happens is that, the bigger the incident packet, the larger the time required to complete the interaction, that is to say to have the scattered packets separated. In the limit of an infinite size, that is to say for the plane wave case, the interaction is never completed, and Pauli is right. But, for realistic packets, Pauli is wrong.

Second Pauli's objection: Theory of transformations

The remark addressed to PEDESTRIANS ($\sim\sim\sim\sim$) at the beginning of the previous subsection still holds for the present subsection.

Pauli has been a strong opponent to causal theories. It is therefore not so much surprising that he found several objections [164] to these theories. The first objection is certainly the most significant one, at least from a historical point of view, if we remember the poisonous effect it had on Louis de Broglie. But the other objections are significant and even emblematic. The second objection concerns the symmetrical status of the theory of transformations, which is broken up by causal theories. See for instance Louis de Broglie [63, 65] or Bitbol [133].

The basic idea of the theory of transformations, introduced by Dirac, is that all observables may be treated on the same footing. We may state that, in relativity, the law of physics should not depend on the observer while, in quantum mechanics, they should not depend on the observable. Each observable generates its own representation (say its own orthonormal basis) and there is no deep reason why a representation or another one should be given a privilege when exposing the formalism (although a given choice may be computationally convenient for a given problem, such as a given choice related to the observable to be measured).

To be specific, let us exemplify the situation by considering only the $|\mathbf{r}\rangle$-representation, that is to say the wave function $\Psi(\mathbf{r})$, that I shall denote for a while $\Psi_r(\mathbf{r})$, in the position space, well suited for location measurements, and the $|\mathbf{p}\rangle$-representation, that is to say the wave function $\Psi_p(\mathbf{p})$ in the momentum space, well suited for momentum measurements. These two wave functions are related by Fourier Transforms, according to [43]

$$\Psi_r(\mathbf{r}) = \left(\frac{1}{2\pi\hbar}\right)^{3/2} \int d^3p \Psi_p(\mathbf{p}) \exp\frac{i}{\hbar}p_j x_j \qquad (8.35)$$

$$\Psi_p(\mathbf{p}) = \left(\frac{1}{2\pi\hbar}\right)^{3/2} \int d^3r \Psi_r(\mathbf{r}) \exp-\frac{i}{\hbar}p_j x_j \qquad (8.36)$$

The densities of probability of presence in location and momentum spaces are then given by $|\Psi_r(\mathbf{r})|^2$ and $|\Psi_p(\mathbf{p})|^2$ respectively (with wave functions being properly normed). We observe that the transformations of Eqs. (8.35) and (8.36) establish a beautiful symmetry between $|\mathbf{r}\rangle$- and $|\mathbf{p}\rangle$-representations, and allow one to evaluate probabilities with similar formulae in both representations. More generally, such transformations do exist between any two representations based on different choices of observables. And, in the words of Bitbol [133], a theory of transformations amounts to pass directly from a list of probabilities associated with the possible results of measurements of an observable A to the corresponding list of probabilities for an observable B, without dealing with the mediation of the state vector $|\Psi\rangle$. In our example, this statement is embodied in the fact that the state vector $|\Psi\rangle$ does not explicitly appear in Eqs. (8.35) and (8.36). What appears is the possibility of expressing $|\Psi_r(\mathbf{r})|^2$ versus $|\Psi_p(\mathbf{p})|^2$ and reciprocally.

But, in contrast, the causal theories give a privilege to the location of the underlying particles. As stated by Louis de Broglie [65], the location of a particle is the result of his mere path in matter, without the physicist having to build a measurement set-up to act on the particle. Location is therefore something much simpler, more elementary than the measurement of a momentum. Then, as a consequence, the principle of localization looked to Louis de Broglie, because of the primordial role played in microphysics by the observable processes of location, as being independent of the principle of spectral decomposition.

Pauli [164] objected that, considering a hidden-variables theory which has no effect on the observed phenomena (same predictions in causal theories than in quantum mechanics), the point of view of the pilot wave introduces an artificial asymmetry in the treatment of two variables of a canonically conjugated pair (such as location and momentum above). He added (referring, as many others, to positivism telling us that the genuine physical reality is only to be found in the set of experimental results) that the pilot wave interpretation appears as *artificial metaphysics*. Heisenberg expressed similar criticisms [146].

Clearly, we can agree that the breaking up of the symmetry involved in the theory of transformations is a much unpleasant feature but, after all, the importance given to this symmetry could be illusory. In science, the search for symmetries, and the importance given to them, has been much useful and fruitful [57]. It is a pillar of modern physics, of the theory of elementary particles for example, and a guide for those who commit themselves to the construction of a Theory of Everything, with the concept of symmetry breaking as another pillar. But, in other cases, it has been misleading with the well known examples of the ancient theory of elements, the idea that planetary motions should be circular, or the use by Kepler of perfect solids to build the System of the World. In some cases, it is productive. In other cases, it could just be a manifestation of the Greek syndrome.

Actually, Bohm was may be not the first one to oppose to the second Pauli's objection, but he did it with eloquence. For him, it is coherent to give to the position of an objective particle the status of a privileged property. It is not an observ-able but a be-able, the hidden variable whose numerical value is well defined. And, for him again, all the other properties receive a precise numerical value only by the act of observation, as in the usual quantum mechanics. More generally, all defenders of causal theories have to disagree with Pauli's demand, and defend the conceptual symmetry breaking down of the equivalence of all observables.

In particular, Louis de Broglie insisted on the issue in many places, complaining with the "elegant, but may be artificial theory of transformations". For instance, to quote him once more, he remarks that, what we always observe when we make a measurement on a particle, is his location [63]. For instance, the action of a particle at a point, such as the darkening of a photographic plate by a photon or by an electron, is a phenomenon which speaks in itself. It does not necessitate a proper active experimental set-up but merely the introduction of a passive photographic plate. In contrast, momentum measurements require a set-up acting on the particle. But, he said, even when we use a prism or a grating to measure momenta, by separating sub-beams associated with different values of the momentum, what we measure is a sub-beam location. And it is therefore from a location measurement that we do deduce a value for the momentum. A more striking example might be spin measurements in a Stern-Gerlach device. This example is particularly striking because spin is a highly non classical property, a genuine quantum mechanical one. However, what we record in such experiments is not directly a spin component, but locations of spots, generated by the interaction between beams emerging from the magnet and for instance a passive detector, such as a screen. Similar considerations are also available from [65]. Yet another similar opinion is available from Bell [142] stating that ... *in physics the only observations we must consider are position observations, if only the positions of instrument pointers. It is a great merit of the de Broglie-Bohm picture to force us to consider this fact. If you make axioms, rather than definitions and theorems, about the "measurement" of anything else, then you commit redundancy and risk inconsistency.* It is known that Bell was in sympathy with hidden variables and the pernickety reader might suspect that Bell could have been victimized by some kind of psychological bias.

But we also have quotations from people not to be suspected of any collusion with or corruption from hidden-variables defenders. For instance, for Feynman [200]): *Indeed, all measurements of quantum-mechanical systems could be made to reduce eventually to position and time measurements (e.g. the position of a needle on a meter or the time of flight of a particle). Because of this possibility a theory formulated in terms of position measurements is complete enough in principle to describe all phenomena.* And we also have Omnès [15] remarking that the base $|x\rangle$ is given a privilege in the decoherence theory ... although this affirmation may be surprising if we compare it with the great lessons from Dirac concerning the invariance of quantum theory through changes of basis.

Very likely, Pauli would not have been convinced by all these considerations, but they are there forever.

Third Pauli's objection. Particles at rest

Again, PEDESTRIANS ($\sim\sim\sim\sim$) are kindly requested to manage with this subsection which contains only three equations.

This third objection is discussed by Heisenberg [146] who mentioned that it had already been pointed out by Pauli. I therefore granted Pauli for this objection. Complementary discussions are available from Louis de Broglie [63] or Belinfante [1]. Following Bohm [29], let us consider an atom in a s-state, that is to say with quantum numbers $l = m = 0$, in other words the ground state. For such a state, the wave function can be written as

$$\Psi = f(r) \exp \frac{i}{\hbar}(\alpha - Et) \tag{8.37}$$

in which α is a constant, E the energy of the state, r the radius taken from the center of the atom, and $f(r)$ is the spatial contribution to the wave function. From Eq. (7.2), with the changes: $a \to R$, and $\varphi \to S$, we see that the quantum Hamilton-Jacobi's function reads as

$$S = \alpha - Et \tag{8.38}$$

identifying with the classical expression of Eq. (2.4), with $S_0(x_j)$ being a constant. Using Eq. (8.7) for the guidance formula, we then find that the guided particle (the electron) has a velocity

$$v_j = \frac{1}{m}\frac{\partial S}{\partial x_j} = 0 \tag{8.39}$$

meaning that the particle is at rest. This happens because there is no coordinate dependent phase term involved in the wave function. More generally, whenever the wave function Ψ is an Eigenfunction of the component of the angular momentum in some arbitrary direction, with Eigenvalue equal to zero, including the case of linear combinations of such wave functions, the electron is standing still.

This result is in obvious contrast with what we learn from quantum mechanics. But, wondering how could this happen, we soon discover that there is no contradiction. The reason why is that, in the objective picture of causal theories, the Coulomb force acting on the objective electron is exactly counter-balanced by the force emanating from the objective quantum potential.

The above argument is sufficient to answer the third Pauli's objection. Let us however provide some complementary points of view. First, we must always keep in mind that there must be some conflicts between the concepts of quantum mechanics and the concepts of causal theories. This is just normal because both theories provide different pictures of the world. In particular, in quantum mechanics, we cannot indeed have a particle at rest at some point because, if it is at rest at some point, its location and velocity are both simultaneously well defined. For objective causal theories, the concept of a particle at rest at a point, conversely, is just something which matches, without any difficulty, the conceptual structure of the theory.

Let us also insist on the fact that such conceptual conflicts can never produce conflicts concerning experimental predictions since, by construction, both quantum mechanics and causal theories must empirically agree. In particular, both theories predict a density of probability of presence (being observed or being found) given by $P(x_j) = (f(r))^2$. We may also complement the point of view of causal theories by returning once more to Louis de Broglie [63] who discussed the consequences of the environment. Let us again consider the fundamental state Ψ_{00} of an hydrogen atom and assume that this atom is not perfectly isolated. For instance, it may be in contact with a sub-quantum mechanical level, possibly the vacuum. Then, Ψ_{00} will be disturbed to $(\Psi_{00} + \delta\Psi_{00})$ so that the associated action is also disturbed from S to $(S + \delta S)$. Then, following Louis de Broglie, the guidance formula implies that the electron will undergo a violent Brownian-like motion. However, the density of probability of presence is not too much modified since $|\Psi_{00} + \delta\Psi_{00}|^2$ remains close to $|\Psi_{00}|^2$. Quantum mechanical predictions are therefore not significantly affected, while the picture of an electron at rest is modified since, now, due to the disturbance $\delta\Psi_{00}$, the electron is no more at rest.

Perrin's objection. Diverging waves

Perrin [63] considered the case of an isotropic diverging wave Ψ propagating from a punctual source S, emitting particles of identical energies. The amplitude of the diverging wave behaves as $1/r$ and, at least far enough from the source, $|\Psi|^2$ behaves as $1/r^2$. Therefore, in the pilot wave interpretation, we have to conclude that the particles are guided by a wave which becomes weaker and weaker when we depart from the source and which, eventually, will become vanishingly small. But, experimentally, far away from the source, we can still produce interferences with an appropriate set-up, as we can do close to the source. Such interferences then have to be the result of the motion of the objective particles guided by a wave Ψ which may be infinitely small. According to Perrin, this cannot be conceived.

In a first step, Louis de Broglie agreed with Perrin [63], at least as far as the pilot wave was concerned, in particular because, once again, the pilot wave is subjective for Louis de Broglie, and therefore, in any case, could not guide objective particles. But, afterward [63], he provided a rebuttal in the double solution framework. Indeed, in this framework, it is easy to view the singularities of the waves u as forming bounded, non spreading, wave packets, particularly if they are stabilized by nonlinear terms. We then have a source isotropically emitting objective particles which preserve their identities when flying away. Therefore, the number of particles per unit of surface of a sphere surrounding the source decreases with an $1/r^2$ behavior, but each particle, remaining intact, preserves its capacity of action all along its motion. Well, this point of view is very close indeed with the one we may have of the behavior of photons emitted by a light source, as viewed by Einstein and as described by quantum mechanics, is not it? And it clearly provides a rebuttal to Perrin's objection. Also, it is worthwhile to point out once more how much, in the mind of Louis de Broglie, the double solution and the pilot wave have to be opposed.

Indeed, let us now return to another mind, the one of Bohm, with the pilot wave Ψ becoming an objective wave. What Louis de Broglie did not forecast is that Perrin's objection could be answered in this pilot wave framework too. For this, let us just consider the expression of the quantum potential, e.g. Equation (8.9) to invoke the simplest expression of the quantum potential among the various expressions we provided. Then, we can immediately see that the quantum potential, as well as the equation of motion 8.10, are not modified when we multiply the wave Ψ by an arbitrary constant, because the amplitude term R appears both in numerators and denominators. This fact is in strong contrast with what happens with a classical wave, and constitutes another striking non classical feature of Bohm's interpretation. It means that hidden particles do not respond to the amplitude (or to the intensity) of the guiding wave, but rather to its shape. Therefore, the quantum potential does not become necessarily small when the wave intensity becomes small, an argument providing a rebuttal to Perrin's objection [29, 68, 179, 190].

For Bohm, it means that the quantum potential plays the role of an active information (e.g. [179]), i.e. the guiding wave is viewed as a source of information, rather than as a source of force. Relying on a metaphoric picture, Bohm compares the action of the guiding wave to a radio message sent to a boat, giving it instructions to modify its trajectory. The content of the message is embedded in the form of the signal, not in the value of its amplitude. Developing further the metaphor, Bohm even later speculated [190], with Hiley, that the treatment of the active information implies that each particle possesses a *subtle and complex structure*. Such a speculation, and others, characterize the later Bohm and, in the eyes of many, weaken his position. They diffuse all around him a strong scent of unfortunate and, I believe, unnecessary metaphysics. They are reminiscent of Louis de Broglie possibly blindly forging ahead. But they do not annihilate the validity of Bohm's rebuttal to Perrin's objection.

The idea of active information just aforementioned is connected to another idea cherished by the later Bohm, namely the one of wholeness which however is not only quite reasonable, but even perfectly fitting the structure of the pilot wave, and also the structure of quantum mechanics which makes the world filled with entangled objects. Indeed, for Bohm and Hiley [190], *the guidance conditions and the quantum potential depend on the state of the whole system in a way that cannot be expressed as a preassigned interaction between its parts. As a result there can arise a new feature of objective wholeness. This is not only the consequence of the non-local interactions but even more it follows from the fact that the entire system of particles is organized by a common pool of active information which do not belong to the set of particles, but which, from the outset, belongs to the whole.*

Einstein's objection. To and fro

We finally come to an objection discussed by Born [201] and which, according to him, is originally due to Einstein, as confirmed by Louis de Broglie [63]. Other sources are from Bohm [30], Freistadt [155], Jammer [24], Holland [40] or Cushing [131]. This objection may be viewed as a particular "particles at rest" objection, but is nevertheless sufficiently different from the one previously examined under this headline to deserve a specific subsection.

Following Louis de Broglie [63], let us consider a particle located on the axis Ox between two perfectly reflecting walls, or mirrors, perpendicular to the axis, and located at $x = 0$ and $x = L$, and let us assume that the particle has a well defined value of energy, say E. We begin with an analysis of this problem in the framework of quantum mechanics. Then, the particle is a quantum object in a stationary state represented by a wave function Ψ. Because the energy E is equal to the square p^2 of the momentum p (within an irrelevant prefactor

$1/(2m))$, there are two possible values for the momentum, so that Ψ is a linear combination of two wave functions, say Ψ_{LR} and Ψ_{RL}. The wave function Ψ_{LR} represents a particle travelling from Left to Right, while Ψ_{RL} represents a particle travelling from Right to Left. To make the problem symmetrical, we may even choose the coefficients of the linear combination in such a way that the probabilities of the *LR-* and *RL-* motions are equal.

This gave Einstein an opportunity for an objection against quantum mechanics, due to the fact that a macroscopic particle actually undergoes only one of the two motions, bouncing *to and fro* between the mirrors, with probability one. But this is an objection concerning the passage from the quantum microscopic world to the classical macroscopic one and, because we already discussed it enough, it should not interest us any more, in the framework of this book. Let us simply mention that we could invoke the decoherence theory according to which, due to the interactions with the environment, one of the macroscopic motions will be selected, with the other one receiving a vanishing probability.

But it also gave Einstein an opportunity to object against the pilot wave theory. Indeed, he observed that the guidance formula implies that the particle (there is only one objective particle between the mirrors, even if Ψ is a linear combination) must be at rest. This results from the fact that one velocity in one direction for one packet annihilates with the velocity in the other direction for the other packet. In the case of a macroscopic particle, this result again conflicts with the observed *to and fro* motion between the walls. Hence, according to Einstein, the pilot wave is untenable.

Louis de Broglie [63] provided two answers to Einstein's objection. According to the first answer, we should depart from a too much idealized problem. In fact, the walls in practice cannot be perfect classical mirrors, because they are actually made out from quantum objects, the constituting atoms. These atoms are subject to thermal fluctuations and, therefore, the boundary conditions are actually not precisely defined. Hence, he argued, if we properly take into account the fluctuating boundary conditions, then the macroscopic velocity of the particle should be close to zero, in agreement with the result of the guidance formula. See the original reference for more details concerning this approach that I do not find particularly convincing, to say the least. For the second answer, Louis de Broglie remarked that the particle should be viewed as a wave packet, rather than being idealized as a combination of plane waves. Since the de Broglie wave length $\lambda = h/p$ decreases when the mass increases, one concludes that, for a macroscopic particle, it will decrease sufficiently to become much smaller than the distance between the walls. Therefore, for a sufficiently large mass, it is no more possible to imagine the particle as being modeled by a stationary wave made out from the superposition of two counter-propagating waves. One should rather imagine a small wave packet oscillating *to and fro* between the walls, in agreement with classical macroscopic physics. This seems to be more relevant but, in any case, it does not reach the heart of Einstein's objection against the pilot wave. It is more a case of dealing again with the frontier between the quantum and the classical worlds, for which we might once more invoke the decoherence theory, and that we decided anyway to dismiss from now on.

Let us now turn to Bohm's answer [30] which does not leave out from the quantum world, and does not deal with macroscopic affairs. As we definitely should, let us think again of our quantum particle between the walls as a superposition of two waves with opposite momenta. This picture is something in the mind, with the events between the walls quite unobservable, and therefore it is not practical from a positivist point of view. It is no more practical that a Schrödinger's cat playing around in a hermetically closed room. To make it practical, let us try to measure the momentum of the particle, an observation which, in quantum mechanical terms, would produce a collapse of the superposition to either Ψ_{LR} or Ψ_{RL}. To look inside the box, we must remove the walls. We may then again infer what is going to happen from the solution to the measurement problem given by Bohm in the framework of the pilot wave theory. Due to the influence of the quantum potential generated by all quantum objects involved in the situation under study, including the ones constituting the boundaries, the removal of the walls, to be viewed as a kind of measurement interaction, will have a dramatic influence on the objective particle previously confined and now liberated. We now have two wave packets propagating without any constraint any more into two opposite directions and, soon, becoming separated in space. One packet is going to contain the objective particle, while the other will remain empty and, from now on, can be ignored. Then, measuring the location of the particle at a time t large enough, we may determine the velocity and therefore, knowing the mass, we have an access to the momentum. In particular, we shall determine whether the momentum is positive or negative, that is to say which wave packet is filled and which one is empty.

Bohm rejected

Rejection does not mean objection. We may answer objections but we are usually hopeless when we are facing rejections, often arising from prejudices and more generally from irrational aspects of the human mind. There are good objective reasons for objections, and for rebuttals. There may also be good reasons for rejections, but they are of a more subjective and evanescent nature, and there is no decent objective rebuttal. We are leaving the realm of pure science, the one of objective knowledge as defined by Popper [25, 61], and entering a world of feelings, often passionate and epidermal, with possibly intolerant attacks. Associated behaviors, unfortunately and deceivingly, are common in the intimate life of laboratories, and would be enough to justify sociological and psychological and psychoanalytical, even sometimes psychiatric, studies on science making people, and on how something objective eventually emerges from a chaos of subjectivities. Regarding causal theories, epidermal excommunications have been openly released in the literature, with some actors seemingly losing their temper. Pleasantly enough, other rejections have been played on a well tempered clavier.

We remember that Louis de Broglie originally badly reacted against Bohm's arrival but that eventually, certainly motivated by Bohm's answers to objections against the pilot wave, he went on starting to develop again his own ideas. Nevertheless, considering that he devoted himself again to the double solution, we might infer that Louis de Broglie remained suspicious regarding the pilot wave. But, in any case, suspicion is not rejection. Furthermore, examining Louis de Broglie and Bohm productions after 1952, we can observe some kind of osmosis between the two minds.

The following quotation from Destouches, in a paper entitled "return to the past" [202] nevertheless expresses mixed feelings concerning this part of the story (my translation from French): The memory of fruitless efforts, fast forgotten, precisely because they have been fruitless, is far from being useless, since it prevents many to start again a work doomed to failure and to commit themselves into blind-alleys where others, before, momentarily got lost ... Prospecting blind-alleys, this is something to which Louis de Broglie devoted several years before his Institute Henri Poincaré time ... His ideas from this period, notwithstanding the so clear judgment addressed to them by their author, came again into discussion these last months. The work of Bohm (1952) revived the attention on the pilot wave theory and on the possibility of building a determinist description of microscopic phenomena.

Next, we already know of the reactions of Einstein regarding Louis de Broglie's version of the pilot wave. As far as we can see, there has been later some contact between Bohm and Einstein and, according to Pinch [154], Einstein even encouraged Bohm to produce an heterodoxy. But, once this was done, Einstein did not like it. In a letter addressed to Born, Einstein wrote [201]: *Have you noticed that Bohm believes (as de Broglie did, by the way, 25 years ago) that he is able to interpret the quantum theory in deterministic terms? That way seems too cheap to me.* This letter is followed by a comment of Born, saying: *The remark he makes about David Bohm's theory is connected with this. Although the theory was quite in line with his own ideas, to interpret the quantum mechanical formulae in a simple, deterministic way seemed to him to be "too cheap".* Born, the father of the probabilistic interpretation, vividly expressed himself his reaction in another letter: *I more or less agree with what you said about de Broglie, Bohm and Schrödinger. Incidentally, Pauli has come up with an idea (in the presentation volume for de Broglie's 50th birthday) which slays Bohm, not only philosophically but physically as well.* More on Einstein versus Bohm in Belousek [162].

Louis de Broglie and Einstein were both eager to eventually reach a satisfactory causal interpretation of quantum mechanics, and their reactions against Bohm, were finally, I would say (my appraisal), not too severe. They have been, both of them, in the inmost depth of their souls, sympathetic (my appraisal again) with Bohm's effort. In contrast, it is now interesting, and illuminating, to turn ourselves to the reactions of the proponents of the Copenhagen interpretation, including Founding Fathers, and more generally to the defenders and to the keepers of the flame of orthodoxy. We then encounter a strong opposition against Bohm's work. In the words of Holland [40], *the reaction to Bohm's work by the Copenhagen establishment was generally unfavorable, unrestrained, and at times vitriolic.*

We already heard from Pauli, considering that Bohm's work was artificial metaphysics [203], and who, in the words of Cushing [131] *is caustic.* And it is Heisenberg, hostile to Bohm's proposal too [53, 146], who denounced it as a superfluous ideological structure. The word "superstructure" is easy to understand: It amounts to the fact that we do not need Bohm's interpretation to achieve correct predictions; quantum mechanics is sufficient to this aim. We may also understand the word "ideological" because the pilot wave formalism, superimposed on and derived from the usual formalism based on Schrödinger's equation, explicitly aims to the restoration of determinism. But the whole expression "superfluous ideological structure" conveys, let us say, a bit of violence; without mentioning that one reason for such a judgment might be an ideological adhesion and adherence to positivism.

Rosenfeld, one of the most faithful advocates of the Copenhagen interpretation, was very acerbic. According to Jammer [24], he regarded Bohm's approach as a *short-lived decay-product of the mechanistic philosophy of the nineteenth century* and stated that it corresponded to an *empty talk.* In Ref. [101], Rosenfeld acknowledged that Bohm's interpretation was very cleverly structured and that it is vainly that we would search for a weak point in its formal layout but, he added: Nevertheless, all this seductive construction is only a trompe-l'oeil. He also asked why we should need to resuscitate old ideas since, in between, the crisis has been solved at a higher level of the theory of knowledge. He added something which may look like a coup de grâce: There is deeply no need to warn that an epistemological criticism of Bohm's work should not point out to the ancient ideas of Louis de Broglie; one can understand that a pioneer penetrating in an unknown territory does not immediately find the right track; it is much less understandable that a tourist got lost when this territory has been explored on a very small scale (my approximate translation from French, however faithful, I believe, to the very meaning in the mind of Rosenfeld). Going on like that, Rosenfeld, the indefatigable herald, stated that "the physicist who still tries to stick at it (the determinism), who refuses to surrender in face of the evidence of complementarity, deserts the rational attitude of the man of science to adopt, whether he wants it or not, the attitude of the metaphysician. Those who want to save determinism are much obliged, effectively, to give up the synthetic position expressed by the idea of complementarity, and to endow again an absolute meaning to one or the other of the wave or particulate aspects of the matter. Naturally, they cannot reach this result without doing violence to logic and they necessarily fall back in contradictions that the complementarity only eliminates". Later on, he stated: "It is the first time, in the history of physics, that one is forced to give up the so convenient distinction between the observer and the object of the observation ... From this to imagine that the very objectivity of the scientific fact is destroyed at its basis, there is only one step that is cheerfully made. On one side, idealists proclaim ... the bankruptcy of science. On another side, some honest defenders of the scientific ideal, get lost in a narrow and erroneous conception of materialism, rant and rave about Bohr and his school, that they accuse to break this ideal". I do not intend to discuss here the scientific content of these sentences (this book offers many opportunities for that), but I just wanted to illustrate how much reactions could be harsh, violent, intolerant.

We may find a summary of Bohm's rejection by referring to Holland [40] stating that *as a whole, Bohm's theory did not enter the mainstream of physics either as a research topic or in textbooks. In fact, in the following 25 years only occasional and sporadic references were made to de Broglie-Bohm theory. Books were written, courses taught and research conducted as if what Bohm has demonstrated to be possible was still in fact impossible. Although de Broglie continued to advertise the idea in his books, and Bohm worked on hidden-variables theories, no development or application of the pilot wave theory was made.* With a bit more of sadness, we may contemplate Cushing's apprehension of the tragedy [131], reminiscent of Greek ostracism, when he states: *In the reconstruction of the history and in the propagation of the new orthodoxy (in, for example, Heisenberg's influential 1929 lectures on quantum mechanics), de Broglie, Einstein, and Schrödinger disappear from Heisenberg's story, which includes as creators of quantum mechanics only those who understood it in the Copenhagen sense.* Of course, in 1929, Heisenberg could not be aware of Bohm's work. But, if he had been aware of it, thanks to some kind of spatiotemporal loop, then, very likely, Bohm would have been rejected from the story too. Cushing also stated that Bohm's approach *was basically ignored, rather than either studied or rebutted. Just as external factors has played a key role in establishing the Copenhagen hegemony, so they once again contributed to keeping this competitor from the field. That a generation of physicists had been educated in the Copenhagen dogma made it all the more difficult for Bohm's theory.*

In the same mood, concerning however Schrödinger rather than Bohm, we have to mention Schrödinger, facing quantum mechanics and in particular the work of Heisenberg, who felt discouraged, or even repelled, by what appeared to him as a rather difficult method of *transcendental* algebra, defying any visualization [47]. Heisenberg wrote to Pauli that the more he was considering the physical part of Schrödinger's theory, the more it looked to him as disgusting. Pauli must have agreed, insofar as he detected, concerning Schrödinger, a neurotic desire for a return to the past. Actually, the rejection suffered by Bohm participated from a more general rejection against all of those who dared to deal with hidden variables. According to Clauser [204], they were considered as quacks, excepted Louis de

Broglie who could not be a quack since he had a Nobel prize. More charitably, he was simply viewed as senile.

Bohm approved

Although Bohm's interpretation (often misrepresented) still nowadays remains marginal (and marginalized), it should never be ignored any more, *eyes wide shut*, if only by its historical virtues, and by telling us, once more, that the human mind should be free of investigating unusual landscapes, outside of any orthodoxy. A good knowledge and understanding of this interpretation and of the issues involved, in any case, should help us to better apprehend the content of quantum mechanics. And, whatever will be our final appraisal, it should forever remain something to remember.

Among the many scientists who delivered Bohm from his solitary underworld, his long-lasting lonesome time spent in the wilderness, a special distinction should be awarded to Bell for whom Bohm's work has been seminal. Bell defended Bohm in many places, as we may check by reading the compilation of Ref. [205]. The influence and reputation of Bell, sometimes considered as the father of the 'second quantum mechanical revolution', made much to restore Bohm good name. In a paper dated 1980 [198], he defended the de Broglie–Bohm version of quantum mechanics, *a sharp one where the usual one is fuzzy, and general where the usual one is special.* In another paper entitled "On the impossible pilot wave" [142], he most eloquently asked: *But why then had Born not told me of this 'pilot wave'? If only to point out what was wrong with it? Why did von Neumann not consider it? More extraordinarily, why did people go on producing 'impossibility proofs', after 1952, and as recently as 1978? When even Pauli, Rosenfeld, and Heisenberg, could produce no more devastating criticisms of Bohm's version than to brand it as 'metaphysical' and 'ideological'? Why is the pilot wave picture ignored in textbooks? Should it not be taught, not as the only way, but as an antidote to the prevailing complacency? To show that vagueness, subjectivity, and indeterminism are not forced on us by experimental facts, but by deliberate theoretical choice?* And in the same paper, he commented: *But in 1952, I saw the impossible done.* Also, Bell [183] acknowledged that Bohm's papers (the ones of 1952) were for him a revelation.

Not only he could see the possibility of eliminating the indeterminism of quantum mechanics but, may be more important, to get rid of *any need for a vague division of the world into 'system' on the one hand, and 'apparatus' or 'observer' on the other.* Concerning proofs of impossibility to be discussed later, he stated that ... *what is proved by impossibility proofs is the lack of imagination.* Bohm's work certainly also drove Bell to the introduction of beables [79, 182, 206], a concept which, in return, pleased Bohm (and Hiley) [180]: *In particular, we will exclude the notion of observables in favour of that of beables. The beables of the theory are those elements which might correspond to elements of reality, to things which exist. Their existence does not depend on 'observation'. Indeed observation and observers must be made out of beables.* The attraction exerted by Bohm on Bell was supplemented by some repulsion concerning quantum mechanics [182, 183]: *I think that conventional formulation of quantum theory, and of quantum field theory in particular, are unprofessionally vague and ambiguous. Professional theoretical physicists ought to be able to do better. Bohm has shown us a way.*

Squires, in his popular account of quantum mechanics, has featured de Broglie-Bohm ideas [13] and remarked: *A much more important advantage of the hidden-variable theory is that it is precise. It is a theory of everything; no no-quantum observers are required to collapse wave functions since no such collapse is postulated.* With the same mood, Cushing, an effective and enthusiastic defender of Bohm, stated [131]: *I suggest that we attempt to generate theories that make the world comprehensible in terms of our inherent patterns of thought,* a sentence which should make the reader thinking of Kant and, in the same vein, Cushing added: *We are able to construct a less incomprehensible, more nearly picturable, representation of the physical universe with Bohm than with Copenhagen.* For Cushing, *it is astounding that there is a formulation of quantum mechanics that has no measurement problem and no difficulty with a classical limit, yet is so little known. One might suspect that there is a historical problem to explain its marginal status.*

Among the many other authors approving Bohm, we also mention Dürr *et al.* [207] stating: *Quantum philosophy, a peculiar twentieth-century malady, is responsible for most of the conceptual muddle, plaguing the foundations of quantum mechanics. When this philosophy is eschewed, one naturally arrives at Bohmian mechanics, which is what emerges from Schrödinger's equation for a nonrelativistic system of particles when we merely insist that "particles" mean particles. While distinctly non-Newtonian, Bohmian mechanics is a fully deterministic theory of particles in motion, a motion choreographed by the wave function. The quantum formalism emerges when measurement situations are analyzed according to this theory. When the quantum formalism is regarded as arising in this way, the paradoxes and perplexities so often associated with quantum theory simply evaporate,* so that, he added: *It is fair to say that Bohmian mechanics is nothing but quantum physics without quantum philosophy.*

Defenders of Bohmian mechanics are still active nowadays and, as a last example in this section, I would like to mention Bricmont who, discussing quantum mechanics for non-physicists in a chapter of a book devoted to the philosophy of quantum mechanics [208], explains that, in a manifest way, quantum mechanics is not complete and concludes that Bohmian mechanics, which is determinist and exhibits a definite ontology (particles possessing a position at any time, and hence *trajectories*), among other attractive properties and features, is a theory which is as natural and neat as any other fundamental theory in physics.

Bohm applied

The most significant set of approvals of Bohmian mechanics is certainly to be found in the fact that it has been successfully applied to various problems. Many people visited the desert where Bohm was wandering and made many oases flourished. The most convincing cases concern the possibility, thanks to the development of computers, and associated softwares, to explicitly compute quantum potentials (and display their charts) and Bohmian trajectories (and exhibit their structure). Such visualizations could vividly illustrate, in a way directly accessible to the human brain, how the hidden world of hidden variables managed to produce the foreground world. It was no more something purely conceptual, embedded in mathematical formulae, but became revealed and apparent as something real, effective, speakable, directly grasped by the eyes and analyzed by the huge capacity of the brain to deal with pattern recognition. These facts should definitively erect Bohm's theory to the status of at least a valuable model.

Dewdney and Hiley [199] provided numerically computed quantum potentials and Bohmian trajectories in the case of one-dimensional scattering by square barriers and square wells, putting these computations in perspective by saying: *One of the less helpful aspects of the Copenhagen school was their insistence on the uniqueness of the usual interpretation which leads to the supposition that what happens to the individual between measurements is somehow inherently indescribable and is not subject to detailed analysis. The ambiguous nature of the individual process has given rise to many speculations as to its origins, some of which ultimately resort to the introduction of human consciousness in an essential way. The calculations based on the quantum potential presented here do not call for such conclusions. Take for example the Schrödinger cat paradox. In terms of the quantum potential approach it is the position of the triggering particle within the incident packet that effectively determines its future, and whatever the final outcome of the process, it will leave the cat in some definite state. In other words the cat will be dead or alive regardless of our knowledge of its final state.* I urge the interested reader to make a detour by examining Dewdney and Hiley's paper and his quite convincing figures.

Similarly, Phillipidis *et al.* [209] explicitly calculated the quantum potential and exhibited a set of Bohmian trajectories in the famous and emblematic case of the two-slit interference experiment, illustrating vividly and pictorially how Bohm's approach provides an objective understanding of this paradigmatic situation. This is an accomplishment of an anticipation of Belinfante [1] when he stated that *an amusing part of Bohm's theory is that in principle it can tell us through which slit a particle came as it hits a screen in one of the constructive-interference fringes behind a system of slits. The Copenhagen version of quantum theory claims that the question through which slit such a particle came has no answer, and therefore should be a (physically) meaningless question. More correctly, quantum theory — according to its statistical interpretation — say that any answer given to the above question is not experimentally verifiable without destroying the interference pattern.* For Philippidis *et al.* [209], *it was not until the late seventies* (about three decades after 1952 Bohm's papers) *that serious interest was rekindled when the trajectories corresponding in the two-slit experiment were explicitly displayed in computer graphics.*

Phillipidis *et al.* [210] provided examples of Bohmian trajectories to illustrate the Aharonov–Bohm effect and demonstrated how a certain fringe shift involved in this effect arises from the quantum potential. In this effect, shifts in interference patterns arise for electrons which travel in field-free regions. The Aharonov-Bohm effect is similarly illustrated by Home and Whitaker [83], while Home and Selleri discussed this effect from the point of view of local realism [211]. See also Tonomura [212]. Dewdney [213] dealt with particle trajectories and interference in a time-dependent model of neutron single crystal interferometry. Bohmian trajectories are also exhibited by Vigier *et al.* [214] for the two-slit experiment, and in connection with neutron interferometry, or by Hiley [215]. Lam and Dewdney [216]) studied correlations in low-particle interferometry, exhibited Bohmian trajectories for this situation, and took this as an opportunity to discuss locality and non-locality. Bozic and Maric [217] is devoted to Bohmian trajectories in Mach-Zender and neutron interferometries, and show how the guiding wave determines the set of possible trajectories, as well as their probabilities. Wu and Sprung [218] studied the ballistic transport of electrons through a quantum wire with a constriction. Shifren *et al.* [219] also simulated Bohmian trajectories in a constricted quantum wire. Courtney and Frasinski [220] used the quantum potential as a tool to understand nuclear physics phenomena. Datta *et al.* [221] provided a Bohmian picture of Rydberg atoms and demonstrated that Bohmian paths of electrons are ellipses to a high degree of approximation. Similarly, Matzkin [222] studied the dynamics of highly excited Rydberg wave packets in terms of Bohmian trajectories. Gonzalez *et al.* [223] is devoted to the behavior of a wave packet traversing a complex potential, and Zheng and Kobe [224] examined the quantum kicked rotor and compared Bohmian results with those of quantum mechanics (of course, they agree). The evaluation of Bohmian trajectories even motivated specific efforts in the study of associated numerical algorithms, such as by Nerukh and Frederick [225].

Early Bohm's theory only considered one wave function Ψ, a way to say that it was meant for spinless particles. Among the subsequent developments of Bohm's pilot wave, after 1952, there has been an extension of pilot wave interpretation to include the concept of spin, still by using well defined objective motions [167, 168, 226]. Relying on such extensions, Dewdney *et al.* [227] provided an objective account of a Stern-Gerlach experiment performed with spin (1/2)-particles, in which the particles have well defined and continuous trajectories, and spin vectors. Dewdney *et al.* [228] similarly exhibited Brownian trajectories for understanding spin superposition in neutron interferometry. Bohmian trajectories with a spin dependent term are also discussed by Colijn and Vrscay [229], [230].

Next, Bohmian trajectories have been found to be relevant to the theory of chaos. The word chaos has been used by Louis de Broglie and by Bohm, in a loose sense, to designate the violent and irregular motions generated in some cases by the quantum potential or, in a slightly different context, by the presence of an underlying sub-quantum mechanical level. As stated by the authors of causal theories, the motions of objective particles can then receive a Brownian-like character, for which words like "random" or "stochastic" may possibly be found more appropriate. See in particular Rafii-Tabar [231] discussing Bohmian trajectories in a stochastic causal version.

But the word chaos may also be given a strong and precise meaning in the framework of the theory of chaos, a sub-field pertaining to the study of the theory of non-linear dynamics. With the effective calculation of Bohmian trajectories, and a careful enough examination of their structures, it became possible to exhibit genuine deterministic chaoticity in Bohmian mechanics. Parmenter and Valentine [232–234] demonstrated that deterministic chaos may occur for Bohmian trajectories in the configuration space, and that the trajectory of a physical system may exhibit deterministic chaos even when the system is non chaotic when treated classically. To diagnose chaos, they relied on the evaluation of the largest Lyapunov exponent (for a review concerning different ways to achieve a diagnosis of chaos, see [59]). Still thinking in the framework of a causal interpretation, Sengupta and Chattarj [235] discussed the signature of chaos in the quantum behavior of a classically chaotic system. Konkel and Makowski [236] exhibited chaotic trajectories for a particle moving inside an infinite two-dimensional rectangular well, relying again on the use of a Lyapunov exponent as a tool for diagnosis of chaos. Wu and Sprung [237] dealt with the fact that the concept of chaos may be applied to Bohm particle trajectories. Bonfim *et al.* [238] displayed chaotic dynamics for a particle trapped in a circular billiard (strange enough, if we remember that the circular billiard is classically integrable), using the magnitude of the Lyapunov exponent to measure the degree of chaoticity. Makowski *et al.* [239] discussed the relationship between the phase of the wave function and the generation of chaotic Bohmian trajectories. He used the Melnikov function method [240] to prove the existence of chaos. Falsaperla and Fonte [241] examined the motion of a single particle near a nodal line, and drew consequences on the possibility of chaotic motion. There is also an interesting series of papers by Chattaraj and Sengupta, or by Chattarj *et al.* using Bohmian or Bohm-like approaches in connection with the topic of quantum chaos [242–248].

Applications to chemistry, or chemistry oriented issues, are available from Levit and Sarfati [249], Prezhdo and Brooksby [250], Gindensperger *et al.* [251] or Guantes *et al.* [252]. More exotic and unexpected (at least unexpected to me) applications of Bohm's ideas can be found in the fields of quantum cosmology and quantum gravity, possibly by using calculations of Bohmian trajectories. In particular, Dewdney *et al.* [253] computed Bohmian trajectories in a cosmological context and concluded that the notion of a wave function of the universe makes sense in a causal framework (it does not make sense in quantum mechanics). See also Hiley [215], and Squires [254], who addressed the question of the smallness of the cosmological constant, and Pinto-Neto *et al.* [255], Pinto-Neto and Colistete [256], Pinto-Neto and Santini [257], Kim [258], Barros and Pinto-Neto [259], Barros *et al.* [260].

Bohm and others after the early Bohm

Bohm did not stop working on the pilot wave after his 1952 seminal papers. Conversely, he developed it further, certainly under the pressure of his own motivations, realizing that his research program was far from being completed, but also likely in reaction to various objections (even irrational) addressed to him. For a courageous man, the fact that the quantum community has been, for a long time, silent, indifferent, or even hostile, must have produced impulses which have forced Bohm to a sustained ramification of his thoughts. One of the most significant steps has been the introduction, in the picture, of a sub-quantum mechanical level.

Geneticity of Born's postulate, and sub-quantum mechanical level

Born's postulate, as indicated by its name, is a postulate, at least in quantum mechanics. We may refer to our postulate numbered 4, in the chapter devoted to the background of quantum mechanics, expressing probabilities, although, when we speak of Born's postulate, we often think of its specification to the density of probability of presence given by $|\Psi|^2$. It is also a postulate in the pilot wave where it serves to help quantum mechanics and pilot wave to produce identical empirical predictions. Since Bohm was well aware of the necessity for the pilot wave to lead to the same predictions as Born's postulate in quantum mechanics, it was indeed expedient to use the same postulate in his causal theory. He really had to borrow it from quantum mechanics. But could we found some deeper justification for this postulate? It could for instance be the necessary consequence of some first principle, at a higher rational level of the organization of the universe that, up to now, we may have failed to identify. Bohm provided a solution corresponding to another possibility, namely that Born's postulate is not the result of a necessary rational demand, but the result of an immanent evolution. Although it is not actually a Darwinian evolution, with a genetic support, the word geneticity may be adopted as a kind of metaphor.

For Bohm, the support of the evolution of quantum systems towards Born's postulate behavior is to be found in the existence of a sub-quantum mechanical level. Bohm was already on this track in 1952 when he remarked that the orbit of a particle in a non stationary state is very irregular and complicated, resembling more a Brownian motion than the motion of a planet around the Sun. However, at this stage, we are facing a Brownian-like motion which is the result of the fact that the amplitude R and the phase S, the quantum potential, and the particle momentum, are the subject of rapid and violent fluctuations, both with respect to space and time. And all this happened in a framework in which Born's postulate is assumed.

But why not making a next step [261–263] and suppose that irregular motions have also an extra-origin, coming from a deeper level? We already have the result, even in quantum mechanics, that, if $P = |\Psi|^2$ holds at a certain time, the equations of motion of the particles will make this relation holds at all later times. But what about the stability of this relation? What happens if, at a certain time, the relation does not hold? Then, Bohm demonstrated that, as a result of random collisions, an arbitrary initial probability density will eventually decay to one exactly expressed by Born's postulate. Such collisions might arise from interactions with other quantum systems, at the ordinary quantum level, but, more fundamentally, they might also arise from a sub-quantum mechanical level which would produce a stochastic behavior of the particles (rather than "stochastic", Bohm and Hiley used the word "chaotic" [68], but I now on prefer to restrain the use of the word "chaotic" to a proper meaning, the one of deterministic chaos). Therefore, Bohm said, ... *all quantum mechanical systems now available in practice have had billions of years in which to come to equilibrium.*

Besides providing geneticity to Born's postulate, the existence of the sub-quantum mechanical level itself is an interesting issue on its own, and Bohm discussed it in several places. In 1957, he insisted that *there is good reason to assume the existence of a sub-quantum mechanical level that is more fundamental than that at which the present quantum theory holds.* Due to the existence of this deeper level, containing new kinds of entities, the quantum theory would not be *complete enough to treat all the precise details of the motion of individual electrons, light quanta ...*

In the early Bohm's version, the guiding wave Ψ influences the objective particles, but without any reciprocal influence, or at least influences which are small enough to be neglected at the quantum level. But such a reciprocal influence could be significant if we take the sub-quantum mechanical level into account. As a consequence, Ψ itself would undergo random fluctuations about an average, and it is this average only which would satisfy Schrödinger's equation. At the sub-level, the laws of physics could remain causal, although, due to their randomness like in Brownian motion, they could necessitate a statistical approach. However, these laws could be different from the ones prevailing at the quantum level and, in particular, they would not necessarily match Schrödinger's equation. Furthermore, they might be non-linear, and such nonlinearities would ensure the existence of discrete energy levels in matter and of electromagnetic energy in quanta. Here and there, we should note the osmotic relationship between Bohm after the early Bohm, and Louis de Broglie after the early Louis de Broglie, a fact which should not surprise us if we keep in mind the historical background of causal theories. The problem of quantum jumps, already discussed with Louis de Broglie, is also similarly commented by Bohm. For him, transitions between discrete energy levels would then be explained by the action of the nonlinearities, and would be continuous at the sub-level, appearing discontinuous at the quantum level due to the very short time required for them, in contrast with the Copenhagen interpretation which considers such transitions as inherently discontinuous, and any attempt to provide a description of them as meaningless. Clearly, we can see here the action of the desire to restore another classical conception, appealing to our mind, according to which nature does not like jumps and therefore does not make jumps (*natura non facit saltus*, of course). Such considerations are also discussed by Bohm, e.g. in Bohm [31] and in Bohm and Hiley [68]. Louis de Broglie favorably commented them, e.g. in [63] and [178].

There is another topic, to which we shall have to devote a bit of time, namely stochastic quantum mechanics (Chapter 14), which also relies on the possible existence of a sub-quantum mechanical level. This topic is well represented by a seminal, often quoted paper, due to Nelson [264]. Differences between the points of view of Bohm and Nelson are discussed by Bohm and Hiley [190]. But the idea of the existence of a sub-quantum mechanical level as discussed by Bohm, Louis de Broglie, Nelson, and many others, actually had a long venerable tradition. Einstein already believed that the success of the statistical interpretation of quantum mechanics could imply the existence of underlying Brownian-like particle motion and that this Brownian-like motion could be represented by a diffusion equation, with a certain diffusion coefficient, which would allow one to recover Schrödinger's equation. As we shall see, this is the basic idea taken up in stochastic quantum mechanics.

The questions raised by the aforementioned new developments of the early Bohm have been thereafter examined by several authors. Valentini [265, 266] introduced a H-theorem to explain the relaxation out from equilibrium to Born's postulate and, later on [267], he discussed the *hypothesis that in the remote past the universe relaxed to a state of statistical equilibrium, at the hidden-variables level.* Potel *et al.* [268] demonstrated that for simple one-dimensional systems, any initial probability distribution of a statistical ensemble relaxes asymptotically to $|\Psi|^2$, if the system is subject to a random noise of arbitrary small intensity. In contrast, Colijn and Vrscay [269] argued that the expected Bohm process of relaxation to $|\Psi|^2$ does not in general occur in some hydrogen systems and therefore that Born's postulate has still to be given the status of a postulate. Complementary discussions may be found in Belinfante [1] or Cushing [131].

Flying away

To go on developing his causal theory, as soon as 1954 [263], Bohm (with Vigier) had to fly away from his early relatively safe ground of spinless non relativistic framework, to more adventurous countries. He discussed the extension of his approach to Pauli's equation for spin, still neglecting relativity effects, e.g. in [226] with coauthors. For Bohm, the electron spin has to be interpreted as the result of a genuine body rotation, giving rise to an intrinsic angular momentum. The idea of the electron as an extended body may be found weird but, in about the same spirit , Vigier [270] discussed an extended charged particle model in the framework of the de Broglie–Bohm interpretation, and mentioned the extension of the theory to handle Dirac's equation. Regarding quantum field theory, Bohm speculated on a deeper nature of fields and particles. He imagined fields behaving as waves but which, due to the actions of nonlinearities, would manifest a tendency to generate particle-like concentrations of energy, always forming and dissolving. Then, particles would not be any more permanent entities but would appear in a random way, as suitable concentrations of energy of fields. Being aware of the difficulties involved in such a research program of extension of his early conceptions, Bohm stated that several lines of research are now open and, being nevertheless optimistic, he added that the achievement of a suitable extended theory was by no means distant.

An approach to relativistic quantum field theories is claimed to be successful in 1984 by Bohm and Hiley [179], an approach which, they said, leads to results which are consistent with all the known experimental implications of relativity, despite non-locality. In his last will book of 1993 [68], Bohm, together with Hiley, insisted on very speculative ideas, such as the conception that a particle possesses a rich and complex inner structure which can respond to information and accordingly direct its self-motion. Obviously, with such a surprising or even provocative vision, we have been driven far away from the original simplicity of the early Bohm. The possibility of internal structures of electrons and photons is also discussed by Hofer [271]. In the same book [68], Bohm and Hiley also considered an ontological interpretation of Pauli's equation, elaborating on his 1957 thoughts on this problem (Chapter 10), dealt with boson fields (Chapter 11), discussed the relativistic invariance of the ontological interpretation (Chapter 12). Chapter 14 focused on the extension of ontological theories beyond the domain of applicability of quantum mechanics. See also [180] for a review by Bohm and Hiley of various theoretical approaches.

Similar efforts and commitments are due to other authors. For instance, in Ref. [62], we may find a note by Vigier concerning the forces exerted on objective particles of spin 0, 1/2, and 1, in the pilot wave framework. Vigier again [272] discussed the de Broglie–Bohm interpretation of quantum mechanics in connection with electromagnetism. See also another earlier review work by Vigier [50]. Takabayasi [167, 168] introduced an interpretation of quantum mechanics which has the feature of being associated with certain classical pictures, generalizing Madelung hydrodynamical model, and criticized Bohm's work, a criticism which is soon after rebutted by Bohm [261]. After a causal treatment of boson fields in quantum field theory by Bohm and Hiley [273], and by Bohm *et al.* [274], Bell [183, 275] provided an extension to the case of fermion fields. Later on, dealing again with such fields, Colin [276] used a stochastic model, due to Bell, to introduce a realistic and deterministic interpretation of any quantum field-theoretic model involving fermion fields. Fenech and Vigier [277] presented a relativistic thermodynamical description of the behavior of the pilot wave where individual micro-objects moving in real space-time are waves and particles simultaneously, a point of view which is named thermodynamical because it is built on sub-quantum mechanical fluctuations. Struyve *et al.* [278] examined the effect of an additional spin-dependent term on the guidance formula in the case of spin 1 particles. Weizel [279–281] attached hidden variables to the existence of a new kind of unobservable particles called zerons, criticized by Heisenberg [53]. Roy and Singh [282] presented a post-Bohm causal theory which, they claimed, is still more realistic than the de Broglie–Bohm mechanics. A variant of Bohm in which the Bohmian particles are endowed with a position-dependent effective mass is given by Plastino *et al.* [283]. Holland [284] developed an hydrodynamical analogy with some relevance to the old Madelung hydrodynamical interpretation, and with de Broglie–Bohm causal theories, allowing one to compute the wave function from the hidden trajectories. For him, *the fact that we can derive the time-dependent wave function from the trajectories indicates that we are dealing here not just with an interpretation (hydrodynamics) but rather with an alternative mathematical representation, or picture, of quantum mechanics.* Yang [285] dealt with a generalization of the de Broglie-Bohm theory by using a complex quantum potential, and applied it to the study of the dynamics of hydrogen atom. The most exotic among exotic applications of Bohm's ideas might be due to Khrennikov [286] who described conscious thinking by a quantum cognitive mechanics generalizing the pilot wave model. Extensive discussions of associated problems and formulations, particularly for quantum field theory, are available from Bohm and Hiley [68], Holland [40, 287], Cushing [166] or d'Espagnat [84]. In particular, the excellent books by Bohm and Hiley [68] and Holland [40] are reviewed by Dickson [288] in a very relevant and well balanced way. A quotation here seems to me compulsory: *However, given their many virtues, these books can hardly be faulted on the ground that they do not say everything. I have raised certain questions and objections not because I find the books lacking in careful discussion of important issues, but because I hope to convince the reader that these books are not completions of the Bohmian program, but inducements to further research.*

Experiments for unfalsifiable pilot wave

In the language of Popper [25, 61], quantum mechanics is well corroborated and has not been falsified. The pilot wave, which exactly makes the same empirical predictions than quantum mechanics is therefore strictly in the same epistemological position than quantum mechanics itself. Rigorously speaking, contrarily to what is suggested in the headline of this section, the pilot wave is not unfalsifiable. It can indeed be falsified. But, if this happened, quantum mechanics would, simultaneously, be falsified too. What I should have said in the headline of this section is that the pilot wave is unfalsifiable *against* quantum mechanics. Hence, rigorously speaking again, there is no sense to invoke experiments to discuss the pilot wave, at least the 1952 version of the early Bohm, *alone*. Heisenberg [146] already remarked that a failure of quantum mechanics would be simultaneously a failure of the pilot wave, saying *it would cut the ground from beneath not only the quantum theory but also Bohm's interpretation.*

The fact that quantum theory and pilot wave make identical predictions has been used against the pilot wave. Holding fast inside the realm of quantum mechanics, the proponents of the Copenhagen interpretation viewed the pilot wave, standing outside of the holy circle, as a *superstructure* or as being *metaphysical*. If we refer to the epistemological language of Popper, the pilot wave indeed is not metaphysical. Being falsifiable (although simultaneously with quantum mechanics), it is on the scientific side, not on the metaphysical side, of the famous Popper's demarcation line. The real epistemological nature of the problem "pilot wave versus quantum mechanics" should better make it to be viewed as a problem of discrimination for which the most adapted framework is the Duhem–Quine under-determination thesis concerning the under-determination of theories by experiments. When the times are ready, we shall eventually deal with it, and propose a solution to it (Chapter 17).

But, in 1952, Bohm was already aware that the lack of discrimination based on empirical facts would be viewed as a weak point of his theory although, from a strict epistemological point of view, at least in the Popper's sense, it was not. Therefore, rather inconsistently, he expressed the hope that his approach could become of crucial importance in a domain of characteristic dimensions equal to 10^{-13} cm (recall, a fermi) where, he said, the present quantum mechanics is totally inadequate. To make Bohm consistent, we have to interpret his mind in a charitable way. He was actually hoping that quantum mechanics was going to fail, and would indeed be falsified. Sure, the pilot wave then would be falsified too, but he believed that the correct track towards a new empirically adequate theory would be found in the spirit of the pilot wave, not in the spirit of Copenhagen.

The situation becomes different if we go on further, visiting again Bohm after the early Bohm, particularly the issue of the geneticity of Born's postulate. This is because we now have the possibility, at least conceptually, to drive hidden variables out of equilibrium, and to observe their relaxation to equilibrium. Phenomena not predicted by quantum mechanics could then be, in principle, detected. In 1953 [262], Bohm reiterated his hope that 10^{-13} cm could be a critical length scale but, this time, he did it by referring explicitly to the opportunity of detecting relaxation to equilibrium ruled out by Born's postulate. As a matter of fact, there has indeed been an experiment carried out by Papaliolios [289] to test this idea, but we shall report on it later, in a more appropriate context. A similar hope is expressed again by Bohm [31], and by Bohm and Aharonov [290], mainly in connection with high energy physics. In his last book, together with Hiley [68], Bohm acknowledged the fact that such hopes have been deceived by the evolution of modern physics, and pushed the frontier of falsification to a more remote location, saying that *between the shortest distances now measurable in physics (of the order of* 10^{-16} *cm) and the shortest distances in which current notions of space-time probably have meaning which is of the order of* 10^{-33} *cm, there is a vast range of scale in which an immense amount of yet undiscovered structures could be contained.* Surely, Bohm is right concerning the ultimate fate of quantum mechanics, which is actually falsified by its inability to deal convincingly with gravitational fields, and this is without any doubt connected with the very notions of space and time which eventually become hopeless, as eloquently discussed for instance by Smolin [9]. But whether the spirit of the pilot wave will find here its accomplishment, or not, this is something we may most reasonably question.

Louis de Broglie [62], in agreement with Bohm, also mentioned the critical length of 10^{-13} cm but, for him, it should be interpreted in the framework of the double solution. At such scales, the singular regions of the waves u associated with particles would cease to be isolated but, conversely, should begin to overlap, expectably producing a new physics. Freistadt [155] also fascinated by the same critical scale commented: *When one of the founders of the quantum theory of fields is ready to express pessimism in public concerning the ability of the quantum field theory, plagued as it is by divergences and other mathematical difficulties, to serve as a reasonable model of physical reality, it is time perhaps to question the foundation upon which the entire edifice of quantum theory is based.* But, whatever the indisputable problems met by theoretical physics, during the last decades, we did not meet any clear clue directing ourselves, in a compulsory way, toward a pilot wave kind of interpretation.

Freidstad also mentioned another possibility, that, at some deep enough level, the objective values (let us call them true values) and the observed values would merge. If this happened, then quantum mechanics could not provide a correct entry to reach this level since it just rejected the idea of true values, in contrast with causal theories which accept both kinds of values, but simply give them different meanings, a true meaning for true values, a coarse-grained (averaging) meaning for observed values. But if they are true values and a level where they merge with observed values, this might open the possibility of non-quantum measurements, that is to say the possibility of measuring true values. Such an idea is already put forward, in 1952, by Bohm, saying: *It is conceivable that we may be able to carry out new kinds of measurements, providing information not about observables ..., but rather about physically significant properties of a system, such that the actual values of the particle position and momentum.*

In a similar mood, Mugur-Schächter [291] defended the possibility of non-quantum measurements, that is to say measurements which, because they would not be of a quantum nature, would allow one to probe the existence of hidden variables. She provided the example of tracks in a cloud chamber from which we would obtain the location of the particle at the origin of the track, together with the momentum by analyzing the development of the track in space. Unfortunately, this was not a good example, but rather the result of a misunderstanding of quantum mechanics. Indeed, quantum mechanics is perfectly able to explain the seemingly classical character of trajectories in a cloud chamber, in its own framework, as the result of a succession of a huge number of forward-peaking events. The track is actually an averaging illusion, from which we may possibly deduce simultaneously some kinds of locations and momenta, but as the result of averaging processes which do not provide simultaneous locations and momenta in the quantum mechanical sense. Obviously, these measurements are non-quantum measurements in the sense that they are not quantum measurements in the proper sense, e.g. they would not satisfy Heisenberg uncertainty relations. But they are not non-quantum measurements in the sense advocated by Bohm and Mugur-Schächter, because they do not give any access to true values. For the discussion of tracks in quantum mechanics, essentially see Mott [292], but also Bohm [41], Bitbol [128], Messiah [293] or d'Espagnat [85].

As an excuse to Mugur-Schächter, let us mention that, in the early ages of quantum mechanics, the understanding of tracks in chambers has not been readily available. For instance, Einstein once asked Heisenberg, in a private discussion : You admit that there are electrons in the atom ... And nevertheless, you intend to completely eliminate orbits or trajectories of these electrons in the atom, although we can directly observe electron trajectories in a Wilson chamber. Could you explain...? In the same discussion, he later on insisted : In a Wilson chamber, we observe the trajectory of the electron going through the chamber. In the atom, conversely, you think that trajectories of an electron do not exist any more. This manifestly looks like absurd. For the narrowing of the space in which the electron is moving cannot make the notion of trajectory disappearing [10].

And, indeed, there has been a time when nor Bohr nor Heisenberg could understand how such a so simple phenomenon as the trajectory of an electron in a Wilson chamber could be reconciled with the mathematical formulations of quantum mechanics or wave mechanics [10].

Appraisals and auto-appraisal

Besides what we have already said concerning the appraisals of Bohm's work, we must end this chapter with a bit of synthesis and recapitulation. I do not intend to provide here definitive conclusions whether or not Bohm is to be eventually accepted or not. This will be tentatively done later (although with a loophole), at the right place (we are not yet equipped enough for doing it now). But we should not leave this chapter without having in mind some of the main relevant issues.

A strong point, already commented by Belinfante [1], is that early Bohm's theory has been formulated in a nonrelativistic framework. In itself, this is not much troublesome because Bohm's aim, at this time, was to provide an alternative to a quantum mechanics which was also built in a non relativistic framework. But further developments of quantum mechanics succeeded to manage fairly well with (special) relativity. Regarding this aspect, what about Bohm? Let us first assume that beables are the objective particles of the early Bohm. If we desire to generalize this picture to the relativistic domain, particularly to high energy physics, where particles can be created and annihilated, there is an immediate problem arising in pilot wave interpretation, namely that the continuity equation (e.g. 8.8) is not valid any more. This equation is a law of conservation of the number of particles, but particles are actually no more conserved in high energy physics. Not only the number of particles may change, but their nature as well. In a quantum mechanical framework, we may deal conveniently with such features. For this, we use an operator of creation for creation, an operator of annihilation for annihilation, and the number of particles becomes an observable too, with an associated operator to deal with the associated measurement. In contrast, the early Bohm is poorly equipped to deal with these kinds of phenomena. We now have to create or annihilate objective particles (with the word "objective" to be taken seriously). Then, worldlines of these particles have to stop abruptly dead, or to start alive from nothing. This certainly motivated Bohm to create new ideas, to become a new Bohm after Bohm. In particular, particles were no more considered as the fundamental entities, a real renegation of the original ideas, letting the stage filled with fields, the new beables, at a still deeper level. Particles then become evanescent appearances, resulting from the formation (creation) and from the dissolution (annihilation) of energy concentrations, under the effect of non-linearities, in a rather *ad hoc* way. Such a scenario has not been supported by any correct mathematics, and even by any mathematics at all. It remained qualitatively programmatic.

Squires [13] took notice of the difficulties encountered by Bohm to satisfy relativity, and similar observations are also available from many authors, such as d'Espagnat [85] or Omnès [88, 97]. D'Espagnat remarked that, to account properly for creations and annihilations, we must give up the idea that the particles would be the basic beables and, instead, the beables should rather be the values of boson or fermion fields at each point [68, 182]. But, he noted that such a shift concerning what beables should be is uncomfortable, stating that so radical changes in the image that we make ourselves of the world lead us to ask when and whether the upheaval will stop. For Omnès, Bohm's theory never successfully solved the problem of quantum fields, or in other words, the passage to quantum field theory has never been accomplished in a convincing way, a fact which did not much plead in his favour, in the eyes of many physicists. In contrast, Bitbol [133] expressed a much more positive appraisal. For him, Bohm's theory [68] has reached such a development that it is possible to affirm that it exactly reproduces the predictions of essentially all known quantum theories, the usual quantum mechanics, the relativistic quantum theory, as well as quantum field theory, and this is done by using only beables. I do not intend to deepen further this point, just because I shall not need it (Chapter 17), but the reader had to be aware of the issue.

Another line of thought may be explored, relying on some ideas of beauty and simplicity that physical theories should exhibit. Indeed, it has been argued that the forms eventually obtained for causal theories, by Bohm and by others after the early Bohm, particularly when flying away, were not convincing for at least a lack of beauty. Some people found that it would be very surprising if quantum phenomena, revealing an undisputed self-coherence and harmony, would find their rational origin in theories which definitely became awkward. Such an argument however, although possibly subjectively convincing to many, is logically very weak (we shall return to this issue in Chapter 17 to refine it). In particular, it does not take into account the fact that the forms questioned are may be not final, that more effort could make them amenable to more appealing shapes. And we should not forget that the alternative, here alternative meaning quantum mechanics, is not free from problems and still spoiled by a precise smell of vagueness, at least in the eyes of many people. For a final motivated rejection, more serious reasons are needed.

And Bohm? What is eventually the auto-appraisal of Bohm by Bohm? It must have obviously been varying and fluctuating all along the story. However, more or less recurrently, since the beginning, likely well aware of difficulties involved in his interpretation, and anticipating more difficulties to come, Bohm stated that he did not regard his proposed interpretation as a final theory but that his principal objective has been to demonstrate that at least one logically self-consistent causal alternative to usual quantum mechanics was possible. This is also the mood conveyed by Bell [142] viewing Bohm's interpretation as an *antidote to prevailing complacency* to quantum mechanics (the word antidote will be much later also used by Dickson in the title of one journal paper [288]). In 1957 [60], Bohm recognized that the theory in its original form, although completely consistent in a logical way, had many aspects which seemed quite artificial and unsatisfactory. He said that his principal purpose was not to propose a definitive new theory, but was rather mainly to show, with the aid of a concrete example, that alternative interpretations of the quantum theory were in fact possible. Discussing subsequent developments, currently developed or anticipated, he commented that *it is clear then that even if none of the alternative interpretations of the quantum theory that have been proposed thus far has led to a new theory that could be regarded as definitive, the effort to find such theories is nevertheless becoming a subject of research on the part of more and more physicists, who are apparently no longer completely satisfied with continuing on the lines of research that are accessible within the framework of the usual interpretation.* He also wrote the following striking comments: *Finally, our model in which wave and particle are regarded as basically different entities, which interact in a way that is not essential to their modes of being, does not seem very plausible. The fact that wave and particle are never found separately suggests instead that they are both different aspects of some fundamentally new kind of entity which is likely to be different from a simple wave or a simple particle, but which leads to these two limiting manifestations as approximations that are valid under appropriate conditions.* Louis de Broglie should have much disliked these comments. What does that mean? Nor wave nor particle again? The temptation of Copenhagen? An attack of lucidity? Or a fever of despondency?

Another auto-appraisal is worth-quoting [179]: *... there was perhaps a misunderstanding as to the intention behind the suggestion of this interpretation. Its purpose was not to suggest that the ideas in it were to be regarded as a final and received version of the ultimate nature of the reality. Rather, it was proposed in the spirit of a provisional point of view, that would help provide further insight into the significance of the quantum theory, a kind of insight that is made possible by the intuitive and imaginative way in which it shows the meaning of the mathematical equations. Thus, with the aid of this and other interpretations, it would be possible to understand the theory more thoroughly and deeply possible than with one interpretation alone.*

When I examined the latest developments of the latest Bohm, I had the same feeling than I had for Louis de Broglie when I was wondering whether he was not blindly forging ahead. I do not intend to mislead the reader, making him believing that I am mastering all these lately meanders. I did not and, at the present time, I am not intellectually equipped to do it. Furthermore, once again, I do not need to do it (otherwise I would have done it before writing this book) because my final appraisal, near the end of this book (Chapter 17), will not depend on these latest developments. But, for the sake of completeness, I had to report on them.

But my strong intimate conviction is that Bohm was becoming more and more embarrassed in his attempt to develop his own ideas up to their *logical conclusion*. When I follow the subjective trajectory of Bohm, in particular from 1980 [31], where he introduced his idea of the implicate order, I found him sometimes very lucid and creative,

and sometimes unduly speculative. A typical example is Chapter 4 of
[31] which provides a much interesting review of hidden variables in
quantum theory, but unfortunately, I felt, embodied in a somewhat
mystical environment. I cannot prevent myself thinking that, most
unfortunately, likely under the influence of various pressures, and
with someone coming over him, Bohm progressively drifted towards
bad philosophy. Such aspects of his work certainly strongly and
unfortunately speak against him, and were counter-productive to
make his ideas studied more carefully, in serenity and even ataraxia.
Favorite themes of the late Bohm with favorite words (implicate and
explicate orders, holomovement or wholeness) was already discussed
in 1971 [294].

I nevertheless deeply feel in sympathy and empathy with Bohm
(and with Louis de Broglie), for his solitary enterprise, his bravery, his
perseverance, his open-mindedness, and I hope that the parts of his
works to be most criticized in an implacable way, could nevertheless
be a source of inspiration for others. He is genuinely a hero of sciences,
to be praised for having explored a territory that someone had to
explore. In any case, his insistence on the concept of wholeness, which
is irrefutable, is much welcome, much in phase with our present
understanding of usual quantum mechanics, and has far-reaching
philosophical consequences. As a testimony, d'Espagnat mentions
that the most plausible ontology is the holistic ontology perceived by
Bohm, and that he himself gives his adhesion to it. The relationship
between the individual and the ensemble, the part and the whole,
concerning the concept of non-separabiltity, was already discussed in
1971 by d'Espagnat in a book [295], reviewed by Hiley [296].

Can we make a final appraisal, from features and reports which are
sometimes contrasted and even contradictory? For sure, we cannot
reject Bohm on the basis of any experimental evidence. For sure,
we cannot reject Bohm on the basis of the fact that the theory
has not been satisfactorily completed: What one solitary man did
not achieve, an army of researchers, like the one who invaded the
now usual quantum world, might have succeeded. And, after all,
there could have been here again a lack of imagination, like the one
which still makes our most advanced physics rather messy. Like Bohm

failing to achieve his story, the story of physics too is not achieved. And also, for sure, we cannot and should not reject Bohm on the basis of subjective and irrational feelings, whatever they could be. Essentially, Bohm has been nearly alone against nearly the whole assembly of experts. But, it already happened that someone, alone at the beginning, eventually revealed himself as the winner.

So, finally, what about Bohm? Can we provide a definitive answer to this puzzling question? Well, may be. But wait a bit, or better: A bit more.

Chapter 9

Other Hidden-Variables Theories

We now have in mind emblematic and paradigmatic examples of hidden-variables theories, namely the double solution of Louis de Broglie, the pilot wave declined into two versions, first the version of Louis de Broglie, then the more achieved version of Bohm, all of these with objective particles as beables, followed by various variants or more elaborated theories, with possibly fields as beables. We are now going to enter a short chapter, short indeed, but sufficient and compulsory, to complement our luggage with a few more suitcases, before attacking the problems of hidden variables from more universal standpoints.

Bohm–Bub theory

PEDESTRIANS ($\sim\sim\sim\sim$): This section is rather made out from technicalities. So, go to the next one.

Bohm and Bub [297] reviewed the basic principles of the usual quantum mechanics, elaborated on the inadequacy of the Copenhagen interpretation, challenged the von Neumann famous proof of impossibility of hidden variables (by now, soon to be discussed), and provided a theory, or better said a model, in which the wave function is represented as a vector in a two-dimensional Hilbert space. To be more concrete and specific, such a two-dimensional Hilbert space may be viewed as generated by the description of a spin 1/2 particle without any translational motion. Then, they supplemented the usual quantum mechanical description with hidden variables.

We are now going to enter the details of this model, with however some changes of notation, due to the fact that the original notations, in the Bohm–Bub's paper, were, I found, a bit disturbing. The state vector $|\Psi\rangle$ of the system under study, assumed to be properly normed, may be written as:

$$|\Psi\rangle = a_1^\Psi |S_1\rangle + a_2^\Psi |S_2\rangle \tag{9.1}$$

in which $|S_1\rangle$ and $|S_2\rangle$ form an orthonormal basis, that some people might prefer to denote as $|+\rangle$ and $|-\rangle$, e.g. see [43]. The quantities a_1^Ψ and a_2^Ψ are expansion coefficients.

The hidden variable is denoted by ξ. Mind however that we do not work any more in the framework of the double solution or of the pilot wave, that it to say we do not claim that ξ is associated with hidden trajectories. On the contrary, we let the nature of ξ unspecified. We just say that it is something hidden which participates to the quantum processes. We now intend to manage in such a way that the ket $|\Psi\rangle$ and the bra $\langle\xi|$ together drive quantum processes in a complete deterministic way. Remember however that, according to the terminology we adopted, Bohm–Bub theory is not causal, although deterministic.

To the hidden variable ξ, we associate another two-dimensional (hidden) Hilbert space, a dual space in the words of Bohm and Bub, in which ξ evolves. Similarly as for Eq. (9.1), we decompose ξ into two terms, however using a bra-relation instead of a ket-relation, according to:

$$\langle\xi| = a_1^\xi \langle S_1| + a_2^\xi \langle S_2| \tag{9.2}$$

in which a_1^ξ and a_2^ξ are the expansion coefficients associated with the hidden variable ξ in the dual space.

We next assume that the hidden variable is randomly distributed on the hypersphere of unit radius in the dual space, and that it satisfies a normalization relation reading as:

$$\sum_i \left| a_i^\xi \right|^2 = 1 \tag{9.3}$$

and, for such a normed state, it may be demonstrated that one can reproduce the usual quantum mechanical averages for all observables.

We now go on, accepting the existence of sub-ensembles which are less dispersed than the aforementioned ones. In a complete deterministic limit, we may even accept sub-ensembles for which the a_i^ξ's are precisely defined so that we only deal with dispersion-free states (another complementary formal definition of dispersion-free states to be found a bit later when we discuss Mugur–Schächter's rebuttal of von Neumann).

Let us now introduce the ratios:

$$R_i = \frac{\left|a_i^\Psi\right|^2}{\left|a_i^\xi\right|^2} = \frac{J_i}{\left|a_i^\xi\right|^2}, \quad i = 1, 2 \tag{9.4}$$

The free evolution, that is to say in the absence of any measurement, of $|\Psi\rangle$ is governed by the linear Schrödinger's equation, as in the Copenhagen interpretation. Let us now consider an impulsive measurement of the observable S underlying the basis $\{|S_1\rangle, |S_2\rangle\}$. As in the Copenhagen interpretation again, this measurement is no more ruled out by Schrödinger's equation (since Schrödinger's equation must be relaxed during a measurement process). But, during such a measurement, we postulate that the evolution of $|\Psi\rangle$ is determined by:

$$\frac{da_1^\Psi}{dt} = \gamma(R_1 - R_2)a_1^\Psi J_2 \tag{9.5}$$

$$\frac{da_2^\Psi}{dt} = \gamma(R_2 - R_1)a_2^\Psi J_1 \tag{9.6}$$

in which γ is some real quantity remaining constant during the measurement, supposed to be positive, and negligible outside of the measurement. From Eqs. (9.4)–(9.6), it is a very simple exercise to establish:

$$\frac{dJ_1}{dt} = \frac{d\left|a_1^\Psi\right|^2}{dt} = 2\gamma(R_1 - R_2)J_1 J_2 \tag{9.7}$$

$$\frac{dJ_2}{dt} = \frac{d\left|a_2^\Psi\right|^2}{dt} = 2\gamma(R_2 - R_1)J_2 J_1 \tag{9.8}$$

Hence,

$$\frac{d}{dt}(J_1 + J_2) = \frac{d}{dt}\left(\left|a_1^\Psi\right|^2 + \left|a_2^\Psi\right|^2\right) = 0 \tag{9.9}$$

From Eq. (9.1), this means that $|\Psi\rangle$ remains normalized during the measurement process. Now, we have taken γ positive and, from their definition in Eq. (9.4), J_1 and J_2 are not negative. Furthermore,

$$J_1 + J_2 = \left|a_1^\Psi\right|^2 + \left|a_2^\Psi\right|^2 = 1 \tag{9.10}$$

So, let us now consider Eq. (9.7). We see that, if J_2 is different from 0, and R_1 larger than R_2, then J_1 will increase up to the time when it reaches its maximum value $J_1 = 1$. In parallel, J_2 decreases down to 0. Therefore, from Eq. (9.1), $|\Psi\rangle$ will tend to $|S_1\rangle$, within an irrelevant phase factor. Similarly, if $R_2 > R_1$ and $J_1 \neq 0$, then $|\Psi\rangle$ will eventually land on $|S_2\rangle$. Therefore, once the measurement is completed, we obtain the values $\hbar/2$ or $-\hbar/2$ respectively. Let us remark that the condition $R_1 > R_2$, using Eq. (9.4) and normalization relations, may be converted to $\left|a_1^\Psi\right|^2 / \left|a_1^\xi\right|^2 > 1$. Similarly, $R_2 > R_1$ may be converted to $\left|a_2^\Psi\right|^2 / \left|a_2^\xi\right|^2 > 1$.

We then have found that the result of the measurement of the spin is determined by the state vector $|\Psi\rangle$ together with the dual vector $\langle\xi|$. What has been developed here is a hidden-variables model which reproduces the usual statistics of quantum mechanics, without any collapse of the wave function or, if there is a collapse, it is not a collapse in the usual quantum mechanical meaning, but the result of a deterministic process, with transients.

In this model, we may take a thermodynamical standpoint. For this, we define a microstate as being given by the couple ($|\Psi\rangle$, $\langle\xi|$). This is not however a classical microstate defined by objective quantities, but a quantum-hidden microstate in which $|\Psi\rangle$ is unobservable and $\langle\xi|$ is hidden. The measurement problem has been solved by introducing a basic irreversible process involving an interaction which couples macrostates $|\Psi\rangle$ and microstates ($|\Psi\rangle$, $\langle\xi|$), showing another difference with classical thermodynamics since the macrostate is in the present case a component of the microstate. As pointed out by Bohm and Bub, this is a new concept.

Let us now remember that, in utmost rigor, there is no solution to the measurement problem in a linear framework, even with decoherence theory. In contrast, the model presented by Bohm and Bub succeeded to provide a solution by introducing non-linearities, this necessary ingredient. To clearly see how these nonlinearities occur, we simply have to rewrite Eqs. (9.5) and (9.6) as:

$$\frac{da_1^\Psi}{dt} = \gamma(R_1 - R_2)a_1^\Psi \left|a_2^\Psi\right|^2 \tag{9.11}$$

$$\frac{da_2^\Psi}{dt} = \gamma(R_2 - R_1)a_2^\Psi \left|a_1^\Psi\right|^2 \tag{9.12}$$

or, better, using the normalization equation for $|\Psi\rangle$, similar to Eq. (9.3), we obtain:

$$\frac{da_1^\Psi}{dt} = \gamma(R_1 - R_2)a_1^\Psi(1 - \left|a_1^\Psi\right|^2) \tag{9.13}$$

$$\frac{da_2^\Psi}{dt} = \gamma(R_2 - R_1)a_2^\Psi(1 - \left|a_2^\Psi\right|^2) \tag{9.14}$$

In contrast with the original description of the measurement process by Bohm, in 1952, which explicitly introduced an interaction between the measured object and the measuring apparatus, the present description is of a phenomenological nature, linking $|\Psi\rangle$ and $\langle\xi|$, without any explicit reference to any measuring device. In the words of Belinfante [1], this description *remains valid through the entire measurement instead of oscillating between descriptions of the object alone and the object as part of a composite system together with the apparatus.*

Bohm–Bub theory for PEDESTRIANS ($\sim\sim\sim\sim$)

Bohm and Bub provided a hidden-variables theory in which the hidden variable, denoted as ξ, does not receive any clear physical meaning, being simply a hidden symbolic quantity (or set of quantities) which is formally incorporated in a symbolic framework. In particular, there is no reason to argue that ξ represents any quantity (velocity, location, or whatsoever) associated with hidden trajectories. According to the terminology that has been chosen in this book, Bohm–Bub theory is therefore not causal, although it is deterministic.

Wiener–Siegel theory

Belinfante dedicated a whole chapter of his famous book [1] to a theory published by Wiener and Siegel in 1953 [298, 299], one year after Bohm's seminal papers. This theory is rather involved and very uncomfortable to explain in a reasonable amount of space. Furthermore, it significantly departs from my final aim in writing this book, in particular it is not much relevant to the final objections I shall draw at the end of our journey. Therefore, I shall be content to inform the reader of its existence and to provide a bit of qualitative discussions on its main features. For more details, the reader might return to the original papers or, better and more economical, to Belinfante which provided a simplified account of it.

Belinfante exhibited a relationship between Bohm–Bub' and Wiener–Siegel's theories, essentially viewing Bohm–Bub's theory as a sub-case of Wiener–Siegel's theory. Therefore, he began to explain the Wiener–Siegel's theory and, afterward, obtained the Bohm–Bub's theory by deleting irrelevant parts of Wiener–Siegel's theory, and also by supplementing the obtained structure with Bohm–Bub's ideas concerning the use of additional terms in Schrödinger's equation, namely those non-linear terms which are only effective during the measurement process.

In both theories, the outcomes of measurements are determined by an algorithm. In Bohm–Bub's theory, the algorithm relies on the behavior of differential equations as we have seen. In Wiener–Siegel's theory, it relies on a somewhat non differential process called the polychotomic algorithm, coming from a Wiener's mathematical generalization of the Brownian motion. But, similarly as in Bohm–Bub's theory, the result of the algorithm is a biased hidden-variables distribution which therefore, due to the occurrence of some kind of relaxation process, can be experimentally tested.

However, as discussed by Belinfante, there are *troubles and paradoxes* in Wiener-Siegel's theory, which however could possibly be treated with adequate remedies. For instance, in this theory, the result of the polychotomic algorithm is non unique when applied to a degenerate operator. This may be diagnosed as an algorithm incompleteness (see also appendix in [299]). *Troubles and paradoxes may also be found when one tries to apply the theory to a composite system with non factorizable state vectors, but paradoxes may already exist for single systems*. Optimistically, Belinfante however stated that one must *end up with a rather simple theory that is easy to comprehend and easy to work with,* although there are shortcomings.

Experiments for unfalsifiable relaxations

Principle

A last issue of importance is that Bohm–Bub's theory (or better: Model) can be a motivation for experiments (although, as we shall see, it is immune to falsification). This comes from the fact that, immediately after the measurement of the spin component in a certain direction, $|\Psi\rangle$ is either $|S_1\rangle$ or $|S_2\rangle$ and, therefore, the expansion coefficients $a_i^\xi (i = 1, 2)$ must satisfy either $R_1 > 1$ or $R_2 > 1$. This provides information on the a_i^ξ's and therefore on $\langle\xi|$. According to Bohm and Bub, a time sequence of successive measurements would then produce a fairly accurate measurement, and revelation, of the hidden $\langle\xi|$. Conversely, if two successive measurements are separated by a time long enough, the a_i^ξ's and the a_i^Ψ's would pursue their own free evolution, without any coupling since $|\Psi\rangle$ must then satisfy the usual linear Schrödinger's equation. Therefore, for a short time, call it a relaxation time, immediately following a measurement, the predicted result for a subsequent measurement would differ from the one predicted by quantum mechanics.

To help understanding, we may present the situation again by considering three cases. In the first case, a first measurement is made at time $t = 0$, and the system is thereafter left to itself. The relaxation process begins to occur and, after a time long enough, say T, we recover the free evolution governed by Schrödinger's equation, with a good enough approximation. The larger T is, the better the approximation is. If we made a second measurement, at time $t > T$, we essentially recover quantum mechanical predictions.

In a second case, a first measurement is made at time $t = 0$, and a second measurement is made immediately after at time $t = \varepsilon$, with ε vanishingly small. Then hidden variables did not had time to relax, and the outcomes of measurements should significantly depart from quantum mechanical predictions. In the last case, we make a first measurement at time $t = 0$, and a second measurement at an intermediary time t_i such as $\varepsilon < t_i < T$. By varying t_i from ε to T, we should then be able to observe the effect of the relaxation process.

So, we may conclude by saying, with Pipkin [156]: *The Bohm–Bub theory gave an instructive example of a modification that in general gives the same predictions as quantum mechanics, but differs in certain cases where measurements are involved,* or with Freedman and Holt [300]: *However, for the short time interval immediately following a measurement, it makes predictions which differ from those of quantum mechanics,* or with Tutsch [301]: *The above results show that the hidden variables need not remain hidden, and it is hoped that the validity of the Bohm–Bub theory may be tested in the laboratory.*

The paper by Tutsch is dated 1968 but, actually, his hope was already fulfilled by a paper reporting on such experiments, dated 1967, due to Papaliolios [289]. Furthermore, such experiments do not really test Bohm–Bub's theory only, but a much larger class of hidden-variables theories in which relaxation processes are involved. This does not concern the early Bohm designed to exactly satisfy quantum mechanical predictions, but it is relevant to a later Bohm, with his relaxation to Born's postulate, or to Wiener-Siegel's theory. In any case, the experimental discovery of relaxation processes in microphysics would be a huge step forward.

Papaliolios experiment

The existence of relaxation processes, which could be associated with existing hidden-variables theories or hidden-variables theories still to be built, can be experimentally tested. We explained above a basic principle to carry out such experiments, requiring successive measurements to detect relaxation phenomena, but this principle might be declined in several ways. One of them has been achieved by Papaliolios [289] providing a direct and simple experimental test.

Papaliolios experiment is performed by sending optical photons through a stack of three polarizers and by measuring how the transmission varies as the final polarizer is rotated. If there is no relaxation process, or a too fast relaxation process, quantum mechanics predicts that the transmission should vary according to Malus law. If there is an effective relaxation process, a departure from Malus law should be observed. The experimental relaxation time in Papaliolios experiment is as short as $7.5.10^{-14}$ s. Experimentally, no departure from quantum mechanical predictions is observed. Therefore, Papaliolios's results support (or better said: Corroborate) quantum mechanics and does not support hidden variables with relaxation processes. It has been estimated that, at room temperature, a relaxation time to hidden variables could be something like 10^{-13} s. This evaluation is invalidated by Papaliolios's experiment. But there is actually no known explicit mechanism to explain hidden-variables relaxation processes, so that any theoretical evaluation of a relaxation time is to be suspected of arbitrariness. Therefore, what actually did the negative result of Papaliolios's experiment is simply to impose a limit upon the relaxation time. All we could say is that, if relaxation processes existed, the Papaliolios's experiments would have not been fast enough to detect them. More in Freedman and Holt [300], Belinfante [1], or Pipkin [156].

Epistemological appraisal

We are then facing a rather interesting epistemological situation. Papaliolios's experiment does not support hidden variables, but it does not falsify them. Since we do not actually know what should be the value of the relaxation time, it is always possible, for the tenants of hidden variables, to state that experiments are simply not fast enough, whatever the number and the quality of experiments done. Therefore, the proposal of hidden variables with relaxation processes is immune to falsification. Remember that we have discussed the fact that the early Bohm's theory is unfalsifiable *against* quantum mechanics. We have here a stronger situation. Relaxation processes are unfalsifiable in a strict sense, that is to say not in a relative sense (with respect to quantum mechanics) but in an absolute sense (corresponding exactly to the original sense given by Popper to the word). Now assume that we possessed an experiment demonstrating the existence of relaxation processes. Then quantum mechanics would be falsified, and hidden variables corroborated. Therefore, we may falsify quantum mechanics, but we may not falsify the existence of relaxation processes.

As far as I know, Papaliolios's experiment is the only one of that sort. Certainly, this experiment should be repeated, or other ones in the same spirit (such as proposed in Belinfante's book) should be made, with more advanced technologies. Valentini too [302], discussing departures from Malus law as a distinctive signature of quantum non equilibrium, suggested new experiments. I understand very well that it might look risky, particularly for those who blindly believe to quantum mechanics, and have more serious things to accomplish. And, in any case, you might have to make a career, if simply to support a family. But the possibility of falsifying quantum mechanics should be able to make some people fairly excited. A researcher who would succeed, even if it seems most unlikely, would be the winner of a huge jackpot.

Chapter 10

Proofs of Impossibility
and Possibility Proofs

There are proofs of impossibility and possibility proofs, a situation which sometimes produced a bit of confusion in the debates on hidden-variables theories, sometimes something worst than a confusion, and that we have to clarify.

Proofs of impossibility with Von Neumann

Brief beginning of the history

In his famous book, providing an axiomatization of quantum mechanics, von Neumann [26] introduced, in 1932, a proof of impossibility of hidden-variables theories, the famous von Neumann's proof. This was a few years after the Solvay Congress of 1927 where Louis de Broglie's pilot wave was defeated. Although Louis de Broglie had already renegaded his work, and rejoined the camp of orthodoxy, the publication of the proof certainly confirmed him that he lost his time with his heretical detour. Furthermore, in the words of Pinch [154], *the proof was accepted and welcomed by the physics elite.* The continuation of the story shows that the acceptation was not based on a careful examination of the proof, but rather on other elements, such as a psychological pressure exerted by the very high reputation of its author, or the unconscious desire of definitely getting rid of an ill-timed and embarrassing issue. Afterward, for a long time, the rejection of hidden variables could be made just by invoking von

Neumann (the man more than the proof). As stated by Belinfante [1], *the truth, however, happens to be that for decades nobody spoke up against von Neumann's arguments, and that his conclusions were quoted by some as the gospel ... the authority of von Neumann's over-generalized claim for nearly two decades stifled any progress in the search for hidden variables theories.* Belinfante also [1] remarked that *the work of von Neumann (1932) was mainly concerned with the axiomatization of the mathematical methods of quantum theory. His side remarks on hidden variables were merely an unfortunate step away from the main line of reasoning.* More than unfortunate however, it was erroneous, and even deeply erroneous, *because of the obviousness of inapplicability of one of von Neumann's axioms to any realistic hidden variables theory* [1].

As far as I know, the first one to challenge the validity of von Neumann's proof has been a philosopher, Grete Hermann, in 1935 [117], three years after von Neumann's book, followed nine years after, in 1944, by another philosopher, Hans Reichenbach [134], according to Pinch [154]. As noted by Bitbol [133], and can be checked in the original reference, recently republished with enlightening complementary discussions by Léna Soler, Hermann correctly identified the weakest point of von Neumann's proof, namely the misleading use of an additivity condition of average values (soon to be discussed), much later on pointed out by Bell again, in 1966 [188]. Furthermore, she charged von Neumann's proof with the accusation of circularity.

However, likely because they were philosophers, Hermann and Reichenbach seemingly did not have much influence on the community of physicists. Very often, they are even forgotten or dismissed. For instance, Pipkin [156] could erroneously write that it was Bell, in 1966, *who first pointed out the axiom by which von Neumann's formulation violated the elementary principles of any realistic hidden variables theory,* a statement strongly and strangely in agreement with Belinfante [1] when he claimed, five years before: *The first to publicly pinpoint the axiom by which von Neumann's formulation violated the elementary principles of any realistic hidden variables theory was Bell (1966).*

As far as I know (again), the first physicist to challenge von Neumann has been Bohm, twenty years after the proof is born, in one of his 1952-papers [30]. The best rebuttal of the early Bohm has obviously been the fact that he produced a logically consistent hidden-variables theory. This was a constructivist rebuttal, explicitly showing that what was demonstrated as being impossible was in fact possible. It definitely established that something was wrong with von Neumann. But what exactly was wrong, this is something which was compulsory to establish. The formal answer to this issue, by Bohm, was rather poor, but he correctly identified the most important point, namely that, in fact, von Neumann implicitly restricted himself, in his proof, to an excessively narrow class of hidden-variables theories. And it just happened that the early Bohm's pilot wave did not pertain to this class. Louis de Broglie then also manifested a reluctance against von Neumann's proof, e.g. in [62, 64]. However, the fact that formal Bohm's rebuttal was unsatisfactory, and required more clarification, was for instance pointed out by Bell [188] saying: *The analysis of Bohm seems to lack clarity, or else accuracy ...*

Bell's rebuttal of von Neumann

Accepting the idea that von Neumann' proof has been rebutted, PEDESTRIANS ($\sim\sim\sim\sim$) are allowed to fly over this sub-subsection as a Buddhist bird would do...

Bell [188] was able to make clear, although in a fairly laconic way, what was unclear in Bohm's attack against von Neumann. In essence, his identification of the reason why von Neumann failed is in agreement with what Hermann had already understood, something like thirty years before. But the influence of Bell's paper, about thirty five years after the proof was announced to the world, a long Middle Age for hidden variables, was top one. The dragon was struck down. But, at that time, that there was a dragon still living and that it had to be destroyed, this was not obvious to all scientists, as we can infer from the following quotation [188]: *The present paper ... is addressed to those who do find the question interesting, and more particularly to those among them who believe that "the question concerning the existence of such hidden variables received an early and decisive answer in the form of von Neumann's proof on the mathematical impossibility of such variables in quantum theory"*. The influence of Bell's paper is acknowledged by Bitbol [133] when he stated that the very sociological turnaround of the community of physicists regarding von Neumann's theorem nevertheless only happened from 1966 on, the date of the publication of Bell's paper... The message being then sufficiently clear, it went on affirming more itself thereafter.

Bell's attack pointed out an additivity postulate used by von Neumann which, when applied to hidden-variables theories, is faulty: *One of the weakest points in the proof*, in the words of Jammer [24]. The essential assumption used by von Neumann, the one we called above the "additivity condition on average values", from now on called the additivity postulate, is contained in the following statement: *Any linear combination of any two Hermitian operators represents an observable, and the same linear combination of expectation values is the expectation value of the combination.* In the words of Mermin [197], this was the *von Neumann's silly assumption.*

The sentence quoted above actually contains two statements. The first one says that any linear combination of any two Hermitian operators represents an observable. This seems something like obvious, but actually is not guaranteed at all. Let us however forget this problem, as irrelevant for our purpose, and focusses on the second statement, the one we properly call the additivity postulate saying that the same linear combination of expectation values is the expectation value of the combination.

Let us consider two Hermitian operators A and B, and let us assume that the sum of the two Hermitian operators A and B is another Hermitian operator, defining an observable, that we denote by C. Let us remark that we have been here a bit more careful than von Neumann. We do not claim that C is an observable for any couple (A, B). We just consider a couple such that C is indeed an observable. We then have:

$$C = A + B \qquad (10.1)$$

The additivity postulate tells us that:

$$\langle C \rangle = \langle A \rangle + \langle B \rangle \qquad (10.2)$$

in which $\langle X \rangle$ is the expectation value of X. In quantum mechanics, the expectation value of X for a given wave function Ψ can be evaluated according to the following rule:

$$\langle X \rangle = \int \Psi^* X \Psi d\mathbf{r} \tag{10.3}$$

in which Ψ^* is the complex conjugate of Ψ. Then, the additivity postulate is seen to be always true in quantum mechanics, since:

$$\langle C \rangle = \int \Psi^* C \Psi d\mathbf{r} = \int \Psi^* (A + B) \Psi d\mathbf{r} = \int \Psi^* A \Psi d\mathbf{r}$$

$$+ \int \Psi^* B \Psi d\mathbf{r} = \langle A \rangle + \langle B \rangle \tag{10.4}$$

In particular, it is true whatever the value of the commutator $[A, B]$, that it so say it does not depend on whether the operators A and B commute or not. Obvious as it seems to be, the additivity postulate is not trivial. This can be exemplified by noting that it does not necessarily apply to Eigenvalues. To see this, let us consider a counter-example taken from Jammer [24]. Let us consider the spin component of an electron in the direction along the bisector line between the x-axis and the y-axis. This observable is represented by the operator:

$$S_{45^0} = \frac{1}{\sqrt{2}} (\sigma_x + \sigma_y) \tag{10.5}$$

The outcome of the measurement of S_{45^0} for the electron, a spin 1/2-particle, may be either $\hbar/2$ or $-\hbar/2$, that is to say the Eigenvalues are ± 1, in units $\hbar/2$. This is different from $(\pm 1 \pm 1)/\sqrt{2}$ that we would obtain if we applied the additivity postulate to the right-hand side of Eq. (10.5). Therefore, indeed, Eigenvalues do not combine linearly. Nevertheless, we still have, as we should:

$$\left\langle \frac{1}{\sqrt{2}} (\sigma_x + \sigma_y) \right\rangle = \langle \pm 1 \rangle = 0 \tag{10.6}$$

$$\frac{1}{\sqrt{2}} (\langle \sigma_x \rangle + \langle \sigma_y \rangle) = \frac{1}{\sqrt{2}} (\langle \pm 1 \rangle + \langle \pm 1 \rangle) = 0 \tag{10.7}$$

that is to say:

$$\langle S_{45^0} \rangle = \frac{1}{\sqrt{2}}(\langle \sigma_x \rangle + \langle \sigma_y \rangle) \qquad (10.8)$$

Although the additivity postulate (which has to be true for commuting as well as for non commuting observables) is satisfied by Eq. (10.8), this example has something specially interesting, namely the fact that σ_x and σ_y precisely do not commute. For commuting variables, the additivity postulate may hold for Eigenvalues having well defined values for common Eigenstates and, in such a case, it follows immediately that it holds too for expectation values. In contrast, we just have exhibited an example, with non commuting variables, where it does not hold for Eigenvalues, although it holds for expectation values.

As stated by Bell [188]: *A measurement of a sum of noncommuting observables cannot be made by combining trivially the results of separate observation on the two terms — it requires a quite distinct experiment. For example the measurement of σ_x for a magnetic particle might be made with a suitably oriented Stern–Gerlach magnet. The measurement of σ_y would require a different orientation, and of $(\sigma_x + \sigma_y)$ a third and different orientation. But this explanation of the nonadditivity of allowed values also established the nontriviality of the additivity of expectation values. The latter is a quite peculiar property of quantum mechanical states, not to be expected a priori. There is no reason to demand it individually of the hypothetical dispersion-free states, whose function is to reproduce the measurable peculiarities of quantum mechanics when averaged over.*

Another example, similar to the one above (picked up from Jammer), but with a different flavour, is available from Bohm and Hiley [68]. Let us take the time to explain it due to its complementary interest. Let us consider a particle whose observables (or operators) for the components of the orbital angular momentum are denoted as L_x, L_y and L_z, with usual notations. Let us also restrict ourselves to the case when the Eigenvalues are $\hbar, 0, -\hbar$. For A and B of Eq. (10.1), let us take $A = L_x/\sqrt{2}$ and $B = L_y/\sqrt{2}$. Let now L_{45^0} be the operator associated to a measurement of the orbital angular momentum component along the bisector line between the x-axis and the y-axis. We have:

$$L_{45^0} = \frac{1}{\sqrt{2}}(L_x + L_y) \tag{10.9}$$

so that L_{45^0} is the operator C of Eq. (10.1). The additivity postulate, valid for expectation values in quantum mechanics, tells us that we indeed may write:

$$\langle L_{45^0} \rangle = \frac{1}{\sqrt{2}}(\langle L_x \rangle + \langle L_y \rangle) \tag{10.10}$$

Now, let us assume the existence of dispersion-free sub-ensembles in which the observables L_x, L_y and L_{45^0} have well defined values $V(L_x), V(L_y)$ and $V(L_{45^0})$ respectively. If these values satisfy the additivity postulate, we should have:

$$V(L_{45^0}) = \frac{1}{\sqrt{2}}\left[V(L_x) + V(L_y)\right] \tag{10.11}$$

Now, let us assume for instance that $V(L_{45^0}) = \hbar$, one of the allowed values indeed. Next, we have $V(L_x) = \hbar, 0, -\hbar$ and also $V(L_y) = \hbar, 0, -\hbar$. We can then see that there is no way to satisfy Eq. (10.11). Hence, there is no dispersion-free states, and hidden variables do not exist. This is substantially von Neumann's proof specified for a convenient example.

The rebuttal, again specified for this example, is as follows. First, in the words of Bohm: *As is well known (and as von Neumann agrees), there is really no meaning to combine the results of non-commuting operators such as L_x, L_y and L_{45^0}. These measurements are incompatible and mutually exclusive.* Nevertheless, Eq. (10.10) for expectation values is still true, due to the validity of the additivity postulate in quantum mechanics, even if it requires the use of three separate mutually exclusive series of experiments. Now, from the fact that Eq. (10.11) cannot be satisfied for dispersion-free sets, we should not conclude that dispersion-free sets do not exist. Conversely, we have to conclude that the additivity postulate does not apply to such sets. Indeed the *true* values V's are not of a quantum nature. The additivity postulate is a peculiarity of quantum mechanics which need not to be satisfied by the true values of hidden-variables theories.

From the previous arguments, the mistake of von Neumann should now be clear and is indeed clearly identified. Von Neumann unduly extended an additivity postulate from quantum mechanics, where it is valid but non trivial, to the realm of hidden variables where it is trivially non valid. The same issue of an undue extension is also put forward by Mugur-Schächter, although in the context of an argumentation which is more logical than physical, and shed complementary light to illuminate the landscape.

Mugur-Schächter's rebuttal of von Neumann

I believe that PEDESTRIANS ($\sim\sim\sim\sim$) are allowed to skip this sub-subsection too, just accepting the fact that Mugur-Schächter provided a rebuttal to von Neumann's proof. They will find however a very short comment, worth visiting, at the end of it. Land on there.

Two years before Bell's rebuttal in 1966 [188], another rebuttal had already been provided, in 1964, by Mugur-Schächter in his thesis [291], more precisely in the first part of her thesis (the second unfortunate part contains speculations on non-quantum measurements, that we already discussed). This work has been less influential than Bell's one, but I must confess, without any disrespect concerning Bell, that I found it particularly impressive and convincing. What did Mugur-Schächter is to analyze the logical structure of von Neumann's proof in order to exhibit that the logic used is inconsistent. We may say that the deep physical content of Mugur-Schächter's rebuttal is similar to Bell's one, but it provides a quite different, but complementary, point of view by insisting on the flawed logical organization of von Neumann's proof. The thesis, interestingly enough, is prefaced by Louis de Broglie who stated that, since now several years, there were already doubts rising against von Neumann's proof and that he already, himself, came to the conviction that von Neumann's reasoning was misleading and circular. The circularity of von Neumann's proof lies in the fact that he implicitly introduced, in his premises, the result that he intended to demonstrate. Louis de Broglie also stated that Mugur-Schächter's work achieved a genuine logical dissection of von Neumann's proof and rigorously demonstrated its fallacious character (For Jammer however [24], the charge of circularity is not justified).

I am now going to present the demonstration of Mugur-Schächter. However, I shall simplify it a bit without, I hope, destroying the gist of it. Also, I shall make a bit of rewording to better match the terminology used nowadays. Mugur-Schächter starts with a logical analysis of von Neumann's proof, which can be summarized in a few steps.

To set down the stage, let us consider a wave function Ψ and N realizations of Ψ, i.e. N copies of Ψ named $\Psi_1, \Psi_2, \ldots, \Psi_N$ (in other words, we consider a statistical set $\{\Psi_1, \Psi_2, \ldots, \Psi_N\}$). Let us also consider an observable R (any observable) that we can measure on Ψ. From the copies, we may then evaluate the expectation value $\langle R \rangle$ of R which actually can also be quantum mechanically evaluated by using Eq. (10.3), or its more general formulation expressed in Dirac notation as:

$$\langle R \rangle = \langle \Psi \mid R \mid \Psi \rangle \tag{10.12}$$

We say that the statistical set is dispersion-free if, for any (the word "any" is very important) R, we have:

$$\langle R^2 \rangle = \langle R \rangle^2 \tag{10.13}$$

Now, the steps to be considered are as follows.

 (i) In quantum mechanics, there is no dispersion-free set (just think of Heisenberg uncertainty relations). This is in particular true for pure states when, and only when, the density operator is a projector [43].

(ii) Assume that we possess hidden variables associated with the behavior of quantum objects. Then, a pure state set can be decomposed into dispersion-free subsets by sorting them out according to the values of the hidden variables.

(iii) But, by (i), there is no dispersion-free set, hence a contradiction which establishes that hidden variables do not exist. As a corollary, quantum mechanics is intrinsically indeterministic.

The rebuttal may proceed as follows. Surely, items (i) and (ii) are correct. But the point is that item (ii) does not apply to quantum mechanics. Indeed, hidden variables are not necessarily of a quantum nature, i.e. they are not necessarily compatible with the postulates of quantum mechanics. In particular, they are not necessarily distributed according to the probability rules of quantum mechanics. For instance, in the pilot wave theory, hidden variables are of a classical kind, insofar as they are associated with classical (or pseudo-classical) deterministic trajectories, whose existence is rejected in the framework of quantum mechanics. They are therefore not distributed according to Born's postulate which applies to the guiding wave Ψ, or say to the observed values of positions, but not to the hidden trajectories, or say to the objective values of positions.

We may also say that the hidden variables, being of a classical kind, are defined *outside* of the logical framework of quantum mechanics in which measured values of observables are of a quantum type. In other words, the theory T_{HV} made out from quantum mechanical theory T_{QM} and from its completion with hidden variables is larger than quantum mechanics ($T_{HV} > T_{QM}$). This is a convenient way to say that T_{HV} does not operate inside T_{QM}, or that it is not internal to T_{QM}. Therefore, item (i) pertains to T_{QM} and item (ii) to T_{HV}. Then, item (iii) is faulty because it unduly transports an element of a smaller theory in the structure of a larger theory.

PEDESTRIANS ($\sim\sim\sim\sim$), start here

Borrowing a question to the title of a paper from Pinch [154], we then may ask: *What does a proof do if it does not prove?* Well, the answer might be that von Neumann's proof indeed did not prove what it was aiming to prove but, yet, it is proving something. It is proving that, if hidden variables do exist, they cannot be of a quantum type because, if they were of a quantum type, then von Neumann's proof would apply to them. This restricts the class of admissible hidden-variables theories, but does not rule them out. The restriction of the class of admissible hidden-variables theories, as we shall see, will require other conditions: They will have to be contextualist and non-local, as is the case for the early Bohm's pilot wave.

Mugur-Schächter went on discussing other proofs against the proof, from Bohm, Weizel, Fenyes, Bocchieri and Loinger, and Louis de Broglie, pointing out their inadequacies. I can, without any harm, leave them out. Complementary discussions on von Neumann's proof are available from Louis de Broglie [62, 63, 65], Pauli [164], Bohm [60], Bohm and Bub [297], Bohm and Hiley [68], Siegel [303], Ballentine [304], Belinfante [1], Jammer [24], Flato *et al.* [305], Pinch [154], Pipkin [156], Bell [142], Wheeler and Zurek [100], Squires [13], Mermin [197], Bitbol [133], d'Espagnat [84]. This rather significant, but nevertheless likely incomplete, list of references is certainly a testimony of the fame of von Neumann's impossibility proof for the debate on hidden variables.

Von Neumann still worrying about hidden variables

Once things are published, they are published. And some authors may feel, soon or later, that they did not make it in the best way. They may find that such sentence is not properly enough phrased, or such demonstration could have been made shorter and more elegant, or even worst, there has been an error, here or there, a real nightmare for theoretical physicists who like perfection. But it is a fact and an ultimate fate that everything which has been produced in theoretical physics is wrong, waiting for the mistakes to be pinpointed and corrected. If there is only one truth, it is a matter of elementary computations in probability theory to demonstrate that the probability of having reached the truth is zero. This paragraph applies in particular to von Neumann's proof, who did publish an erroneous proof.

As far as I know, we do not possess any information concerning what von Neumann eventually thought of von Neumann's proof. But it seems well established that he went on worrying about hidden-variables theories. In any case, according to Wigner [74, 306], von Neumann possessed another objection, although unpublished, against hidden variables. To explain this objection, let us for instance consider a Stern-Gerlach experiment, or more precisely an indefinite number of repetitions of Stern-Gerlach experiments. We begin, for instance, with a measurement of a spin component along the z-direction, followed by a measurement of a spin component in the x-direction, followed by still another measurement of a spin component along the z-direction, followed again by another measurement of a spin component along the x-direction, and so on ... Let us now assume again, for instance, that we are dealing with spin 1/2-particles.

The first measurement is used to prepare the system in a pure state, say spin up in the z-direction. Once this is done, the outcomes of the second measurement can be spin forward or spin backward along the x-direction, both with probabilities 1/2. If these values are deterministically determined by hidden variables, the result obtained (say $+$ or $-$ along the axis $0x$) provides a restrictive information on the values of the hidden variables. Next measurement, now with probabilities 1/2 for both spin up or spin down (along the z-direction) will restrict further the values of the hidden variables. Eventually, after N measurements (N very large but finite), the hidden variables will become restricted to a very narrow range. Yet, this narrow range should still be able to completely determine all further measurements, after the N first measurements, even if the number of further measurements is assumed to be infinite (we do not worry with the fact that it is in practice impossible to carry out an infinite number of measurements). Is this possible? Well, for von Neumann, it seemed unreasonable.

According to Wigner [306], it seems that von Neumann was thinking in terms of a variety of hidden variables to determine the outcomes of the measurements, not in terms of a single variable. Then von Neumann would have pointed out *to the unreasonable large variety of hidden variables which must be assumed if one wishes to account for the postulate (implicit in quantum mechanical theory) that no matter how many successive measurements we undertake of a system, the distribution of the hidden variables remains sufficiently unsharp so that the outcomes of measurements are as unpredictable as they were to begin with.*

It is difficult to know why von Neumann did not publish his proof. A likely possibility is that, although the proof seems intuitively convincing, von Neumann did not succeed to give it a satisfactory mathematical shape. Also, we may advance as a guess that, when trying to make it mathematically safe, he found that the proof was basically flawed. I am daring to propose a rebuttal that, may be, von Neumann eventually discovered. Let us begin by assuming that each hidden variable in the variety of hidden variables is continuous. Then, we may invoke a Cantor-Bernstein theorem, telling us that $\mathbf{R}^n$ is equipotent to $\mathbf{R}^m$, whatever n and m being positive integers [307]. Let n denote the number of hidden variables in the von Neumann variety of hidden variables, and m be equal to 1. Then, the Cantor-Bernstein theorem implies that we may reduce the von Neumann variety to a single hidden variable ranging over $\mathbf{R}$. This reduction is not strictly necessary for the rebuttal, but it is useful to support the intuition. Now the cardinal of $\mathbf{R}$ is infinity (not a countable infinity however, but the infinity of the power of continuum). Therefore, there is plenty of room in $\mathbf{R}$ to deal with an infinite number of successive measurements, and von Neumann-still-worrying is rebutted.

The full real line is even not necessary to develop the argument. For instance, let us assume that the single hidden variable does not spread over $\mathbf{R}$ but is instead distributed on the open segment $]0, 1[$. After N successive measurements, the hidden variable may have been restricted to a number of open segments located inside the original unit open segment, with children segments narrowing more and more when the number N of measurements increases. It is however a property of the power of continuum that, whatever N, whatever the number of children segments, and whatever the narrowness of these segments, the cardinal of the set formed by all children open segments remains strictly equal to the cardinal of the original unit open segment.

Next, as another possibility, let us assume that each hidden variable in the variety of von Neumann may take an infinite countable number of values. The argument used above for continuous variables may be, *mutatis mutandis*, repeated to this new case by relying on the equipotence between $\mathbf{N}^n$ and $\mathbf{N}^m$. Finally, the rebuttal may be made complete, in a similar way, if we assume that the variety of von Neumann is made out from a mixture of continuous and countable hidden variables. I believe that von Neumann, although he had a new objection in mind, soon realized that it could not be mathematically defended.

Early other proofs of impossibility

There are also early other proofs of impossibility which however have never been so famous and disputed as von Neumann's proof. One year after von Neumann's proof appeared in the literature, Solomon [308] also examined the possibility of hidden-variables theories and concluded that any deterministic hidden-variables theory is incompatible with the intrinsic indeterminacy of quantum mechanics. Another demonstration of the fact that quantum mechanics is intrinsically indeterministic was given by Destouches-Fevrier in 1945 [309, 310], with the conclusion that it is no more possible to return to determinism in the microphysics realm. Later on, one year after the publication of the 1952 early Bohm's papers on pilot wave, Destouches [202] took as guaranteed the results of von Neumann and Solomon, gave a rebuttal to a recent objection from Louis de Broglie [152], and concluded that quantum mechanics is essentially indeterministic. He affirmed that any future theory which would replace the present quantum theories would have to be indeterministic too.

I do not know any attempt to refute these works, a kind of indifference which might be easy to explain. First, these new proofs of impossibility, dated 1933, 1945 or 1953, when von Neumann was still largely dominating the stage, were likely found to be neither interesting nor worthwhile to be deeply examined. Second, later on, and nowadays, it may be simply sufficient to state that we possess a convincing rebuttal of von Neumann's proof of impossibility and, more important, that we possess a consistent example of deterministic hidden-variables theories, namely the early Bohm's pilot wave. Therefore, without any doubt, these early other proofs of impossibility must be flawed somewhere. I shall be content in this book with this expedient point of view, although a more careful examination of these early other proofs, to identify where the shoe pinches, could deserve a bit of effort.

Late other proofs of impossibility

Even after 1952 Bohm's pilot wave, and after the criticisms already addressed against von Neumann, other proofs of impossibility have been published. In dealing with these new proofs, some authors considered themselves in the filiation from von Neumann, and were aiming to improve an imperfect proof. After all, may be von Neumann's demonstration was wrong, but this does not mean that the conclusion was wrong too. It could be right, and an improved demonstration could be able to establish this point correctly. The persistence of such an effort to get rid of hidden variables might seem to be welcomed, particularly from those who were not aware of the existence of Bohm's pilot wave ... or simply ignored it. Indeed, if these damned hidden variables do not exist, we better should be able to prove it, and to get rid of a long-lasting and irritating issue. But Bohm was living on the stage and, simply, could not be ignored. The effort to find new proofs of impossibility which, taking account for the existence of Bohm, was doomed to failure, seemed weird to Bell. For instance, he could ask [142]: ... *extraordinarily, why did people go on producing impossibility proofs, after 1952 ...?*, and he could also speak of *the strange story of the von Neumann impossibility proof* and of *the even stranger story of later impossibility proofs*.

The topic of late other proofs of impossibility is however not very easy, not easy-reading and not easy-reporting, particularly when they need to invoke logical-mathematical approaches, the propositional calculus of quantum mechanics, and when they use swearwords such as lattice of propositions or ortho-complemented lattice, or worst (swearwords which are more familiar to mathematicians than to physicists). Furthermore, as we shall illustrate, the literature is often confusing, polemical and contradictory. The reader wanting to quickly have a flavour of the topic may refer to a collection of papers on the logico-algebraic approach to quantum mechanics, edited by Hooker [136]. The second paper, by Strauss [311], leads to the result that *all attempts at interpreting the quantum mechanical formalism in terms of classical probabilities ... are doomed to failure*, a way to dismiss hidden variables. Conversely, in a subsequent paper of the same collection, Kamber [312] concluded that *the existence of hidden parameters ... cannot ... be excluded ... mathematically.* These two quotations illustrate the contradictions that we may have to face.

The story goes on in the same style. An early work among the late works is by Gleason [313] on the basis of which Jauch and Piron [314] thought that they had dismissed the possibility of hidden variables. The works of Gleason, and of Jauch and Piron, are discussed in a simplified manner by Belinfante [1]. Also, other comments on these works are available from Shimony [93]. However, Belinfante, considering that Gleason's proof was rather abstract, preferred to derive Gleason's result from a later more understandable work by Kochen and Specker, to which we shall return soon. Bohm and Bub [315] provided a rebuttal of Jauch and Piron, rebutted by Jauch and Piron, followed by a rebuttal of the rebuttal of the rebuttal by Bohm and Bub again [316]. The refutation of Jauch and Piron by Bohm and Bub is also discussed by Gudder [317]. In 1966, Bell, in the same paper in which he pointed out the faulty axiom in von Neumann's proof [188], also criticized the proofs from Gleason, and Jauch and Piron, but made clear, relying on Gleason's work, that a class of hidden-variables theories, possibly called non-contextualist theories (we shall discuss them), are inconsistent, more specifically inconsistent for quantum systems whose Hilbert spaces are more than two-dimensional (dimension three or greater). In his paper, Bell took the opportunity to provide a new proof of Gleason's result, which is independent of the scheme used by Gleason. This is similar to the result to be soon later obtained by Kochen and Specker [28]. These impossibility proofs are actually no-go theorems (do not go to the classes of forbidden hidden variables). In the words of Mermin [197], the results obtained *by demonstrating that a hidden-variable program necessarily requires outcomes for certain experiments that disagree with the data predicted by the quantum theory, are called no-hidden-variables theorems (or, vulgarly, "no-go theorems").*

Anyway, it should however be now clear to the reader that the so-called proofs of impossibility cannot prove the impossibility of something which, with Bohm, has been constructively shown to be possible. What they can however do, when properly examined, is to exclude some classes of hidden-variables theories. Focusing on this point of view, Bell examined which classes of hidden-variables theories are actually ruled out by these proofs. He emphasized that these classes form a rather small subset of all possible hidden-variables theories. Jauch and Piron's proof, as von Neumann's proof, required the additivity of expectation values of non-commuting variables which, as we have seen, is valid in quantum mechanics, but need not to be applied to hidden variables. Gleason considered theories in which the outcome of a measurement is independent of which compatible observables were simultaneously measured. Such theories are called non-contextual theories. But here again, this was an undue restriction. Hidden variables theories need not to be non-contextual theories. Conversely, they have to be contextual. The rather subtle issue of contextuality will be examined and discussed in the next chapter. As stated by Shimony [93]: *Bell (1966) noted however that Gleason's theorem does not preclude a more complex type of hidden-variables theory, called "contextual" according to which a complete state assigns a definite truth value to a proposition only relative to a specific context.* Bohm and Bub [315] also contradicted Jauch and Piron, saying that *they actually prove nothing at all* and that the argument of Jauch and Piron is circular, an accusation already addressed previously to von Neumann's proof. In a subsequent paper, Jauch and Piron [318] provided a rebuttal to this rebuttal, something like turning down the objection point-blank, being content with a *restatement of our result in a nontechnical language.* See also Jauch [319] for a book whose chapter 7 is dedicated to hidden variables. In this chapter, Jauch maintains a theorem according to which the existence of hidden variables (*of a certain kind*) is in contradiction with empirical facts. This is soft enough due to the cautious expression: *Of a certain kind.* Indeed, what happens most generally for proofs of impossibility is that they refer to certain definitions or premises which imply the impossibility of hidden variables when these definitions or premises are accepted. However, it happens that the use of other definitions or premises might be allowed, so that proofs of impossibility only refer to *a certain kind* of hidden variables.

A second somewhat parallel line of exposition starts with Kochen and Specker [28] who *give a proof of the nonexistence of hidden variables*. Actually, once again, they did not prove what they claimed to prove, but they proved that hidden-variables theories must be contextual, a result already obtained ten years before by Gleason, although by different means. As stated by Freedman and Holt [300], they *considered only theories in which the outcome of a measurement was independent of which compatible observables were simultaneously measured*, that is to say non-contextual theories. It is however acknowledged that the proof of Kochen and Specker is complicated (indeed, it is). For an easier path, the interested reader should better turn himself to Belinfante [1] which provided a simple enough discussion concerning contextuality. In plain terms, Belinfante remarked that the argument used by Kochen and Specker (and also by Jauch and Piron) simply denies that the way in which a measurement is made could have an effect on the result of the measurement, a very strong hypothesis indeed! A conclusion may be borrowed to Jammer [24]: *Kochen and Specker have proved once again that noncontextual hidden variables do not exist in quantum mechanics.*

For another line of exposition, let us consider a work from Gudder [320]. Relying on previous works, in particular on Jauch and Piron, Gudder examined proofs of impossibility, in the framework of formal logic, using the swearword concept of an ortho-complemented partially set which is complete with respect to compatible elements. Although the content is frightening-reading, the conclusion can fortunately be easily delivered as follows: A quantum system admits hidden variables if and only if it acts physically as an ideal classical system. Hence, because it is well known that quantum systems do not behave like classical systems (just think of Heisenberg uncertainty principle), hidden variables must be excluded from quantum mechanics. We identify here again something which is like a circle, as in von Neumann's proof. An heuristic rebuttal can be done by referring to the example of Bohm's pilot wave. In this pilot wave, there are two levels, the higher level of quantum mechanics, and the lower level of hidden variables. The higher level, the one of quantum mechanics, being not of a classical nature, indeed does not receive hidden variables. The lower level is the one of the hidden trajectories. What did Bohm is to succeed to connect the two levels which, somehow, can be considered each of them in their own right. At least, this is the result of my effort to dismiss Gudder in a simple, and I hope essentially, correct way, avoiding any technicality. A bit later on, Gudder [321] himself remarked, taking into account the fact that Bohm's theory had never been shown to be inconsistent, that *clearly, there is something wrong here. One obviously cannot have a HV* (Hidden Variables) *theory if it is impossible.* Going on further, Gudder eventually demonstrated that hidden variables are always possible.

Bub [322], facing such a confusion, took the expedient but rather arbitrary point of view that *the term hidden-variables theory is justifiably used to denote the kind of theories rejected by von Neumann, Jauch and Piron, Kochen and Specker,* but *it is suggested that the term should not be used as a label for the theories considered by Bohm and other workers in the field. Such theories should be regarded as fundamentally compatible with the original Copenhagen interpretation of the quantum theory, as expressed by Bohr,* adding: *The conclusion of this paper will therefore be that there are no hidden variables theories of quantum phenomena in the usual sense, that the term hidden variables theory for the kind of theory considered by Bohm and his collaborators is unfortunate and misleading, and that this latter approach might well be characterized as an extension of Bohr's conception of wholeness as opposed to von Neumann philosophy.* This is a kind of normative attitude introducing a dichotomy between what deserves and what does not deserve to be called hidden variables, if not a kind of propaganda to shield Bohm from attacks from hidden-variables opponents. In any case, even if the concept of wholeness is somehow common to Bohr and Bohm, it is extraordinarily difficult to see Bohm as a continuator of Bohr.

And now, a fairly funny but clever thing. Greechie and Gudder [137] considered *two hidden variable proofs in the quantum logic framework, one a proof that they do not exist and one a proof that they do.* This is indeed reminiscent of a Jesuitical exercise in rhetoric, in which the same man can prove that God does not exist and, just after, with the same cleverness in persuasion, that He does exist.

At this point, is the reader confused by a literature which is indeed confusing? There are however a small number of simple key points to make him comfortable (i) proofs of impossibility can never prove impossibility (ii) they can only prove that some classes of hidden variables are impossible (iii) among the set of impossible hidden-variables theories, there are those theories which are non-contextual. This last item needs clarification. It will need a chapter, after this chapter.

More in Jauch [319], Belinfante [1], Jammer [24], Freedman and Holt [300], Flato *et al.* [323], Gréa [324], Pinch [154], Fine and Teller [325], Peres [326], or Bohm and Hiley [68].

Possibility proofs

Once again, the easiest possibility proof is simply the very existence of Bohm's pilot wave, a constructivist proof in which an edifice is constructed so that we can simply say: Just look at it. It is logically consistent, and this consistency has never been contradicted. Once we understand Bohm's theory, and understand that it is logically consistent, there is really no need to investigate further the issue. If we just want to know whether hidden-variables theories are possible or not, the issue is indeed closed. We are however going to dig a bit further, with formal approaches, to provide a still better understanding. Actually, we have many other things to learn.

PEDESTRIANS ($\sim\sim\sim\sim$) may skip the next subsection, being content with the conclusive last lines, but they should be able to deal with the next one after it, at least most of them.

A simple particular formal approach

In this subsection, we are going to provide a simple example due to Bell [188], an example which is also discussed by Flato *et al.* [323]. The reader might find that this example is too particular to be convincing. In the words of Flato *et al*, it *may be too simple*. But, even if it is too simple, it is enlightening and sufficiently illustrating to support later discussions.

Following Bell, we consider a system with a two-dimensional state space, say to be specific a particle of spin 1/2 without any translational motion, or more generally a two-dimensional system which can be reduced to a spin 1/2 (several examples in [43]). In such a case, any quantum mechanical state may be represented by a spinor Ψ which is a two-component wave function. Any observable R in the two-dimensional state space may be represented by a square Hermitian matrix of rank 2, which can always be decomposed under the form:

$$\mathbf{R} = \alpha\mathbf{I} + \boldsymbol{\beta}.\boldsymbol{\sigma} \qquad (10.14)$$

in which $\mathbf{I}$ is the unity matrix, the super-vector $\boldsymbol{\sigma}$ has for components the Pauli matrices σ_x, σ_y, and σ_z, α is taken to be a real constant, and $\boldsymbol{\beta}$ is taken to be a real vector (Bell indeed takes α and $\boldsymbol{\beta}$ as real). Explicitly, Eq. (10.14) can be rewritten as:

$$\mathbf{R} = \alpha \begin{pmatrix} 1 & 0 \\ 0 & 1 \end{pmatrix} + \beta_x \begin{pmatrix} 0 & 1 \\ 1 & 0 \end{pmatrix} + \beta_y \begin{pmatrix} 0 & -i \\ i & 0 \end{pmatrix} + \beta_z \begin{pmatrix} 1 & 0 \\ 0 & -1 \end{pmatrix}$$
$$(10.15)$$

Let us consider the rather simple case in which $\beta_x = \beta_y = 0$. Then:

$$\mathbf{R} = \alpha \begin{pmatrix} 1 & 0 \\ 0 & 1 \end{pmatrix} + \beta_z \begin{pmatrix} 1 & 0 \\ 0 & -1 \end{pmatrix} \qquad (10.16)$$

from which we see that Eigenvalues are

$$E_v = \alpha \pm |\beta_z| \qquad (10.17)$$

Physics should not depend on the orientation of the vector $\boldsymbol{\beta}$. Therefore, in the most general case of Eq. (10.15), Eigenvalues should be:

$$E_v = \alpha \pm |\beta| \qquad (10.18)$$

According to the basic rules of quantum mechanics, associated expectation values are:

$$\langle \alpha\mathbf{I} + \boldsymbol{\beta}.\boldsymbol{\sigma} \rangle = \langle \Psi, (\alpha\mathbf{I} + \boldsymbol{\beta}.\boldsymbol{\sigma}) \Psi \rangle \qquad (10.19)$$

Let us now complement quantum mechanics with a real hidden variable, denoted λ, pertaining to the interval $[-1/2, 1/2]$. A microstate is then defined by the couple (Ψ, λ) in which Ψ is the spinor. By a rotation of coordinates, the spinor can always be given the simple form

$$\Psi = \begin{pmatrix} 1 \\ 0 \end{pmatrix} \tag{10.20}$$

For such a spinor, using Eq. (10.19), quantum mechanics computation rules allow one to find

$$\langle \alpha \mathbf{I} + \boldsymbol{\beta}.\boldsymbol{\sigma} \rangle = (1\ 0)\,(\alpha \mathbf{I} + \boldsymbol{\beta}.\boldsymbol{\sigma}) \begin{pmatrix} 1 \\ 0 \end{pmatrix} = \alpha + \beta_z \tag{10.21}$$

Indeed,

$$(1\quad 0) \left[\alpha \begin{pmatrix} 1 & 0 \\ 0 & 1 \end{pmatrix} + \beta_x \begin{pmatrix} 0 & 1 \\ 1 & 0 \end{pmatrix} + \beta_y \begin{pmatrix} 0 & -i \\ i & 0 \end{pmatrix} + \beta_z \begin{pmatrix} 1 & 0 \\ 0 & -1 \end{pmatrix} \right] \begin{pmatrix} 1 \\ 0 \end{pmatrix}$$

$$= (1\quad 0) \left[\alpha \begin{pmatrix} 1 \\ 0 \end{pmatrix} + \beta_x \begin{pmatrix} 0 \\ 1 \end{pmatrix} + \beta_y \begin{pmatrix} 0 \\ i \end{pmatrix} + \beta_z \begin{pmatrix} 1 \\ 0 \end{pmatrix} \right]$$

$$= \alpha + \beta_z \tag{10.22}$$

Now, let us set

$$\left. \begin{array}{ll} \text{(i)} & X = \beta_z \text{ if } \beta_z \neq 0 \\ \text{(ii)} & X = \beta_x \text{ if } \beta_z = 0, \beta_x \neq 0 \\ \text{(iii)} & X = \beta_y \text{ if } \beta_z = 0, \beta_x = 0 \end{array} \right\} \tag{10.23}$$

and also

$$\left. \begin{array}{l} signX = +1 \text{ if } X \geq 0 \\ signX = -1 \text{ if } X < 0 \end{array} \right\} \tag{10.24}$$

Now, let us introduce microstate Eigenvalues denoted as $E_{(\Psi, \lambda)}$ defined by

$$E_{(\Psi, \lambda)} = \alpha + |\boldsymbol{\beta}|\, sign \left(\lambda |\boldsymbol{\beta}| + \frac{1}{2} |\beta_z| \right) signX \tag{10.25}$$

in which the vector $\boldsymbol{\beta}$ is still given by $(\beta_x, \beta_y, \beta_z)$, but in the new coordinate system in which the spinor Ψ reduces to the form of Eq. (10.20). This vector being given, the dispersion-free microstate

(Ψ, λ) completely determines the microstate Eigenvalues $E_{(\Psi,\lambda)}$. If we were able to measure the microstate, particularly the value of λ, the probability for the associated microstate Eigenvalue would be exactly equal to 1, in contrast with the probabilities associated with quantum Eigenvalues.

The philosophy of hidden-variables theories is that quantum expectation values are obtained by averaging over hidden variables. Using an uniform averaging over λ, we are then led to the evaluation of an integral over the microstate Eigenvalues, according to

$$\langle \alpha + \boldsymbol{\beta}.\boldsymbol{\sigma} \rangle = \int_{-1/2}^{+1/2} [\alpha + |\boldsymbol{\beta}| \, sign \left(\lambda \, |\boldsymbol{\beta}| + \frac{1}{2} \, |\beta_z| \right) signX] d\lambda \quad (10.26)$$

After a simple evaluation (but better start with case (iii) of Eq. (10.23) to check), we obtain

$$\langle \alpha \mathbf{I} + \boldsymbol{\beta}.\boldsymbol{\sigma} \rangle = \alpha + \beta_z \quad (10.27)$$

that is to say the same result than with quantum mechanics.

PEDESTRIANS ($\sim\sim\sim\sim$), reconnect here.

We therefore possess a deterministic hidden-variables model, based on a hidden variable λ, which reproduces quantum mechanical predictions. There is however an important difference with respect to Bohm's pilot wave, namely the fact that λ does not receive any physical meaning, so that this hidden-variables model does not allow one to propose any new interpretation of quantum mechanics. Such hidden variables, we may call them dummy hidden variables (some people might prefer to call them artificial hidden variables, and this is actually the terminology we shall later retain).

A simple general formal approach

The reader might worry that the previous model has been developed in a very special case (spin 1/2) and that, may be, it would be impossible to generalize it. As it is, it might even look a bit *ad hoc* and contrived. But it is a fact that one formal example is sufficient to demonstrate that hidden variables are possible, simultaneously ruining all pretensions of proofs of impossibility. But we may have a stronger result: Not only hidden variables are possible, but they are *always* possible. Bell [327] stated this result as follows: *If no restrictions whatever are imposed on the hidden variables, or on the dispersion-free states, it is trivially clear that such schemes can be found to account for any experimental results whatever. Ad hoc schemes of this kind are devised every day when experimental physicists, to optimize the design of their equipment, simulate the expected results by deterministic computer programs drawing on a table of random numbers,* adding however: *Such schemes ... are not very interesting. Certainly what Einstein wanted was a comprehensive account of physical processes evolving continuously and locally in ordinary space and time.* The same argument has been served again in 1976 [328]: *That the apparent indeterminism of quantum phenomena can be simulated deterministically is well known to every experimenter. It is now quite usual, in designing an experiment, to construct a Monte Carlo computer programme to simulate the expected behavior ... Every such programme is effectively an ad hoc deterministic theory, for a particular set-up, giving the same statistical predictions as quantum mechanics.*

According to Wigner [306], *it is rather obvious that, given any quantum mechanical measurement represented by the operator Q one can introduce a hidden-variable model.* This can be done as follows. Together with any operator Q, and associated measurements, let us introduce a hidden variable denoted q. Let us demand that the statistical distribution of q reproduces the probabilities for the various possible (Eigenvalues) $\lambda_1, \lambda_2, \ldots$ of the quantum measurements for Q, and examine whether this is always possible. The answer is positive. In order to demonstrate this, we biunivocally associate domains $D(\lambda_1), D(\lambda_2), \ldots$ of q with each possible measurement outcomes $\lambda_1, \lambda_2, \ldots$ Let us now consider a state $|\Psi\rangle$ and a distribution $P_\Psi(q)$. We then postulate that the distribution $P_\Psi(q)$ assigns a probability to the domain $D(\lambda_i)$ which is equal to the probability that the measurement of Q on Ψ yields the value λ_i. We are done.

This may be generalized to the case when we are dealing with several operators $Q_1, Q_2, \ldots$ It suffices to introduce a hidden variable

q_j for each operator Q_j. Assume for instance that the spectra of all operators are discrete. Hence, let us call λ_{ij} the ith Eigenvalue ($i = 1, 2, ..., N_j$) of the jth operator ($j = 1, 2, ..., N$). To each discrete value λ_{ij}, we can associate a hidden variable q_{ij} in such a way that the value of q_{ij} determines λ_{ij}. If we associate a probability $P_\Psi(q_{ij})$ to each q_{ij}, we may adjust the $P_\Psi(q_{ij})$'s to recover the various probabilities for the λ_{ij}'s. Of course, for each j, the sum of all probabilities over i must be equal to 1.

Let N_{hv} be the number of hidden variables required for the process. We may use N_j hidden variables for the jth operator having N_j Eigenvalues. And the summation of all the N_j's over $j = 1, 2, ..., N$, that is to say for all operators, will provide a value for N_{hv}. We then have a list of hidden variables q_{ij} which may be arranged by using a single integer s ranging from 1 to N_{hv}. This means that the number of hidden variables can actually be reduced to 1, denoted by s, taking N_{hv} possible different discrete values. If we are dealing with continuous spectra, j still takes on discrete values, but the discrete index i is to be replaced by a continuous index. We are then dealing with a number N of continuous segments which, by invoking again the Cantor-Bernstein theorem, can be merged into one single segment. Therefore, once again, the number of hidden variables may be reduced to 1. Similar considerations may be used for mixed spectra. Then, one can reduce all discrete spectra to one discrete hidden variable, and all continuous spectra to one continuous hidden variable. The number of hidden variables has then be reduced to 2. Whether it is possible to achieve a further reduction from 2 to 1, this is something that I am not able to answer.

The previous discussion is similar to the one we used when dealing with von Neumann's consecutive measurements. Concerning also von Neumann, let us remark that the procedure we used here is the one also used for the construction of dispersion-free statistical sets which, according to von Neumann's proof, are forbidden. Indeed, in his book [26], von Neumann was already aware of such a possibility for a decomposition of a statistical set into dispersion-free subsets. But, ironically enough, he did not use this fact to conclude that hidden variables are always possible. Instead, as we know, he ended with the claim that such decompositions are actually not possible, and that hidden variables must always be excluded.

Wigner [306] stated that the number of hidden variables increases enormously when we assume an increasing number of operators and of consecutive measurements. I hope having correctly proven that such is not the case. It remains however true that hidden variables introduced as above do not receive any physical interpretation. Therefore, they do not permit alternative interpretations of quantum mechanics, and look *ad hoc* and artificial. Certainly, this is a kind of hidden variables that most physicists would not like to consider. Whether this difficulty can be overcome, this is something I do not know. It might simply be *a lack of imagination,* in the words of Bell [142]. But, anyway, as stated by Flato *et al.* [323]: ... *it is quite trivial to construct theories containing hidden variables which do not have any physical meaning. The real problem will of course be to construct theories with hidden variables having physical meaning, capable of reproducing all known quantum-mechanical results and possessing also predictions in domains not yet covered by quantum mechanics. No example of such a theory is known to our day* (and, as far as I know, still today).

Gudder [321] explains the contradiction between proofs of impossibility and possibility proofs in the following terms: *The proponents of hidden variable theories have an idea of what these theories should be and have given examples of such theories. The antagonists have a different idea of what a hidden variable theory should be and have proved that such theories are impossible in the present general framework of quantum mechanics. These proofs are irrelevant since they do not refer to the hidden variable theories as formulated by the advocates of these theories.* To avoid any ambiguity, he then provided a definition of what is a hidden-variables theory, which is in Bohm's spirit. Now that we have in mind several examples of hidden-variables theories, it is appropriate to precisely define what a hidden-variables theory is, and this can be done in a concise and perfect way, following Gudder, as follows: *The state m of a quantum mechanical system is not complete in the sense that another variable ξ can be adjoined to m so that the pair (m, ξ) completely determines the system. That is,* *a knowledge of* (m, ξ) *enables one to predict precisely the outcome of any single measurement. Furthermore, an average of* (m, ξ) *over the values of* ξ *gives the usual quantum state* m. Gudder afterward demonstrated that, in a certain sense, a hidden-variables theory is always possible (although the usual, even clever, physicist might feel distressed by some mathematical statements, heavily relying on a calculus of propositions, used to establish the result). Gudder's work is acknowledged by Holland [40], saying: *It proved possible to show that quantum mechanics can always be supplemented by hidden variables.* The paper of Gudder has been complemented by Greechie and Gudder [137]. See also Fine [329] who agreed with the statement that we may always build a deterministic hidden-variables model which reproduces quantum mechanical results.

Chapter 11

Contextuality

We already met the concept of contextuality in several places, when we stated and loosely explained that the Bohm's pilot wave is a contextual theory, but also regarding the proofs of impossibility of Gleason, and Kochen and Specker, when we restated their results, saying that, actually, what they proved is that admissible hidden-variables theories have to be contextual. This result may be viewed as a theorem, often named KS-theorem, or for reasons that we shall make clear Bell–KS theorem. There is another important result, that admissible hidden-variables theories must be non-local, a result which may also be viewed as a theorem, named Bell's theorem (much more famous than KS-theorem). Non-locality will be discussed in the next chapter. In the present chapter, we are going to deal with contextuality (a concept that we have now to understand clearly), and with the proof that hidden-variables theories must be contextual, that is to say: The properties assigned to hidden variables must be determined by the experimental context of the measurement.

Kochen and Specker for everyday cyclists

Cyclists go faster than PEDESTRIANS ($\sim\sim\sim\sim$) but climb lower than mountaineers. I believe that cyclists can do with this section, but I am not sure about the fate of PEDESTRIANS ($\sim\sim\sim\sim$). Anyway, they may try. Otherwise, considering that the next section is somehow in the same mathematical mood than the present one, they better have to go directly to the next chapter, just taking a bit of time to gain some amount of information, here, or there, from a few useful comments.

As we stated, Kochen and Specker's proof is difficult. Belinfante produced a simplified proof but it is still too involved for an easy presentation, in a reasonable number of pages (at least in the framework of this book). Fortunately, we also possess a still simpler version exposed by Bitbol in the Annex III of one of his books [133]. This is like a kind of green lane for everyday cyclists. You do not need to be a champion having to climb the Galibier or Tourmalet passes.

Following the green lane, let us consider spin-1 particles. When we measure a component of the spin in any direction, the possible outcomes are: $-1, 0, +1$ (with $\hbar$-units) producing three spots in a Stern–Gerlach experiment. This is in particular true if measurements are made along the directions x, y and z, with associated observables denoted S_x, S_y and S_z respectively. Hence, if we measure the squares of the components in any direction, in particular S_x^2, S_y^2 and S_z^2, there are only two possible outcomes, namely 0 and 1. Let us now consider the observable:

$$\Sigma = S_x^2 + S_y^2 + S_z^2 \qquad (11.1)$$

For a spin $s = 1$, the result of the measurement of Σ is $s(s + 1) = 2$. If we admit hidden variables, these hidden variables must determine the values (a concise way to designate: The values of outcomes of measurements) of S_x, S_y and S_z. Therefore, they must also determine S_x^2, S_y^2 and S_z^2. Then, because each component S_i, $(i = x, y, z)$, may have the values $-1, 0, 1$, and S_i^2, $(i = x, y, z)$, may have the values $0, 1$, we conclude that two of the values of S_i^2 must receive a value equal to 1, and the third one must receive a value equal to 0, so that, as required, the sum of the three values (Σ) is equal to 2.

Let us now rotate our coordinates about the x-direction and, instead of S_x, S_y, S_z, let us consider $S_{xnew}, S_{ynew}, S_{znew}$, with $xnew, ynew, znew$, the new directions after rotation. We then may consider the new observable:

$$\Sigma_{new} = S_x^2 + S_{ynew}^2 + S_{znew}^2 \qquad (11.2)$$

Here, the reader should not worry: In the right-hand side of Eq. (11.2), the first term is indeed S_x^2, not S_{xnew}^2. We can now make the same kind of reasoning as above, leading to the following observations. If we measure Σ with an apparatus M, or Σ_{new} with an apparatus M_{new}, then, in both cases, the outcome of the measurement must be equal to 2. And again, among the three observables $S_x^2, S_{ynew}^2, S_{znew}^2$, two of them must receive a measurement value equal to 1, and the third one a measurement value equal to 0.

We are now going to introduce an assumption, that we call the non-contextual assumption, which looks most reasonable, and from which we shall draw dramatic consequences concerning quantum mechanics and hidden-variables theories possibly underlying quantum measurements. To begin with, let us take notice of the fact that the observable S_x^2 pertains to the right-hand side summations of both Eqs. (11.1) and (11.2). Now, let us assume that S_x^2 receives one of its possible values, say 1, determined by the hidden variables. The most reasonable non-contextual assumption tells us the following: If we have 1 with apparatus M for Σ, then we also have 1 with apparatus M_{new} for Σ_{new}. In other words, the value 1 does not depend on the context, that is to say whether the experimental set-up is M or a rotated set-up M_{new}. Similarly, if S_x^2 yields its other possible value 0, it must be 0 independently of the experimental set-up being M or M_{new}. Our most reasonable assumption therefore provides an additional constraint concerning the observable S_x^2 which is shared by both Σ and Σ_{new}. However, we have the following extraordinary result: This constraint is not compatible with quantum numerical predictions, meaning that the non-contextual assumption must be rejected. Therefore, hidden-variables theories have to be contextual, as is indeed Bohm's pilot wave.

We now complete the proof of the KS-theorem by demonstrating that, indeed, quantum mechanical predictions cannot be recovered from our aforementioned non-contextual hidden-variables theory. For this, we make a geometrical transposition of our constraint (two constraints actually, one for the value 1, the other for the value 0).

Let us consider a set of three orthogonal vectors in Newtonian space, called a trio. Let us next consider a family of trios. Let us assign the value 1 or 0 to each vector of a trio. Our algebraic constraints may then be converted to geometrical constraints as follows.

(i) For any trio, one vector is assigned the value 0 and the two other vectors are assigned the value 1.

(ii) Let us consider the value assigned to a given vector. This vector may pertain to several trios. Then, the value assigned to the vector does not depend on the trio to which it pertains.

Now, let us consider two orthogonal vectors $\mathbf{a} = (x_a, y_a, z_a)$ and $\mathbf{b} = (x_b, y_b, z_b)$ with components taken from a Cartesian coordinate system (x, y, z). Let us assign the value 1 to each of these vectors, and denote this as:

$$[\mathbf{a}] = [(x_a, y_a, z_a)] = 1 \tag{11.3}$$

$$[\mathbf{b}] = [(x_b, y_b, z_b)] = 1 \tag{11.4}$$

There is one and only one vector $\mathbf{c}$ orthogonal to both $\mathbf{a}$ and $\mathbf{b}$. Item (i) implies that this vector must be assigned the value 0 and, according to item (ii), any vector orthogonal to $\mathbf{c}$ must be assigned the value 1. Any vector of this kind lies in the plane defined by $\mathbf{a}$ and $\mathbf{b}$ and therefore can be written as $(\alpha\mathbf{a} + \beta\mathbf{b})$ with $\alpha, \beta \in \mathbf{R}$. Hence, we have the rule: If $\mathbf{a}$ and $\mathbf{b}$ are two orthogonal vectors and $[\mathbf{a}] = [\mathbf{b}] = 1$, then $[\alpha\mathbf{a} + \beta\mathbf{b}] = 1$, whatever α and β pertaining to $\mathbf{R}$.

Let us now consider the trios $(1, 0, 0)$, $(0, 1, 0)$ and $(0, 0, 1)$ and let us assume $[(1, 0, 0] = 0$. By item (i), we then must have: $[(0, 1, 0)] = [(0, 0, 1)] = 1$. We next examine the vector $(2, 1, 0)$ which lies in the plane defined by $(1, 0, 0)$ and $(0, 1, 0)$, and makes an acute angle with respect to $(1, 0, 0)$. We now show that this vector cannot receive the value 1, by establishing that it would lead to a contradiction.

For this, we therefore have $[(1, 0, 0)] = 0$, $[(0, 1, 0)] = [(0, 0, 1)] = 1$, and assume $[(2, 1, 0)] = 1$. Applying several times the aforementioned rule, we deduce a set of relations reading as:

$$[(2, 1, 0) + (0, 0, 1)] = [(2, 1, 1)] = 1 \tag{11.5}$$

$$[-(0, 1, 0) + (0, 0, 1)] = [(0, -1, 1)] = 1 \tag{11.6}$$

$$[(2, 1, 0) - (0, 0, 1)] = [(2, 1, -1)] = 1 \tag{11.7}$$

$$[-(0,1,0)-(0,0,1)] = [(0,-1,-1)] = 1 \qquad (11.8)$$

$$[(2,1,1)+(0,-1,1)] = [(2,0,2)] = 1 \qquad (11.9)$$

$$[(2,1,-1)+(0,-1,-1)] = [(2,0,-2)] = 1 \qquad (11.10)$$

as can be seen by noting that the two vectors present in the left-hand side of any of these relations are orthogonal.

Now, the vectors $(2,0,2)$, $(2,0,-2)$ and $(0,1,0)$ are three orthogonal vectors, i.e. they form a trio. From one of our starting relations, namely $[(0,1,0)] = 1$, and from the results obtained in Eqs. (11.9) and (11.10), each of the vectors in the above trio is assigned the value 1. By item (i) which demands that one of the vectors should be assigned the value 0, this is impossible. Therefore, if we have $[(1,0,0)] = 0$, implying $[(0,1,0)] = [(0,0,1] = 1$, we cannot simultaneously have $[(2,1,0)] = 1$. Hence, we must have $[(2,1,0)] = 0$. So, we have the following easily stated result: $[(1,0,0)] = 0$ implies $[(2,1,0)] = 0$.

In other words, what we have obtained is the following: Given items (i) and (ii), if a certain vector is assigned the value 0, then there exists at least another vector, making an acute angle with respect to the previous one, which must be assigned the value 0 too.

Let us now convert back this result to our original problem concerning spin 1 particles. For this, we consider a component of the spin in a certain direction. If the square of this component receives the value 0, a true value determined by hidden variables, then there is at least another component, in a direction making an acute angle with the previous one, whose square must receive the value 0 too.

However, this is in deep contradiction with the predictions of quantum mechanics. Indeed, if S_u in direction u is measured to be 0, that is to say S_u^2 is measured to be 0 too, then quantum mechanics predicts that S_v, in direction v making an acute angle with u, possesses a certain probability to be measured as $S_v = \pm 1$, so that S_v^2 possesses a certain probability to be measured as $S_v^2 = 1$. In other words, among the particles having $S_u^2 = 0$, there is a certain number of them which will be measured as $S_v^2 = 1$. This does not agree with our previous result telling us that there exists at least a direction v such as all particles must receive the value 0.

Therefore, we must reject the non-contextual assumption. Hence, we have the Kochen–Specker theorem: Any admissible hidden-variables theory must be contextual.

Mermin, just in himself

We now examine another argument published by Mermin in 1993 [197], about 25 years after the work by Kochen and Specker. Such a large time gap is due to the fact that this argument is a result of a long historical development, starting, more or less arbitrarily but significantly, with Bell in 1964 [189] on the concepts of locality and non-locality, discussed in the next chapter. All over this period, many things have been learned, organized and reorganized, and, due to a better understanding gained by many efforts of several contributors, both deepened and often simplified. As a result, we do not need to find a version of the Mermin argument for everyday cyclists, as required with Kochen and Specker. We may expose Mermin, just as it is, in himself (but for minor rewordings), although it may still be fairly hard for PEDESTRIANS ($\sim\sim\sim\sim$). Before doing this, let us however mention another paper by Mermin [330], providing two examples which significantly simplify the no-go theorem of Kochen and Specker. Preferably, both papers should be examined together.

Following Mermin [197], we now consider a four-dimensional state space generated by two spin-(1/2) particles, particle 1 being associated with Pauli matrices $\boldsymbol{\sigma}^1_\mu$ and particle 2 with Pauli matrices $\boldsymbol{\sigma}^2_\nu$. Any observable in the state space may conveniently be represented in terms of these Pauli matrices (and of unity matrices), e.g. [43]. They enjoy a certain number of properties worth recalling.

Property 1: The squares of each matrix are unity, and therefore each of them have Eigenvalues equal to ± 1:

$$\left(\sigma^1_\mu\right)^2 = \left(\sigma^2_\nu\right)^2 = 1 \tag{11.11}$$

Property 2: Any component of $\boldsymbol{\sigma}^1_\mu$ commutes with any other component of $\boldsymbol{\sigma}^2_\nu$.

Property 3: Let μ and ν specify orthogonal directions ($\mu = x, y, z$ and $\nu = x, y, z$), then $\boldsymbol{\sigma}^i_\mu$ anticommutes with $\boldsymbol{\sigma}^i_\nu$ for $i = 1, 2$, and for $i = 1, 2$:

$$\boldsymbol{\sigma}^i_x \boldsymbol{\sigma}^i_y = i \boldsymbol{\sigma}^i_z \tag{11.12}$$

It is convenient, for either $i = 1$ or 2, omitting the superscripts, and simplifying the matrix notation from bold $\boldsymbol{\sigma}$ to normal σ, to recapitulate the properties in a series of formulae as follows [14]:

$$\left.\begin{aligned}
\sigma_x^2 = \sigma_y^2 = \sigma_z^2 &= 1 \\
\sigma_x \sigma_y = -\sigma_y \sigma_x &= i\sigma_z \\
\sigma_y \sigma_z = -\sigma_z \sigma_y &= i\sigma_x \\
\sigma_z \sigma_x = -\sigma_x \sigma_z &= i\sigma_y
\end{aligned}\right\} \tag{11.13}$$

One can then build nine observables conveniently arranged in the following table (coming back to bold notations, with superscripts for particles 1 and 2):

$$\begin{matrix}
\boldsymbol{\sigma}^1_x & \boldsymbol{\sigma}^2_x & \boldsymbol{\sigma}^1_x \boldsymbol{\sigma}^2_x \\
\boldsymbol{\sigma}^2_y & \boldsymbol{\sigma}^1_y & \boldsymbol{\sigma}^1_y \boldsymbol{\sigma}^2_y \\
\boldsymbol{\sigma}^1_x \boldsymbol{\sigma}^2_y & \boldsymbol{\sigma}^2_x \boldsymbol{\sigma}^1_y & \boldsymbol{\sigma}^1_z \boldsymbol{\sigma}^2_z
\end{matrix} \tag{11.14}$$

These observables exhibit the following properties:

Property 1: The observables in each of the three rows and each of the three columns are mutually commuting. This is immediately evident for the top two rows and for the first two columns from the left. It is also true for the bottom row and rightmost column, because, in every case, we can use a pair of anticommutations.

Here is an example concerning the first row:

$$[\boldsymbol{\sigma}^1_x, \boldsymbol{\sigma}^1_x \boldsymbol{\sigma}^2_x] = \boldsymbol{\sigma}^1_x \boldsymbol{\sigma}^1_x \boldsymbol{\sigma}^2_x - \boldsymbol{\sigma}^1_x \boldsymbol{\sigma}^2_x \boldsymbol{\sigma}^1_x = \boldsymbol{\sigma}^1_x \boldsymbol{\sigma}^1_x \boldsymbol{\sigma}^2_x - \boldsymbol{\sigma}^1_x \boldsymbol{\sigma}^1_x \boldsymbol{\sigma}^2_x = 0 \tag{11.15}$$

Here is an example using anticommutations:

$$\begin{aligned}
[\boldsymbol{\sigma}^1_x \boldsymbol{\sigma}^2_x, \boldsymbol{\sigma}^1_y \boldsymbol{\sigma}^2_y] &= \boldsymbol{\sigma}^1_x \boldsymbol{\sigma}^2_x \boldsymbol{\sigma}^1_y \boldsymbol{\sigma}^2_y - \boldsymbol{\sigma}^1_y \boldsymbol{\sigma}^2_y \boldsymbol{\sigma}^1_x \boldsymbol{\sigma}^2_x \\
&= (\boldsymbol{\sigma}^1_x \boldsymbol{\sigma}^1_y)(\boldsymbol{\sigma}^2_x \boldsymbol{\sigma}^2_y) - (\boldsymbol{\sigma}^1_y \boldsymbol{\sigma}^1_x)(\boldsymbol{\sigma}^2_y \boldsymbol{\sigma}^2_x) \\
&= (i\boldsymbol{\sigma}^1_z)(i\boldsymbol{\sigma}^2_z) - (-i\boldsymbol{\sigma}^1_z)(-i\boldsymbol{\sigma}^2_z) = 0 \tag{11.16}
\end{aligned}$$

Property 2: The product of the three observables in the column on the right is (-1). The product of the three observables in the other two columns and all three rows is $(+1)$.

Here is the demonstration for the first product:

$$\sigma_x^1\sigma_x^2\sigma_y^1\sigma_y^2\sigma_z^1\sigma_z^2 = \sigma_x^1\sigma_y^1\sigma_z^1\sigma_x^2\sigma_y^2\sigma_z^2 = i\left(\sigma_z^1\right)^2 i\left(\sigma_z^2\right)^2 = -1 \tag{11.17}$$

And, for the second product, here is an example:

$$\sigma_x^1\sigma_y^2\sigma_x^2\sigma_y^1\sigma_z^1\sigma_z^2 = \sigma_x^1\sigma_y^1\sigma_z^1\sigma_y^2\sigma_x^2\sigma_z^2 = i\left(\sigma_z^1\right)^2 (-i)\left(\sigma_z^2\right)^2 = +1 \tag{11.18}$$

Now, if we consider mutually commuting observables, then any identity satisfied by the observables must also be obeyed by the values they receive. This means in particular that *property 2* may be converted to a corollary in which "observables" is replaced by "values of the observables". Hence, the product of the values assigned to the three observables in each row must be $(+1)$, and the product of the values assigned to the three observables must be $(+1)$ for the first two columns, and must be (-1) for the rightmost column.

But, Mermin stated: *This is impossible to satisfy, since the row identities require the product of all nine values to be 1, while the column identities require it to be -1.* The impossibility may also be shown from Eqs. (11.17) and (11.18). Indeed, let $V(\mathbf{X})$ denote the value of the observable $\mathbf{X}$. Then, recalling that values are numbers, that is to say commuting quantities, Eq. (11.17) may be converted to:

$$V(\sigma_x^1)V(\sigma_x^2)V(\sigma_y^1)V(\sigma_y^2)V(\sigma_z^1)V(\sigma_z^2) = -1 \tag{11.19}$$

and Eq. (11.18) to:

$$\begin{aligned} V(\sigma_x^1)V(\sigma_y^2)V(\sigma_x^2)V(\sigma_y^1)V(\sigma_z^1)V(\sigma_z^2) \\ = V(\sigma_x^1)V(\sigma_x^2)V(\sigma_y^1)V(\sigma_y^2)V(\sigma_z^1)V(\sigma_z^2) = +1 \end{aligned} \tag{11.20}$$

Hence a contradiction. Afterward, Mermin similarly examined, with a similar conclusion, the case of an eight-dimensional state space (instead of a four-dimensional state space) made out from three independent spin-$(1/2)$ particles (instead of only two particles), a case which is even simpler and more versatile, as stated by Mermin.

What has been proven here is a Bell–KS theorem on the impossibility of some kinds of hidden variables, if these hidden variables have to satisfy quantum mechanical predictions. To see that hidden variables are indeed concerned, we just need to imagine that they determine the values $V(\mathbf{X})$ discussed above. The theorem has something to do with locality since the particles are far apart, but, to justify its discussion in this chapter, it must be noted that it also has something to do with contextuality.

To understand this, let us make Mermin speaking: *In all these cases ... we have tacitly assumed that the measurement of an observable must yield the same value independently of what other measurement must be made simultaneously ...* In particular, in the four-dimensional example above, and also in the eight-dimensional example that we did not discuss, *...we required each observable to have a value in an individual system that would give the result of its measurement, regardless of which of two sets of mutually commuting observables we chose to measure it with. But since the additional observables in one of those sets do not commute with the additional observables in the other, the two cases are incompatible. These different possibilities require different experimental arrangements: There is no a priori reason to believe that the results ... should be the same. The result of an observation may reasonably depend not only on the state of the system (including hidden variables) but also on the complete disposition of the apparatus.* But, here is now the explicit introduction of the concept of contextuality by Mermin: *This tacit assumption that a hidden variable theory has to assign to an observable A the same value whether A is measured as part of the mutually commuting set $A, B, C, ...$ or as part of a second mutually commuting set $A, L, M, ...$even when some of the $L, M, ...$ fail to commute with some of the $B, C, ...$ is called non-contextuality.* Therefore what has been proven is that the tacit assumption must be rejected: Quantum mechanics is contextualist, and hidden-variables theories have to be contextualist. This result might be felt as strange, at least sufficiently strange to motivate asking: *Is the Bell–KS theorem silly?*

Mermin further commented: *It is surely an important fact that the impossibility of embedding quantum mechanics in a noncontextual hidden-variables theory rests not only on Bohr's doctrine of the inseparability of the objects and the measuring instruments, but also on a straightforward contradiction, independent of one's philosophic point of view, between some quantitative consequences of noncontextuality and the quantitative predictions of quantum mechanics.* Also, according to Mermin, the previous remark may be declined in another way, in connection with the fact that the KS- or Bell–KS theorem is less famous than the Bell's theorems discussed in the next chapter, as follows: *One reason the Bell–KS theorem is ... less celebrated ... is that the assumptions made by the hidden-variables theories it prohibits can only be formulated within the formal structure of quantum mechanics.* This statement can only be fully appreciated when we have in mind a sufficient material, concerning non-locality, EPR-paradox, Bell's inequalities and theorems, all topics to which we now turn.

Chapter 12

Non-locality

In the previous chapter, we have demonstrated the Kochen–Specker theorem, according to which any admissible hidden-variables theory (reproducing the predictions of quantum mechanics) must be contextual. The present chapter is devoted to a second, more famous theorem, namely Bell's theorem, according to which any admissible hidden-variables theory must be non-local. This issue, in a book devoted to hidden variables, is inescapable. However, to do justice to it, this chapter has to be fairly big and, I am afraid, this could distract the reader. A possible reading strategy is then to skip this chapter (or to read it quickly) and, because it nevertheless contains a significant material, to return to it later. The reader who would choose this strategy can do it without any significant damage if only he keeps in mind (i) the content of Bell's theorem and (ii) that Bohm's pilot wave is indeed non-local. This is indeed the strategy I intend to recommend because it would allow one, in a first reading, to avoid loosing contact with the essential story I have in mind in writing this book. This chapter is certainly compulsory but, I believe, it cannot be made easy-reading. So, if you do not feel motivated enough, just jump to the next chapter. If you are motivated, not too much, but still significantly enough, you may try to pick up everything you can. This statement is valid for PEDESTRIANS ($\sim\sim\sim\sim$) too, although many parts of this chapter are actually quite readable by them. Furthermore, concerning the most technical issues, there is somewhere below a subsection devoted to these PEDESTRIANS ($\sim\sim\sim\sim$).

EPR-paradox, reality and completeness

Einstein–Podolsky–Rosen (EPR) paradox

We already know that Einstein defended an objective view of physics and of the attached reality, with events taking place in the arena of space and time (more generally space-time), governed by the principle of causality, in compliance with deterministic rules. This vision of the world is in deep conflict with quantum mechanics and, as a result, Einstein remained, all along his life, a permanent opponent to the Copenhagen orthodoxy. This opposition is brilliantly crystallized and expounded in what is certainly one of the best known and most influential papers in quantum mechanics, a paper which, in the mind of Einstein, I guess, was intended to shot a final stroke to the devil. This paper has been published in 1935, in the Physical Review, in collaboration with Podolsky and Rosen, under a title by now most famous: *Can quantum mechanical description of physical reality be considered complete?* [143].

Not surprisingly, there has been a huge number of expositions and re-expositions of the EPR-argument, some of them even easier to follow, I found, that the original argument and which, in any case, may provide complementary hints and insights. Here is a sample, I hope not being too long for the patience of the reader: Bohm [41], [60], Bohm and Hiley [68, 331, 332], Belinfante [1], Jammer [24], Flato *et al.* [305], Bell [328], Hiley [296], d'Espagnat [85, 104], Pais [147], Mermin [333], Howard [334], Holland [40], Bitbol [133], or Omnès [88]. Rather than exposing the original argument from EPR, I shall essentially provide two complementary simple re-expositions, one from Omnès, and the other one from Bitbol, plus a brief allusion to Pais, all of them already quoted in the above sample.

Following Omnès [88], let us consider a physical system made out from two particles, numbered 1 and 2. To simplify the discussion, let us assume without any loss of generality that these particles move along the direction x. Let us denote by x_1 and x_2 the possible values (that is to say if we measure them) of the positions of particles 1 and 2 respectively. Similarly, let us denote by p_1 and p_2 the possible values (again if we measure them) of the momenta of particles 1 and

2 respectively. We assume that the initial state of the system is given by a wave function Ψ reading as

$$\Psi = \varphi(x_1 + x_2)\phi(p_1 - p_2) \qquad (12.1)$$

in which φ and ϕ are narrow Gaussian functions. We can take these functions as narrow as we want so that, eventually, we may think of them as delta functions. In this limit, both $(x_1 + x_2)$ and $(p_1 - p_2)$ are strictly determined. Let us remark that we may express $\phi(p_1 - p_2)$ as a Fourier Transform depending on the argument $(x_1 - x_2)$ so that Ψ becomes $\Psi(x_1 + x_2, x_1 - x_2)$ which is a wave function in a configuration space spanned by the coordinates $(x_1 + x_2)$ and $(x_1 - x_2)$.

Now, let us make a measurement of the location of the first particle and assume that we find $x_1 = a$. Therefore, with an extreme accuracy, we know that the other particle is located at $x_2 = -a$ (the extreme accuracy is a perfect accuracy if we use delta functions instead of narrow Gaussian functions). Next, immediately after the first measurement (that is to say before any free evolution of the wave function had time to occur), we measure the momentum of particle 2 and, let us assume, we find $p_2 = b$. Therefore, we simultaneously know the values $x_2 = -a$ and $p_2 = b$ of the location and momentum of particle 2. This is in complete contradiction with quantum mechanics, in particular with Heisenberg uncertainty relations which prohibit the simultaneous precise determinations of a location and of the associated conjugate momentum, without even having to recall the fact that these quantities are complementary in the language of Bohr, and require mutually exclusive kinds of experiments. This looks like a paradox and has therefore be named the EPR-paradox. However, EPR did not intend to produce a paradox. In their mind, they indeed did not provide a paradox but an argument, that we may call the EPR-argument. The aim of this argument was to prove (their conclusion) that quantum mechanics is not complete. EPR in their paper set down a dilemma : *Either (1) the quantum mechanical description of reality given by the wave function is not complete or (2) when the operators corresponding to two physical quantities do not commute the two quantities cannot have simultaneous reality.* The EPR-argument annihilated the second term of the alternative.

Clearly, an obvious way to completeness is then to introduce hidden variables.

For gaining another flavour of the state of affairs, let us present again the EPR-argument, developed in a slightly different way by Bitbol [133]. We shall expose it using a few steps.

(i) We again prepare a quantum system made out from two subsystems (more later on the concept of system in quantum mechanics), or better said: Particles, denoted particles 1 and 2. The two particles interacted in the past and, as a consequence, they are now correlated. The preparation of the system is carried out in such a way that we can predict with certainty the values of two specific observables. Let these observables be the distance D separating the two particles and the sum P of the momenta of the particles (this is possible because the operators D and P commute, as we can readily check)

$$D = x_1 - x_2 \tag{12.2}$$

$$P = p_1 + p_2 \tag{12.3}$$

(ii) Let us make a measurement of the location of the first particle. If the outcome of the measurement is $x_1 = a$, then, by using Eq. (12.2), we have determined that

$$x_2 = a - D \tag{12.4}$$

(iii) In the original EPR-paper, we have the following most important definition: *If without in any way disturbing a system, we can predict with certainty (i.e. with a probability equal to unity) the value of a physical quantity, then there exists an element of physical reality corresponding to this physical quantity.* In item (ii), we have determined x_2 without having measured it, that is to say, let us put it that way, without having disturbed it. Indeed there has been a disturbance, but this disturbance concerned particle 1, not particle 2 which has been left free. Therefore, the location x_2 possesses an *element of reality*. Using other stronger words, this location is an objective property.

The clever reader might contest this conclusion. He might argue that we indeed disturbed particle 2, through the measurement of particle 1, since both particles are correlated. Such an interpretation however asks many formidable questions. For instance, particle 1 could be located in my laboratory, in Rouen, France, and the other particle 2 could be located in an alien laboratory, for instance somewhere in Messier 31, better known under the name of Andromeda galaxy. Which kind of instantaneous interaction would allow us to understand the fact that, if we measure $x_1 = a$ in Rouen, then "instantaneously" x_2 is $(a - D)$ somewhere in Andromeda galaxy? This is so unreasonable that we just decide not to listen the clever reader any more, at least for the time being, and go on further.

In deciding to go on further, we also ignore a remark from Harré [118] concerning the EPR-definition of what is an element of reality, according to which the above EPR-principle *is surely mistaken.* For example, *the value of the diameter of a Ptolemaic epicycle can be determined with probability equal to unity, but that is clearly no ground for accepting epicycles.*

(iv) We can also make a measurement of p_2 and find, say, $p_2 = b$. Hence, the particle 2 now possesses both a location $(a - D)$ and a momentum p_2.

(v) But this is again in contradiction with the formalism of quantum mechanics. Therefore, according to EPR, quantum mechanics is not complete.

I would also like to briefly mention a third source with a variant. Following Pais [147], we can measure x_1 and know x_2 by using Eq. (12.2). Therefore, x_2 is *an element of reality.* We then can measure p_1 and know p_2 by using Eq. (12.3). Therefore, p_2 is also *an element of reality.* This means that x_2 and p_2 can be viewed as objective quantities, again in complete contradiction with what indeed seems to be an incomplete quantum mechanics. In the words of Pais, *this simultaneous predictability makes the reality of x_2 and p_2 depend upon the process of measurement carried out on the first system which does not disturb the second system in any way. No reasonable definition of reality could be expected to permit this.*

Now, that we have different examples in mind, the diagnosis of incompleteness can be summarized as follows: (i) we may simultaneously know the values of two specific observables (ii) in contradiction with quantum mechanics. Therefore, something is lacking in quantum mechanics, something which should allow us, if we knew it, to simultaneously determine the values of the two specific observables considered. This "something", let us call it "hidden variables". In the words of Bell [189]: *The paradox of EPR was advanced as an argument that quantum mechanics could not be a complete theory but should be supplemented by additional variables. These additional variables were to restore to the theory causality and locality.*

With again different examples that we now have in mind, let us return to the terminology: EPR-paradox versus EPR-argument (for complementary discussions on this point of terminology, see also Jammer [24] or Laloë [335]). Actually, there is no paradox, in the sense that there is no logical inconsistency. But a way to state EPR-conclusion is that, if we hang on an objective view of reality, in Einstein's spirit, then quantum mechanics indeed is not complete. In other words, quantum mechanics does not permit the reunion of the values of two conjugated observables in an unique causal description in space and time.

The Einstein's spirit underlying EPR-paper is vividly testified in an exchange of letters with Born ([201], letters 88 from Einstein to Born, and 89, from Born to Einstein, with a comment from Born, dated 1948, that is to say 13 years after the EPR-paper). Einstein summarized again his philosophy : *The concepts of physics relate to a real outside world, that is, ideas are established relating to things such as bodies, fields, etc... which claim a 'real existence' that is independent of the perceiving subject - ideas which on the other hand, have been brought into as secure a relationship as possible with the sense-data. It is further characteristic of these physical objects that they are thought of as arranged in a space-time continuum.* After this preliminary introduction, he restated the EPR-argument, and added: *I am therefore inclined to believe that the description of quantum mechanics ... has to be regarded as an incomplete and indirect description of reality, to be replaced at some later date by a more complete and direct one.* This was four years before Bohm published his pilot wave, that Einstein did not like.

But we do not need to hang on an objective view of reality *à la Einstein* and, more important, Nature does not necessarily behave in this way and, furthermore, religious people could ever insist that God is not obliged to behave as we believe he should behave. As Bohr would have said, who are we to tell God what He had to do? All these comments look like driving us toward metaphysics in the bad sense of the term (there is however a good sense of the term). But actually not: Such questions will soon enter the realm of experimental metaphysics.

Bohr's rebuttal

Bohr reacted very quickly to EPR-attack, producing the same year, in 1935, two papers to extinguish the fire, one in the Physical Review with the same title than for the EPR-paper [144], the other one in Nature [145], two pieces of work to be attached to the Bohr–Einstein debate. However, the EPR-debate between Einstein and Bohr, concerning EPR-argument, was not exactly a debate, but a dialogue of the deaf. And Bohr's reply was not exactly a reply but a blunt refusal. He turned down EPR-argument point-blank. For, what Bohr essentially did, was simply to reassert his Copenhagen message: There is no objective reality in the sense of Einstein.

What is meaningful indeed is to consider the whole system, particles under investigation plus experimental set-up, as an indivisible unit, producing a holistic structure of wholeness (these words so praised much later by Bohm). More specifically, the particles 1 and 2 are in an entangled state (a notion introduced by Schrödinger in the last century thirties) and, in such a meaningful state, it is meaningless to refer to their individual properties, even if they are far remote one from the other. Furthermore, while EPR exhibited quantum non-measurement determinations (determinations of quantities on non-measured quantum systems, not to be confused with non-quantum measurements), Bohr insisted on the determinations of quantities by quantum measurements, writing: *It is only the mutual exclusion of any two experimental procedures, permitting the unambiguous definition of complementary physical quantities, which provides room for new physical laws.* For him, the choice of different observables to be measured on particles 1 and 2 implies mutually exclusive experimental arrangements. Then, the notion of separation of two sub-systems of a single quantum system is meaningless.

Bohm, that man who liked the concept of wholeness, commented Bohr's reply as follows [31]: *The paradox of EPR was resolved by Niels Bohr in a way that retained the notion of indeterminism in quantum theory as a kind of irreducible lawlessness in nature. To do this, he used the indivisibility of a quantum as his basis. He argued that in the quantum domain the procedure by which we analyze classical systems into interacting parts breaks down, for whenever two entities combine to form a single system (even if only for a limited period of time) the process by which they do this is not divisible... Consequently, one must regard what has been previously been called the 'combined system' as a single, indivisible, overall experimental situation. The result of the operation of the whole experimental set-up does not tell us about the system that we wish to observe but, rather, only about itself as a whole.*

Once more, the philosophical postures behind Einstein on one hand, and Bohr on the other hand, two mutually exclusive individuals, are simply incompatible (in the words of Kuhn [115], they are incommensurable). Not surprisingly, Bohr's rebuttal did not convince Einstein who, in 1948, as we have seen, was maintaining his position in a letter to Born. Also, in 1949, in a reply to criticisms in Schilpp's compilation [123], he wrote: *I am, in fact, firmly convinced that the essentially statistical character of contemporary quantum theory is solely to be ascribed to the fact that this theory operates with an incomplete description of physical systems*[121].

However, the issue addressed by EPR, and the associated Bohr–Einstein incompatibility, did not really interest the physicists. For most of them, it was a matter of taste, of personal epistemological and philosophical options, without any effective influence on the practice of quantum mechanics which was going on flying, jumping from success to success. We shall have to wait about thirty years before Bell's bell sound, and was listened to.

What is a quantum system?

We may partly clarify the situation with a bit more of rigor concerning our terminology, particularly concerning the notions of system and sub-system that I have (on purpose) used and possibly misused up to now. Let us do it in the frame of reference imposed by quantum mechanics.

Let us assume that we have two particles numbered 1 and 2 that have never interacted in the past. It might be extraordinary difficult, in practice, to exhibit two such particles if everything strongly interacted with everything during the pre-Planck age of the universe. But we may instead assume that our particles have forgotten all consequences of their interaction (say by decoherence), with probability so close to 1 that the complementary probability is vanishingly small and, invoking the zero Borel axiom, can be set to 0. Then the wave function Ψ of the combined system may be factorized under the form $\Psi(1)\Psi(2)$ in which $\Psi(j)$ is the wave function of the particle j, $(j = 1, 2)$. In such a case, there is no problem if we say that the combined system is a system, that particle 1 is a system, that particle 2 is a system, and that the combined system is made out from two sub-systems, because each system or sub-system possesses its own wave-function. Then, in agreement with postulate 1 of quantum mechanics, each system is defined by a wave function (more generally, by a ket).

However, if the particles have interacted in the past, the wave function of the combined system cannot be factorized any more but, on the contrary, takes the form $\Psi(1, 2)$. Furthermore none of the particles can be any more characterized by an independent wave function. We say that particles 1 and 2 are in an entangled state. Then, according to postulate 1, particles 1 and 2, which cannot be defined by a wave function (or more generally by a ket) are not quantum systems. There is one system, and only one system, the so-called combined system. There is no sense any more, in this entangled state, to discuss individual properties of the particles. For instance, a photon entangled with another photon is not a system. To say it strongly, an entangled photon *does not exist as an entity*. From now on, when using the concept of system, it is this rigorous definition that we must have in mind: "something" is a system if, and only if, it possesses a wave function (or, more generally, a ket). Then, the so-called combined system previously discussed is not a combined system because it is not a combination of systems. Strictly speaking, we should have adopted this point of view as soon as postulate 1 was stated. But I preferred waiting a more appropriate place (now) to insist on this most fundamental point.

This also gives us an opportunity to comment once more on Bohr versus Einstein. Bohr would have agreed with the above precise definition concerning what a system is. On the contrary, Einstein (at least the Einstein of 1935) would have disagreed. For him, particles 1 and 2 are systems, and objectively exist as independent entities. For Bohr, viewing particles 1 and 2 as systems is meaningless. Who is right? We provide the answer now, and shall comment it in the sequel : Einstein is wrong, in orthodoxy. Therefore, we are now on dragged down into the mysteries of entangled states.

To do completely justice to Einstein, let us provide a later quotation, dated 1949, showing that his mind had evolved by accepting the idea that entangled particles could possibly not form a system: *But on one supposition we should, in my opinion, absolutely hold fast: The real factual situation of the system S_2 independent of what is done with the system S_1, which is spatially separated from the former ... One can escape from the conclusion only by either assuming that the measurement of S_1 (telepathically) changes the real situation of S_2 or by denying independent real situations as such to things which are spatially separated from each other,* that is to say (my comment) by denying that S_1, as well as S_2, forms a system.

These mysteries are expressed by Belinfante [1] as follows: *We thus meet the paradoxical result that quantum mechanics prescribe correlations between measurement results at spatially separated posi-tions on separate particles that previously by interaction formed a composite system ... but does not provide any formalism by which we could understand how the separate particles upon reaching the separate apparatus could acquire the information that seems neces-sary to make the two apparatus find results that show the quantum-mechanically predicted correlations ... we are mystified by an absence of a deterministic mechanism in quantum theory for enforcing its own predictions.*

Bohm–EPR argument

Bohm is acknowledged for having provided a new version of the EPR-argument (hence the given name of Bohm–EPR argument), a version which is usually found conceptually simpler and, more important, which constituted a significant step for driving the debate, out from metaphysics or academic commitments, to the possibility of experimental falsification. The exposition of the argument is available from [41, 336], the former being an extract of the latter. It is then mathematically demonstrated that, in the framework of quantum mechanics, if we consider a so-called combined system made out from two atoms (none of these atoms constitute a system), having a total spin 0, the spin components of each atom in arbitrary directions (obviously, *if* we measure these components) are correlated. In other words, we are facing a system of two entangled atoms, with total spin 0. Let us insist on the fact that these correlations are predicted by quantum mechanics although, in orthodoxy, before the measurements, the values of the spin components do not exist (they are only potential). Instead of using observables (positions, momenta) having continuous spectra, as done by EPR, Bohm used observables (spin, spin components) having discrete spectra. The change from continuous to discrete spectra is the essential source of simplification in the argument. Let us now make this general discussion a bit more specific, following Bohm and Aharonov [290].

We consider a molecule of total spin 0 made out from two spin-1/2 atoms. This is a combined system whose wave function is not factorizable but, instead, reads as (e.g. [43], Chapter 10):

$$\Psi = \frac{1}{\sqrt{2}}(|+,-\rangle - |-,+\rangle) \tag{12.5}$$

in which $|\epsilon_1, \epsilon_2\rangle$ is a ket with the first particle having an Eigenvalue ϵ_1 and the second particle having an Eigenvalue ϵ_2. Also $\epsilon_1 = \pm\hbar/2$ is abridged to $\pm$, and similarly for ϵ_2. The state of Eq. (12.5), of total spin 0, is a singlet state.

Let us now suppose that the molecule is disintegrated through a process which does not modify the spins of the individual atoms. The two atoms (entities, but not systems) are now flying apart, in opposite directions, and although they do not any more interact significantly (via any kind of forces known to us), and although they do not share any information channel known to us, and although the values of their individual spins do not exist since we have not measured them, being potential and not actual, the total spin remains equal to 0. Now, let us measure a spin component of an atom in a given direction, say x. We may find any of the allowed values, that is to say $\pm\hbar/2$. But, because the total spin is zero, we immediately know, without making any measurement, that the spin component of the other atom, in the same direction, is $\mp\hbar/2$, exactly the opposite. Therefore, by measuring any component of the spin of one atom, we obtain the value of the same spin component of the other atom, without any measurement, that is to say without having directly interacted with this other atom in any way. This is an example of quantum correlation between entities (not systems) widely separated after having interacted in the past. In the present case, we may say that the correlation is perfect, although in a more precise language, it is actually a perfect anticorrelation. In general, for spin components which are not measured along the same directions, correlations still exist, but they are not perfect any more.

In some sense, the anti-correlation observed seems obvious, and, if we are indulgent enough (that is to say if we do not search for any explanatory mechanism) without any mystery. Indeed, let us invoke a law of conservation of the spin. Since the total spin is 0, it is indeed clear that, if the spin component of one atom *is* equal to (+*something*), the same spin component of the other atom *is* equal to (−*something*). But this reasoning is completely misleading, just the result of a not careful enough use of the language because, in orthodoxy, a spin component *is not* something if it has not been measured. Therefore, in quantum mechanics, the situation we have to face is indeed more puzzling. This puzzle may be expressed by the following question, may be too intuitive and meaningless: Once a measurement is made on one atom, how the other atom *knows what he has to do?* Furthermore, we can indeed check, for the time being in principle, that the other atom knows what he has to do by carrying out a measurement on this other atom, and observing that he actually behaves as it was supposed to behave. The situation is still more puzzling if we remember that the two separated entities may indeed be separated by, in a strict sense, astronomical distances, from here to Andromeda, or even from here to the "end of the world". Furthermore, we may freely decide to choose the common orientation (or more generally the different orientations) of the Stern–Gerlach apparatus while the particles are still in flight. How particles to be measured *know*, in advance, while they are in flight, what will be our eventual choices for the orientation (or the orientations)?

The EPR-argument concerning objectivity can be translated, *mutatis mutandis*, to Bohm's version: If we can know the value of a spin component without measuring it, without in any way disturbing the associated atom, then it is *an element of reality*. It has to be predetermined and, therefore, quantum mechanics is not complete.

It is indeed difficult to make simpler than this Gedanken experiment. However, Bohm and Aharonov [290] admitted that this experiment, using spins, is not easily practicable. They stated that we should rather study the polarization properties of entangled (correlated) photons. As an example, they suggested to use entangled photons produced in the annihilation radiation of a positron-electron pair. In this process, two photons are simultaneously emitted with opposite momenta. It can then be shown that each photon is always emitted in a state of polarization orthogonal to that of the other, or more properly said: If we measure the states of polarization, one is found to be orthogonal to the other. This is true independently of the choice of the axis with respect to which the state of polarization is defined. Such experiments using photons, and other ones, more and more sophisticated, more and more convincing, will eventually be performed, with a kind of culmination with Aspect's experiments. The verdict of these experiments, teaching us who is guilty, whether it is Bohr or Einstein, will be pronounced later.

More in Bohm and Aharonov [337], Bohm [31], Bell [189] or Flato *et al.* [305], and also in Feynmann [14] who, very pedagogically, instead of using spin components, discussed the EPR-argument in terms of polarization of photons.

Rietdijk under the banner of Einstein

Rietdijk, both a theoretical physicist and a sociological philosopher, may be viewed as a defender of Einstein and of his four-dimensional reality in space-time. In particular, he insisted on the fact that, independently of our human perception, both past and future realistically exist, implying determinism, a consequence of the theory of relativity recently well popularized by Greene [70], and which made Einstein claiming that time is an illusion, albeit a very convincing one. For Rietdijk, the actual existence of the whole space-time bread would make no surprise in facing retroactive influences, such as the ones discussed regarding delayed-choice experiments (or zigs and zags of Costa de Beauregard, to be discussed fairly soon), since future already exists: The universe as a whole is a deterministic and coherent existing thing, in which time, at some deeper level of understanding, does not matter.

Such a summary of Rietdijk'vision is well exemplified in his book on waves, particles and hidden variables [192]. The first chapter of this book is devoted to a rigorous proof of determinism derived from the special theory of relativity. But, because special relativity is deterministic, the proof can only be circular. From this, he however concluded that *the indeterministic conception of 'orthodox Copenhagen' quantum mechanics has been disproved,* with the consequence that *there is determinism, also, in microphysics.* Another consequence, as he stated, is that *there is no free will.* Another subsequent statement, from Rietdijk, is that *if quantum mechanics is correct ... special relativity cannot be, because of the paradox of Einstein, Podolsky, and Rosen ...* The author went on with several Gedanken experiments in the spirit of the EPR-argument, or in the spirit of delayed-choice experiments, with several conclusive statements such as: *The usual quantum mechanics has* *serious shortcomings ... the current view of quantum mechanics cannot be right ... an experiment is discussed that should make it possible to transmit signals with a velocity greater than c, if we do not revise some fundamental conceptions about micro-processes ... the famous 'hidden variable' does in fact exist ... be in accordance with De Broglie's and Bohm's views that matter waves are 'guiding waves', particles remaining corpuscular ... the orthodox tenet that all information about a stream of particles is contained in Ψ, is invalidated ... instantaneous action at a distance conflicts with relativity ...* and so on.

Rietdijk's discussions are closely connected with classical arguments, including relativistic concepts, that is to say with arguments accepting space and time or space-time, in contrast with the fact that many quantum phenomena cannot be understood in such a framework. However, lucidly enough, Rietdijk realized *that some of the reasonings and images brought out here are very provisional and will fail at several points.* Published a bit after Bell's inequalities and one year before the experimental test of Freedman and Clauser (to be discussed below), Rietdijk's book may be viewed as a testimony of the difficulties which arise if we intend to stick to a classical conception of things in the quantum mechanical realm. It is a condensation of all EPR-like arguments and consequences that we are going to investigate in this chapter.

Possible solutions to the EPR-paradox

If we consider the EPR-argument as a paradox, which solutions to it are possible? Stated otherwise, which kinds of explanation could we invoke to make unmysterious the correlations exhibited by EPR-like arguments and experiments? According to Van Fraassen [21], there are six kinds of explanation available, namely: Chance, coincidence, coordination, pre-established harmony, logical identity, and common cause. The interested reader is kindly requested to refer to the original book by Van Fraassen to examine how this author dealt with the issue. In this subsection, I am adopting another kind of classification which, I feel, better matches the aim of this book. Discussion of the validity of the items in this classification is postponed to a later time, when we shall be aware of relevant experimental results.

Hidden variables

When the whole EPR-story is finished to be exposed (we are not yet done on this point), it is sometimes or even often claimed that hidden variables in quantum mechanics are definitely ruled out, on the basis of experimental results (to be discussed later). However, this is not true. A perfect example is the realistic and deterministic solution to the EPR-paradox already proposed by Bohm in his early work [30]. This should not surprise us since we know that Bohm's pilot wave and quantum mechanics yield identical predictions for the outcomes of experiments. Also, the EPR-paradox relies on the interpretation of some quantum measurements and is therefore somewhat akin to the measurement problem, to which Bohm proposed a solution. Actually, in the words of Bohm [31]: *This aspect of the problem* (EPR) *is evidently one that any acceptable theory of hidden variables must somehow manage to deal with in a satisfactory way.*

Let us then return to Bohm, and consider observables having continuous spectra, as in the original EPR-argument (although not as in the Bohm-EPR's version of the argument which was dealing with spin). Since we consider two particles, we have to describe the situation with an objective trajectory in an objective six-dimensional configuration space. Instead of using narrow Gaussian wave functions as in Eq. (12.1), we use delta functions as in the original EPR-paper. Then, Bohm wrote the wave function under the form:

$$\Psi = \delta(x_1 - x_2 - a) \tag{12.6}$$

In this wave function, the phase term S is zero and, therefore, by using the guidance formula, we conclude that particles are at rest. Furthermore, Eq. (12.6) implies that the possible locations of the particles must satisfy:

$$x_1 - x_2 = a \tag{12.7}$$

Let us now carry out measurements. In the framework of Bohm's pilot wave, we remember that measurements are objective deterministic processes. However, the details of these processes are essentially unpredictable, particularly if we assume sub-quantum mechanical fluctuations. Indeed, let us measure the position of the first particle. The measurement process introduces uncontrollable fluctuations in the wave function of the composite system, made out from the two individual particulate sub-systems (in the present objective framework, each particle deserves the name of system). Due to the quantum mechanical forces generated by the quantum potential, these fluctuations are transferred to the second particle and, as a result, we also have uncontrollable fluctuations concerning the objective momenta of each particle. Similarly, let us measure the momentum of the first particle. Then, other uncontrollable fluctuations induce uncontrollable changes in the position of each particle. In other words, we may also say that quantum mechanical forces instantaneously transmit uncontrollable disturbances between the two particles (this is again the non-locality aspect of the pilot wave interpretation). Therefore, due to these instantaneous transmissions, the two particles are correlated. Without dealing with any quantitative specific analysis, extremely difficult if not impossible to perform, requiring at least the use of numerical analysis and computers, we may then however qualitatively understand the mystery of correlations at large distances.

This analysis is somehow akin with Bohr's analysis, at least in the mind of Bohm, when, later on, he wrote (with Bub) [297]: *Consequently, the paradox of EPR does not arise ..., because a composite system must always be regarded as an indivisible totality which in principle* (we should better add: Also in practice, since we can imagine to perform the corresponding calculations) *cannot be subdivided into independently existing unities.* Well, if we put it that way, then it is the essence of Bohr's refutation of the paradox.

In the sequel of one of his early papers [30], Bohm commented on a possible conflict with relativity, due to the existence of instantaneous transmissions of disturbances between the sub-systems. If there is a conflict on this point, then we would also have a similar conflict with quantum mechanics, connected with such key-expressions as: Instantaneous correlations at long distances, delayed-choice experiments, Bohr's wholeness, quantum mechanics outside of space and time, and so on. Bohm however argued that there could not be any conflict because the uncontrollable disturbances induced by the quantum mechanical forces do not permit any signalling. We shall return to this most important issue, and refine it, in due place.

No need to say, Bohm's explanation, as well as, more generally Bohm's pilot wave, cannot be ruled out by experiments, even the ones of Aspect, since once again the pilot wave has been designed in such a way that it leads to the same predictions as quantum mechanics. Therefore, hidden variables cannot be dismissed on experimental grounds. More by Bohm and Hiley in [331], and see also a causal account of non-local EPR-spin correlations by Dewdney *et al.* [338].

Common causes

Bohm's pilot wave is an objective realistic theory and we just have seen that, in this framework, there is no real paradox with the EPR-paradox. More generally, let us take another example of an objective realistic point of view borrowed from our everyday experience, relying on the existence of common causes in the past. Indeed, if we invoke common causes in the past to explain the EPR-paradox, then the paradox vanishes. Our example will rely on hidden postcards, involved in a realistic story.

This story will show the performances of two heroes (the word "hero" is admittedly a bit too much emphasizing, but it is just a rhetoric way to stress the relevance of the story), one man named Gérard and one woman named Monique. On a certain day, say on 18th Mai, 2011, both of them are located at Charles de Gaulle airport, Paris. They are going to fly apart, the particle named Gérard to Narita airport, Tokyo, and the particle named Monique to J.F.Kennedy airport, New York. While still together, they bought two post-cards from Paris, one representing Eiffel tower and the other showing the pyramid of the Grand Louvre museum. With the deterministic complicity of a third traveller, one of the post-card is hidden in Gérard's luggage, and the other in Monique's luggage, without any of them knowing which post-card he (she) has received.

On his arrival in Tokyo, Gérard searches for his post-card in his luggage, finds it, observes that he has been flying with the pyramid and, *instantaneously*, knows that Monique is in possession of the tower. Similarly, on her arrival in New York, Monique observes that she has been flying with the tower and, *instantaneously*, knows that Gérard is in possession of the pyramid. As Bell said [328], *the person who first looked was just the first to know.*

The observations at any place (Tokyo or New York) by no means disturb or interact with the observations at the other place. Both the pyramid and the tower have been strongly objective elements of reality since the beginning of the story. We are indeed facing a perfect correlation (say spin up for the pyramid, spin down for the tower) at long distances, between what happens here and what happens there, even far away. But there is nothing mysterious in such a correlation. It has its origin in Paris, with a common cause at the French source, at the moment of the departures of two particles flying away.

There are actually many variants of this story in the literature, all however sharing the same fundamental ingredient: A common cause. If this worked in quantum mechanics, it would be very comfortable indeed. The results of the future measurements on separated particles would then be simply written somewhere in each particle luggage. We could also say, in a more sophisticated algorithmic picture that, from the source, each particle is flying with a deterministic tabulation, or manual, of instructions for its future, the two manuals for the two particles being arranged, from the beginning at their common place of birth, in such a way that they have to play a coherent and consistent game.

But the question is: Can we explain long range correlations, in quantum mechanics, by invoking common causes, at the source? The answer is: Not in quantum mechanics. This is because, in quantum mechanics, when we open a luggage to observe its content, in particular which postcard it contains, we do not just observe the postcard, the pyramid or the tower. Opening and observing is a quantum measurement process. It does not show something which already existed, but creates the result of the observation. Before any observation, both the pyramid and the tower are potentially present in both suitcases, but none of them is actually present in none of the suitcases. Once an observation is made, say the observation of the pyramid, in one luggage, we know that the tower is present in the other luggage, that is to say, far away, the tower instantaneously comes in actuality. The opening of the luggage containing the tower then does not create the tower any more. It reveals something which became actual inside that luggage, as soon as the content of the other luggage has been observed.

And this is certainly to be found mysterious, even difficult to swallow, by many. Teller [339], referring to Fine [329], tried to make the mystery less mysterious when he writes: *But, in addition, the pair of objects displays an objective probability for correlation. Such 'random devices in harmony', as Fine called them, together have a disposition to manifest their display properties in a correlated way, without benefit of any common cause or direct causal chain. Why, he challenged, should this be any more mysterious than the objective, undetermined chance behavior of the individual objects?* Putting away the use of the word 'objective' which has not the same acceptation than the one we used in this book, excepted if it refers to the weak objectivity of quantum mechanics, I just ask the reader to wonder whether his feeling of strangeness indeed just evaporated under the influence of this Teller's quotation.

It is often stated that experiments verifying quantum mechanics rule out the invocation of common causes to explain long range correlations. In other words, common causes are dismissed by experiments. What is true is that common causes are dismissed by quantum mechanics. But it is not true that common causes are dismissed by experiments : We still have this irritating although fascinating Bohm's theory, with deterministic trajectories, still leading to the same predictions than quantum mechanics, including the existence of long range correlations. The Bohmian trajectories are deterministically governed from the beginning of their stories, when particles depart from their common source, therefore deterministically submitted to a common cause. But something else, seemingly mysterious too, occurs in this conception : The (long range) non-locality of the forces arising from the quantum potential.

Spontaneous factorization

From the point of view of quantum mechanics, particles participating to any EPR-experiment are entangled, a physical fact which is mathematically expressed by the non-factorized, and non-factorizable, form $\Psi(1,2)$ of the wave function describing the pair of particles. Such wave functions have been called wave functions of the second kind, in contrast to wave functions of the first kind which are simple products, such as $\Psi(1)\Psi(2)$. According to Furry [340, 341], one could avoid the EPR-paradox by assuming that, when flying and becoming macroscopically separated, the entangled particles described by a wave function of the second kind would spontaneously (or under the action of some mechanism, possibly to elucidate) disentangle, producing a system which can eventually be described by a wave function of the first kind, with individual particles deserving again the name of systems. This idea has also been later on further discussed by Bohm and Aharonov [290, 337]. We shall call it Furry's hypothesis in the sequel.

Without entering into mathematical details, Furry's proposal may be reasonably grasped by intuition: If entanglement is the source of the EPR-paradox, disentanglement could solve it. Also, non-locality in quantum mechanics would then always be a transient phenomenon, with the return to locality as a rule. Locality being the essence, it would also explain why our everyday vision of the world is so classical. But it would also imply that the accepted rules of the usual quantum mechanics have to be broken up when dealing with many-body systems. Well, *a priori* why not? However, this may be *a posteriori* tested, that is to say by experiments. As we shall see, such experiments falsify Furry's hypothesis which is then to be dismissed. More precisely, long range correlations predicted with the aid of Furry's hypothesis do not agree with long range correlations predicted by quantum mechanics, and experiments will be strongly in favour of quantum mechanics. Also, Furry's vision of quantum phenomena being more local than non-local, the reader will not be surprised to hear that it is empirically close, or even identical, with the results of local hidden-variables theories, those local hidden-variables theories examined by Bell, related to Bell's inequalities (the topic of our next section) and dismissed by experiments. Actually, we shall see that precise tests of Bell's inequalities provide simultaneously a way to rule out both local hidden variables, and Furry's hypothesis.

Zigs and zags

One knows that an electron propagating forward in time is equivalent to a positron propagating backward in time. Then, the legend (I have no reference to prove whether this legend is correct or not; I just found it in my personal neuronal network system) says that, Feynman, wondering why all the electrons of the universe are the same, realized that the answer might be that there is one and only one electron in the universe. The picture behind this idea is that this electron propagates forward in time, turns backward as a positron, reappears forward as an electron, and so on. Using Feynman-like diagrams, it takes the form of a succession of zigs and zags, visually generating what may be called a Feynman zigzag. In a series of papers, Costa de Beauregard [342–346], inspired by the above idea, proposed a solution to the EPR-paradox relying on a particular symmetry property of time, some kind of interaction with the past akin to the one we already discussed regarding delayed-choice experiments.

Let us again consider an EPR-experiment with measurements possibly made at far apart locations A and B on a pair of entangled particles. Let us make a measurement at location A. Following Costa de Beauregard, we now assume that some kind of signal, or may be better said influence, is flowing from A at the moment of the measurement, toward the past, and reaches the source at the moment of the emission of the pair. Afterward, particle B, on which we make a measurement just after the measurement carried out at A, is informed of this measurement at A, by a forward time connection from the moment of the emission of the pair to the moment of the measurement at B. In the words of Costa de Beauregard, *the Einstein correlations between presently separated systems are tied by Feynman zigzags.* We may also view such a process as a variant of the common-causes-in-the-past-explanation made possible by a zig (or a zag) from location A at the moment of the measurement at A to the location of the source at the moment of the emission of the pair.

The zigzag process may be described by using another language, closer to the one of relativity. Let us call S the source. Then *the correlations found to exist between measurements at A and B is not tied, in space-time, along the spacelike vector AB (which is physically empty) but necessarily along the Feynman-style zigzag ASB made of two timelike vectors (which is physically occupied).* Also, according to Jammer [24], who used still another language, the time symmetry invoked by Costa de Beauregard may be viewed as a symmetrical principle of retarded and advanced actions: *As soon as advanced actions are admitted, the measurement process performed on one of the two particles may be conceived as producing an effect on the other particle at a time when the two particles were still interacting with each other, a conception which obviously would resolve the "paradox".* In some sense, even loose, the vision of Costa de Beauregard shares some similarity with the Wheeler interpretation of delayed-choice experiments because, in some sense, but not in the spirit of the Copenhagen interpretation viewing a measurement process as an indivisible process, both Costa de Beauregard and Wheeler are saying that, if the future is uncertain, the past is not completely fixed, but still fuzzy.

The point of view developed by Costa de Beauregard is in deep contrast with a well-known prohibition from Einstein forbidding to telegraph into the past (this issue, telegraphing into the past, will be later revisited and deepened in another context). According to Costa de Beauregard, this prohibition *does not hold at the level of the quantal stochastic event (the wave collapse).* This is a way to say that, at the level of the basic quantum mechanical phenomena, space-time, as viewed by Einstein, is no more relevant, again an issue that will be later revisited and deepened.

We are here facing an interesting and rather original point of view which is intermediary between the may be too psychologically rigid commitment of the classical trio, and the may be too mystical and flamboyant vision of the Copenhagen interpretation. From the classical trio, Costa de Beauregard retains the synthetic idea of causality acting in an arena made out from space and time. But he made a step away from it by accepting a different organization of the arena, namely by accepting that causality might flow both forward and backward in time. This acknowledges the fact that the classical trio, *stricto sensu*, as perceived and defended by Einstein, cannot provide an explanation for EPR-experiments, and anticipates the other fact that local hidden-variables theories, akin in spirit and motivation with the classical trio, is sentenced to failure. What Costa de Beauregard then tried to do is to start from the classical trio, and to demand an amendment, as weaker, as conservative, and as parsimonious as possible: Preserve the arena in which so many wonderful performances took place, but give it an extended topology. From an epistemological point of view, this was a fairly safe attitude. We should not impulsively jump into any revolutionary action if we are not forced to do it (a historian of sciences would provide us with many examples, although others with counter-examples).

As for a counter-example indeed, let us leave the amended classical trio and prepare us to join the Copenhagen interpretation. When a theoretical structure is amended, it may happen that the amendment is sufficient, not only to explain a specific problem, but also as a fruitful way to go forward, and to encompass more and more phenomena, possibly making new predictions, such new predictions being possibly afterward corroborated by experiments. Then, the amendment will be retained, and the whole cathedral will be preserved. This did not happen with the proposal of Costa de Beauregard. The silence which accompanied this proposal (see however Selleri and Tarozzi [347] and Stapp [90, 348, 349]) has been oxymoronic (i.e. eloquent). Likely, the classical trio could not be amended in such a *cheap* way. The whole cathedral, others believed, long ago before Costa de Beauregard, had to be destroyed and reconstructed anew, even with all the ancient stones rejected. This had been the Copenhagen commitment, whether you like it or not. The classical trio is no more amended but rejected : Exit the space and time arena, and welcome to the intrinsic indeterminacy. This last sentence is reminiscent of the Bohr-Einstein debate. In this spirit, Rietdijk [350] indeed discussed retroactive influences *à la Beauregard* versus the famous debate.

According to Costa de Beauregard, non-locality in EPR cannot be understood otherwise than by invoking the kind of time symmetry (*an intrinsic past-future symmetry*) that he introduced. In view of the above discussion, this was a rather definitive or pessimistic statement...

Bell's inequalities and theorems

With EPR-type arguments, we vividly encountered the issue of non-locality (after the one of contextuality in the previous chapter). Indeed, the most important issue in EPR was not the one of determinism, but the one of non-locality, for which Einstein has been speaking of a spooky action at a distance. This is expressed in a letter (letter 84) from Einstein to Born [201] in which he wrote: *I cannot seriously believe in it because the theory cannot be reconciled with the idea that physics should represent a reality in time and space, free from spooky actions at a distance,* a point of view which is challenged by Bohr in his reply to Einstein, regarding EPR. But, whatever our interpretation (non-local hidden variables, or orthodox quantum mechanics), it really seems that a measurement in a certain region of space may instantaneously change "something" in another far-off region, even if this "something" is only, in a very subjective point of view, the knowledge we have of what happens in this far-off region. And this certainly produces an uneasy feeling. There is a connected issue, the one of non-separability which is sufficiently similar to the one of non-locality to allow us, for the time being, to confuse them, before refining later. We also heard previously of the non-locality of causal theories and of quantum mechanics, and of local or non-local hidden-variables theories. With Bell's inequalities and

Bell's theorems, we are rising the curtain on a new act, to possibly improve our understanding of these issues.

Belinfante's inequality. Locality implies inequalities

Before explicitly discussing Bell's inequalities, I intend to prepare the mind of the reader to these inequalities by showing that locality indeed implies inequalities, whatever their explicit forms could be. For this, I refer to Belinfante [1], with some amount of rewording, and also omitting various subtleties which are irrelevant for my purpose.

Let us consider a non-factorized wave function $\Psi(1,2)$ representing an entangled state of two particles, in the usual quantum mechanical framework. On particle 1, we decide to measure R different observables, and we assume that each measurement can yield M different values. When measuring these R different observables, we may use an orthonormal set of basis functions for each observable and therefore, as a whole, R orthonormal sets denoted as $\{\Phi_i^\rho(1)\}$ in which $\rho = 1, 2, ..., R$ identifies one of the R orthonormal sets, and $i = 1, 2, ..., M$ denotes the Eigenfunctions for the set ρ. The symbol (1) recalls us that we are dealing with particle 1. Next, on particle 2, we decide to measure S different observables, and we assume that the dimension of the Hilbert space for these observables is N (instead of M). We then have to deal with S orthonormal sets $\left\{\Lambda_j^\sigma(2)\right\}$ with $\sigma = 1, 2, ..., S$ and $j = 1, 2, ..., N$.

We can now combine each of the R measurements on particle 1 with each of the S measurements on particle 2. Therefore, if we make all possible measurements on the pair $(1, 2)$, we have to make $P = RS$ different measurements on the pair, which may return $L = MN$ different possible outcomes. We label the measurements by using an index $\pi = 1, 2, ..., P$ formed from the couples (ρ, σ), and the possible outcomes by using an index $k = 1, 2, ..., L$ formed from the couples (i, j). The number of possible measurement outcomes for the pair $(1, 2)$ and all of the P measurements is therefore equal to $L^P = (MN)^{RS}$.

In a hidden-variables theory, the outcomes of measurements are determined by microstates $(\Psi(1,2), \xi)$ formed from the quantum state $\Psi(1,2)$ and from hidden variables, say, to simplify, one hidden variable, denoted by ξ. Since we have L^P different quantum mechanical outcomes, we may arrange the values of ξ in L^P different domains, each domain for ξ being mapped to one of the L^P measurement outcomes. The reader is expected to remark that this argument is reminiscent of a similar argument (a decomposition argument, producing dispersion-free states) that we already invoked to demonstrate that hidden-variables theories are always possible.

But we are now going to make a special, but seemingly reasonable, extra-demand, that we shall call the locality demand and which will completely change the state of affairs. According to this demand, what happens here concerning particle 1 should not depend on what happens there concerning particle 2 (and reciprocally). In particular, what happens here when *measuring* something on particle 1 should not influence what happens there concerning particle 2, nor what happens there when *measuring* something on particle 2. In the language of quantum mechanics, we may state the locality demand by saying that particles 1 and 2 behave independently, that is to say they form proper systems and each particle may receive a wave function, $\Psi(1)$ for particle 1 and $\Psi(2)$ for particle 2. The wave function for the combined system then reads as $\Psi(1)\Psi(2)$.

Under such circumstances, let us start a new measurement campaign. We again consider the measurement of R observables on particle 1 (with M possible outcomes for each observable) and the measurement of S observables on particle 2 (with N possible outcomes for each observable). The number of outcomes for particle 1 is therefore equal to M^R and, similarly, the number of outcomes for particle 2 is N^S. Hence, the total number of outcomes for the pair $(1,2)$ is $M^R N^S$.

But, in general

$$M^R N^S < (MN)^{RS} \tag{12.8}$$

This, I shall call it Belinfante's inequality. It is sufficient to conclude that locality is not compatible with quantum mechanics. In particular, if we possess a hidden-variables theory satisfying the locality demand, it cannot reproduce all the predictions of quantum mechanics, nor all the predictions of a hidden-variables theory, called non-local, which is not submitted to the constraint of the locality demand. Therefore, any hidden-variables theory making the same predictions than quantum theory must be non-local. This is essentially Bell's theorem, here deduced from Belinfante's inequality. Historically however, Bell's theorem is the consequence of other inequalities, known as Bell's inequalities.

Bell's inequalities, new stars in the sky

For a long time, from 1935 to 1964, the issues concerning the EPR-argument (non-locality, non-separability, associated hidden-variables theories) have been essentially considered as academic (or even scholastic), without any relevance to the practice and to the development of quantum mechanics. Not only the new doctrine (quantum mechanics) was flying from success to success, but the number of new problems, automatically generated by the new paradigm, was more than great enough to occupy an army of researchers, both experimentalists and theoreticians. What was fashioned was to make experiments, experiments, and to calculate, calculate, in order to garner more successes, invading the new kingdom (this is obviously reminiscent of a witticism granted to Feynman, saying that you have to remain silent, and calculate). For this alone, there has been a kind of tacit consensus to ignore the problems raised by the EPR-argument, seemingly of a metaphysical nature insofar as they could not be grasped by experiments and, in any way, in any case, very far from the prevailing, even moderate, positivist posture. Cushing [48] expressed a similar opinion, however in a somewhat more ruder way : *There was an emphasis, by most of the rapidly growing band of practitioners, on computation as opposed to thinking about foundational questions. It is simplest and most efficient for disciples to follow the path of the master and use the scheme to calculate.*

Nevertheless, in 1964, in a reasonable number of lines, Bell [189] demonstrated that, in the case of two entangled particles, no local hidden-variables theory is compatible with quantum mechanics. Even if it is going to take a bit of time, let us here reproduce the demonstration of Bell. It is so fundamental that I want the reader to study it now, rather than kindly requesting him to refer to the original work. Also, a bit of rewording may be useful. PEDESTRIANS ($\sim\sim\sim\sim$) are allowed to directly jump to the PEDESTRIAN ($\sim\sim\sim\sim$) flag below, in the next section, with the headline: Bell completed, or more Bell without Bell.

Bell's demonstration refers to the Bohm's version of the EPR-argument, the one with spins, rather than to the original argument, the one with continuous spectra. Therefore, once again, let us consider a pair of spin $(1/2)$-particles forming a composite system in the spin singlet state. After some initiating event, both particles move freely into opposite directions. We measure spin components by using Stern-Gerlach instruments. If the measurement of a spin component of particle 1 in a given direction defined by the vector **a** yields the value $(+1)$ or (-1), then the measurement of the same spin component (in the same direction) of particle 2 yields (-1) or $(+1)$, respectively, i.e. we are facing a perfect anti-correlation. This situation was in essence the basis of the Bohm–EPR argument.

We now consider a more general situation with direction **a** for particle 1, and direction **b** for particle 2. In general, the anti-correlation does not hold any more but there will still be a correlation, although not perfect. This new situation will be analyzed under a constraint, called the locality assumption. According to this assumption, if the two spin component measurements (one on particle 1, the other on particle 2) are carried out when the particles are far apart from each other, then the orientation of the magnet in a Stern–Gerlach experiment for a measurement does not influence the result of the other far apart measurement, made with another Stern–Gerlach device. In a loose way, we may again state that what happens here does not influence what happens there. Or, in a looser way, we may state: What happens around here does not depend on what happens around there. For a non quantum mechanical post-Einsteinian physicist, this is indeed a very reasonable assumption, at least for many people. Furthermore, in the case of a perfect correlation (or anti-correlation) once we have measured the spin component of the first particle, we can predict with certainty the result of the measurement of the spin component of the second particle, even if this measurement is not performed.

An easy way to explain these features is to imagine that the results of the measurements are predetermined, and therefore to rely on a hidden-variables theory. Let us denote by $\boldsymbol{\lambda}$ the hidden variables. I used a bold notation for $\boldsymbol{\lambda}$ to insist on the fact that it does not matter whether it denotes a genuine single variable, or a set of continuous or discrete variables, or a mixed set, or whatsoever. The result of measuring the spin component of particle 1 in direction $\mathbf{a}$ is then completely and objectively determined by the direction $\mathbf{a}$ and by the hidden variable $\boldsymbol{\lambda}$. Let us call this result $\boldsymbol{\sigma}_1.\mathbf{a}$ and simply take it as a notation, independently of its precise meaning in quantum mechanics. Similarly, the result $\boldsymbol{\sigma}_2.\mathbf{b}$ of measuring the spin component of particle 2 in direction $\mathbf{b}$ is completely and objectively determined by $(\mathbf{b}, \boldsymbol{\lambda})$. Let us call A and B the outcomes of measurements along $\mathbf{a}$ and $\mathbf{b}$. For spin $(1/2)$-particles, we have, in units $\hbar/2$

$$A(\mathbf{a}, \boldsymbol{\lambda}) = \pm 1, B(\mathbf{b}, \boldsymbol{\lambda}) = \pm 1 \qquad (12.9)$$

Eq. 12.9 furthermore indeed satisfies the locality assumption. We explicitly see that the measurement B does not depend on $\mathbf{a}$, nor A on $\mathbf{b}$. Next, with the hidden variable $\boldsymbol{\lambda}$, we associate a probability distribution $\rho(\boldsymbol{\lambda})$. Then, the objective expectation value of the product of the two components $\boldsymbol{\sigma}_1.\mathbf{a}$ and $\boldsymbol{\sigma}_2.\mathbf{b}$ is given by:

$$P(\mathbf{a},\mathbf{b}) = \int A(\mathbf{a}, \boldsymbol{\lambda}) B(\mathbf{b}, \boldsymbol{\lambda}) \rho(\boldsymbol{\lambda}) d\boldsymbol{\lambda} \qquad (12.10)$$

We now demand that this expectation value, based on the hidden variable $\boldsymbol{\lambda}$, be equal to the expectation value predicted by quantum mechanics for the same situation, reading as

$$P_{QM}(\mathbf{a},\mathbf{b}) = -\mathbf{a}.\mathbf{b} \qquad (12.11)$$

In particular, if $\mathbf{a} = \mathbf{b}$, we recover the perfect anti-correlation $(P_{QM} = -1)$ of the Bohm-EPR argument. Now, the fact is that Eqs. (12.10) and (12.11) are not compatible as we are going to demonstrate.

We use a normalized probability distribution $\rho(\boldsymbol{\lambda})$:

$$\int \rho(\boldsymbol{\lambda})d\boldsymbol{\lambda} = 1 \tag{12.12}$$

Due to Eq. (12.9), and using the above normalization condition, we readily see that $P(\mathbf{a},\mathbf{b}) = \pm 1$. Therefore, we have

$$P(\mathbf{a},\mathbf{b}) \geq -1 \tag{12.13}$$

We obtain (-1) only if $A = -B$, excepted for a discrete set of points $\boldsymbol{\lambda}$ (of Lebesgue measure zero). Under such circumstances, we may rewrite Eq. (12.10) as:

$$P(\mathbf{a},\mathbf{b}) = -\int A(\mathbf{a},\boldsymbol{\lambda})A(\mathbf{b},\boldsymbol{\lambda})\rho(\boldsymbol{\lambda})d\boldsymbol{\lambda} \tag{12.14}$$

Let $\mathbf{c}$ be another unit vector. We then have

$$P(\mathbf{a},\mathbf{b}) - P(\mathbf{a},\mathbf{c}) = -\int \left[A(\mathbf{a},\boldsymbol{\lambda})A(\mathbf{b},\boldsymbol{\lambda}) - A(\mathbf{a},\boldsymbol{\lambda})A(\mathbf{c},\boldsymbol{\lambda})\right]\rho(\boldsymbol{\lambda})d\boldsymbol{\lambda} \tag{12.15}$$

which, by using Eq. (12.9), may be rewritten as

$$P(\mathbf{a},\mathbf{b}) - P(\mathbf{a},\mathbf{c}) = \int \left[-A(\mathbf{a},\boldsymbol{\lambda})A(\mathbf{b},\boldsymbol{\lambda})\right]\left[1 - A(\mathbf{b},\boldsymbol{\lambda})A(\mathbf{c},\boldsymbol{\lambda})\right]$$
$$\times \rho(\boldsymbol{\lambda})d\boldsymbol{\lambda} \tag{12.16}$$

But the quantity $-A(\mathbf{a},\boldsymbol{\lambda})A(\mathbf{b},\boldsymbol{\lambda})$ is ± 1, so that its absolute value is $+1$. And the quantity $[1 - A(\mathbf{b},\boldsymbol{\lambda})A(\mathbf{c},\boldsymbol{\lambda})]$ is non negative and therefore is always equal to its absolute value. Then, taking the absolute value of Eq. (12.16), we deduce

$$|P(\mathbf{a},\mathbf{b}) - P(\mathbf{a},\mathbf{c})| \leq \int \left[1 - A(\mathbf{b},\boldsymbol{\lambda})A(\mathbf{c},\boldsymbol{\lambda})\right]\rho(\boldsymbol{\lambda})d\boldsymbol{\lambda} \tag{12.17}$$

In the r.h.s. of the inequality 12.17, we readily identify $P(\mathbf{b},\mathbf{c})$ by invoking Eq. (12.14), and therefore, with the aid of the normalization Eq. (12.12), we obtain

$$1 + P(\mathbf{b},\mathbf{c}) \geq |P(\mathbf{a},\mathbf{b}) - P(\mathbf{a},\mathbf{c})| \tag{12.18}$$

which is the first of our Bell's inequalities (*the* Bell's inequality), the first one actually of a series of variants.

Afterward, Bell rather laconically concluded that this inequality can be violated in quantum mechanics (see also [328]). I shall rather follow Greenberger *et al* and take one example [351]. Let us take the vectors $\mathbf{a}, \mathbf{b}, \mathbf{c}$ in the (xy)-plane, and define them in this plane by azimuthal angles ϕ_1, ϕ_2, ϕ_3. Let us now more specifically take $\phi_1 = 0$, $\phi_2 = \pi/3$ and $\phi_3 = 2\pi/3$. Using Eq. (12.11), we then have

$$P_{QM}(\mathbf{a}, \mathbf{b}) = P_{QM}(\mathbf{b}, \mathbf{c}) = -\cos(\pi/3) = -1/2 \qquad (12.19)$$

$$P_{QM}(\mathbf{a}, \mathbf{c}) = -\cos(2\pi/3) = 1/2 \qquad (12.20)$$

From these equations, we readily see that the quantum P_{QM}'s do not satisfy *the* Bell's inequality. Hence, the hidden-variables approach developed above is not compatible with quantum mechanics and we have to conclude that the locality assumption is faulty. Therefore, local hidden-variables theories are not admissible (if quantum mechanical predictions are correct). To avoid any misunderstanding, let us insist a bit. We have not proven the following claim: Local hidden variables cannot reproduce quantum mechanical predictions, but the following one: Local hidden-variables theories cannot reproduce *all* quantum mechanical predictions.

In his 1964-paper, Bell made another step that is worth-reporting. Instead of $P(\mathbf{a}, \mathbf{b})$ in the hidden-variables framework, and of $(-\mathbf{a}.\mathbf{b})$ in the quantum mechanical framework, we consider quantities $P_{av}(\mathbf{a}, \mathbf{b})$ and $(-\mathbf{a}.\mathbf{b})_{av}$ which are obtained by averaging $P(\mathbf{a}', \mathbf{b}')$ and $(-\mathbf{a}'.\mathbf{b}')$ over vectors $\mathbf{a}'$ and $\mathbf{b}'$ within small angles around $\mathbf{a}$ and $\mathbf{b}$. Because we do not expect that these quantities are equal, we may suppose that, whatever $\mathbf{a}$ and $\mathbf{b}$, we may introduce a difference bounded by ϵ, according to the inequality

$$|P_{av}(\mathbf{a},\mathbf{b}) + (\mathbf{a}.\mathbf{b})_{av}| \leq \epsilon \qquad (12.21)$$

We are now going to show that ϵ cannot be made arbitrarily small. For this, we also suppose that, whatever $\mathbf{a}$ and $\mathbf{b}$, we have

$$|(\mathbf{a}.\mathbf{b})_{av} - \mathbf{a}.\mathbf{b}| \leq \delta \qquad (12.22)$$

From inequalities 12.21 and 12.22, we deduce

$$|P_{av}(\mathbf{a}, \mathbf{b}) + \mathbf{a}.\mathbf{b}| = |P_{av}(\mathbf{a}, \mathbf{b}) + (\mathbf{a}.\mathbf{b})_{av} - (\mathbf{a}.\mathbf{b})_{av} + \mathbf{a}.\mathbf{b}|$$

$$\leq \delta + \epsilon \qquad (12.23)$$

Next, from Eq. (12.10), we have

$$P_{av}(\mathbf{a}, \mathbf{b}) = \int A_{av}(\mathbf{a}, \boldsymbol{\lambda}) B_{av}(\mathbf{b}, \boldsymbol{\lambda}) \rho(\boldsymbol{\lambda}) d\boldsymbol{\lambda} \qquad (12.24)$$

Due to inequalities 12.9 for $A(\mathbf{a}, \boldsymbol{\lambda})$ and $B(\mathbf{b}, \boldsymbol{\lambda})$, we deduce

$$|A_{av}(\mathbf{a}, \boldsymbol{\lambda})| \le 1, |B_{av}(\mathbf{b}, \boldsymbol{\lambda})| \le 1 \qquad (12.25)$$

Using Eqs. (12.23) and (12.24), specified for $\mathbf{a} = \mathbf{b}$, we obtain

$$|P_{av}(\mathbf{a}, \mathbf{b}) + \mathbf{a}.\mathbf{b}|_{\mathbf{a}=\mathbf{b}} = \left| \int A_{av}(\mathbf{b}, \boldsymbol{\lambda}) B_{av}(\mathbf{b}, \boldsymbol{\lambda}) \rho(\boldsymbol{\lambda}) d\boldsymbol{\lambda} + 1 \right|$$
$$\le \delta + \epsilon \qquad (12.26)$$

that is to say, recalling Eq. (12.12)

$$\int \left[A_{av}(\mathbf{b}, \boldsymbol{\lambda}) B_{av}(\mathbf{b}, \boldsymbol{\lambda}) + 1 \right] \rho(\boldsymbol{\lambda}) d\boldsymbol{\lambda} \le \delta + \epsilon \qquad (12.27)$$

Now, from Eq. (12.24), we have

$$P_{av}(\mathbf{a}, \mathbf{b}) - P_{av}(\mathbf{a}, \mathbf{c})$$
$$= \int \left[A_{av}(\mathbf{a}, \boldsymbol{\lambda}) B_{av}(\mathbf{b}, \boldsymbol{\lambda}) - A_{av}(\mathbf{a}, \boldsymbol{\lambda}) B_{av}(\mathbf{c}, \boldsymbol{\lambda}) \right] \rho(\boldsymbol{\lambda}) d\boldsymbol{\lambda}$$
$$= \int A_{av}(\mathbf{a}, \boldsymbol{\lambda}) B_{av}(\mathbf{b}, \boldsymbol{\lambda}) \left[1 + A_{av}(\mathbf{b}, \boldsymbol{\lambda}) B_{av}(\mathbf{c}, \boldsymbol{\lambda}) \right] \rho(\boldsymbol{\lambda}) d\boldsymbol{\lambda}$$
$$- \int A_{av}(\mathbf{a}, \boldsymbol{\lambda}) B_{av}(\mathbf{c}, \boldsymbol{\lambda}) \left[1 + A_{av}(\mathbf{b}, \boldsymbol{\lambda}) B_{av}(\mathbf{b}, \boldsymbol{\lambda}) \right] \rho(\boldsymbol{\lambda}) d\boldsymbol{\lambda}$$
$$(12.28)$$

We next use the inequalities 12.25 and deduce, from Eq. (12.28):

$$|P_{av}(\mathbf{a}, \mathbf{b}) - P_{av}(\mathbf{a}, \mathbf{c})| \le \int \left[1 + A_{av}(\mathbf{b}, \boldsymbol{\lambda}) B_{av}(\mathbf{c}, \boldsymbol{\lambda}) \right] \rho(\boldsymbol{\lambda}) d\boldsymbol{\lambda}$$
$$+ \int \left[1 + A_{av}(\mathbf{b}, \boldsymbol{\lambda}) B_{av}(\mathbf{b}, \boldsymbol{\lambda}) \right] \rho(\boldsymbol{\lambda}) d\boldsymbol{\lambda}$$
$$(12.29)$$

With the aid of Eq. (12.24) and inequality 12.27, this becomes

$$|P_{av}(\mathbf{a}, \mathbf{b}) - P_{av}(\mathbf{a}, \mathbf{c})| \le 1 + P_{av}(\mathbf{b}, \mathbf{c}) + \delta + \epsilon \qquad (12.30)$$

in which we used the normalization condition of Eq. (12.12).

We now invoke inequality 12.23 and, after a number of lines (although this can be intuitively felt by the hurried reader), we obtain

$$|\mathbf{a}.\mathbf{c} - \mathbf{a}.\mathbf{b}| - 2(\delta + \epsilon) \le 1 - \mathbf{b}.\mathbf{c} + 2(\delta + \epsilon) \qquad (12.31)$$

that is to say

$$4(\delta + \epsilon) \ge |\mathbf{a}.\mathbf{c} - \mathbf{a}.\mathbf{b}| + \mathbf{b}.\mathbf{c} - 1 \qquad (12.32)$$

which may be viewed as a second Bell's inequality.

Now, let us take the following example

$$\mathbf{a}.\mathbf{c} = 0, \mathbf{a}.\mathbf{b} = \mathbf{b}.\mathbf{c} = 1/\sqrt{2} \qquad (12.33)$$

Then, from the inequality 12.32, we obtain

$$4(\delta + \epsilon) \ge \sqrt{2} - 1 \qquad (12.34)$$

This shows that, for small finite δ, ϵ cannot be arbitrarily small. Therefore, P_{QM} cannot be arbitrarily closely represented by the hidden variables expectation value of Eq. (12.10). This implies that, *a fortiori*, such a representation cannot be exact. Hence, once more, we have to reject our hidden-variables approach, that is to say the locality assumption on which it was based.

Before going further, it is likely worthwhile to summarize what we have recently learnt (a bit of redundancy, even if not strictly necessary from a logical point of view, is always useful from a pedagogical point of view). EPR-argument, and later on Bohm–EPR argument, analyzing long-range correlations between spatially separated particles, concluded that, let us say seemingly, quantum mechanics is incomplete. A possibility to reach completeness is to introduce hidden variables. Bell demonstrated not that such a completion is impossible, as often stated or believed, but that such a completion is not possible by invoking local hidden-variables theories. The demonstration relies on the exhibition of inequalities, thereafter called Bell's inequalities, which are not necessarily contradicting quantum mechanics but which, under certain circumstances, do violate quantum mechanical predictions. The experimental violation of Bell's inequalities would then definitely rule out local hidden-variables theories, and this, based on experimental evidences, would remain true even if quantum mechanics, eventually, was falsified, because such a violation would falsify the locality assumption, independently of the existence of the quantum mechanical formulation. Therefore, hidden-variables theories have to be non-local. More than a matter of fact, it is a fact of nature.

Bell revisited by Wigner

The reader might have found that the derivation of Bell's inequalities in the previous subsection has been rather abstract (indeed it has been fairly abstract), that it was not illuminating enough (I agree), that we can follow the steps of the demonstration but that we cannot grasp its essence (yes). It is a fact that introducing a new idea is something extraordinarily difficult, and that it is exceptional that such a novelty can be introduced in the best clarity of the mind. But Bell did it, likely in the best clarity of his mind, even if not necessarily in the best clarity of the mind of his readers. *My* readers might have preferred the use of Belinfante's inequality to reach Bell's theorem (I prefer it) but, from Bell, 1964, to Belinfante, 1973, there is a gap of nearly ten years. And there is no escape to the fact that Bell lighted a new star in the sky, in a way that I find most brilliant, or even incredible. I cannot imagine how he found its path to his result. Respect.

This being heartedly, honestly, and carefully said, I am now happy to introduce the reader to Bell's inequalities by following another path, available from Wigner [306] who claimed making Bell's argument somewhat simpler (I agree again, although it could just be a matter of taste) and more specific. A variant is available from Pipkin [156] (see also Freedman and Holt [352]). As with Bell, we consider a pair $(1, 2)$ of spin $(1/2)$-particles. A spin component may be measured along one of three different directions denoted as $\omega_1, \omega_2, \omega_3$. Let us make a simultaneous measurement of spin components on the pair $(1, 2)$, that is to say a measurement of the spin component of particle 1 along direction ω_i and a measurement of the spin component of particle 2 along direction ω_j. We have 3 components for each spin and, for each component of one spin, we have 3 components for the other spin. As a whole, we therefore have 9 measurements for the pair, that we may denote as $[1\omega_1, 2\omega_1], [1\omega_1, 2\omega_2], [1\omega_1, 2\omega_3], [1\omega_2, 2\omega_1], \ldots, [1\omega_3, 2\omega_3]$, in which $i\omega_j$ denotes a measurement of the jth component (in direction j) on the particle numbered i.

For any spin component, the outcome of a measurement can only be $\pm\hbar/2$, more conveniently denoted as ± 1, or even $\pm$. Therefore, for a pair of directions (ω_i, ω_j), we only have four possible outcomes of measurements, namely $(+, +), (+, -), (-, +)$ and $(-, -)$ in which (ϵ_1, ϵ_2) denotes an outcome ϵ_1 on particle 1 in direction ω_i and an outcome ϵ_2 on particle 2 in direction ω_j. Since we have nine pairs (ω_i, ω_j) and four outcomes for each pair, we have as a whole 4^9 possible outcomes.

Let us now assume that everything is predetermined by a hidden variable λ. Let us take λ continuous, just to simplify the language, the exposition, and the functioning of the mind. We may then partition the range of λ into 4^9 domains (again, this kind of decomposition...) and biunivocally associate each of the 4^9 domains with each of the 4^9 possible outcomes. We may also arrange ourselves in such a way that some averaging over λ in each domain allows one to recover the probabilities of the associated quantum measurement outcomes. This is particularly obvious if we just think that we may adequately and possibly in an *ad hoc* way define the averaging process in each domain. We then obtain a consistent hidden-variables theory in agreement with the predictions of quantum mechanics. Up to now, nothing new: We have just proven that, at least for the case under study, we can always construct a hidden-variables theory, whether it is local or non-local.

We now introduce a locality assumption, in Bell's spirit. We state it as follows: The hidden variable λ must determine the spin component of one particle in any direction ω_k, independently of the direction ω_l in which the spin component of the other particle is measured. Again, this is what we called a reasonable assumption, particularly if we remember that the two particles are spatially well separated, that is to say if we remember that the apparatus measuring one spin is much remote from the apparatus measuring the other spin.

But this new assumption dramatically decreases the number of domains in which we can partition the range of the underlying hidden variable λ, precisely from 4^9 to 2^6. Indeed, instead of nine pairs (ω_i, ω_j) with four measurement outcomes (ϵ_1, ϵ_2) for each pair, we now have three measurements on one particle, independently of the measurements on the other particle, making only six relevant measurements, and, for each of these measurements, we only have two outcomes, namely $(+)$ or $(-)$. This is a version of Belinfante's inequality.

The 2^6 domains we are now facing may be denoted as $(\sigma_1, \sigma_2, \sigma_3 \mid \tau_1, \tau_2, \tau_3)$ in which the σ's refer to the first particle, the τ's refer to the second particle, $\sigma_i = \pm$, $\tau_i = \pm$, and the vertical bar is used to insist on the fact that what happens for one particle does not depend on what happens for the other particle. This bar also reflects the fact that the quantum mechanical wave function can be factorized as $\Psi(1)\Psi(2)$ that we could also write under the form: $\Psi(1 \mid 2) \neq \Psi(1, 2)$. To clarify the notation, we take an example. Let us consider a domain denoted as $(+ - - \mid - + -)$. For this domain, the measurement outcomes are: (i) $+, -, -$, in directions $\omega_1, \omega_2, \omega_3$ respectively, for particle 1, independently of the outcomes for particle 2 and (ii) $-, +, -$, in directions $\omega_1, \omega_2, \omega_3$ respectively, for particle 2, independently of the outcomes for particle 1.

We now consider the specific case of a singlet state of the two spin system. Let $P(\sigma_1, \sigma_2, \sigma_3 \mid \tau_1, \tau_2, \tau_3)$ denote the probability for the hidden variable to be located in the domain $(\sigma_1, \sigma_2, \sigma_3 \mid \tau_1, \tau_2, \tau_3)$. Also, let $\theta_{12}, \theta_{23}, \theta_{31}$ denote the angles (all between 0 and π) between the directions ω_1 and ω_2, ω_2 and ω_3, ω_3 and ω_1, respectively. We now have the following results:

(i) The probability that the measurement of the spin component of particle 1 in direction ω_i be equal to the measurement of the spin component of particle 2 in direction ω_j ($+$ or $-$ in both cases) is:

$$P(\omega_i : \pm \mid \omega_j : \pm) = \frac{1}{2} \sin^2 \frac{1}{2}\theta_{ij} \tag{12.35}$$

with an obvious notation.

(ii) The probability that the measurement of the spin component of particle 1 in direction ω_i be the opposite of the measurement of the spin component of particle 2 in direction ω_j is:

$$P(\omega_i : \pm \mid \omega_j : \mp) = \frac{1}{2} \cos^2 \frac{1}{2}\theta_{ij} \tag{12.36}$$

Equations (12.35) and (12.36) can be obtained by making direct calculations in the framework of quantum mechanics or, for the case under study, by using the fact that a singlet state is spherically symmetric [306]. But, as we are going to show, these results cannot be reproduced by any local hidden-variables theory.

We begin with the observation that, as a consequence of Eq. (12.35), $P(+, \sigma_2, \sigma_3 \mid +, \tau_2, \tau_3) = 0$, and more generally

$$P(\omega_i : \pm \mid \omega_i = \pm) = 0 \tag{12.37}$$

This implies that $P(\sigma_1, \sigma_2, \sigma_3 \mid \tau_1, \tau_2, \tau_3)$ is different from 0 only when $\sigma_1 = -\tau_1, \sigma_2 = -\tau_2, \sigma_3 = -\tau_3$. There are therefore only eight such possibilities which are different from 0.

Let us now consider the following event : The measurement of the spin component of particle 1 in direction ω_1 and the measurement of the spin component of particle 2 in direction ω_3 are both positive. Let us denote $P(\omega_1 : + \mid \omega_3 : +)$ the probability of this event. This can be obtained by summing up the probabilities of 16 events (2 directions are left free for particle 1, 2 directions are left free for particle 2, and we have 2 possible outcomes for each of these four directions). For the sake of clarity, these 16 events are listed below (omitting comas in the notation)

$$(+++\mid+++), (++-\mid+++), (+-+\mid+++), (+--\mid+++),$$

$$(+++\mid+-+), (++-\mid+-+), (+-+\mid+-+), (+--\mid+-+),$$

$$(+++\mid-++), (++-\mid-++), (+-+\mid-++), (+--\mid-++),$$

$$(+++\mid--+), (++-\mid--+), (+-+\mid--+), (+--\mid--+).$$

Many events in the above list have probability 0, for instance $(+ + + \mid + + +)$ or $(+ + - \mid + + +)$. Upon a systematic inspection, we may then see that only two events have probabilities different from 0, namely: $(+ + - \mid - - +)$ and $(+ - - \mid - + +)$. Therefore, we obtain

$$P(\omega_1 : + \mid \omega_3 : +) = P(+ + - \mid - - +) + P(+ - - \mid - + +) \tag{12.38}$$

which, from Eq. (12.35), can be written as

$$P(\omega_1 : + \mid \omega_3 : +) = \frac{1}{2} \sin^2 \frac{1}{2}\theta_{13} = \frac{1}{2} \sin^2 \frac{1}{2}\theta_{31} \tag{12.39}$$

Let us now investigate the right-hand side of Eq. (12.38). In this right-hand side $P(+ + - \mid - - +)$ is the probability of an event in which the measurement of the spin component of particle 1 in direction ω_2 is positive and the measurement of the spin component of particle 2 in direction ω_3 is positive too. But we have

$$P(\omega_2 : + \mid \omega_3 : +) = \frac{1}{2} \sin^2 \frac{1}{2}\theta_{23} = P(+ + - \mid - - +)$$
$$+ P(- + - \mid + - +) \tag{12.40}$$

in which two terms of the kind $P(\sigma_1, \sigma_2, \sigma_3 \mid \tau_1, \tau_2, \tau_3)$ only are different from zero.

Hence, the inequality

$$P(+ + - \mid - - +) \leq \frac{1}{2} \sin^2 \frac{1}{2}\theta_{23} - P(- + - \mid + - +) \tag{12.41}$$

Similarly, considering now the term $P(+ - - \mid - + +)$ in the r.h.s of Eq. (12.38), we have another inequality

$$P(+ - - \mid - + +) \leq \frac{1}{2} \sin^2 \frac{1}{2}\theta_{12} - P(+ - + \mid - + -) \tag{12.42}$$

Inserting inequalities 12.41 and 12.42 in Eqs. (12.38), (12.39), we obtain a new Bell-kind inequality reading as

$$\sin^2 \frac{1}{2}\theta_{31} \leq \sin^2 \frac{1}{2}\theta_{23} + \sin^2 \frac{1}{2}\theta_{12} \tag{12.43}$$

in which we have taken into account the fact that probabilities are non negative. This inequality must necessarily be always satisfied if we want our local approach predictions to agree with the predictions of quantum mechanics.

Let us however consider the case of three coplanar directions ω_i such that ω_2 bisects the angle between ω_1 and ω_3, that is

$$\theta_{12} = \theta_{23} = \frac{1}{2}\theta_{31} \tag{12.44}$$

The above inequality becomes

$$2\sin^2\theta_{12} \geq \sin^2\frac{1}{2}\theta_{31} = 4\sin^2\frac{1}{2}\theta_{12}\cos^2\frac{1}{2}\theta_{12} \tag{12.45}$$

implying

$$\cos^2\frac{1}{2}\theta_{12} \leq \frac{1}{2} \tag{12.46}$$

This is violated for a range of θ_{12}, for instance for $\theta_{12} = \pi/6$, leading to $\cos^2(\pi/12) \sim 0.93$. Other violations may be found, even if the directions ω_i are not coplanar. Therefore, once again, we have to reject the locality assumption and conclude that local hidden-variables theories cannot reproduce all predictions of quantum mechanics.

Bell revisited by Bell

Bell went on working on the topic. In particular, in 1971 [327], he demonstrated that any stochastic theory satisfying the locality condition is also incompatible with quantum mechanics. Therefore, not only deterministic local hidden-variables theories do not agree with quantum mechanics, but also those local theories which are not deterministic. Indeed, determinism, after all, was not the main issue. The fact that determinism was not the main issue is also emphasized by Squires [13] when he states that locality, rather than determinism, is the key ingredient for proving Bell's inequalities.

As far as I know, the last work of Bell on hidden variables is dated 1990 [353], the sad year when he passed away. This 1990-paper from Bell is reproduced in the unspeakable compilation [205] and discussed in a book by d'Espagnat [85]. It relies on simple calculations, although tedious, boring, as admitted by Bell. The inequality exhibited by d'Espagnat has actually been published for the first time, in 1969, by Clauser, Horne, Shimony and Holt, known as CHSH, [354] and, at this time, already provided a generalization of the original Bell's inequality. This original inequality only concerned phenomena for which, when the two directions **a** and **b** identify, the theory predicts a perfect anti-correlation. On the contrary, the content of 1990 Bell's paper is very general and, in particular, nothing is postulated concerning the entities involved, whether they are particles, waves, or something else. Rather than discussing Bell in its generality, we shall discuss a simpler derived proof available from Appendix 1 of the book by d'Espagnat [85]. The interest of this proof, besides its elegance, is that it deals with photon polarizations used for experiments to be discussed later. For complementary discussions and points of view, see Annex II in a book by Bitbol [133], and Omnès [15].

In this simplified proof, we return to a somewhat less general case than in 1990 Bell' paper, namely a case in which a perfect correlation (here not an anti-correlation) is predicted when directions **a** and **b** are the same. This may be experimentally achieved via some relaxations of atoms, producing photon pairs. Specifically, we now assume that we are dealing with photons and that we measure photon polarizations A, B, C along three directions **a**, **b**, **c**, respectively. Any measurement can only provide two outcomes (two polarizations) that we shall conveniently denote as $(+)$ and $(-)$. We next invoke a hidden-variables theory in which the outcomes of measurements are predetermined by the value of a hidden variable (or a set of hidden variables, again it does not matter).

Let us now focus on what happens on a detector managing with photons which, for instance, travelled to the right (R). In quantum mechanics, the orientation of the detector plays a most significant role in the outcome of the measurement. Then, it is for instance meaningless, in quantum mechanics, to think of a simultaneous measurement of polarizations along three different orientations on a given photon if only due to the fact that, once a polarization measurement along one direction has been carried out, the quantum state of the photon has been modified. However, in the framework of our hidden-variables theory, since the results of the measurements are predetermined, it is meaningful to think in that way.

In particular, it is meaningful to consider the number of photons $R(+ + +)$, associated with predetermined values of hidden variables, which would provide the result $(+)$ if the detector was oriented along **a**, the result $(+)$ too if it were oriented along **b**, and the result $(+)$ again if it were oriented along **c**. More generally, we are allowed to consider the number of photons $R(\epsilon_1, \epsilon_2, \epsilon_3)$, $\epsilon_i = \pm$, which would provide the result ϵ_1 if the detector was oriented along **a**, ϵ_2 if it was oriented along **b**, and ϵ_3 if it was oriented along **c**. Let us denote by $R(+ + \forall)$ the number of photons which would provide the result $(+)$ if the detector was oriented along **a**, the result $(+)$ too if it was oriented along **b**, and this whatever $(\forall)$ the outcome of measurement with orientation along **c**. We then obviously have

$$R(+ + \forall) = R(+ + +) + R(+ + -) \tag{12.47}$$

and, also, with obvious similar notations

$$R(+\forall+) = R(+ + +) + R(+ - +) \tag{12.48}$$

$$R(\forall + -) = R(+ + -) + R(- + -) \tag{12.49}$$

Adding Eqs. (12.48) and (12.49), and using afterward Eq. (12.47), we obtain

$$R(+\forall+) + R(\forall + -) = R(+ + \forall) + R(+ - +)$$
$$+ R(- + -) \tag{12.50}$$

Now, $R(\epsilon_1, \epsilon_2, \epsilon_3)$ is a number of photons, and hence cannot be negative. It then follows that Eq. (12.50) implies the inequality

$$R(+\forall+) + R(\forall + -) \geq R(+ + \forall) \tag{12.51}$$

Similarly, as for $R(\epsilon_1, \epsilon_2, \epsilon_3)$, we may define a new number of photons $L(\epsilon_1, \epsilon_2, \epsilon_3)$ for measurements of the polarizations of photons by the left (L) detector. Due to the perfect correlations we assumed between photons travelling to the right and photons travelling to the left, we have for instance $R(+ + \forall) = L(+ + \forall)$, and so on. We can then define another new number of photons $N(+ + \forall)$ according to

$$N(+ + \forall) = R(+ + \forall) = L(+ + \forall) \tag{12.52}$$

Let us assume that we intend to measure $R(+ + \forall)$ by using the right detector. This requires two simultaneous measurements, one with an orientation of the detector along direction **a**, and a second one (immediately after the first one) with an orientation along direction **b**. But this would not be meaningful because the first measurement would have modified the state of the photon.

But, in contrast, invoking the assumed strict correlation between right and left channels, the number $N(+ + \forall)$ can be retrieved from measurements. For this, it is sufficient to measure along **a** for one particle and along **b** for the other particle, two measurements which are now meaningful. If we now return to Eq. (12.51), interpret the numbers R as measurable numbers N, and convert numbers to probabilities, we obtain a new Bell-kind inequality reading as

$$P(+\forall+) + P(\forall + -) \geq P(+ + \forall) \tag{12.53}$$

which may be experimentally tested. Then, experimental results allowed researchers to observe violations of the inequality under certain circumstances, agreeing with quantum mechanical predictions of correlations at a distance. Once more, we therefore have to reject the locality assumption, not only because it disagrees with quantum mechanics (after all, quantum mechanics could be erroneous and, without any doubt, it is indeed erroneous at a certain more sophisticated level of the description of the world), but because it is in disagreement with experimental facts. Whatever will be the ultimate fate of quantum mechanics, non-locality then appears as an experimental manifestation of the phenomenological world, and may be of the Thing in Itself, even veiled. The classical trio has definitively exploded, without any foreseeable possibility of renovation.

It is sometimes believed or stated that such results rule out hidden variables, and that we must then renounce to them. Clearly, such is not the case. Non local hidden-variables theories are still admissible, and we still have Bohm's theory wandering on the stage. For those who do not like hidden variables, the shoe should still be hurting.

Bell completed, or more Bell without Bell

PEDESTRIANS ($\sim\sim\sim\sim$), land on here.

Bell has been the primary and, as far as I know, the sole precursor of the idea that locality implies inequalities. We have seen some variants of them, and we are now well equipped enough to go on further without any need to rely again on physico-mathematical formulations. If we had to remember only the essence, it would be that Bell's or Bell-type inequalities generally provide upper bounds on the strength of correlations which may be observed on two spatially separated particles.

But, actually, many other Bell-kind inequalities, providing extensions of the original one to more general situations, with less restrictive assumptions, are available from the literature, by various authors, and with various purposes, just as when dealing with proposed or effective measurements. In this section, we are trying to do justice to them, in a qualitative way, without however claiming exhaustibility. Actually, an exhaustive story would be very complicated and lengthy to report, if not impossible to properly investigate. In the words of d'Espagnat [69], *it is not possible to give here a systematic survey -not even a schematic one- of all the debates the ... inequalities give rise to.* A difficulty for such a systematic survey is due to the fact that the story has many facets, and exhibits some kind of fractal ramification. In particular, in the words of Jarrett [75]: *Bell-type inequalities can be derived from different sets of premises; some versions of Bell's theorem employ one or another premise (some formulation of determinism, for example) which other versions do not; in some treatments certain premises are consolidated and others not made explicit, and, in general, things just formulated differently in different expositions.* Therefore, let us be content with a bit more of information, completing without completeness our knowledge of the issue, a knowledge which in any case is already sufficient for the sequel.

The original paper from Bell [189] historically opened the way to experimental investigations but, although seminal in a strong sense, was not enough developed by itself to clearly see how such experimental investigations could be made. Thereafter, several people have carried out the work further, up to the point where how to achieve actual experiments was indicated in detail, among them, to start with, Clauser, Horne, Shimony and Holt (CHSH again) [354]. This paper has been much influential and, as a consequence, people often speak of the Bell-CHSH's inequalities. Its aim was to cover actual systems, and led to the introduction of a terminology using the expressions: Objective local theories, or realistic local theories. It has been completed by further works from Shimony [355], from Clauser and Horne [356] who discussed experimental consequences of objective local theories, from Clauser [357, 358], and from Clauser and Shimony [359]. The latter paper by the way contains an extensive review of the subject of Bell's inequalities and hidden variables.

Next, the quantum mechanist who dislikes hidden variables cannot be completely satisfied with Bell's inequalities which deal only with local hidden variables. Some steps have been provided toward a more satisfactory conclusion. For instance, in 1973, Mc Guire and Fry [360] derived an inequality, similar to Bell's ones, for non-local hidden-variables theories in which the hidden-variables density distribution of the particles measured is non-local, but where the detectors are contained in some finite volume. Non-local models (besides obviously Bohm's one) have been discussed for instance by Edwards [361], and by Edwards and Ballentine [362], who constructed an example of a non-local model which satisfies a generalized Bell inequality, even for perfect analyzers and detectors. Selleri and Tarozzi [347], who quote them, and provided a pedagogical review on Bell's inequalities and non-locality, emphasized the fact that locality is by no means a necessary condition to obtain inequalities, but that, on the contrary, inequalities may be obtained for non-local models too. Non-local models which satisfy an inequality are also discussed by Garuccio and Selleri [363] who have shown that a natural extension of local hidden-variables theories to include non-local effects still imply inequalities. See also Stapp [90]. The issue of non-local hidden-variables theories is therefore ready to become of an increasing interest. For instance, in 2003, Leggett [364] demonstrated that a class of non-local realistic theories is incompatible with the predictions of quantum mechanics, this having been followed by an experimental test in favour of quantum mechanics, by Gröblacher *et al.* [365].

In short, local variables imply Bell's inequalities but Bell-kind inequalities are compatible with non-local variables. This means that a violation of Bell's inequalities rules out local hidden-variables theories. But a violation of Bell's inequalities does not rule out non-local hidden-variables theories, because the non-local hidden-variables theories which are actually ruled out always rely on particular assumptions, not always explicit, and not necessarily relevant to definitively close the issue. The reader who would like a conclusion without having to spend his time on the associated literature only needs to remember an inescapable fact: Bohm is still alive. The quantum mechanist who dislikes hidden variables must remain dissastified.

For the reader who would like to spend more time on these issues, here is a list of relevant complementary references: Bohm and Hiley [331, 332], Bohm [366], Hiley [296], Stapp [90, 349, 367], Eberhard [368], Pipkin [156], Wheeler and Zurek [100], or Cushing and McMullin [369].

Bell-kind inequalities for PEDESTRIANS ($\sim\sim\sim\sim$)

The locality assumption, behind Bell's inequalities or Bell-kind inequalities, drives us back from quantum mechanics to a world more objective, and more realist. We may then wonder whether we could not find inequalities too in the most objective and most realist world of our everyday macroscopic experience. The answer to this question is : Yes, we can. These inequalities, I call them Bell-kind inequalities for PEDESTRIANS ($\sim\sim\sim\sim$). To understand these inequalities, we do not need any elaborated physics nor any mathematics (or just a small amount of them). We just need to concentrate a bit. Furthermore, everything is clearly exhibited and described with plain words, all of them having a precise meaning, and nothing is hidden. In particular, there is no hidden variables, a way to illustrate that hidden variables are not necessary to derive Bell-kind inequalities. We then provide a way to extract the gist of them. PEDESTRIANS ($\sim\sim\sim\sim$) may then go on up to the next alert.

Here is an example taken from d'Espagnat [126]: In a population of individuals, whatever the population, the number of women younger than forty years is smaller than or equal to the number of smoking women augmented by the number of non smoking individuals younger than forty years. This is a theorem. Variants with cats or nails are available from [98].

At first sight, the theorem may look a bit puzzling, but it is illuminating to think about it up to the forthcoming moment when it becomes fully obvious. Here is a demonstration. Let us denote the class of women of less than forty years as YW (Y for Young and W for Women), and let $N(YW)$ be the number of young women in the population. Such a young woman (may be actually a child, or even a baby) may pertain to the class of smoking women. Let us denote the class of Smoking Women as SW, and let $N(SW)$ be the number of smoking women in the population. But a young woman may also pertain to the class of Non Smoking Young people. Let us denote the class of non smoking young people (man or woman) by NSY and let $N(NSY)$ be the number of non smoking young people. A non smoking young woman pertains to the class NSY. But, in this class, there are individuals which are not woman. Therefore a young woman (class YW) may pertain to the class SW of smoking women (when she smokes) or to the class of non smoking young people (when she does not smoke), the latter class containing also men. Therefore, we have

$$N(YW) \leq N(SW) + N(NSY) \tag{12.54}$$

The above theorem can be given a statistical flavour. Instead of counting our different N-numbers in a given population, let us pick up three subsets of same size, called samples, and numbered $1, 2$ and 3. Then, we obtain a new version of the theorem: If the common size of the samples is large enough, then the number of young women in sample 1 is smaller than or equal to the number of smoking women in sample 2, augmented by the number of non smoking young individuals in sample 3. This can be written as

$$[N(YW)]_1 \leq [N(SW)]_2 + [N(NSY)]_3 \qquad (12.55)$$

which is inequality 12.54 supplemented by indices referring to the different samples used.

The two theorems given above are of a strict (elementary) mathematical nature. They use some properties (young or not, smoking or non smoking, man or woman), but the theorems themselves actually do no depend on the nature of these properties. We solely need the fact that each property is dichotomous. It could as well be "this or that" or, with another notation, "+ or −".

Let us now return to physics and consider again, as we have done previously, the case of two particles, perfectly correlated, flying apart. Let us analyze measurements carried out on the two particles, with a locality assumption, according to which a measurement on a particle does not affect a measurement on the other particle. We also assume that the properties of the particles are predetermined and are not affected by their journeys to the detectors. We may measure a property A on a particle and a property B on the other particle, on a large enough number of pairs of particles, called sample 1. On another sample of same size, called sample 2, we may measure property B on one particle and property C on the other particle. And, in a last sample, called sample 3, again of the same size, we may measure property A on one particle and property C on the other particle.

Let us call the property A the age, property B the sex, and property C the "smokiness", a neologism for being smoker or non smoker (names of properties do not matter, so why not using these names?). Let us also assume that the measurement of property A can only yield two values: + also named Young or − (not Young). Similarly, we assume that property B can only yield two values: + (Woman) or − (Man), and that property C may also only yield two values: + (Non Smoker) or − (Smoker). Then, inequality 12.55 can be translated to

$$[N(A:+,B:+)]_1 \leq [N(C:-,B:+)]_2 + [N(C:+,A:+)]_3$$

$$(12.56)$$

Inequality 12.56 is the Bell's microphysical version of the inequality 12.55 for PEDESTRIANS ($\sim\sim\sim\sim$).

In a paper by Bell [120], we have another PEDESTRIAN ($\sim\sim\sim\sim$) version, not concerning shoes for PEDESTRIANS ($\sim\sim\sim\sim$), but socks, the famous Bertlmann's socks. The story of Bertlmann's socks, to begin with, is similar to the one concerning Gérard and Monique, that we used to discuss common causes as a possible solution to the EPR-paradox. It happens that Dr. Bertlmann, a somewhat eccentric physicist (there indeed has been a motivating Dr. Bertlmann), likes to wear socks of different colors, everyday. For instance, one sock is always pink, and the other sock non-pink. By assumption, it is impossible to predict what will be the color of the right foot sock when you welcome Dr. Bertlmann, in the morning (say that the right foot is hidden by some useless and weird piece of trousers). But, you can watch the left foot sock. If you observe that this left sock is pink (non-pink), then you immediately know, without any measurement, that the right sock is non-pink (pink). This a perfect (rather short distance) correlation between (not too much) spatially separated systems, actually a fairly exact translation to the everyday experience of what has been a possible solution to the quantum EPR-paradox, also a vivid way to expose the EPR-paradox issue to your baker and to your butcher.

Now, we may have a PEDESTRIAN ($\sim\sim\sim\sim$) inequality, a macroscopic Bell's inequality, if we consider pairs of Dr. Bertlmann socks being washed. Here it is : *The probability of one sock passing at 0° and the other not at 45° plus the probability of one sock passing at 45° and the other not at 90° is not less than the probability of one sock passing at 0° and the other not at 90°.* This wearing socks inequality, as in the previous example, may be given a microphysical translation which, however, does not agree with quantum mechanical predictions.

Another simple form of such macroscopic inequalities may be found in an Appendix of a book by Squires [13]. In about the same spirit, I would also recommend the reader to examine the behavior of a magic, holy or devil, device described by Mermin [333], inspired by Hardy [370], with variants [75, 371–374]. According to Van Fraassen [21], one of these devices has already been used for military purposes. Here is Van Fraassen's report of the story. *There are two generals, Alfredo and Armand, who wish to strike a common enemy simultaneously, unexpectedly, and very far apart. To guarantee spy-proof surprise, they ask a physicist to construct a device that will give them a simultaneous signal, whose exact time of occurrence is not predictable. (This is a science-fiction story - their physics is like ours but their technology is much advanced. It happened in a galaxy long ago and far, far away...). The physicist gives each a receiver with three settings, and constructs a source which produces pairs of particles at a known rate, travelling towards those receivers. In each receiver is a barrier; if a received particle passes the barrier, a red light goes on, and otherwise a green one is on. The probability of this depends on the setting chosen. But when the two generals choose the same setting, one member of the pair of particles passes if and only if the other does not. Alfredo and Armand agree to choose a common setting and agree to turn on their receivers for 1 minute every other morning at eight o'clock (starting on a certain day), and then Alfredo will strike the first time his light is green while Armand will strike as soon as his light is red.*

Bell's theorems

Bell's theorem may basically be expounded as follows: A local hidden-variables theory cannot reproduce the whole set of quantum mechanical predictions. This is a consequence of Bell's inequalities.

The formulation above is the formulation we shall retain in this book. However, in the same way that there are many Bell's inequalities which may correspond to the different premises used to start the demonstration (for instance deterministic or stochastic hidden variables), we actually possess several Bell's theorems. For instance, d'Espagnat [84] distinguishes three Bell's theorems. The first theorem, called Bell 1 by d'Espagnat, is essentially the one which results from the original 1964 Bell's paper [189]. Premises are identified as follows (i) the locality assumption according to which particles do not interact any more when they are experiencing the measurements and (ii) a criterion of objective reality. This criterion of objective reality requires the use of hidden variables since Ψ alone is inoperative to define an objective reality. Therefore, from this point of view, as pointed out by d'Espagnat, hidden variables have not to be postulated but are required as a consequence of the premises. By the way, this explains why our PEDESTRIANS ($\sim\sim\sim\sim$) Bell-kind inequalities did not use hidden variables: The everyday world is classical, not quantum mechanical, and the second premise of objective reality is automatically satisfied. Then, we may also say, as pointed out by Aspect [375], that the locality assumption is crucial to derive Bell's inequalities.

I shall not discuss Bell 2 and Bell 3 theorems (in the terminology of d'Espagnat) which involve some subtleties not required for this book. But the interested reader who would like to dig further may refer to the book by d'Espagnat. Also, Peres [326] compared the logical structures underlying the theorems of Bell, for non-locality, and of Kochen and Specker, for contextuality. There are also some subtleties associated with the concept of counterfactuality which are discussed by d'Espagnat [84, 85, 104, 126]. Furthermore, I recall that Ref. [205] is a compilation of Bell's papers. Also, Kafatos [376] is another valuable compilation of papers devoted to Bell's work. In particular Kafatos [377] discusses the implications of Bell-type correlations to the early universe. Hooker [378] comments Bell's work in the perspectives of the Bohr–Einstein debate. Cushing and McMullin [369] deepen the concepts of locality and separability associated with Bell's theorems. A book edited by R.A. Bertlmann (the physicist with strangely paired socks) and A. Zeilinger [379] provides many complementary insights concerning Bell (the man and the work). An early history of Bell's theorem is also available from J.A. Clauser [204].

Experimental corroborations and falsifications

With the words "corroborations" and "falsifications", I explicitly refer again to the epistemology of Popper [25, 61]. We shall first discuss experiments which are tests for quantum mechanics, without however testing the locality issue. These experiments are in agreement with quantum mechanics. But an agreement between a theory and experiments never proves the validity of the theory. In the words of Popper, we have simply to say that the theory is corroborated. These corroborative experiments, I shall call them zeroth generation experiments. Next, we shall examine other experiments explicitly aiming to decide whether hidden variables are to be accepted or rejected. The results of these experiments is that we have to conclude that Bell's inequalities are violated by experimental results. Hence, local hidden-variables theories have to be rejected. In the words of Popper, they are falsified. These experiments are designated as higher-order generation experiments. Before the experiments are exposed, the reader should not make any prognostication on their results. For instance, Selleri [380], relying on a certain conception of realism, stated that *it would be strange if the experiments being made by Clauser, Holt, Horn and Shimony were to confirm quantum mechanics.* Well, we are going to see...

Zeroth generation experiments

Bell's inequalities emphasize the importance of experiments on pairs of particles which are spatially separated and correlated. However, such experiments do not necessarily violate Bell's inequalities, even if quantum mechanics is complete and correct. Indeed, evidencing Bell's inequalities requires specific experimental configurations. Therefore, if quantum mechanics is correct, experiments on correlated particles should satisfy quantum mechanics, but also, very often, should not violate Bell's inequalities. Such corroborating experiments have been carried out rather early, even a long time prior the advent of the first Bell's inequality in 1964, and therefore obviously did not discuss such inequalities. Nevertheless, the experimentalists might already have in mind the EPR-paradox, and the desire to check quantum mechanics on correlated pairs. There was also certainly the idea that, if quantum mechanics were falsified by such experiments, then a hidden-variables theory might be necessary, or at least such a theory could have been a possible solution to explain the observed discrepancies (which have actually not been observed). Besides, these experiments have also been valuable because they opened the way to more sophisticated investigations pertaining to higher-order generation experiments, explicitly dedicated to Bell's inequalities tests. Let us however remark that the aim of the EPR-argument, at least in the mind of the authors, was to show that quantum mechanics is incomplete, but not that it is incorrect.

The story starts with Wheeler [381] who suggested experiments to study correlated pairs of photons, more precisely of gamma ray photons emitted during electron-positron annihilation processes. These correlations concern the states of polarizations of the photons and can be evaluated into the quantum mechanical framework. Following this suggestion, effective theoretical evaluations have been carried out in 1947 by Pryce and Ward [382] and in 1948 by Snyder *et al.* [383]. According to Pryce and Ward, if the two gamma quanta emitted on the annihilation of a slowing moving (ideally at rest) electron-positron pair both undergo a Compton scattering event, then, in view of the fact that they are polarized in perpendicular directions, it is reasonable to expect that the scatter distributions will be correlated in azimuthal angles. Theoretical evaluations are carried out under such circumstances. The relevance of Compton scattering is due to the fact that gamma detection is made by means of this kind of scattering. This was compulsory because no effective polarization filters were available for hard photons, such as the ones produced by annihilation radiation, so that polarizations had to be measured indirectly through the effectiveness of Compton scattering at chosen angles [1].

The first experiment implementing this scheme is due to Bleuler and Bradt, in 1948 [384]. They measured, using Compton scattering, maximum and minimum rates of coincidence between counters detecting annihilation quanta for perpendicular and parallel orientations of the two polarizations associated with the pair of emitted quanta. They claimed that experiments are in agreement with quantum mechanical predictions. Similar experiments have also been published, the same year, by Hanna [385]. In contrast with Bleuler and Stadt, Hanna found significant discrepancies between quantum mechanical predictions and experimental results, but remarked that, although the results from Bleuler and Stadt were in agreement with quantum theory, they were spoiled by a sufficient range of uncertainty to make them compatible with their own results. So, at this stage, the results were rather inconclusive.

Other similar experiments have afterward been reported by Vlasov and Dzehelepov, in 1949, yielding a qualitative agreement with the theory (see [156]) and by Wu and Shaknov in 1950 [386]. Wu and Shaknov's results were in agreement with quantum mechanics too. Furthermore they were in disagreement with Furry's hypothesis, proposed as a possible solution to the EPR-paradox, concerning spontaneous factorization of entangled wave functions. Hereford's results, published in 1951 [387], were rather inconclusive. As far as I can see, results from Bertolini *et al.*, in 1955 [388], were only marginally in agreement with quantum mechanics, or even slightly in disagreement: A certain ratio ρ of coincidence counts is theoretically evaluated to $\rho = 2.32$, while experiments yield $2.09 < \rho < 2.27$.

Up to now, we therefore have a set of experimental results leading to an overall fair corroboration of quantum mechanics but with some marginal, inconclusive, or even contradictory results. Borrowing words from Wheeler and Zurek [100], let us say that the agreement is only qualitative. However, Bertolini *et al.*'s experiments have been improved by Langhoff, publishing in 1960 [389], with an improved angular resolution, agreeing well with quantum mechanics. Still more precise measurements of this kind have been published by Kasday or Kasday *et al.* [390–392], after Bell's inequalities were announced to the world.

These last experiments from Kasday and Kasday *et al.* have been conveniently exposed under the headline of zeroth generation experiments, because they deal with annihilation radiation. But we are already encroaching a bit on the territory of first generation experiments (although such first generation experiments will be devoted to the use of optical photons). Indeed, the interpretation of these experiments, with the aid of reasonable additional assumptions, concludes to an agreement with quantum mechanics, and to a violation of Bell's inequalities. Furthermore, they falsify once more Furry's hypothesis. However, in an Appendix to his 1971-paper, Kasday explained, with a counter-example, that these experiments (and their interpretation) cannot definitively rule out local hidden-variables theories.

Next, still using annihilation radiation and gamma rays, Faraci *et al.*, in 1974 [393], exhibited a significant disagreement with quantum mechanics, although, more moderately, they conclude: *The results do not give a decisive answer for the existence of local hidden variables because, they stay, within the experimental uncertainty, just on the Bell's limit.* Furthermore, in Faraci's experiments, some of the annihilation photons used had a coherence length of 7 cm, others of 47 cm, allowing one to observe a decrease of correlations with an increase of the distance between measurement locations, therefore supporting Furry's hypothesis.

These results have been contradicted by Kasday *et al.* [392], already cited above, Wilson *et al.* [394] and Bruno *et al.* [395] who, in contrast, found a satisfactory agreement with quantum mechanics. Kasday *et al.* [392] stated that their results are evidence against local hidden-variables theories. However, in a fairly contradictory way, they also wrote: *It would be pleasing to be able to say that the results of this experiment rule out local hidden variables theories. We cannot say that, and in fact, it appears that no experiment done with currently available techniques could lead to such a definite conclusion* (in particular due the need of at least one additional assumption required for the interpretation), adding: *It has been pointed out ... that unless measurements are made without polarizers, and without making such an assumption, a hidden variables theory could be constructed which reproduces the predictions of quantum mechanics for such experiments.*

Furthermore, in Wilson *et al.*'s experiments, the separation between the photon source and the detectors could vary by as much as 2.5 m, and the difference in separation between detectors by as much as 1 m, to be compared with a coherence length of photons of 12 cm. No dependence of polarization correlation upon either kind of separation was observed, contradicting Furry's hypothesis. Bruno *et al.'* experiments, once more, were not compatible with Furry's hypothesis. The relevance of these works, and others, for testing Furry's hypothesis is discussed by Bohm and Hiley [332].

The use of gamma rays and hence of Compton scattering has been recognized as an inefficient way to deal with correlated photons, certainly explaining the various contradictory results obtained. As a consequence, an experimental test of Bell's inequalities, and of local hidden-variables theories, after the advent of the second quantum mechanical revolution of Bell, could not be made convincing enough with such means. We should also have in mind the Kasday's counter-example of 1971: It is possible to construct an *ad hoc* local hidden-variables theory that reproduces all the results obtained by using Compton scattering of annihilation photons. These are supplementary reasons to incorporate all annihilation photon experiments under the headline of the present subsection, even when they deal with Bell's inequalities. More convincing experiments can be made with lower energy photons, say optical photons, where, instead of indirect detections via Compton scattering processes, we may use more conventional and efficient polarizers and detectors, more readily available.

An example of such a work (the first one actually) is by Kocher and Commins, in 1967 [396], who investigated correlations in linear polarizations of two photons emitted in a calcium atomic cascade. This paper appeared only three years after Bell's work of 1964 (published in a rather obscure journal, we must say, a journal that, as far as I understood, only had one issue), and Kocher–Commins did not make any reference to Bell's inequalities. Their experimental results agree with quantum mechanical predictions, but the data were taken under such circumstances that a latter re-interpretation, with Bell's inequalities and local hidden variables as the goal, could only be inconclusive (recall that the violation of Bell's inequalities requires some specific experimental configurations). Furthermore, the polarizers used were not efficient enough to make their results reliable. Nevertheless, with these experiments, the next step drives up to first generation experiments.

More by Belinfante [1], Jammer [24], Pipkin [156], or Wheeler and Zurek [100]. Let us also mention Freedman and Holt [300] who, while already dealing with low energy photon correlations, quoted a series of experiments concerning polarization correlations of annihilation photons stopped in matter.

First generation experiments

Some of the zeroth generation experiments already demonstrated that the issues raised by Bell's inequalities could be brought into the experimental domain, and became really physical rather than academic or metaphysical. But the use of annihilation radiation was poorly adapted to the purpose. With the introduction by Kocher and Commins of experiments using low-energy photons, a new avenue was opened for more sophisticated and convincing experiments, that I call first generation experiments, relying on the study of the correlations of polarized photons emitted by atomic cascades. I shall report on these experiments using a nearly perfect chronological order.

Clauser *et al.* [354], five years only after 1964 Bell's paper, have recognized the significance of Bell's inequalities and theorems to an issue which has been propagating forward in time since the EPR paradox, and which could, thanks to Bell, from this time on, be experimentally investigated. Analyzing the previous literature, they pointed out the inadequacy of experiments based on annihilation radiation and Compton scattering, the one of Wu and Shaknov in particular [386], and also pointed out that the experiments with low energy photons by Kocher and Commins [396] used polarizers which were not sufficiently efficient. Furthermore, measurements were made only for two particular values of the relative orientations of polarizers, namely $0°$ and $90°$, far from being sufficient to track Bell's inequalities. Accounting for this analysis, they went on proposing an improved experiment to study the polarization correlation of a pair of optical photons generated by an atomic cascade, that was expected to provide a definitive discrimination between quantum mechanical predictions and the ones derived from local hidden-variables theories, a discrimination relying on Bell's inequalities. However, for doing this, as we already mentioned, the authors had to present a generalization of the original Bell's theorem which could be applied to realizable experiments, namely the CHSH-version.

The test proposed by Clauser *et al.* has thereafter been carried out by Freedman and Clauser [397]. I would like to report the results by quoting Bohm [366] because such a quotation gives us the opportunity of a valuable connection with Bohm's mind and view: *The result definitively fits in with the notion that the quantum system behaves as an indivisible whole, not relevantly analyzable into independent parts, each of which would exist in a separate region of space. And so, the need for new forms of insight into this question is made even sharper and clearer than it was before.* We may also let Freedman and Clauser speaking with more heart cooling terms: *We have measured the linear polarization correlation of the photons emitted in an atomic cascade of calcium. It has been shown by a generalization of Bell's inequality that the existence of local hidden variables imposes restrictions on this correlation in conflict with the predictions of quantum mechanics. Our data, in agreement with quantum mechanics, violate these restrictions to high statistical accuracy, thus providing strong evidence against local hidden-variables theories.* To be specific, the result from Freedman and Clauser was expressed as an experimental correlation factor equal to 0.300 ± 0.009, agreeing with quantum mechanical predictions giving a value equal to 0.301, but in contradiction with the relevant Bell's inequality implying a value smaller than 0.25.

But bad news are coming. Soon after, results by Holt, and by Holt and Pipkin, however not released in the archival literature [398, 399], were strongly in favour of local hidden-variables theories. They obtain 0.216 ± 0.013, not violating Bell's inequalities, and differing flagrantly from the quantum mechanical value equal to 0.301. Holt and Pipkin, in spite of a careful analysis, did not succeed to detect systematic errors which would be sufficient to explain their result, nor later exegetes. For some reason, Holt and Pipkin certainly remained skeptical about their findings, likely explaining why they have not been published. Furthermore, the authors recommended that the experiment should be repeated by someone else with different configurations of apparatus (this will be done).

Although the reader might feel that I am going into a digression, I would like to comment more on these experiments, from methodological and epistemological points of view. The results from Holt and Pipkin will remain embarrassing, even if other convincing evidences were afterward obtained. Usually, as I could find in the literature, people reassured themselves by remarking that, generally, all possible systematic errors would tend to wash out the correlations. Therefore, we may argue, experiments in favour of local hidden variables, with correlations smaller than the expectable quantum mechanical values, are less convincing that other experiments falsifying local hidden variables. But I hope not to shock anyone if I dare say that this is a shabby-looking loophole. It is nice that the experiments of Holt and Pipkin have indeed be repeated with different configurations, as recommended but, considering the importance of the issue at stake, it would have been compulsory to reproduce it exactly with the same configuration, in order to make experimental investigations on this experimental investigation. We can imagine that this has been impossible at this time, due to contingencies and, nowadays, it is likely not possible any more because, considering technological advances, the experimental required ingredients are possibly not available any more. Therefore, we have here a falsifying experiment which cannot be falsified, that is to say, in a Popperian sense, which has become metaphysical. We have likely to do with this forever.

Let us return to the chronology of first generation experiments. We then have to mention Fry [400] who, in a very technical paper, derived a general formula for two-photon coincidence rates, worked out theoretical predictions for different experimental configurations, and consequently discussed various experiments to test hidden-variables theories. In particular, he analyzed a mercury cascade, the kind of cascade used by Holt and Pipkin. Three years later, the same author, together with Thompson [401], relying on the previous analysis dated 1973, had repeated the two-photon correlation experiment with such a cascade, and obtained results in excellent agreement with quantum mechanics, and in clear violation with a Bell's inequality.

Next, Clauser and Horne [356], as Fry, went on with a theoretical analysis. After a critical review of the previous literature, since Bell ten years before, the authors defined a class of hidden-variables theories that they called objective local theories, relevant to experimental arrangements in which, as already discussed, two correlated particles emitted by a source fly apart to be eventually detected by detectors, each detector being preceded by an analyzer. The positions of the two analyzers are defined by externally adjustable parameters **a** and **b** which can be taken (but not necessarily) as being the angles specifying the orientations of the analyzers. A significant step is that the notion of locality is refined by explicitly introducing the concept of space-like separation, just as defined in relativity: The interaction between one particle and its associated apparatus (analyzer plus detector) has no time, considering the speed of light limitation, to affect the interaction between the other particle and its apparatus (this refinement will prove to be useful for later discussions). As a consequence, the probability for an apparatus to be triggered does not depend whether the other apparatus is triggered or not, or upon the choice of the parameter used for this apparatus, or even whether this apparatus is installed or not. Then, an objective local theory, whether deterministic or stochastic, is defined by the factorization of probabilities or associated numbers. Thereafter, the authors derived a consequence of any objective local reality, which is ready for experimental tests, taking the form of a Bell-kind inequality, possibly (under some circumstances) violated by quantum mechanical predictions. This Bell's inequality had not yet been presented elsewhere, and suggested new experiments. A subsequent discussion insisted on the fact that testing inequalities requires drastic conditions on the efficiencies of analyzers and detectors. This difficulty may be circumvented by relying on an assumption, called non-enhancement assumption, according to which the probability of a count with a polarizer in place is less than or equal to the probability with the polarizer removed. Reanalyzing the aforementioned paper by Freedman and Clauser [397], it is then found that, using an auxiliary assumption (the non-enhancement assumption, or a similar one), the results obtained concerning the violation of a Bell-kind inequality were confirmed or even reinforced.

In 1975, Freedman and Holt [352] were motivated by the fact that *recent experiments to test the predictions of local hidden-variable theories have led to conflicting conclusions* (remember the experiments by Holt and Pipkin). They therefore reviewed the theoretical foundations of the experiments, and the experiments as well. To provide a bit more of information on the contradiction between experiments, let us state that Freedman and Clauser [397] used an atomic cascade in calcium, and exhibited a violation of the corresponding Bell's inequality by more than six standard deviations, while Holt and Pipkin used an atomic cascade in mercury. The authors also stated, in their review: *Because there are many more systematic errors that reduce the strength of the correlation than increase it, we would be inclined to favor the calcium results ... but the issue is too fundamental and too important, to allow any experimental discrepancies to remain unexplained.*

In 1976, with Clauser [357], we return to actual experiments. The aim was to revisit, in some way, the experiments by Holt and Pipkin. Indeed: *An experiment was performed to search for the anomalous two-photon polarization observed earlier by Holt and Pipkin using a cascade of atomic mercury. Although the present experimental arrangement differed only slightly from theirs, the anomalous results were not observed.* Fortunately (for defenders of the usual quantum mechanics): *The discrepancy with the quantum-mechanical predictions observed by Holt and Pipkin was not found in the present data,* but, unfortunately: *However, the cause of their discrepancy was not pinpointed either.* The same year, complementary results have been published by Clauser [358]. His results exhibited a reasonable agreement with quantum mechanics, without however violating the appropriate Bell-kind inequality (recall once more that inequalities are not necessarily violated by quantum mechanics). Furry's hypothesis is nevertheless clearly refuted by the data obtained.

In 1978, Clauser and Shimony [359] provided a detailed survey of experiments up to this date, and also a variant for a proof of Bell's theorem. According to these authors, essentially all local theories formulated within a realist framework could be tested by using a single experimental arrangement. Besides evaluating the experimental results already obtained, they gave prospects for future experiments. After the use of annihilation radiation and of cascade-photon experiments, they also consider a third class of experiments, namely experiments based on proton-proton scattering.

Such a proton-proton scattering experiment has been reported by Lamehi-Rachti and Mittig, in 1976 [402]. Bell's inequality is tested by using the measurement of spin correlations (the kind of situation discussed in the Bohm-EPR argument, no more photons but spins again) in low-energy proton-proton scattering. The experimental results are in good agreement with quantum mechanics, and violate Bell's inequality. According to Pipkin however [156], *the experiment does not give an unambiguous test of the Bell inequality because of the low efficiency of the polarization analysis. It does, however, provide a strong argument against the local hidden variable theory and for quantum mechanics.*

More information on first generation experiments are available from Belinfante [1], Freedman and Holt [300], Bohm and Hiley [68], [332], Aspect [403], [404], Pipkin [156], d'Espagnat [104], Selleri and Tarozzi [347], Wheeler and Zurek [100], Wigner [74], Squires [13] or Shimony [93]. Also, Clauser [204] delivered a real-life history of these experiments.

Non-locality and non-separability

Before dealing with second generation (Aspect) experiments on correlated photons, implementing drastic improvements of the exper-imental arrangements, some kind of culmination made accessible by all the mountaineers who paved the track, it is useful to make a break to discuss a bit more the issues of non-locality and non-separability, two notions that we decided up to now to identify. I do not intend, in this subsection, to tell everything about these issues but simply to prepare our mind, to inflict to the reader a first inoculation. In the next subsection, a booster injection will be delivered.

We already have in mind the notion of locality just as discussed by Bell in 1964, with other opportunities for refinements provided by the derivations of various variants of the original Bell's inequality. It is now important to remark that the original locality assumption used by Bell never referred to relativity. Since the beginning of our discussion on non-locality, we had to wait Clauser and Horne, in 1974, just above, before introducing the expression of space-like separation, so typical of a relativist framework. Conversely, the invocation of a locality requirement by Bell, in 1964, was simply reasonable and natural, fitting rather well our physical instincts. But no theoretical reason, no fundamental principle, and none of our available theories, was used to support the idea. As it was introduced, it rather looked as a supplementary principle to be introduced as an extra-physical principle.

Nevertheless, we could for instance imagine that there existed a kind of interaction, still unknown to us, some para-tachyonic effect or some time zigzags *à la Beauregard*, which would indeed allow one detector to have an instantaneous influence on the other one, or more generally to instantaneously connect what happens here and what happens there. But, we possess indeed a well established theory, relying on two postulates, one of them (the relativity principle) being a first principle: A theory which imposes the limitation of the speed of light to the transmission of information between two separated pieces of space, namely relativity. Then, with this framework in mind, let us again consider two correlated photons, emitted by a source, flying apart into opposite directions. Assume that a first detector is triggered by a photon, and analyzed by using an analyzer whose orientation is defined by a direction **a**. Let the second detector situated at a distance X from the first detector. The time required for any information, satisfying relativity, to travel from one detector to the other is denoted as t_{travel} and must satisfy: $t_{\text{travel}} \geq X/c$, in which c is the speed of light, and the equality holds if, and only if, the information propagates at the speed of light. Let us assume that the orientation **b** of the second detector is chosen at a time t, after the triggering of the first detector, satisfying $t < t_{\text{travel}}$. Then no information satisfying relativity had time, starting from the first detector, to reach the second detector. We therefore have a refined definition of locality, that we shall now call separability: What happens here cannot be connected with what happens there, by any information propagating at a speed larger than the speed of light. With the kind of experimental set-up just discussed above in this paragraph, the separability assumption might be tested. But this specific example we discussed is not convenient for actual experiments. More generally, and in a better way for experimental implementation, as proposed by Bell, we should modify quickly enough, and ideally in a random way, the orientations of both polarizers during the time of flight of the photons, in such a way that no influence, satisfying relativity, had time to put the detectors into communication. This is the concept used by Aspect in his second generation experiments. If the correlations are washed out when the orientations of the polarizers are switched quickly enough, then the world of quantum events is a separable world. Otherwise, the world of quantum events is a non-separable world. In the last case, the possibility of a conflict between relativity and quantum mechanics would have to be examined.

The above concept of separability is well in resonance with Einstein's mind. As stated by d'Espagnat [405]: *For the sake of convenience let us call the following principle the principle of separability, as formulated by Einstein: If S1 and S2 are two systems that have interacted in the past but are now arbitrary distant, the real, factual situation of system S1 does not depend on what is done with system S2 which is spatially separated from the former. This principle can be somewhat widened or understood in a broad sense so as to mean: No finite influences can propagate arbitrarily far away. Alternatively, it can be given the more restrictive sense that there are no space-like propagation of influence.*

There is a huge literature devoted to the concepts of locality (non-locality) and separability (non-separability), sometimes contradictory, not necessarily because the authors do not agree on the essence, but because they may have chosen different definitions, and/or ramified sub-definitions. Compare for instance the definitions above (to which we shall hold fast) and the definitions of Bub and Clifton [184]. For them, locality is defined as follows: *If two systems are spatially separated, then the determinate properties ('real state') of one system cannot be directly influenced by any measurement on the other system,* and separability as follows : *The determinate properties ('real states') of spatially separated systems are independent of each other,* definitions which however must be best understood in a hidden-variables framework.

D'Espagnat [84] confirmed that such words as locality, separability, and their negative counterparts, do not always have the same meaning in the writings of different authors, and that also, sometimes, several such names may be used by different authors to designate the same concept. See also d'Espagnat [69] for a review on the concept of separability (together with Bell's inequalities). Cushing [131], in a rather monistic vision, agreed that there are various types of locality, but decided to use interchangeably the terms non-local and non-separable, although *logical distinctions do exist between them.* Locality and separability are also discussed by Bell, e.g. in 1966 [188], with other complementary discussions available from Clauser *et al.* [354], d'Espagnat [405], Aspect [403, 404], Garuccio and Selleri [363], Pipkin [156], Wheeler and Zurek [100], Bub and Clifton [184] Furthermore, a book edited by Cushing and McMullin [406] contains many relevant papers, in particular those by Jarrett [75], Stapp [407], Hughes [408] or Howard [334].

Second generation (Aspect) experiments

The usefulness of the discussion given in the previous subsection is attested by Aspect [403] when he stated: *The words locality and separability, and likewise local and separable, are sometimes taken as synonymous. However, the specific definitions given here are natural and, in the context of the present paper, useful.* To fully understand this quotation, and better see the connection between the previous subsection and the present one, let us summarize the work of Aspect, published in 1976, even if this introduces a bit of redundancy. Such a redundancy could be eliminated but, considering the subtleties involved, it is a good idea, for pedagogical purposes, to accept it and even, deliberately, to reinforce it.

Following Aspect [403, 409] who proposed new experiments to test Bell's inequalities, the crucial point to the derivation of inequalities is not determinism, as proved by the fact that stochastic theories may lead to inequalities too (recall Bell [327], the local objective theories of Clauser and Horne [356], and see also Selleri and Tarozzi [347], Bohm and Hiley [68]). Indeed the crucial point is the locality assumption and it is this locality assumption which has been tested, and as a whole rejected, by previous experiments. But it remained to test separability, or more precisely Einstein's separability.

The locality assumption, let us recall it, tells us that the setting of a measuring device does not influence the results obtained with another remote measuring device (and also does not influence the emission process at the source). Invoking relativity theory, we have to introduce separability which may be stated as follows [295]: *The setting of a measuring device at a certain time (event A) does not influence the result obtained with another measuring device (event B) if the event B is not in the forward light cone of event A (nor does it influence the way in which particles are emitted by a source if the emission event is not in the forward light cone of event A). In other words, all these events are space-like separated.*

Then, according to Aspect, we can conceive separable theories that do not fulfill Bell's locality. Such theories take into account the possibility of interactions between remote measuring devices (this means that they do not satisfy Bell's locality) but these interactions do not propagate faster than light (they therefore satisfy Einstein's separability). Hence, we may possess theories satisfying Einstein, without satisfying Bell, a way to show that Einstein's separability does not imply Bell's locality. Another way to say this, borrowing an expression to Aspect, is to state that the principle of separability is a *weaker assumption* than the principle of locality. But (this is an important issue), it may be connected with an existing theory, namely relativity.

What Aspect proposed to achieve is then to test Einstein, rather than Bell. In an EPR-experiment, if the orientations of the polarizers are set in advance and kept fixed, we may imagine that some kind of information may have been exchanged between the particles, even at a velocity less than or equal to that of light, so that eventually both particles had enough time to become correlated. As stated by Bell [189], *the settings of the instruments are made sufficiently in advance to allow them to reach some mutual rapport by exchange of signals with velocity less than or equal to that of light.* Then, what happens for one apparatus may depend on what happens for the other apparatus. Therefore, following an idea already previously put forward by some people, such as Aharonov and Bohm [410], Bell [189], Shimony [355] or Clauser and Horne [356], in order to test separability, Aspect [403, 409], and see also [404], proposed that the orientations of the polarizers should not be set in advance and kept fixed. The second generation of experiments discussed and started by Aspect then would, at least in an ideal version, use *versatile polarizers whose orientations are changed rapidly and repeatedly in a stochastic manner,* and also independently, in a time comparable with the time of flight of the photons (time varying analyzers). The setting of one detector is then made at a time when it is already too late to permit its influence, even propagating with the speed of light, to reach the other detector.

To this second generation also pertain less ideal configurations in which a complete stochasticity in the manner to change the orientations of the polarizers is not achieved. In particular, in the 1976 proposal of Aspect, as he said: *We believe that the experimental scheme we are proposing, although it is not an ideal one, is interesting in that it embodies a device for changing the orientations of the analyzers in a time comparable to the time of flight of the photons.* The paper also discussed generalized Bell's inequalities adapted to such experimental arrangements, inequalities which *can be derived from the principle of separability, with no further locality assumption made. Since these inequalities still conflict with the quantum-mechanical predictions, such modified experiments would be able to discriminate between quantum mechanics and separable hidden variable theories.* Furthermore, *such experiments could also allow one to reject Furry's assumption, without the locality assumption.*

Separability versus locality is also discussed by noting that *the impossibility of an influence between two distant polarizers appears as a consequence of the principle of separability, which is a less stringent assumption than the principle of locality tested in previous works.*

Experiments (without however using time varying analyzers) were first published in 1981 by Aspect *et al.* [411], by studying the linear polarization correlation of the photons emitted in a radiative atomic cascade of calcium. The results are in excellent agreement with quantum mechanical predictions, exhibit a strong violation of generalized Bell's inequalities, and rule out a whole class of realistic local theories. They are also in disagreement with Furry's hypothesis. They provided the clearest evidence hitherto available against local hidden-variables theories. The consequences of such experiments are also drawn by Aspect [404] according to which long distance correlations between space-like separated events cannot be simply interpreted in terms of common causes in the past. The choice of the orientation of a polarizer therefore seems to influence instantaneously, that is to say faster than allowed by the speed of light, the results of the measurements carried out by means of the other polarizer. However, he added, from this point up to the idea that we could telegraphy, that is to say transmit usable signals faster than light, there is only one step ... that we should avoid to make.

In 1982, Aspect *et al.* [412] reported new experimental results still based on the study of correlated photons generated by a calcium cascade, with an improved set-up, although still without using time varying analyzers. These results yield *the strongest violation of Bell-kind inequalities ever achieved, and excellent agreement with quantum mechanics.* Also, the authors, still giving some hope to the opponents to the Copenhagen interpretation, state that *only two loopholes remain open for the advocates of realistic theories without action at a distance. The first one, exploiting the low efficiencies of detectors, could be ruled out by a feasible experiment. The second one, exploiting the static character of all previous experiments, could also be ruled out by a timing experiment with variable analyzers now in progress.*

Eventually, after several years of theoretical and experimental training, including the previous experiments without using time varying analyzers, experiments *with* time varying analyzers could be reported for the first time [413]. The results were in agreement with quantum mechanics, and violated Bell's inequalities by more than five standard deviations. However, *the ideal scheme has not been completed since the change is not truly random, but rather quasi-periodic ... A more ideal experiment with random and complete switching would be necessary for a fully conclusive argument against the whole class of supplementary-parameter theories obeying Einstein's causality.* Note that this sentence uses the terminology 'supplementary parameters' rather than the one of 'hidden variables', in relation with the fact that the expression 'hidden variables' has often been denounced as a misnomer (as an example, recall that Ψ is unobservable, and therefore can be viewed as hidden too). A very pedagogic, enlightening, and convincing review of such experiments, and of the issues involved, is available from Aspect [375].

Furthermore, we must from now on keep in mind that the intriguing features of non-locality and/or non-separability have nothing to do with any form of quantum theory, or with any available interpretation. These features occur in the phenomena themselves, and pertain to the stuff of the world. It just happens that any empirically adequate theory, like quantum mechanics, must accommodate them. Complementary discussions are available from Ballentine [36], d'Espagnat [405], Pipkin [156], or Bohm and Hiley [68].

Criticisms, counter-criticisms, and a brief romance

Should we be perfectly convinced by experiments violating Bell's inequalities, in particular by Aspect experiments? Nearly so. Indeed, it would look incredible that such an accumulation of experimental results, in agreement with quantum mechanics, and in disagreement with local (and separable) hidden-variables theories, would eventually be found misleading. This would really require us to invoke some mysterious and diabolical conspiracy, possibly forged by the

evil demon of Descartes, this evil demon that Descartes himself
eventually rejected on the basis that (although not in the words of
Descartes) God could be subtle, but not malicious (a link between
Einstein and Descartes spirits). However, once again, the issue is so
fundamental and far-reaching that we should not allow us to take
any risk, like omitting a detail that could possibly turn upside down
the whole landscape. It is indeed a Darwinian agent of progress to
criticize and possibly falsify, in Popper's spirit. And the advocates of
local/separable hidden-variables theories do their job properly, even
if desperate, when they vigorously search for loopholes. Therefore, in
this subsection, I shall discuss criticisms, counter-criticisms, and take
the reader to a brief romance with third generation experiments.

As mentioned by Aspect himself, the ideal theoretical scheme
of time-varying experiments has not been perfectly completed
because the switching of the variable polarizers was not genuinely
random, but rather quasi-periodic. Nevertheless, Aspect remarked
that switches on the two sides were driven by different generators
at different frequencies and that, therefore, it is very natural to
assume that they functioned in an uncorrelated way. Furthermore,
later on, an experiment has been constructed unifying the aspects
of large separation distance (about 360 m) and fast switching, with
randomness implemented [414, 415].

Another criticism is that the photon detectors were not perfect,
that is to say they did not have an efficiency equal to, or at least
sufficiently close to one. As stated by Bohm and Hiley [68], *with
the actual rather low efficiencies of these detectors, there seems
to be room for assumptions concerning the hidden variables of the
apparatus which could still preserve locality in spite of the experimental results. However, these assumptions seem rather arbitrary and
artificial and, in fact, they give the impression of being contrived just
to 'save the appearances'.* In fact, the experimental demonstration
of the violation of Bell's inequalities is only effective if we make an
extra-assumption, namely that, on average, the photons which have
not been detected would behave, if they were detected, in the same
way that the photons which have effectively been detected [85, 375].
Obviously, as a consequence of the low efficiencies of the detectors,

not only all pairs of photons are not detected but, also, one member of a pair may be detected, without the other member being detected.

Pipkin [156], discussing the issue, pointed out what would be the obvious ideal situation: *It would be preferable to have a system with unity detection efficiency that was sensitive to both members of each correlated pair* and, I dare to add, which would be sensitive to each pair. Pipkin also refers himself to a paper by Pearle [416] who *has shown how agreement with quantum mechanics can be obtained for a deterministic local hidden variable model by rejecting 'anomalous' data in which only one particle is detected.*

We however have the reassuring statement that all experimental inefficiencies lead to a decrease of the correlation function [375]. That is to say, if we consider an experiment in disagreement with quantum mechanics, not violating Bell's inequalities (exhibiting a too poor correlation in the data), and therefore being consistent with the existence of local or separable hidden variables, then we may found them questionable, arguing that experimental defects could have biased the results toward a decrease of the observed correlations. Conversely, if we consider an experiment in agreement with quantum mechanics, violating Bell's inequalities, and therefore leading to the rejection of local or separable hidden variables, then it is more difficult to invoke experimental defects to contradict the results obtained. We have here a fortunate lack of symmetry between success and failure.

So, asking again the question : Should we be perfectly convinced by experiments violating Bell's inequalities?, we may refine our answer, and state it as: More than nearly so. But some people remain skeptical. For instance, in 2005, Genovese [417] could write that *a conclusive experiment falsifying in an absolutely uncontroversial way local realism is still missing.* He also stated that, after forty years of experiments, no conclusive experiments have yet been performed, mainly due to the low efficiencies of the detectors used that demand for additional assumptions. He also expressed the hope that, he believed, an ultimate experiment could not be far in the future. He also nevertheless acknowledged that this was a personal opinion, that this opinion is not generally shared, and that indeed most of the physicists are convinced in face of the large amount of experimental data in disfavor of local hidden variables. There is also the opinion that, for instance after the resistance of detection loopholes to be eliminated, there could be a practical impossibility of falsifying local realism. Hence, if only to please Genovese, and other similarly skeptically oriented scientists, the interest of third generation experiments is warranted. Another discordant voice is by Marshall *et al.* [418] who claimed that local realism has not been refuted by atomic quantum experiments, but these authors proposed a local realistic model which does not perfectly agree with quantum mechanical predictions.

Aspect [375] provided a rather recent review of third generation experiments, developed since the late 80's. According to this reference, these *experiments, with new sources and schemes, have lead to a series of Bell's inequalities which have all confirmed quantum mechanics. They can close most of the loopholes still left open in the second generation experiments.* Genovese [417], although skeptical as we could see him just above, provided a review with more than 500 references.

For the brief romance, let us mention Aspect-like experiments by a group of Genova, with Gisin [419], where quantum correlations have been observed over more than 10 km, and by a group of Vienna, with Zeilinger [420], where entanglement-based quantum communication has been achieved over 144 km.

I would like to end this subsection by giving my own appraisal of the situation. This is not the opinion of an expert having carried out by himself any experiment of that sort in the field. Rather, it is the opinion of a theoretician (although not, for most of his life, a quantum mechanist) who examined the literature and had carefully to make up his mind. It is also the opinion of an author, discussing the foundations of quantum mechanics, with an unassailable fallback position. Indeed, for him, this author, it does not matter whether quantum mechanics is experimentally correct or not (agreement with experiments form the crowning of a theory, not its origin). But what does matter, in the framework of the present book, is the relationship between the concepts of quantum mechanics (and its formalism) with the concept of hidden variables (and their formalisms). Here, the messages from Kochen and Specker, and from Bell, and from others too, are clear: Hidden-variables theories, if they have to agree with quantum mechanics, must be contextualist and non-local (non-separable).

This being said as a preliminary to an end, here is what I adhere to regarding experiments: I am convinced.

Teleportation

Bennett *et al.* [421] demonstrated in 1993 that the property of quantum entanglement could be used to achieve a quantum teleportation of states. Some authors like to refer to Aristotle to explain the issue. Indeed, in quantum teleportation, the substrate that is to say the matter (a photon, an atom ...) is not teleported. What is teleported is the form, that is to say the quantum state which, in some sense, informs the substrate of what it has to be. The distinction discussed by Aristotle between matter and form finds here an unexpected justification and actual implementation.

I do not intend to report on the successful experimental implementations which followed, but only to mention an issue relevant to our purpose. We could believe naively that the success of teleportation of states could be another way to prove that Bell's inequalities have been violated, but the issue is more subtle, as discussed by Popescu, in 1994 [422], one year after the paper from Bennett *et al.* Popescu indeed remarked that Bell's correlations and the capacity for teleportation seem equivalent. This comes from the fact that if the two particles constituting the non-local channel are maximally entangled, as for a singlet state, then a state can be teleported from a location (where Alice is operating) to another location (where Bob is operating) with perfect accuracy. Therefore, it is easy to draw the following conjecture: *Whenever two systems which are separated in space violate some Bell inequality, they can be used for teleportation, and vice versa.* But Popescu actually shew that *surprisingly enough, the above conjecture is wrong.* Indeed, he established *that there are situations in which two systems separated in space do not violate any*

Bell inequality but still are useful for teleportation. This subsection may be viewed as a necessary side remark, however outside of the main stream of this book.

Relativity challenged?

So, the world, as far as we can see, is contextual, non-local/non-separable, as taught to us by the formalism of quantum mechanics, by Bohm's pilot wave and its hidden variables, and also as nearly perfectly checked by a huge amount of more and more sophisticated experiments. Sure, there are still skeptical people. But, you know very well, it is nearly impossible to find any statement on which every one would agree. You may always take the point of view of the systematical doubt as expressed by Descartes, and decide to hold fast on it, even if Descartes did not eventually do this. There are still individuals, or sects, or various kinds of communities, believing that dinosaurs are still alive, hidden in huge African caves, that American heroes did not land on the Moon in 1969, that God created the universe as it is, a small number of millenniums ago (something that is most strictly impossible to refute from a pure logical point of view), and what else? There is also the frightening possibility that some of them may be right, as we might argue by remembering how many people believed that the earth was flat, located at the center of the universe, that there were different laws of physics for the sub-lunar and celestial realms (even Aristotle, this genius, and a majority of learned doctors of the brightest universities of the Middle Age), and what else?, nevertheless eventually knocked down by the evolution of beliefs, and of objective science. So, yes, however, as far as we can see (most of us), the world is contextual, non-local/ non-separable. The reader who does not agree might decide to stop reading this book, if even he started to read it. I am going on writing for the enlightened (or illuminated?) ones who share this conclusion.

Then, we are facing this highly non classical participatory universe of wholeness, to use this word so praised by Bohm and that Bohr would not have rejected, if we are allowed to guess. Entangled objects are deeply connected, outside of space and time, in agreement with the vision of Bohr. When we say, in EPR-kind experiments, that what happens here is instantaneously influenced by what happens there, we are compulsively thinking, trained by everyday life, and educated as we have been by our study of classical physics, in terms of instantaneous influences between here and there. Thinking that way, however, is thinking in terms of spatial connections, through influences which may need more or less time to establish these connections. That is to say, we are thinking in terms of space and time pictures which are meaningless if things happen outside of space and time. The lesson we are learning, a dramatic long-range shot, is likely that, to make more progress, we certainly have to go beyond our intuitions, even the ones concerning the existence of space and time. Whatever these concepts would eventually point out to, they must be misleading. But then, what about the relativity of Einstein? The simple fact to ask this question: Relativity challenged?, demonstrates how much our present understanding of quantum mechanics, gained by this weird story on hidden variables, seemingly conflicts with our conceptions of everyday experience of space and time, even refined to the space-time conception of Einstein. What is still hidden in the hidden worlds? But, to begin with, is there any conflict between relativity (gravity excluded) and quantum mechanics?

We essentially had in mind, since the beginning of the book, the non relativistic version of quantum mechanics. In particular, we enunciated postulates valid for this version and, if we wanted to deal with a relativistic version, these postulates should be amended. Clearly, we should not expect any *a priori* dangerous conflict between a nonrelativistic version of quantum mechanics and relativity. Any conflict of that sort would not be *a priori* dangerous because we would simply conclude, before deciding whether we are facing a severe problem or not, that we have to extend our non relativistic version to the relativity domain (even if it is not simple). In other, complementary, and more general words, there should not be any conflict between a quantum mechanics outside of space and time, and relativity glued to space and time representations, giving two deeply different visions of the world. Any conflict of that sort should be solved by a search for unification which should be eventually successful because, if we may have different visions of the world, there is only one world, at least one world relevant to us.

But, thinking more about it, we can quickly enough discover that we are nevertheless facing a strange situation indeed: The facts and features which are bothering us are not limited to the non relativistic version of quantum mechanics. True, everything may be charged on the existence of entangled states, and it happens that the concept of entangled states, as we have seen, originally pertains to the non relativistic version of quantum mechanics. We could then imagine that this troublesome concept should vanish in a relativistic version, and therefore that we then should have not to bother with it any more. But this is not the case. Entangled states are stubborn. In particular, we find them and check them with correlated photons, these messengers of light flowing at the speed of light, the most relativist realm experimentally revealed to us. We find them and check them, not in our theories, but in the very experimental world of phenomena exhibiting its non-separable character, whatever our theories would be.

Therefore, we have to deal, without any possible escape, with non-locality, and non-separability, which certainly apparently violate special relativity, because, in the language of space and time, something has the ability to produce instantaneous influences. This "something" might be some non-local hidden variables (as in Bohm's pilot wave), some kind of unknown fields not yet identified, which could possibly, but not in the space-time framework, provide an adequate mechanism to explain the phenomena, or even simply save the appearances. Such kinds of explanation could generate theories which however might not be Lorentz invariant.

Before going further and put forward other possibilities, let us discuss the problem of Lorentz invariance raised by our questioning, see for instance Hughes [408] inspired by Shimony. Let us consider an EPR-kind experiment with correlated photons. We assume that the measurement of a component of polarization along direction **a** at a motionless location A and the measurement of the other component of polarization along direction **b** at a motionless location B are simultaneous, *in the frame of reference of the laboratory.* This is easy to arrange and check. For example, to this aim, we may locate a photon detector at a third motionless location C with the distance AC equal to the distance BC. When a measurement is made at location A, a photon signal is sent toward C, and the time t_A of arrival at C of this signal is recorded. Similarly, when a measurement is made at location B, a photon signal is sent toward C, and the time t_B of arrival at C of this signal is recorded too. If we observe that $t_A = t_B$, we know that measurements at locations A and B have been simultaneous. Clearly, in a convenient setting, the location C may be midway between locations A and B.

But simultaneity is not a Lorentz invariant property. Actually, there is at least another frame of reference F_A in which the measurement event associated with the particle located at A (called A-event) happens before the similar event (called B-event) for the particle located at B, and there is also at least another frame of reference F_B in which the measurement event (B-event) associated with the particle located at B happens before the similar event (A-event) for the particle located at A. Now, within F_A, an A-event has occasioned a change of state at B (because the transmission of influence is quantum mechanically instantaneous) prior to the occurrence of the B-event, but, within F_B, if the change of state at B has occurred, it has occurred as the result of the B-event, this event occasioning instantaneously a change of state at A, that is to say a change of state occurring before the A-event. We are then facing a strange state of affair indeed, even paradoxical. Nevertheless, this paradoxical report of the situation is the consequence of the fact that we applied the concept of separability to a non-separable situation. That, we should not do.

The main interest of this Gedanken experiment is to insist once more on the consequences of separability in relativity and non-separability in quantum mechanics. In the words of Pipkin [156], *some individuals find the apparent nonlocality present in quantum-mechanical measurements and in particular in EPR-type experiments particularly unsatisfying. They are disturbed by the fact that measurements made by observer B at a space-time point outside the light cone of observer A can influence the results of measurements made locally by observer A.*

But we are actually fortunate enough because no real conflict with relativity can happen. This is because quantum non-locality does not provide any means to transmit a signal faster than light by using measurements on entangled states. In other words, we cannot use EPR-type particles, quantum mechanically correlated, to achieve any transfer of information from here to there or from there to here. A way to view this is to consider a pair of quantum particles in a non factorized state $\Psi(1,2)$. Then, as already pointed out by the orthodox Bohm, in 1951 [41], before his conversion to hidden variables heterodoxy, the expectation value of any operator concerning only particle 1 (particle 2) does not depend on any interaction undergone only by particle 2 (particle 1). This feature is a direct consequence of the formalism of quantum mechanics, and does not require any extra-consideration. More qualitatively, and sufficiently enough, quantum interactions between entangled particles 1 and 2, spatially separated (measured at separated locations), automatically occur, in an uncontrollable way. But, if they are uncontrollable, they cannot be used for signalling, just because, for sending signals, we must be able to exert a control on the signals sent (and received). Therefore, we cannot use quantum correlations to invent any kind of what is sometimes called: Superluminal telegraphy. This is a no-signalling theorem, that is to say a theorem about no faster than light transmission by using EPR-correlations. The expression passion-at-a-distance, introduced by Shimony, has been used to express the non-separability features exhibited by the quantum mechanical formalism, and observed in the experiments, e.g. Redhead [423]. But the no-signalling theorem implies that there is a *peaceful coexistence* between quantum mechanics and relativity, an expression again due to Shimony [424]. The impossibility of superluminal signalling is also discussed in many other places, for instance by d'Espagnat [69]. Hence, strange as it might look, even the non relativistic version of quantum mechanics, which has not to deal with relativity, is already protected, by some kind of mysterious anticipation, from any deep conflict with relativity. In the words of Fenyes [425], *one arrives at the bizarre conclusion that a theory (quantum mechanics) which is itself not relativistically formulated, nonetheless seems uncannily prevented from coming into conflict with relativity theory.*

The no-signalling theorem using quantum correlations can be vividly illustrated by returning to actual experiments in the following way. Let us assume that we achieve a series of measurements at two separated locations on pairs of entangled particles, and, for instance, that these measurements concern spin components. At the first location, the observer records a series of outcomes and makes a list out of them, say: $+ + - + + + + - +...$ Similarly, the observer at the second location records a series of outcomes and makes a list out of them, say: $- - + - - - - + - ...$ Each list must reveal a random character agreeing with the laws of probability of quantum mechanics. To check the entanglement, the two lists must be compared. When this is done, we indeed observe a perfect correlation (here an anti-correlation) between the two series of outcomes. But this checking requires to transfer each of the lists, from the first and second locations to a third common location (possibly this common location can be the first or the second location), something which cannot be done faster than light. Quantum channeling outside of space can only be observed by channeling in space. Therefore, since the entanglement cannot be detected before the comparison is made, it follows that no signal, faster than light, can be exchanged at a superluminal speed.

Another important complementary point of view is that quantum correlations between entangled particles require us to postulate a superluminal communication between these particles only if we assume that these particles are physical quantum mechanical systems. Then, we are indeed facing spatially separated systems, existing in separable states, and these systems can affect each other only by invoking some kind of communication between them, a communication which then has to be superluminal. But, here, we should remember a previous discussion on the notion of system in quantum mechanics, a discussion that the reader might have found too pernickety, but which now reveals its significance: Particles in entangled states are not systems, because they simply do not have states.

We then are driven back to a vision of quantum mechanics, outside of space and time (and of space-time) and to the issue of wholeness. This last word may be applied to both conceptions of Bohm and Bohr. From the point of view of wholeness, both visions are effectively much akin, but there is also a drastic difference which is best pointed out in the present context. For Bohm (discussing the compatibility between his interpretation and relativity in his last book, with Hiley [68]), the quantum potential is an objective entity, and therefore the property of objectivity must be conceptually transferred to the instantaneous actions at a distance between the different systems (yes: Systems, in Bohm's framework) related by the forces deriving from the potential. Therefore, we may have to state that Bohm's wholeness, or Bohm's non-separability, indeed violates Einstein's relativity (this is not exactly the point of view of Bohm however). But this conflict, possibly unbearable to us, is hidden. It is hidden, removed from our eyes, because what we eventually are allowed to observe, the quantum mechanical predictions, do not exhibit it. For Bohr, the whole experimental set-up is an indivisible unit that we are not allowed to break into pieces, even in the mind and, therefore, thinking of these pieces as entities in space and time, is meaningless. Bohr denied the classical idea that objects play their games in themselves in a space-time arena, independently of the measurement interactions which eventually make objects objective. With Bohm, the conflict is hidden. With Bohr, it does not exist. Einstein did not like non-locality, but the one of Bohm in which objective particles are instantaneously and objectively connected by the objective forces deriving from the objective quantum potential was for him certainly more repugnant than the one of Bohr, for which there is no objective particles, no objective forces, no objective quantum potential, but simply quantum correlations transcending space and time. This price to be paid before accepting Bohm's theory is found too much by many.

This may also be viewed as a last word to the long-lasted Bohr-Einstein debate. Concerning EPR-experiments (specifically

EPR-experiments, not necessarily for all aspects of quantum mechanics), Bohr is victoriously triumphant (or triumphantly victorious), and Einstein defeated. There should be still another last word behind this last word, still hidden in the mysteries of the universe, but this lastest word, we do not know it. In the words of Jarrett [75], *providing an adequate elucidation of this inseparability, and of the measurement problem itself, an elucidation which gives us a more wholehearted reconciliation between relativity theory and the "passion-at-a-distance" associated with this inseparability - this is the grand, unfinished task.* This grand, unfinished task might *simply* be the complete unification of all of our physical theories, the fantastic and fantasized Theory of Everything.

More in Bohm and Hiley [426], Aspect [404], Selleri and Tarozzi [347], Squires [13], Teller [339], Folse [427], Shimony [93], Holland [40], Cushing [131], or Genovese [417].

Bell's theorems without inequalities

PEDESTRIANS ($\sim\sim\sim\sim$): Just read this section in a loose way, or even skip it.

As we have seen, the violation of Bell's inequalities for correlated two-particle systems implies Bell's theorems concerning the rejection of local (or separable) objectivity. These theoretical achievements, completed by experimental falsifications, could have been the end of the story. But new very spectacular advances were provided when it was discovered that we could argue against such objectivities without using any inequality. For this, the "only thing to do" was to consider entangled systems with more than two particles. This opened a new line of research initiated by Greenberger, Horne, and Zeilinger, known as GHZ, in 1989 [428].

To start decorating the stage, following GHZ, let us again consider the case of an EPR-experiment (with two particles), in which *a measurement on one particle allows one to predict what happens to the other particle with 100% certainty. This is the case when one measures the spin of one particle in direction* **a**, *and then measures the other either in the same or opposite direction, i.e.* **b** $=\pm$**a**. *This* is the consequence of Eq. (12.11), yielding the expectation value $P_{QM}(\mathbf{b} = \pm\mathbf{a}) = \mp 1$, that is to say perfect anti-correlation or perfect correlation, respectively. Such cases are called superclassical cases by GHZ [428].

Superclassical cases lead to tautologies when we use Bell's inequality of Eq. (12.18). For instance, let us consider an extreme easy case for which $\mathbf{b} = \mathbf{a} = \mathbf{c}$. Then Eq. 12.18 leads to

$$1 + P(\mathbf{a}, \mathbf{a}) \geq |P(\mathbf{a}, \mathbf{a}) - P(\mathbf{a}, \mathbf{a})| = 0 \qquad (12.57)$$

This is indeed a tautology since P is an expectation value which cannot be smaller than 0.

GHZ then asks the question to know whether, facing a super-classical case when one can make a definite prediction, and when Bell's inequalities are tautological, it is possible to build a single deterministic local hidden-variables model agreeing with quantum mechanical predictions. Here is now a counter-example obtained by using a four-body problem.

We consider, to begin with, one particle of spin $s = 1$, initially in a state $m = 0$, that is to say : $|\Psi\rangle = |s, m\rangle = |1, 0\rangle$, in a magnetic field oriented along the z-axis. We then assume that this particle decays into two other spin-1 particles, one travelling along the positive z-axis, and the other one travelling along the negative z-axis. Afterward, each of these two particles experience another decay to two spin-$(1/2)$ particles, the first two also moving along the positive z-axis, and the other two moving along the negative z-axis. Therefore, eventually, the initial particle has decayed into four entangled spin-$(1/2)$ particles.

The state vector of the initial particle can be written as

$$|1, 0\rangle = \frac{1}{\sqrt{2}} \left(|+ + --\rangle - |- - ++\rangle \right) \qquad (12.58)$$

which may be viewed as a generalization of Eq. (12.5). The quantum mechanical expectation value to find the four spins in directions $\mathbf{a}, \mathbf{b}, \mathbf{c}, \mathbf{d}$ respectively is given by

$$P_{QM}(\mathbf{a}, \mathbf{b}, \mathbf{c}, \mathbf{d}) = -\cos(\alpha + \beta + \gamma + \delta) \qquad (12.59)$$

in which $\alpha, \beta, \gamma, \delta$ are angles defining the directions $\mathbf{a}, \mathbf{b}, \mathbf{c}, \mathbf{d}$ respectively, in the (xy)-plane, that is to say in the plane orthogonal to the directions of propagation of the spins. Eq. (12.59) may be viewed as a generalization of Eq. (12.11).

Let us now consider specific cases when

$$\alpha + \beta + \gamma + \delta = 0, \pi \tag{12.60}$$

Then, the cos-term in Eq. (12.59) is equal to (-1) or $(+1)$ and, therefore, if we make measurements for three angles, we can predict the result of a measurement for the fourth angle with 100% certainty. This defines a superclassical case with four particles.

Let us now try to build a local, deterministic, hidden-variables model for this case, under the constraint that the predictions from this model must agree with quantum mechanical predictions. If we measure a spin component of one of the four spin-$(1/2)$ particles resulting from our two-decay process, we can only get two returns, say ± 1, as in Eqs. (12.9). Then, instead of Eq. (12.10), we should now have

$$P(\mathbf{a}, \mathbf{b}, \mathbf{c}, \mathbf{d}) = \int A(\mathbf{a}, \lambda) B(\mathbf{b}, \lambda) C(\mathbf{c}, \lambda) D(\mathbf{d}, \lambda) \rho(\lambda) d\lambda \tag{12.61}$$

in which however

$$A(\mathbf{a}, \lambda) B(\mathbf{b}, \lambda) C(\mathbf{c}, \lambda) D(\mathbf{d}, \lambda) = \pm 1 \tag{12.62}$$

which, in the superclassical case, must be valid together with Eq. (12.60). Recall also that the outcomes (± 1) of each measurement $X(\mathbf{x}, \lambda)$ must be determined by the values of $\mathbf{x}$ and of λ, that is to say by the microstate $(\mathbf{x}, \lambda)$. Then, let us suppose that the angles $\alpha, \beta, \gamma, \delta$ and the value of λ are such that we have

$$A(\mathbf{a}, \lambda) B(\mathbf{b}, \lambda) C(\mathbf{c}, \lambda) D(\mathbf{d}, \lambda) = +1 \tag{12.63}$$

There is an obvious possible way to satisfy this relation, namely

$$A(\mathbf{a}, \lambda) = B(\mathbf{b}, \lambda) = C(\mathbf{c}, \lambda) = D(\mathbf{d}, \lambda) = +1 \tag{12.64}$$

that is to say, in short

$$A = B = C = D = +1 \tag{12.65}$$

Now, the values of A, B, C, D, when they change, can only change abruptly from $(+1)$ to (-1) since ± 1 are the only possible outcomes. But, GHZ argued, such abrupt changes are in conflict with the continuous changes allowed for the angles $\alpha, \beta, \gamma, \delta$. Therefore, we conclude that we cannot, via these continuous changes, switch from the value $(+1)$ of Eq. (12.63) to the other value (-1) allowed by Eq. (12.62). Hence, we cannot produce a consistent hidden-variables model for the superclassical case under study.

After the 1989-paper, actually an outline in conference proceedings, the next year, in 1990, GHZ together with Shimony [351] published a somewhat more elaborated version of the 1989 argument in a much more extended paper. They demonstrated that the premises of the EPR-paper are inconsistent when applied to quantum systems consisting of at least three particles (that is to say not four particles as in 1989, but more than two particles). They considered again superclassical cases, both for three- and four-particle systems, discussing Gedanken experiments and their real implementation.

Let us revisit the four-particles system discussed previously. We are then dealing again with four spin-$(1/2)$ particles generated by a two-decay process. Particles 1 and 2 propagate freely along the positive z-axis and particles 3 and 4 propagate freely along the negative z-axis. Particles 1 and 2, spatially separated, are analyzed by using Stern–Gerlach devices with orientations $\mathbf{a}$ and $\mathbf{b}$ respectively. Particles 3 and 4, spatially separated, are analyzed by using Stern–Gerlach devices with orientations $\mathbf{c}$ and $\mathbf{d}$ respectively. For the sake of convenience, we explicit rewrite the initial state vector as

$$|\Psi\rangle = \frac{1}{\sqrt{2}}(|+ + --\rangle - |- - ++\rangle) \qquad (12.66)$$

The demonstration of this expression is given in an appendix to the paper. The expectation value of the product of the outcomes is again denoted by $P(\mathbf{a}, \mathbf{b}, \mathbf{c}, \mathbf{d})$. To generate a superclassical case, the orientations $\mathbf{a}, \mathbf{b}, \mathbf{c}, \mathbf{d}$ are taken in the (xy)-plane. With a notation slightly different from the one used in Eq. (12.59), we have

$$P_{QM}(\mathbf{a}, \mathbf{b}, \mathbf{c}, \mathbf{d}) = -\cos(\alpha + \beta - \gamma - \delta) \qquad (12.67)$$

Using the expression "perfect correlation" for both correlation and anti-correlation, we may consider two cases of perfect correlations.

In case (i), we have

$$\alpha + \beta - \gamma - \delta = 0 \tag{12.68}$$

$$P_{QM}(\mathbf{a}, \mathbf{b}, \mathbf{c}, \mathbf{d}) = -1 \tag{12.69}$$

and, in case (ii)

$$\alpha + \beta - \gamma - \delta = \pi \tag{12.70}$$

$$P_{QM}(\mathbf{a}, \mathbf{b}, \mathbf{c}, \mathbf{d}) = +1 \tag{12.71}$$

In these two superclassical cases, if we measure spin components for any three particles, then we can predict with certainty the outcome for the fourth particle. We next introduce a locality assumption: When we make measurements on the particles, because they are arbitrarily far apart, they are not supposed to interact any more, and hence we should assume that nothing happens to any of them as the result of what happens with the other three particles.

We are now going to try building a local hidden-variables theory for our four spin-$(1/2)$ particles, in the same spirit than in Bell [189], a theory which should allow one to recover quantum mechanical predictions. Instead of dealing only with $A(\mathbf{a}, \boldsymbol{\lambda})$ and $B(\mathbf{b}, \boldsymbol{\lambda})$ as in Bell [189], we now have to deal with four quantities, namely : $A(\mathbf{a}, \boldsymbol{\lambda})$, $B(\mathbf{b}, \boldsymbol{\lambda})$, $C(\mathbf{c}, \boldsymbol{\lambda})$ and $D(\mathbf{d}, \boldsymbol{\lambda})$. These functions A, B, C, D represent the outcomes of spin measurements along directions $\mathbf{a}, \mathbf{b}, \mathbf{c}$ and $\mathbf{d}$ respectively for a microstate (Ψ, λ) and, therefore, each one can only receive one of the two values $(+1)$ or (-1).

We now restate Eqs. (12.68) to (12.71) in the framework of the hidden-variables model. Eq. (12.59) is still valid. From this equation, in case (i), i.e. $\alpha + \beta - \gamma - \delta = 0$, that is to say for

$$A(\mathbf{a}, \boldsymbol{\lambda})B(\mathbf{b}, \boldsymbol{\lambda})C(\mathbf{c}, \boldsymbol{\lambda})D(\mathbf{d}, \boldsymbol{\lambda}) = -1 \tag{12.72}$$

we have

$$P(\mathbf{a}, \mathbf{b}, \mathbf{c}, \mathbf{d}) = -\int \rho(\boldsymbol{\lambda})d\boldsymbol{\lambda} = -1 \tag{12.73}$$

in which we have used the normalization condition for ρ. This result agrees with quantum mechanical predictions. Similarly, for case (ii), i.e. $\alpha + \beta - \gamma - \delta = \pi$, that is to say for

$$A(\mathbf{a}, \boldsymbol{\lambda})B(\mathbf{b}, \boldsymbol{\lambda})C(\mathbf{c}, \boldsymbol{\lambda})D(\mathbf{d}, \boldsymbol{\lambda}) = +1 \qquad (12.74)$$

we have

$$P(\mathbf{a}, \mathbf{b}, \mathbf{c}, \mathbf{d}) = \int \rho(\boldsymbol{\lambda})d\boldsymbol{\lambda} = +1 \qquad (12.75)$$

agreeing again with quantum mechanical predictions. Nevertheless, let us examine Eq. (12.72) for four different, well chosen, values of $\mathbf{a}, \mathbf{b}, \mathbf{c}, \mathbf{d}$ that is to say of $\alpha, \beta, \gamma, \delta$. We then may write

$$A(0, \boldsymbol{\lambda})B(0, \boldsymbol{\lambda})C(0, \boldsymbol{\lambda})D(0, \boldsymbol{\lambda}) = -1 \qquad (12.76)$$

$$A(\phi, \boldsymbol{\lambda})B(0, \boldsymbol{\lambda})C(\phi, \boldsymbol{\lambda})D(0, \boldsymbol{\lambda}) = -1 \qquad (12.77)$$

$$A(\phi, \boldsymbol{\lambda})B(0, \boldsymbol{\lambda})C(0, \boldsymbol{\lambda})D(\phi, \boldsymbol{\lambda}) = -1 \qquad (12.78)$$

$$A(2\phi, \boldsymbol{\lambda})B(0, \boldsymbol{\lambda})C(\phi, \boldsymbol{\lambda})D(\phi, \boldsymbol{\lambda}) = -1 \qquad (12.79)$$

From Eqs. (12.76) and (12.77), we obtain:

$$A(\phi, \boldsymbol{\lambda})C(\phi, \boldsymbol{\lambda}) = A(0, \boldsymbol{\lambda})C(0, \boldsymbol{\lambda}) \qquad (12.80)$$

and, from Eqs. (12.76) and (12.78)

$$A(\phi, \boldsymbol{\lambda})D(\phi, \boldsymbol{\lambda}) = A(0, \boldsymbol{\lambda})D(0, \boldsymbol{\lambda}) \qquad (12.81)$$

The two last equations imply

$$\frac{C(\phi, \boldsymbol{\lambda})}{D(\phi, \boldsymbol{\lambda})} = \frac{C(0, \boldsymbol{\lambda})}{D(0, \boldsymbol{\lambda})} \qquad (12.82)$$

Since $D(\phi, \boldsymbol{\lambda})$ is equal to ± 1, and hence is equal to its inverse, the last equation may be rewritten as

$$C(\phi, \boldsymbol{\lambda})D(\phi, \boldsymbol{\lambda}) = C(0, \boldsymbol{\lambda})D(0, \boldsymbol{\lambda}) \qquad (12.83)$$

From Eqs. (12.83) and (12.79), we then obtain

$$A(2\phi, \boldsymbol{\lambda})B(0, \boldsymbol{\lambda})C(0, \boldsymbol{\lambda})D(0, \boldsymbol{\lambda}) = -1 \qquad (12.84)$$

This last equation, together with Eq. (12.76), yields

$$A(2\phi, \boldsymbol{\lambda}) = A(0, \boldsymbol{\lambda}) \qquad (12.85)$$

meaning that A does not depend on ϕ, i.e. it is a constant. Let us now invoke the not yet used Eq. (12.74) and rewrite it for $\alpha = \theta + \pi$, $\beta = 0$, $\gamma = \theta$, $\delta = 0$. These values satisfy $\alpha + \beta - \gamma - \delta = \pi$, and yield

$$A(\theta + \pi, \boldsymbol{\lambda})B(0, \boldsymbol{\lambda})C(\theta, \boldsymbol{\lambda})D(0, \boldsymbol{\lambda}) = +1 \qquad (12.86)$$

This last equation, together with Eq. (12.77), implies

$$A(\theta + \pi, \boldsymbol{\lambda}) = -A(\theta, \boldsymbol{\lambda}) \qquad (12.87)$$

Let us now consider Eq. (12.85) for $\phi = \pi/2$. This yields

$$A(\pi, \boldsymbol{\lambda}) = A(0, \boldsymbol{\lambda}) \qquad (12.88)$$

that we compare with Eq. (12.87) specified for $\theta = 0$, that is to say with

$$A(\pi, \boldsymbol{\lambda}) = -A(0, \boldsymbol{\lambda}) \qquad (12.89)$$

Then the two last equations show that, since A cannot be 0 (it must be ±1), we have reached a contradiction. This contradiction holds for the four-particle system we have studied. But, throughout the core of the demonstration, the argument β contained in $B(\mathbf{b}, \boldsymbol{\lambda})$ has been kept equal to 0. Therefore, the contradiction also holds for a three-particle system. We must then conclude that we cannot build a local hidden-variables theory in agreement with quantum mechanical predictions for systems of three or four particles (more generally, for systems of more than three or four particles). This is even impossible for superclassical cases where, in the case of two particles, Bell's inequalities are tautological. This important result is obtained without using any inequality. Afterward, the paper we are discussing goes on with a specific similar study for a system with three particles, without spin, and discusses the possibility of experimental tests. A history of the GHZ approach is reviewed by Greenberger [429].

Bell's theorems and non-locality without any inequality have also been discussed in several other places, in particular by Mermin

[197, 330], Hardy [370], Goldstein [430], Boschi *et al.* [431]. However, the 1993-paper by Mermin [197], rather than concerning a Bell's theorem *stricto sensu*, that is to say concerning locality, better concerns a Bell–KS theorem, that is relevant to contextuality. It has therefore been examined in the previous chapter. More generally, Bell's theorems without any inequality, in a broader sense, may designate Bell's theorems or Bell–KS theorems. Peres [432] did a job similar to the one of Mermin, in a four-dimensional space, using only six operators, but his work depended on the choice of a particular quantum state, namely a singlet state. Mermin, by using a larger number of operators, avoided any restriction to any particular state, a fact which is acknowledged by Peres in an addendum to his paper. Hardy [433] demonstrated non-locality for two particles without using inequalities for almost all entangled states. He compared his work to the ones of GHZ and Bell as follows: *If nature really is local such that quantum mechanics is wrong then in a GHZ-type experiment it is necessary that there would be individual events that violate the predictions of quantum theory. In Bell's proof it would only be necessary to have a statistical violation of quantum mechanics The proof presented in this paper falls halfway between these two extremes.* This means that the formalism of Hardy may lead either to a statistical violation or to a violation for a single event (or both). Laloë [335] commented GHZ in a vivid way as follows : *Roughly speaking, while the two-particle systems can lead to violations of locality of the order of 25% and require the evaluation of a combination of suitable correlation rates, 100% violations become possible in theory with three particles (or more) in one single experimental result,* and went on proposing experimental schemes.

Non-locality and contextuality

The previous chapter on contextuality and the present one on non-locality discussed two notions which present some similarities, two notions which are actually so similar that they can easily be confused and even (incorrectly) subsumed when Bohr explained us the indivisible character of a whole experimental set-up, or when Bohm insisted on the wholeness prevailing in quantum mechanical phenomena. However, as stated by Bohm and Hiley [68], *it is important not to confuse the question of context dependence and that of locality even though it is difficult to separate the two.* The attention of the reader has already been attracted to the fact that the two notions should indeed be distinguished, simply because we used two separate chapters to study them. I now would like to end the corpus formed by these two chapters to refine the issue (still leaving however some questions unanswered).

Let us first come back to Mermin [197] and examine a case in which locality implies non-contextuality. As in the chapter on contextuality, let us again consider two sets of observables $\{A, B, C, ...\}$ and $\{A, L, M, ...\}$ sharing the same observable A. We assume that all observables $A, B, C, ...$ in the first set are commuting observables, and we also assume that all observables $A, L, M, ...$ in the second set are commuting observables too. But it is assumed that not all the observables $B, C, ...$ are commuting with all the observables $L, M, ..$, for instance B and M do not commute, or C and L do not commute ...

The measurement devices which measure the commuting observables $A, B, C, ...$ are named $M_A, M_B, M_C, ...$ respectively and they are supposed to be all far apart. Similarly, the measurement devices which measure the commuting observables $L, M, ...$ are named $M_L, M_M, ...$ respectively and they are supposed to be all far apart too. Furthermore, these measuring devices are also supposed to be all far apart from M_A which pertains to both sets of commuting variables. Let us make a locality assumption: The absence of any action (spooky or unspooky) at a distance between the different pieces of apparatus which are far apart. Then the outcome of the measurement of A should not depend whether we simultaneously measure the other commuting observables $B, C, ...$, or the other commuting variables $L, M, ...$, or even whether we simultaneous measure a mixture of observables all commuting with A, but which do not all commute together such as $B, C, L, M, ...$ in the case where, for instance, B and M do not commute. In other words, this measurement is non-contextual and locality implies non-contextuality (in this example). More generally, Mermin wondered: *Can we prove a Bell-KS theorem in which we assume non contextuality only when it can be justified by locality?*, and confessed that he does not know any way to accomplish this trick that could work for arbitrary states.

But we can show that non-contextuality does not necessarily reduce to locality. To give one sufficient example, let us serve again the table of nine observables used in the chapter on contextuality 11.14:

$$
\begin{array}{ccc}
\sigma_x^1 & \sigma_x^2 & \sigma_x^1\sigma_x^2 \\
\sigma_y^2 & \sigma_y^1 & \sigma_y^1\sigma_y^2 \\
\sigma_x^1\sigma_y^2 & \sigma_x^2\sigma_y^1 & \sigma_z^1\sigma_z^2
\end{array}
\tag{12.90}
$$

This table exhibits four local observables, namely: σ_x^1, σ_x^2, σ_y^2, and σ_y^1. It also exhibits five non-local observables, namely: $\sigma_x^1\sigma_x^2$, $\sigma_y^1\sigma_y^2$, $\sigma_x^1\sigma_y^2$, $\sigma_x^2\sigma_y^1$, and $\sigma_z^1\sigma_z^2$. The local observables are obviously local since, if they are measured, then they are measured at a given location, either at the location of particle 1 or at the location of particle 2. The other observables are obviously non-local since they are products of two observables, one measurable at the location of particle 1 and the other at the far apart location of particle 2. Therefore, the demonstration of the associated Bell-KS theorem, relying on the above table, given in the chapter on contextuality, and explicitly using non-local variables, shows that non-contextuality does not reduce to locality. In the words of Mermin, *if we intend to rely only on locality, we are blocked for the assignment of noncontextual values to the five coupled observables.*

In his 1993-paper, Mermin also commented Bohm's pilot wave, up to now our most cherished hidden-variables theory, reminding us that we shall have, soon or later in this book, to come back to it : *Bell's favorite example of a hidden-variables theory, is not ... explicitly contextual but explicitly and spectacularly nonlocal, as it must be in view of the Bell-KS theorem and Bell's theorem*, adding : *Bohm theory defies all impossibility proofs*. Reciprocally, Bohm had time, just ultimately, to discuss Mermin [68], and to show that his pilot wave is not in contradiction with Mermin's results, something we understand very well by now.

The issue of locality and non-contextuality is also discussed by Brown and Svetlichny [434].

Chapter 13

Classification of Hidden-Variables Theories

Everything we had to learn concerning hidden-variables theories, what has been accomplished, what has been reported in the literature, what has been gained after so many decades (nearly one century of effort), is now behind us. We should fasten seatbelts and prepare ourselves to land on. But, for an optimized strategy, we need to clean and clear our mind, and to organize it. The most efficient way is to provide a careful classification of hidden-variables theories, so that we shall be able to separately consider each item in the classification, and to draw specific conclusions for each of these items. After that, we shall be well equipped and in good shape to dare trying to close the issue. This is particularly important because, as is revealed by the literature, there exist many definitions, possibly contradictory, concerning what are, or what should be, hidden variables, a situation which, very often, obscured the debates and recurrently make them unduly confusing. One of the aims of a satisfactory classification is to accept all these available definitions, but to sort them out.

In 1973, Belinfante [1] already proposed such a (venerable) classification. It was rather well adapted to the knowledge available at this time and is still illuminating and worth reporting, and this not only from a pure historical point of view. Nevertheless, nearly forty years later, I must confess that I find it fairly old-fashioned, or even obsolete. Therefore, I shall afterward propose a new classification which seems to me better matching our present knowledge and understanding of the issue to which this book is devoted.

Belinfante's classification

Belinfante distinguishes between zeroth-, first-, and second-kind theories. I shall review these different kinds of theories with however a somewhat critical point of view, helping to the introduction of a renewed classification.

Zeroth-kind theories

We possess proofs of impossibility, and yet we possess at least one consistent hidden-variables theory (Bohm's one again), an obvious discrepancy. But, as we have seen, each author having provided a proof of impossibility actually admitted some postulate(s) which should be satisfied by hidden variables theories they had in mind, for instance the additivity postulate of von Neumann, and afterward demonstrated that the set of postulates used was self-contradictory (more precisely, with admittedly a somewhat different meaning: Was not compatible with quantum mechanics). We are then facing, not the impossibility of hidden variables, but merely the impossibility of hidden-variables theories satisfying the self-contradictory set of postulates accepted as premises to the demonstration. Such impossible theories (assuming that quantum mechanics is correct) are called zeroth-kind theories by Belinfante.

As we know clearly now, these theories comprise theories for which hidden variables satisfy the additivity postulate, non-contextual theories, local and separable theories (Belinfante, in 1973, did not provide this list: It would have been retrodictively anachronistic). We can also make more specific the list of zeroth-kind theories, at least those considered by Belinfante, by recalling a list of authors, such as von Neumann (additivity postulate), Jauch and Piron (supposed to be an improvement of von Neumann), Gleason (contextuality), or Kochen and Specker (contextuality).

Furthermore, according to Belinfante, because they are self-contradictory, *the impossibility of theories of the zeroth kind can be shown without performing new experiments.* This is however something I would like to criticize, or better to refine. For example, non-contextual hidden-variables theories are taken as zeroth-kind theories but, if quantum mechanics is indeed contextual, we may ask the question to know whether quantum mechanical happenings in the experimental world are indeed contextual or not, that is to say whether quantum mechanics is experimentally corroborated or not. However, this requires experiments. As another and better example, I could have similarly discussed non-locality rather than contextuality, with all the different generations of experiments we have reviewed in the previous chapter but, in 1973, when Belinfante proposed his classification, the issue of non-locality had not been developed as it is now, and the associated hidden-variables theories were not viewed by him as zeroth-kind theories. By incorporating the issues of locality and separability to build the list of zeroth-kind theories, I am somewhat misrepresenting Belinfante's classification, but, also, I am anticipating the new forthcoming one.

It is interesting to remark that the authors of proofs of impossibility usually did not understand what they were doing, like the sleepwalkers of Koestler [8]. They indeed believed that they have produced proofs of impossibility, rather than proofs of impossibility of certain classes of hidden-variables theories. For instance, discussing von Neumann, Belinfante stated that von Neumann *should have concluded that the kind of hidden variable-theory defined by him was an impossible theory, i.e. a theory of the zeroth kind. This, however, he did not realize and rather, erroneously, conclude that hidden variable-theories are, as a whole, impossible.*

The interest of impossible theories is however significant. As stated by Belinfante: *One might think at first that hidden variables theories of the zeroth kind are of no interest. What good is a theory that is impossible?* But, *the interest of hidden variables theories of the zeroth kind is that they are a warning. They are a warning for what one can never expect any realistic hidden variables theory to accomplish.* Or also, *Mathematicians can disprove any theory by adopting axioms that contradict the theory. We have seen how several authors have taken this approach and have constructed so-called theories of the zeroth-kind. These theories were useful as warnings what axioms should be avoided if one try to construct a theory of the first kind* (soon coming).

We have to add that the study of these theories indeed helped us to dramatically better understand the strange properties of the quantum world. This is obtained by creating a contrast, the kind of contrasts or alternatives which always help us to better grasp concepts, by using a couple of opposite meanings, like black and white, luminous and dark, or up and down. In the present case, the couple is formed by: Quantum mechanics satisfying quantum mechanics, and zeroth-kind theories which do not satisfy quantum mechanics. They do not satisfy quantum mechanics because none of them *can assign a set of truth values for all propositions of quantum theory to any hidden-variable state.*

First-kind theories

Fist-kind theories are the consequence of the quest for determinism. For Belinfante, they have to share a conceptual framework which is defined as follows: Hidden variables in first-kind theories possess a distribution of values which, if out of equilibrium, returns irreversibly and rapidly to an equilibrium distribution. When the hidden variables are out of equilibrium, theories of the first kind make experimental predictions which do not agree with quantum mechanical predictions. But quantum theory follows when, after a relaxation process, the hidden variables have reached their equilibrium distribution.

We should expect that, after all the huge period of time having elapsed since the beginning of our universe, equilibrium distributions should have been reached everywhere, and hence quantum mechanics should be valid everywhere. But, it is possible *to witfully and predictably perturb such an equilibrium distribution*. When this is done, it should be possible to observe deviations from quantum mechanics. Therefore, first-kind theories can be experimentally tested. We remember that this has been indeed done by Papaliolios [289], and that the results of his experiment perfectly agreed with quantum mechanics. We should also however remember that such results, certainly not corroborating theories of the first kind, nevertheless did not falsify them. As we put forward, such theories are immune to falsification. This must be what Belinfante had in mind when he states: *Theories of the first kind are not that easily distinguishable experimentally from quantum theory.*

A paradigmatic example is the Bohm–Bub's theory (associated by Belinfante with the theory of Wiener and Siegel) that we have exposed previously. Another example is Bohm after Bohm, when he introduced a genetic point of view for Born's postulate, and his speculations on the sub-quantum mechanical level. Nevertheless, the definition given by Belinfante for theories of the first kind does not apply to the pilot wave of Louis de Broglie, nor to its more elaborated version of Bohm. Actually, Belinfante seems to view Bohm's pilot wave as a first-kind theory (page 312 of his book), but this is inconsistent with his definition, since this pilot wave does not refer to the notion of an asymptotic equilibrium. Indeed, the concept of a relaxation process is not included in the early version of the Bohm's pilot wave approach. Therefore, as we already discussed, this early approach of Bohm is immune to discrimination and, if it were falsified, quantum mechanics would be simultaneously falsified too. Similar considerations are valid for the unachieved double solution too. Yet, both the double solution and the pilot wave were the consequences of a quest for determinism, even if other issues (non-locality, contextuality) were raised as by-products. This certainly points out to a weakness of Belinfante's classification. It could possibly be repaired by introducing a dichotomous subdivision of theories of the first kind, but I shall prefer to take a more decisive action, that is to say to introduce another classification.

Second-kind theories

Instead of being motivated by the quest of determinism, theories of the second kind are motivated by the EPR-argument, and all the resulting efforts concerning entangled states. Nowadays, we can view such a basis for classification as rather arbitrary and inconvenient, insofar as, as we can see, with the early Bohm, that the concepts of determinism, non-locality, contextuality, are rather entangled too.

But, in 1973, considering the state of affairs prevailing at this time, it was certainly reasonable.

However, Belinfante was aware of the entanglement of concepts underlying his classification: ... *that the predictions of hidden-variables of the first kind, like pure quantum theory, have aspects of nonlocality that are not easily answered and that lead people to the invention of hidden-variables theories of the second kind...* Nevertheless, the basic idea for his classification may be expressed as follows: *... we define hidden-variables theories of the first kind as hidden-variables theories constructed with the following two primary purposes in mind (1) the theory shall not be self-contradictory and (2) the theory shall yield all results of pure quantum theory for all ensembles of physical systems in which the distribution of the hidden variables values is a certain equilibrium distribution. By (1) these theories differ from theories of the zeroth kind; by (2) they differ from theories of the second kind.* The origin of these last theories (the second kind theories) is *the quest for theories that look like causal theories when applied to spatially separated systems that interacted in the past.* These theories lead *to deviations from quantum theory. In fact, these theories were constructed with the purpose of contradicting quantum theory, where quantum theory does not explain the behavior of composite systems in a way that makes the individual elementary particles or quanta behave individually locally causally.* The acceptance of deviations from quantum mechanics is partly due to the fact that *there are ... people so much dissastified with quantum theory that at the time this is written they are looking for deviations from quantum theory that would exist even if the hidden variables distribution were an equilibrium distribution.* One reason for such a dissatisfaction, according to Belinfante, is that *we must meet the paradoxical result that the principles of quantum mechanics prescribe correlations between measurement results at spatially separated positions on separate particles that previously interacted,* but *quantum mechanics does not provide any formalism by which one could understand how the separate particles upon reaching the separate apparatus would acquire the information that seems necessary to "make" the two apparatus find results that show* *the quantum-mechanically predicted correlations.* In other words: *We are mystified by an absence of a deterministic mechanism in quantum theory for enforcing its own prediction.*

To complete this with a brief review of specific hidden variables of the second kind, let us just recall Furry's hypothesis or the theories, much comfortable for the mind, which would explain quantum correlations between separated particles by invoking common causes at the source. But as we know, Furry's hypothesis is not supported by experiments, and common causes have to be dismissed. For a review, see Belinfante or return to the subsection on the possible solutions to the EPR-paradox.

And now, again, in this more precise context, what about Bohm's theory? Being not self-contradictory, but conversely logically consistent, it does not pertain to the class of zeroth-kind theories. If we adhere strictly to the above definition for theories of the first kind, it does not pertain to this class since relaxation processes are absent from it (in his earlier version). It would remain the possibility of putting it in the file of second-kind theories but this is much unsatisfactory because, although Bohm had in mind the EPR-experiments, this was not his primary motivation (or at least not his single one) but rather a by-product. Therefore, the entanglement of concepts of Belinfante's classification goes on making problems. Furthermore, another reason why Bohm's theory does not pertain to the second-kind class, at least in the mind of Belinfante, is that, as he stated: *A hidden variables theory of the second kind ... denies the existence of nonlocal quantum mechanical correlations.* Therefore, eventually, there is definitely no room for the early Bohm in Belinfante's classification.

I do not need to tell much more on these second-kind theories. Nearly forty years after Belinfante's book, we know much more on these issues than Belinfante could know. In particular, we know that second-kind theories, as viewed by Belinfante (they intended to get rid of non-locality), disagree with quantum mechanics, with Bohm's pilot wave, and with experiments. We may therefore, without any scruple, ask them to leave the stage.

Big Book classification

The Big Book, according to the famous mathematician Paul Erdös [435], is the huge book (presumably much larger than the size of the universe) in which God listed all mathematical perfect proofs. One pregnant hope of Erdös was that, after his death, he would be allowed to have a look to this huge book. I do not know whether the Big Book also contains perfect classifications produced by physicists, embedded in the sensible world, so far away from the Paradise of Plato ideals. And, even if it is so, it might be presumptuous to hope that the classification I am going to propose is good enough for the Good Big Book. But, as Carmen would have said (in Bizet's opera) if she knew mathematics, physics, and the science of classifications: It is sweet to hope ...

I introduce a first dichotomy between Hidden-Variables theories which are Impossible (IHV-theories) and Hidden-Variables theories which are Possible (PHV-theories). IHV-theories are the ones which are self-contradictory, and/or the ones which conflict with experimentally corroborated quantum mechanical predictions. To simplify, I shall make a risky step further and assume that all quantum mechanical predictions will be, in the future, corroborated by experiments. In saying so, I obviously do not deeply and completely pretend that quantum mechanics is genuinely correct. We know that *it is incorrect*, not only because it is surpassed by more elaborated theories, such as quantum field theory, but more important and decisive because its marriage with gravity is still, may be for a long time, in jeopardy. Planck scales are still waiting for us. I just refer to the domain of validity of quantum mechanics, and to the possible adequacy of hidden variables in this domain.

Therefore, the risky step being done, I may redefine IHV-theories as being the ones which are internally contradictory, and/or the ones which conflict with quantum mechanics. If the above assumption is not correct, then the Big Book knows it, and my classification is not perfect, but is nevertheless listed as a possibility most reasonable if we consider the state of affairs at the beginning of the new century. Examples of IHV-theories are those who assume the additivity postulate of von Neumann, locality, separability, and non-contextuality. The class of IHV-theories is the union of the class of zeroth-kind theories and of the class of second-kind theories, in the terminology of Belinfante.

In contrast, PHV-theories, in agreement with Bell–KS and Bell's theorems, must be contextualist, non-local (non-separable). Also, they must agree with quantum mechanics. Otherwise, they would be impossible. We may however refine by distinguishing between physical and non physical (artificial) theories, and also, from another point of view, distinguish them by examining how they behave with respect to experiments.

Non physical PHV-theories are the ones which used hidden variables of a mathematical nature, designed to allow mathematical tricks, but which cannot receive a clear physical meaning. The hidden variables, introduced for such theories, I call them artificial. Examples of such artificial hidden variables are the ξ of the Bohm–Bub's theory, and the λ used by Bell and GHZ, and others. The creation and the study of non physical PHV-theories have been proven to be much valuable, as we have seen. In particular, they allowed us to deepen our understanding of quantum mechanics by shedding new light on the concepts of non-locality or non-separability. Therefore, the study of non physical PHV-theories should not be discouraged. It is indeed impossible to forecast what we could still possibly learn by playing with them. Nevertheless, a non physical PHV-theory, that is to say a hidden-variables theory using artificial hidden variables, and agreeing with quantum mechanical predictions, would be of little use to propose a new physical interpretation of quantum mechanics and, therefore, I shall not consider them any more, independently of their intrinsic interest. As a physicist, I am therefore giving a due privilege to physical PHV-theories, that is to say, by definition, which are not non physical.

Now, what about experiments? In utmost rigor, according to the (may be provisional) classification introduced here, the experimental issue should not be relevant because, according to our definitions, a PHV-theory must agree with quantum mechanics, and with quantum-mechanical predictions, supposed to remain valid in the future. But additional discussions are nevertheless necessary.

Let us in particular consider Bohm–Bub's theory. Is it a IHV- or a PHV-theory? Here, the matter is a bit subtle. We know that Bohm–Bub's theory, as well as theories considered by Bohm after Bohm, pointing out to the occurrence of relaxations processes, indeed make predictions which do not agree with quantum mechanical predictions, when hidden variables are out of equilibrium. Therefore, seemingly, Bohm–Bub's theory should be considered as pertaining to the class of IHV-theories, and should be straight away rejected. But, in practice, as we have seen, Bohm–Bub's theory is immune to falsification. It is impossible to experimentally demonstrate that relaxation processes do not exist, because it is always possible to argue that such relaxation processes are so rapid that they have not been detected. What can be falsified, by means of experiments *à la Papaliolios*, is quantum mechanics, but not hidden-variables theories with relaxation processes. If this happened, the Big Book classification would have to be revisited, without however any severe damage. But I made a risky assumption, *the* risky assumption (risky from a pure logical point of view) because I do not believe that this will happen. Therefore, in practice, it is quite reasonable to consider Bohm-Bub's theory as a PHV-theory. Nevertheless, such a theory (and similar ones) is from now on dismissed, not because it is impossible, but because it is artificially non physical.

So, after having applied a filter process implied by the Big Book classification, that is to say after having rejected IHV-theories and non physical PHV-theories, we are left, from now on, with the residual class, the one of physical PHV-theories. Obviously, it is hardly possible to discuss physical PHV-theories which do not exist. We are therefore left with only two possibilities: The unachieved double solution and the (unachieved too) pilot wave, plus something else to be discussed now: stochastic quantum mechanics.

Chapter 14

Stochastic Quantum Mechanics

Stochastic quantum mechanics shares many important and specific features with the hidden-variables theories which are still playing their performances on the stage, namely two physical PHV-theories: The double solution and the pilot wave. Eventually, even if I may regret it, I shall have to dismiss these theories, although there will be a loophole for defenders. But the reasons why these theories will be dismissed can simultaneously be applied to stochastic quantum theory. Then, with one stone, why should we refuse to kill several birds?

Following a definition from Wheeler and Zurek [100] and using a bit of rewording, stochastic quantum mechanics is an interpretation which takes note of the fact that Schrödinger's equation is analogous to a diffusion equation (this equation which is basic in the theory of diffusion processes by Brownian motion), and turns this analogy to an ontology. For doing this, it is assumed that the process of diffusion underlying Schrödinger's equation is driven by some kinds of forces.

A first question to be asked is then as follows: If quantum mechanics is the consequence of a diffusion process, we want to know which kinds of entities are actually diffusing? The answer, such as expressed by Nelson in a paper often viewed as seminal [125], is that the diffusing entities are actually particles in the Newtonian sense. We might call them *objective* particles as well, so that we are facing an ontology in the sense of Bohm. From this point of view (and others), stochastic quantum mechanics is much akin to causal theories (double solution, pilot wave). It introduces once more the idea of objective entities evolving behind the masquerade of the wave function.

The second question to be asked is the one to know which kinds of forces are acting on the particles. Similarly as in Brownian motion, the motion of particles is a stochastic process (hence the name of stochastic quantum mechanics), which, as already discussed, does not conflict with determinism, and whose origin is to be found in a sub-quantum mechanical level (in the same way that Brownian motion is the result of a sub-macroscopic molecular level). This reference to a sub-quantum mechanical level establishes another conceptual connection with causal theories, more particularly with Louis de Broglie resuscitated and Bohm after Bohm.

Entities of the sub-quantum mechanical level interact with the objective particles by some kind of random impacts, in the spirit of the old physics of Descartes where the transmission from causes to effects required effective causal contacts. In such a framework, we can see that we must again introduce hidden variables, actually two layers of hidden variables. The first layer, just below the quantum mechanical level, is made from true values associated with the objective particles, as in the early Bohm's pilot wave. The second layer, below the first layer, is made from true values associated with the entities of the sub-quantum mechanical level, as in Bohm after Bohm. Hence, we can view stochastic quantum mechanics as a new case, not yet explicitly discussed before, of hidden-variables theories. Since this scheme agrees with Schrödinger's equation acting at the uppest level, and must therefore make the same predictions than quantum mechanics, it pertains to the class of PHV-theories. Furthermore, since it relies on physical objects (objective particles for the first layer, other objective entities for the second layer), it pertains to the more restricted class of physical PHV-theories, the only class that we decided to go on discussing.

Although some proponents and defenders of stochastic quantum mechanics like to refer to the 1967-Nelson's paper [125], the origin of this interpretation can be traced back to a long time ago, in a 1910-paper by Einstein and Hopf [436], even before the advent of usual quantum mechanics, and its associated Copenhagen interpretation.

The reader might prefer to refer to another paper published in 1913 by Einstein and Stern [437], or more conveniently to Bergia *et al.* [438] who presented an English translation of the 1913 Einstein's paper, originally published in German, together with illuminating discussions and comments.

The main issue of these works from Einstein, relevant to our purpose, is an hypothesis according to which there would exist a background stochastic field even at zero temperature (a precursor of our modern quantum vacuum). The zero-point energy is then viewed as the consequence of a mechanism of interaction between oscillators and radiation fields. This significantly departs from the treatment of zero-point energy in quantum mechanics where it appears naturally, even without referring to any radiation field, for instance when we study an harmonic oscillator governed by Schrödinger's equation (although the harmonic oscillator is the prototypical system used afterward to quantize the electromagnetic field). Nevertheless, we can see the seed for stochastic quantum mechanics in this old idea of Einstein, at least as far as the sub-quantum mechanical level is concerned.

Concerning the other ingredient of stochastic quantum mechanics, namely the analogy between Schrödinger's equation and a diffusion equation, it is also not new. Actually this analogy has already been pointed out by Schrödinger himself, in 1931 [24, 131, 439]. However, a real outlook of stochastic quantum mechanics has been introduced for the first time, in 1933, by Fürth [440], followed twenty years later by Fenyes [425] who provided a careful attempt to interpret quantum mechanics as a theory of Markov processes in configuration space, although little is said concerning the origin of the external random mechanism generating the Brownian motion. Mugur-Schächter [291] commented on the fact that Fenyes demonstrated that there exists a complete analogy between the formal structure of quantum mechanics and the one of the theory of Markov processes (and of the theory of diffusion), and used this context for a complementary contestation of von Neumann's pseudo-theorem on the impossibility of hidden variables. Takabayasi, already quoted in connection with Madelung's and Bohm's works, also discussed stochastic quantum mechanics. Kalitsin [441], inspired by Einstein and Hopf [436] made a step further by viewing the above mentioned quantum forces of stochastic quantum mechanics as originating from the fluctuating electromagnetic field of the vacuum (see complements in Wheeler and Zurek [100]). Papers on the same topic have been published, essentially at the same time by Weizel [279–281].

With Bohm and Vigier [263], and Bohm [60], the line of research initiated and developed by Einstein, Schrödinger, Fürth, Fenyes... on stochastic quantum mechanics made some kind of contact with another line of research, the one of Louis de Broglie and Bohm on causal theories. In these works, opening a new period of his commitment to a new interpretation, Bohm went to the consideration of the *very interesting and suggestive possibility ... of a sub-quantum level containing hidden variables.* He recognized that the postulation of such a level could be unpleasant to many because, if such a level actually existed, *we could never have direct experience with the entities in it* and therefore would always ignore which kind of entities they would be. Here, Bohm supposed that it could receive attacks from positivists for which the existence of such hidden entities should not be considered. The counter-attack to the supposed attacks is worth serving. He first conceded to the positivist Zeitgeist that, *may be we should not postulate the existence of entities which cannot be observed by methods that are already available, but it happens that entities which cannot be observed by methods already available could possibly later be observed with new methods not yet already available.*

It is ridiculously simple to illustrate such a point of view using, once more, the history of atomism. Actually, this counter-attack is worshipping at the altar of the positivist deities: If positivists are not pleased today, they just have to wait, and might well possibly become pleased to-morrow. But the counter-attack did not stop there and went on overthrowing and destroying the altar. As Bohm claimed, it is not true that our concepts only originate from our everyday experience. Well, this is one of the Great Debates for philosophers and I do not intend to try doing justice to it in a few lines. I shall here just mention that I agree with Bohm for at least one reason: What we are is not the result of our everyday experience only, but it is also the result of a huge evolution since the beginning of the life on earth, and even also since the beginning of the universe. All this is somehow embedded in the matter we are formed from, in our very bones, in our very genes. No more here on the issue of positivism that we already discussed but, what I intended to do, was to show how much the epistemological status of stochastic quantum mechanics is similar to the one of the late causal theories of Louis de Broglie or of Bohm. *Mutatis mutandis*, what has been discussed above for Bohm after Bohm applies to stochastic quantum mechanics too. We have two birds to hunt, but they are from the same family. It will make things easier.

For the sake of convenience, our birds should receive given names. Let us call one SQM (Stochastic Quantum Mechanics) and the other one LBB (Louis de Broglie–Bohm). Essentially, they have been flying independently, on their own, with however, sporadically, a bit of tiny cross-talks. It is us, watching the flights of these birds, who can clearly see that, although separated, they are correlated, in the present case by a common cause: The invocation of sub-levels.

We left SQM with Weizel in 1953, and 1954. A bit more than a decade later, Nelson [264] published a paper which is acknowledged as being a seminal paper. We could believe that Nelson's work had been prepared by the efforts of his precursors (we all need precursors) but it seems that, when completing his work, Nelson was not aware of the previous works by Fenyes or Weizel. Soon after, his paper was complemented by a whole book on the dynamical theories of Brownian motion [125]. In his 1966-paper, Nelson provided a re-derivation of Fenyes results, but extended them significantly as follows.

Let us take the example of an electron which is objectively viewed as a point particle of mass m, "objectively" here meaning: In the sense of Newtonian mechanics (therefore, the particle of mass m could also be possibly discussed in a Hamilton-Jacobi's framework). Nelson thereafter made the basic assumption that *any particle of mass m constantly undergoes a Brownian motion with diffusion coefficient inversely proportional to m.* The diffusion coefficient, if written as $\hbar/(2m)$, allows one to establish a connection between the diffusion equation and Schrödinger's equation. More precisely, in this framework, Schrödinger's equation may be directly derived from Newtonian mechanics, with the wave function Ψ being related to a density of probability of the objective position $\mathbf{x}(t)$ of the objective particle versus time. As in the pilot wave theory, but in a different way, the wave function Ψ is something which is overhanging an underlying objective reality.

The expression "Newtonian mechanics" used above must be understood as meaning what it means, that is to say: Newtonian mechanics. In particular, the Newton's law $\mathbf{F} = m\mathbf{\Gamma}$ is assumed to hold but, because we are facing a stochastic Brownian process, $\mathbf{\Gamma}$ denotes the mean acceleration of the particle. Furthermore, in contrast with the usual macroscopic Brownian motion, whose equation of motion exhibits a term for friction, the motion of objective particles in SQM is frictionless (this is required to preserve Galilean covariance, and allows us not to bother with the introduction of a concept of temperature at the sub-level). Forces acting on the particles are of two kinds (i) usual macroscopic forces, such as Coulomb forces, for instance in the case of electrons interacting with a nucleus and (ii) sub-level random forces generating a very irregular Brownian-like motion. The theory obtained is not strictly deterministic, in the absence of a specific deterministic mechanism to explain the origin of the random forces, but it could be made deterministic (although stochastic) if such a mechanism were implemented. And also, being stochastic, probability concepts have to be used, but these probabilities are of a quite classical nature (associated with our lack of knowledge of details), not of an intrinsic quantum nature.

In his 1967-book, Nelson lucidly wondered whether it is an accident that Markov processes, with the dynamical law $\mathbf{F} = m\mathbf{\Gamma}$, are formally equivalent to Schrödinger's equation. In other words, do we have here a meaningful or a misleading meaningless analogy? Indeed, analogies may be misleading and meaningless and my thesis is that it is indeed here the case. This will be proven later when SQM will be eventually dismissed (for good reasons), excepted for the existence of a possible loophole.

Nevertheless, Nelson's work had great influence and many successors. La Pena-Auerbach [442] generalized the work of Nelson, still in the framework of an extension of Newtonian mechanics. In a particular but interesting case, his generalization reduces to the equations first proposed by Nelson. This author too obtains Schrödinger's equation, but as the simplest equation, from a certain mathematical point of view, in a generalized framework. See also further works by La Pena-Auerbach and Cetto [443] and by La Pena-Auerbach [444] in which the authors dealt with the case of identical particles and with particles possessing a spin. Maric and Zivanovic [445] discussed the fact that *the underlying physical picture is that of a perpetual motion of the ultimate constituents of matter, a kind of Zitterbewegung which, although undetectable* (but remember that Ψ too is undetectable), *might be taken as the starting point in the elaboration of the description of physical phenomena*. In the same spirit, they also commented that *instead of starting from the Schrödinger equation as a fundamental postulate and then simply reinterpreting it*, the approach followed allows this equation to be *derived from postulates which are inherited from the logical structure of classical physics*. They also made another comment much relevant for us: *In Nelson's approach almost all trajectories are continuous but they are not differentiable. Therefore, the velocities do not exist ... the image of the physical process which follows consists of an independent diffusion process for each particle but the mean (forward) velocity of any particle depends on the position of all others. The theory is strongly non-local and the configuration space description becomes necessary.* What is relevant for us here, is the non-locality feature. Its appearance in this context should not surprise us. Since Schrödinger's equation, and therefore quantum mechanical predictions, are recovered in Nelson's framework, we could have indeed expected that the non-locality of quantum mechanics would have rubbed off on, and somehow already existed at, the underlying level made out from hidden entities, as is also the case for Bohm's pilot wave. Later on, Davidson [446] provided another extension of Fenyes' and Nelson's approaches in which the diffusion constant of the Markov process is turned to a free parameter, and still later on, Fritsche and Haugk

[447] provided a new look at the derivation of Schrödinger's equation from Newtonian mechanics. We could pursue this story, insofar as researches in SQM are still very active. Already, in 1983, more than twenty years ago, in a supplementary section in Wheeler and Zurek [100], there was a mention of a bibliography on stochastic quantum mechanics, containing over 200 papers. Nowadays, on a certain search engine, "stochastic quantum mechanics" may return you about one million pages. But we do not need to learn more, in the context of this book.

Now, while we have been watching the SQM bird flying, what happened with the other bird, the LBB one? In 1976, Bohm [366] discussed some of his work, carried out together with Vigier, concerning the *statistical features of a quantum system as originating in perturbations from a random background of motions at a deeper sub-quantum-mechanical level*, traced back to a work by Bohm and Vigier [263]. In a monograph [448], Vigier himself dealt with the modeling of quantum statistics in terms of a fluid with irregular stochastic fluctuations, and with a derivation of Nelson's equations. In 1989, Vigier provided a complementary discussion of the concept of sub-quantum mechanical level, in connection with the behavior of particle-like singularities in the double solution [176]. The same year, Bohm and Hiley [190] reviewed the stochastic interpretation of quantum mechanics and demonstrated that, like in the causal interpretations, non-locality is necessarily involved, an important feature that we already mentioned not long ago. They also compared their approach (Bohm's one actually) with Nelson's one, both of them implying non-locality and, interestingly enough, claimed priority. For Bohm indeed, the stochastic interpretation of the quantum theory was first introduced by Bohm and Vigier, in 1954 [263]. The claim for priority is correct with respect to Nelson, but ignored all the precursors in this line of research, these precursors that Nelson ignored too. Although the claim for priority of Bohm could simply indeed indicate a lack of knowledge of the previous literature, an unfortunate case unfortunately too common, it could also mean, at least in his mind, that the stochasticity features he pointed out in quantum mechanics are strongly and inherently conceptually tied with the very existence of the pilot wave. An extension of the stochastic approach to the relativistic domain is also discussed in the reference [190] and, in his last book, with Hiley [68], Bohm provided an ultimate account of the stochastic version of his ontological interpretation. Then, with the death of Bohm, we let the curtain come down on the story of the LBB bird.

With Sidhart [449], both lines of research (QSM and LBB) are considered together, discussing converging points between the stochastic approach of Nelson and the de Broglie–Bohm approaches. The same issue of convergence and confluence is also discussed by Santos and Escobar [450], and by Santos [451]. Indeed, both the QSM- and the LBB-approaches share the idea of the existence of classically objective trajectories followed by objective particles, submitted to a stochastic motion, that is to say the existence of an objective sub-level from which quantum mechanics, with its pseudo-intrinsic indeterminacy, would emerge at the foreground level of the experimentally accessible phenomena. The stochastic ingredient is lacking in the early Bohm's theory, but the objective sub-level made out from objective particles, is already there. At the end of the previous chapter, we were left with an interest for only two non artificial PHV-theories: The double solution and the pilot wave. We now have a third candidate enlarging a bit the family: Stochastic quantum mechanics. And the question we still have to answer is the following: Is this family decent enough to still deserve some visits?

Chapter 15

When the Heart Hurts

An ode for and against irrationality

The issue of hidden-variables theories, and more generally of the possible interpretations of quantum mechanics, is or should be, in some kind of ideal rational world, a purely (a word guilty of being suspicious) scientific matter. We already had the opportunity, in particular when discussing Bohm rejected and Bohm approved, to comment that such has not been the case. Passions, when the heart hurts, when the soul is soul-feared, agitated themselves as soon as they could find any opportunity to mix things up. In this chapter, I intend to discuss again the same topic, but from a somewhat more general and less specific point of view. Furthermore, our previous encounters with this issue, scattered in several places, were *a priori* (before we had a good enough overview of the hidden-variables landscape) while, in contrast, the present chapter, forming one unity at a single place, is *a posteriori* (after the landscape being explored finely enough).

The aspects to be discussed in this chapter must be considered as irrational. Surely, we might fight for years and years, or even centuries and centuries (as some philosophers indeed did), or worst, to define and determine what is rational and what is irrational, and above all what is reason (this Natural Light, as stated by Spinoza [130]). I do not intend to embark myself on such a galley, and shall be content with a somewhat naive and fuzzy, but sufficient picture, that the charitable reader will accept.

In this may be too naive picture, mathematics are rational: Starting from axioms, we use demonstrations, following rules, in particular logical rules, to reach theorems, on which every one, clever enough, should agree (but what is exactly logic, or in other words: Where does it come from?). Also, physics is rational. It does not exactly demonstrate (in the stronger sense of the word, something like : Establishing definitely), but it uses mathematical languages to hypothetico-deductively deduce results from hypotheses, to be tested with experiments. The hypotheses have to be structured in consistent theories, in the framework of which the hypothetico-deductive game may play. All people do not necessarily agree with the results obtained, as we have seen with the various interpretations of quantum mechanics we have reviewed. However, experiments must agree (but how exactly do we invent theories, and where the *laws of nature* come from?). With philosophy and metaphysics (including parts of theology and ethics), we are beyond the demarcation line of Popper, leaving the scientific domain. Yet, we are still in a rational domain: Philosophers do not demonstrate, but they rely on argumentations, using all the power, even spoiled by weaknesses, of the Reason (but what is Reason?). Obviously, as we know, philosophers do not agree together (the consequence of the fact that a philosophical argument is not a demonstration) and, in a sense, any human being makes its own philosophy, even if he is not a philosopher. But refined argumentations imply rationality. Indeed, a good argumentation uses good reasons, even pieces of logic, but the problem raised may be too ill-defined, or too under-determined to possibly produce an agreement. We may exchange good reasons and still disagree. Furthermore, if philosophers eventually succeed to agree, it could be that the problem has crossed back the demarcation line, to land on the scientific side. From this point of view, Philosophy could be the Great Mother whose children are recurrently snapped up by the Big Jaws of Science.

After rational investigations relying on pure reason (mathematics), hypothetico-deductive demonstrations subjected to falsifications (physics) and well formed argumentations (philosophy), we now make one step further, may be tiny, but which leads us to irrationality, made out from fears and feelings, hopes and desires, joys and sorrows, faith and disbelief..., and discussed by many so-called human or social sciences: Study of religions, psychology, psychoanalysis, sociology, politics, ... or even history, which, although not irrational in themselves, have to grasp on and deal with the irrational aspects of the human minds. It would be meaningless to consider such aspects of the life as meaningless: They are, on the contrary, most meaningful for each of us, and they pertain to the stuff of the universe, since we vividly experience them, even if they are out of the grasp of our physics.

The above convenient classification, with a few discrete points on a 1D-space, makes discontinuous something which is certainly multifaceted and much more continuous (if the mathematicians allow me to use this expression) in a space with a dimension larger than 1. Even in an one-dimensional continuous picture, we should imagine, at any point, a mixture of intertwined rational and irrational aspects, in various proportions, flowing from the most rational statement at one left end (such as, may be: 1 is 1), to the most irrational feeling at the other right end (such as, may be: The love of a saint woman for an exterminating dictator). Where should we locate the hidden-variables issue on this line?: Somewhere in between, more on the left, but not too far away from the right. We therefore have to examine how much right ingredient is poured in the left pot. However, to do this, we shall avoid too much redundancy with what we already have said. But we have to do this, even in a book devoted to scientific subjects.

One good reason (one reason is enough if it is good) is that implicit passions, beliefs, prejudices, can dazzle the scientists by their excess of dark light, and drive them away from the right track, even when they are cautious and open-minded. Passions, beliefs, prejudices, should better be correctly identified, as far as possible, to recover an acute vision. Indeed, unfortunately (or may be fortunately), scientists are not those perfect cold-blooded machines sometimes popularized, according to legendary but erroneous pictures, in some books or movies. As any one, scientists think not only with their brain, but with their whole body. They may use the brain (this organ without any nerve), but they feel they are right or not with the belly. They move hands to grasp the concepts, and must better have a nose for having a flair. They put their heart and soul into it, even when they perform according to the "rules" of science. All this gives us much energy to go forward, but we are always in danger of rushing into blind-alleys. The most influential passions may be those raised by religions (including atheism).

Religious beliefs and disbeliefs

The most significant and pregnant aspects of the history of human kind are certainly to be found in religious beliefs and disbeliefs. Religions deal with the most essential issues wandering in and making wondering the human mind: Birth, life, death. In particular, death is definitely outside of any possible human experience in a human life, like a crunching black hole, in the most tautological way available: If you are living and experiencing the life, then you are not dead and do not experience death, and if you are dead, then you do not experience *this* life any more. From a positivist point of view, any question concerning death is therefore meaningless and yet, human beings will always keep on asking meaningful questions on this issue. Being aware of our awareness, and of its possible dissolution, is a trademark of the so unified human race and, when archeologists discover old graves from beyond the graves, the skulls are stamped: *Homo Sapiens Sapiens.*

The issue is so deadly vital that it generated an endless skull-bone farandole of carnages, pogroms, wars of religion, Holy and Devil Crusades, Holy and Devil Offices, terrorisms of all sorts and kinds, manifesting this fanaticism and intolerance stamped, not only on our skulls, but also in our minds. Atheism, largely motivated by such excesses, this religion of the counter-religions, did not escape from the fatal fate, as dramatically exemplified many times, in many places, during the last century. Nietzsche announced that God was dead, but it is impossible to kill God in the human psyche, and the corpse of God, as has been said, is still moving. And even a simple trembling of It generates formidable earthquakes.

Historically, science started its startling development, in Western countries, as a consequence, at least according to some historians of science, of the belief in a God who created the Universe and, therefore, the laws of nature. More generally, according to Einstein, *all the finer speculations in the realm of science spring from a deep religious feeling* [2]. Even the idea that nature possessed laws is not so obvious as it may look to us now, and some philosophers even contest the existence of such laws, telling us that they are illusions or more moderately of little importance. However, progressively, these laws came to express a necessity so inescapable and automatic that God faded away. For Newton, he was still necessary to make the laws obeyed and to prevent perturbations to affect the legal Order of the World but, for Laplace, he was already an unnecessary hypothesis. At least this was the kind of answer made by Laplace to Napoleon when he was urged to tell where was the place of God in the mechanistic System of the World inherited from Newton. But He (God) is still creeping below.

And even when people are not deeply religious, they may behave in a religious way, believing, believing, believing ... to "something", for instance to the free will of individuals, if only because they do not want to be simply marionettes agitated by the laws of nature. Also, we should not forget the possible allegiance of some of us to these modern avatars of religions, called ideologies, with their religious and intolerant manifestations. Hence, the search for hidden-variables theories, especially causal theories (in this context, determinism is a more important issue than non-locality or contextuality), and their rejection, could not be a cool- or cold-blooded affair.

As stated by Pinch [154], ... *the discussions which surround the quest for hidden variables in quantum mechanics have, on both sides of the camp, often been conducted in a spirit of aggressiveness which resembles more the defence of orthodoxy of one ideology than a spirit of scientific objectivity.* Belinfante already had previously stated [1]: ... *the merits of hidden variable theories from an "engineer's" point of view would be nihil, just like the merits of kinetic theory are nihil for describing a thermodynamical measurement. The only merits of hidden variable theory in such a case would be for the person "believing" in determinism, in reassuring him in his faith.* In the same spirit, Belinfante also spoke of the religious belief that *nature must be deterministic* and that *everything happening in nature must be predetermined by previous happenings in the physical world.* In contrast with the faithful defenders of determinism, there are those who defended the opposite position, the one who accepted with pleasure and even relief quantum mechanics and its intrinsic indeterminacy (because it gives birth to a more wide opened and oxygenated vision of the world), regarding the commitment to determinism of their opponents as a mere superstition. As expressed by Belinfante again, although with a bit of indecisiveness: *We then can take one of several possible approaches. We can retreat to positivism and refuse to consider questions meaningful to which the answers cannot be verified directly by observation ... Another possibility is that we resign to the fact that nature in principle acts in whatever way it likes to act, and that the so-called laws of nature observed in the past are merely statistical averages over a large number of free-will acts of nature. This point of view can easily be extended in a way to make it attractive to those inclined to ascribe much of what is happening to the existence of some principle called God.*

There are indeed good religious reasons to praise indeterminism rather than determinism (see Popper [61], the chapter on clouds and clocks, for an interesting discussion of this issue). If determinism is right, we may eventually land on the Laplace vision of the world, in which God is no more necessary, with the necessity embodied in the laws of nature. But this rules out free will and, more than that, if determinism is erected as a supreme authority, then even God Himself should be governed by it too. But what is a God without free will, determined by determinism? If there is no place any more for free will, there is also no place any more for God. Such a vision is pregnant in the evolution of our modernity, and the development of a materialist scientism, who invaded the western culture, and the western civilization, and generated so much despair and so many tragedies. The issue is a matter of instinct, of passion, of faith, playing with the nerves, with hearts and souls, but it is not a matter of logic. Indeed, if we go on trying to think that way, in compliance with logical rules, we have the mystery of knowing where the laws of nature are coming from, and what is the nature of the laws. We are rushing to aporias, and meet antinomies, similar to the ones exhibited by Kant.

Conversely, the concept of indeterminism gets on well with the concept of God: Free will possible everywhere, for Him, for us. Theologians would be very pleased with the vision that we can then encompass: God, with His free will, created the Universe, and the laws of the Universe, but, in His Great Wisdom, He poured in It enough indeterminacy to allow Him to interact with It. With the laws of nature, He made His fingerprints visible with naked eyes for those having a good enough vision to see them and, with indeterminacy in the laws of nature, He allowed Him to have His fingers playing the Cosmic Harpsichord, to manifest His Omnipresence and His Omnipotence.

In the words of Belinfante, *the difference in opinion between the convinced believers in determinism and the convinced believers in God is a religious strife in which rationality may be of little help.* Founding Fathers also, not only deacons and preaching friars, have been explicitly involved in such a religion oriented debate. In a letter (letter 83) to Einstein [201], Born expresses the fact that he was unable to understand how Einstein could combine an entirely mechanistic universe with the free will required for an ethical individual. In contrast, for him (Born), a deterministic world is abhorrent. He noticed that Einstein, with his philosophy, somehow managed to harmonize the automata of lifeless objects with the existence of responsibility and conscience, but it was something that he (Born) was unable to achieve.

Einstein, the believer to the God of Spinoza, who acknowledged Spinoza as the philosopher who exerted the most significant influence on his Weltanschauung, could have used him as a shield. Indeed, for Spinoza, the deity is viewed as impersonal. Although believing to a God, another name for Nature, Spinoza also believed in a deterministic universe, that all things in nature proceed from necessity, in a perfectly tight way. The belief of human beings to free will is a belief, but it is not founded. Actually, the will is not free at all. It is determined by external causes which lead us to such or such decision, or to accept or deny something [129] As a complementary point of view, Spinoza also stated that, regarding the freedom of human will, ... no man wants or makes anything which has not been decreed by God since ever, adding that, how this can be made compatible with human freedom is beyond our understanding [113]. Everything is actually determined by the necessity of the divine nature, non only regarding existence, but also to exist and produce some kind of effect in a certain way, so that there is no contingency [130]. The will cannot be called a free cause, but it is solely a necessary cause [130]. However, it is somehow as an *a priori* for Spinoza, because it is the nature of the Reason to consider that things are not contingent, but necessary [130].

But, we could argue, if there is no free will, we have to conclude that there is no responsibility. In particular, for Spinoza, good and evil, or sin, are nothing else than ways of thinking, and not at all things or whatever of this kind possessing existence [129]. Furthermore, all objects in Nature are things or effects. But good and evil are neither things nor effects. Therefore, good and evil do not exist in Nature [129]. Everyone is attracted or repulsed by what he judges as good or evil, but this is made necessary by the laws of nature [130]. The philosophy of Spinoza is then somehow beyond good and evil. As a result, not surprisingly, Spinoza was highly regarded by Nietzsche, in particular in a letter dated 1881 in which he acknowledged Spinoza as a precursor who denied the existence of the freedom of will or of the moral order of the world [452] although, elsewhere, Nietzsche was comparing the way of making philosophy of Spinoza as the one of a spider [453]. It is then no wonder that Spinoza has often been viewed as an atheist. In any case, Born, very likely, would not have liked Einstein using Spinoza as a shield.

But Einstein could have used better shields, namely Kant [103] and certainly more convincingly Schopenhauer [105]: Determinism pertains to the Phenomenon, and free will to the Thing in Itself. The conflict between determinism and free will is then dissolved by the claim that the two conflicting notions pertain to two different levels of reality, that is to say: The conditioned level of human knowledge, and the unconditioned level of the indescribable.

The above discussion shows that we may associate determinism and the atheist idea that there is no sense in life, on one hand, and indeterminism and the idea that God does exist, on the other hand. But, although strange it may look to many, there is also the possibility of associating God and determinism, in the manner of Spinoza. Also, in about the same manner, in a book by Rietdijk already discussed [192], we may read, at the end of it: *I feel that the only possibility that Universe in general, and human life in particular, might have a deeper sense and purpose, consists in that the entire history of the universe, of mankind and of every individual, is proceeding according to a great design, however opaque it may be for us. An indispensable condition for the possibility that such a design might exist is that, indeed, Einstein was right in saying: 'Der Herrgott würfelt nicht' (God doesn't gamble). If God did, if indeterminism and chance did exist as current quantum mechanics maintains, life and the world would be a lottery. Therefore, our only hope of survival, in the deepest sense of the word, the only hope of the truly religious man, has to be set on determinism, on hidden variables.*

The issue of the relationship between science and religion, in view of the existence of quantum mechanics, is also discussed in the seventh chapter of a book by Heisenberg [10], in which various testimonies are collected. For Pauli, the narrowness of the scientific ideal of an objective world, where everything had to evolve in time and space, in compliance with the law of causality, provoked the conflict with the spiritual forms of various religions. But it is science itself which destroyed this narrow framework, with the theory of relativity, and more significantly with quantum mechanics. Therefore, the relationship between science and the content of the spiritual forms of the various religions takes a different aspect. In particular, for Pauli, the principle of complementarity of Bohr allows us to understand that the idea of a material object, completely independent of the way used to observe it, is only an abstract extrapolation, which does not correspond to any precise kind of reality. In contrast, Dirac developed the thesis that the religion is only the opium of the people, offered to individuals to make them dreaming of happiness and to console them of the injustice they experience. This led Pauli to conclude that Dirac too possessed a religion having a first commandment, that God does not exist and Dirac is his prophet. The interested reader may refer to Heisenberg for more details [10].

We also have to discuss secular religions: Marxism, or dialectical materialism, and whatsoever, which have been given concrete expressions by various kinds of communisms. Although they are not proper religions in some strict sense, they manifested many features which are typical of the worst features of dogmatic approaches to religions. By trying to commute the Heaven from the Sky to the Earth, and unfortunately often turning It to nightmares, they properly became improper religions.

Such ideological constructions are discussed by Heisenberg [146], regarding works by Blotkhintsev and Alexandrov. In particular, Blotkhintsev considered that, among the different idealistic trends in contemporary physics, the (so-called) Copenhagen interpretation is the most reactionary. Lenine is invoked as an authority to confirm the validity of the dialectical materialism, and as a rescue worker of the materialistic ontology. The feature of quantum mechanics most alien to this trend of philosophy is the introduction of the observer. According to Alexandrov, we must deal with objective conditions and objective effects. On this point, the dialectical materialism converges with the demands of the proponents of hidden-variables theories, a fact which is historically and philosophically far from being absurd since both dialectical materialism and hidden-variables theories flourished in the wake of classical physics.

Actually, any enlightened enough thinker would tell us straight away that all these discussions were rubbish, that science is one thing, religion another thing, and that they should not be confused nor mixed together. Einstein would agree. As he said, science does not need mysticism, mysticism does not need science ... but human beings need both of them (*science without religion is lame, religion without science is blind* [122]). Kant would have agreed too: Science pertains to the Phenomenon, and God is outside of any experience. And Popper could have said (although he did not) that science is immune (should be immune) to religious arguments and that religion is immune (should be immune) to scientific arguments with, in between, a demarcation line. But, separated as they could be in the logic of epistemology, they are not separated in the human mind. The debate did exist, and still does exist.

Societal influences

The researcher is not a perfect robot or a perfect thinking machine isolated from his environment, from his colleagues, collaborators or competitors, from supervisors or students, from supporting institutions, from the society, from the prevalent cultures and ideas nebulized and diffused in the air, from all sorts of societal influences. And any researcher knows how it could be tributary, boosted or destabilized, helped or destroyed, by benevolent or malevolent attitudes, good appraisals or gossips. We may possibly adequately or inadequately react to such influences and pressures, but much more insidious and pernicious is the Zeitgeist that impregnates as a spooky smoke the very air we breathe. The question asked is then to know, without any desire to justify any scientific relativism, what have been the influences of such evanescent happenings on the development of quantum mechanics.

Nagel [141] discussed the reciprocal interaction between the environment influencing the history of discoveries, with, in return and reaction, the influence of discoveries, born and growing, which exerted their action on the mind of researchers. As he stated: *Indeed, theories that have been notably fruitful in the natural sciences have also been repeatedly used for ideological purposes, to endorse or to condemn current social institutions and practices, to support or to oppose proposed changes in public policy, or to serve as a foundation for some philosophical or theological systems.* With regard to this issue, it could be a good idea to read again Kuhn [115], with his concept of paradigm, his distinction between normal and revolutionary sciences, or even to revisit the nefarious Feyerabend [139].

Regarding more specifically quantum mechanics, Forman [454] developed an interesting thesis according to which there has been a stream of work, in Germany, to take distance with the idea of causality. This would have happened just after the end of World War I in which Germany was defeated. Certainly, the end of the war was a relief for Germans, but however without any real satisfaction. Actually, it is not hard to understand the big amount of despair and humiliation that the people of the beaten country had to suffer. The bitter defeat, with the sanction of the too much unfair and irresponsible Treaty of Versailles, in 1919 (loss of colonies, huge amount of money asked for reparations from a destroyed country, under the burden of its own reconstruction ...) would have awakened temptations to reject the old philosophical views which, presumably, could have provoked the final failure (and we know how much Hitler and the Nazism were to be fed later again by such motivations).

Among these philosophical views, and above all of them, the causality was to be rejected, due to his deleterious dissolving effect on the feelings of destiny and will and pagan desires. According to Forman, at this time, a large number of German physicists, *for reasons only incidentally related to developments in their own discipline, distanced themselves from, or explicitly repudiated, causality in physics.* This trend away from causality was embodied in the Zeitgeist of the Weimar culture. German physicists, in this dark post-defeat age, had to recover a prestige that they have lost and for this, to improve their public image, they had to depart from a rigorous deterministic view of the universe that was becoming an abhorred feature of the physical world picture. Forman also stressed the possible relationship between the development of these ideas and the influence of the logical positivism of the Vienna Circle, but such an influence is not firmly founded and even doubtful (I do not know any evidence of it). Let us simply remember, as already remarked, that the arrival of Carnap in Vienna is dated 1926, the year of Schrödinger's equation, one year before the beginning, with Heisenberg, of what has been called quantum mechanics. But, even if Forman is not right in all aspects, it remains that positivism, existing long before the Vienna Circle, was an incitation to accept the intrinsic indeterminacy of quantum mechanics, and a motivation for accusing the belief to determinism of being of a metaphysical nature. But, if determinism could be branded as a no-go-any-more area by some people under such influences, an attitude giving birth to some kind of new "Sturm und Drang" movement, indeterminism was viewed by others as irrational. The irrational appeal of indeterminism was an opportunity for Einstein to speak of *his fellow physicists ... rushing to embrace a failure of causality without having made any serious attempt to explore the possibilities for a causal solution.*

Another significant development in Forman's paper concerns the influence of a famous book by Spengler [119], entitled: *The Decline of the West (Der Untergang des Abendlandes)*, published in 1918. About one hundred thousands copies have been released, and the book has been widely read in academic circles, including by physicists. Forman provided us with a synthetic account of Spengler views, in relationship with Weimar culture, and the reader is kindly requested to turn to Forman's paper, if he is interested enough by this aspect of the issue. Here, I intend to provide a few extracts to exemplify how much Spengler's book constituted an attack against the principle of causality. Spengler opposed mathematics and the principle of causality on one hand, and the destiny on the other hand. The former pertains to the dead nature of Newton and the second to the living nature of Goethe. It is then a very serious mistake to transfer the principle of causality, the concepts of law and system, that is to say the structure of the rigidified being, in the realm of historical problems which is the domain of the problem of destiny. Rigidified forms repudiate the life. Formulae and laws spread over a crystallization on the image of nature. Numbers kill. The exact procedures used by the physics (the old causal one) lead to death. But the pure phenomena of real life are far from any causality with Goethe, as a precursor, who thoroughly rejected the corresponding principle. The causality corresponds to what is rational, to the law, to what can be expressed ... while destiny is the name for an internal certainty that we should not describe. Causality requires a discrimination that is to say a destruction while destiny is a creation. Therefore, destiny is related to life, causality to death. The dead space and the dead objects are the facts of the physicist. It remains to outline the end of the western science, an end that we can safely perceive. After two centuries of scientific orgies, we have enough of them. It is not the individual who is tired of these orgies, it is the soul of the culture. Therefore, Spengler predicted, the western science, tired of its efforts, will one day return in its psychic homeland.

I believe that the previous extracts are eloquent enough. Spengler, exhibiting a kind of radical relativism, is also a paradigmatic representative of what is called "Lebensphilosophie", certainly a successor of Nietzsche who also disliked science, to better emphasize the instinct of life and its will to power. We know how much, disastrously, such forces to life may become forces to slaughter.

When we refer to Goethe as we have made above, and as Spengler actually did, we also refer to what has been called the "philosophy of nature", a poor translation of "Naturphilosophie", a general philosophical stream which expressed a rebellion against a dominant mechanistic vision of nature, propagated by the triumph of Newtonian dynamics, a triumph that, as far as I understand, would have been rejected by Newton himself. Goethe is usually considered as an emblematic representative of this stream. For instance, in his theory of colors, Goethe advocated the idea that the understanding of colors could not be reduced to Newtonian optics, but that the perceiving subject should be, in a deep way, taken into account. Nowadays, we know that, in some sense, Goethe was right. There is a huge gap indeed between physical colors defined by the physicists and psychic colors felt by an observer. However, may be, this remark is scientific and too much causal (we may study how physical colors are physiologically analyzed by the perceiving organs, and somehow converted to signals fed to the brain). As far as I understand "Naturphilosophie" and Spengler's point of view, it would be more relevant to state that the beauty of a beautiful landscape is not to be found in the landscape itself (not in the radiations emitted by the landscape), but in the mind of the observer. Can the reader find any way to disagree with this evidence?

Obviously however, even if we agree with the thesis of Forman which, in the worst case, may generate a bit of antipathy, and which, in a better case, could contain a part of truth, we are not forced to conclude from this to a scientific relativism. We could be simply content to state that the emergence of indeterministic concepts was facilitated by the spirit of Weimar culture, in about the same way that the belief in God and the correlated belief that, when He created the universe, He equipped it with laws of nature, facilitated the emergence of modern science, about four centuries ago.

And this may be contrasted with the behavior of the French scientists, pertaining to the victorious family of Allies who won World War I, and who, isolated on the other side of the Rhine river, not impregnated by the Deutsche Zeitgeist, resisted to the song of the sirens singing the indeterminism aria. As winners of World War I, French "air du temps" was free of any philosophical questioning, still flagrant with the influences of Descartes, Rousseau, Voltaire, the Age of the Enlightenment and the determinism of Laplace. Such a point of view, at least, gives us the feeling to understand Louis de Broglie, the recipient of a proud, or even arrogant, unchallenged tradition, when he committed himself to the development of causal theories.

In any case, whether there was or not a relationship between the Weimar culture and the development of quantum mechanics and its intrinsic indeterminacy, whether Spengler has been influential or not, the question of the acceptance of causality in classical physics, of its impact on the decline of the West as viewed by Spengler, remains a relevant question for historians of science. In 1949, 26 years after the failed attempt by Hitler to take over the Bavaria government, 16 years after Hitler was made chancellor and, notwithstanding Versailles agreement, began immediately a rearmament program, Heitler [455] could write: *Some have taken it for granted that a living organism is nothing more than a complicated mechanical and chemical system that is entirely subjected to the laws of classical physics and therefore itself deterministic in the same sense. Functions of the mind can, in such a picture, only be regarded as by-products of a deterministic mechanism, and must therefore be precisely predictable themselves. Clearly such views would destroy entirely the concepts of free will and ethical behavior, and indeed of "life" itself. And perhaps they have gone some way to destroy both. In the decline of ethical standards which the history of the past fifteen years has exhibited, it is not difficult to trace the influence of mechanistic and deterministic concepts which have unconsciously, but deeply, crept into human minds. But what is clear is that the new situation, with which we are confronted in quantum mechanics, has created room for an approach to the problem of life ... which is no longer chained to the deterministic views of classical physics.*

From a societal point of view, but less dramatically, we may also discuss the outcome of the battle between causal theories and the Copenhagen interpretation as a result of the inability of people such as Louis de Broglie or Bohm to efficiently communicate with their opponents or also with those who were simply indecisive. At the Solvay Congress of 1927, a turning-point of the story, Louis de Broglie, having essentially worked alone and defended himself alone in face of a somewhat small but unified community of "mencheviks" ready to take the power as "bolcheviks", was defeated. He did not had time, alone, to polish enough his arguments, and could not decently provide a rebuttal to Pauli's objection. What would have happened if Louis de Broglie were not alone, if instead of exploring mainly the double solution, he had a school of collaborators, and if the pilot wave was developed up to his logical conclusion, possibly with the aid of this school? What would have happened if, instead of giving up, Louis de Broglie produced the Bohm's pilot wave version as a de Broglie's version, in 1928? What would have happened if, when von Neumann published his erroneous proof in 1932, this proof, so much used by the defenders of the Copenhagen interpretation, was immediately struck down? This is the issue of contingency to which we shall dedicate our next section.

But, before of doing so, we may say that Louis de Broglie unfortunately turned back at the turning point without any tunneling. Later on, for Bohm, the situation was already much more delicate and desperate. He was not facing any more a small community gathered in a lecture room, but he was facing a huge community of orthodox quantum mechanists spread all over the world. This issue is discussed by Pinch [154]. Pinch was trying to understand communication problems which arise in the development and practice of science, more precisely in the *actual* development and not, as he said, *in the third world* (this is an implicit reference to the third world of Popper). For Pinch, an obvious example for this issue is that, when Bohm published his pilot wave version in 1952, 25 years after the Solvay Congress of 1927, quantum mechanics was in a dominant position. It was taught everywhere and students, professors, researchers, engineers ... were used to it. Experts discussed and published their results, exchanged information, were working and thinking, by using what has become the orthodox language. Much room was over-filled and the possibility for an outsider to make him heard was excessively small. Therefore, the fate of any alternative interpretation is to be forgotten, particularly if it does not provide any new experimental predictions, to be corroborated or falsified. It was one of the merits of Bell to perceive the signal lost in the noise, to amplify it, and to make it audible by the orthodox community.

Furthermore, the pregnancy of an orthodoxy always generates heterodoxies, a fact which is also of a societal nature. Majorities generate minorities and minorities have to oppose to majorities. Belinfante [1] mentioned those who participated to the search for hidden variables for purely polemic reasons, without however really believing that nature was governed by such variables, and/or those who just wanted to show that they could not been brainwashed by the Copenhagen interpretation.

Societal features are also extensively discussed by Cushing [131, 166] but we shall convoke again soon this author in a somewhat related context. Another paper of interest is the one by Bonk dated 1994 [177], in which *the arguments offered for and against various interpretations of quantum mechanics at the Solvay conference on physics in 1927 are examined, and the Bayesian framework of "logical learning" is applied in an attempt to elucidate the fast and widespread acceptance of the Copenhagen interpretation in the contemporary scientific community.* For Bonk, *the Copenhagen view finally succeeded, ..., because a few, but highly representative theories in the realistic programme had been eliminated, discrediting strongly the whole programme. This in turn effectively blocked further pursuit of realistic alternatives for some time.* Pinch [154], already quoted above, also reminds us of the influence of a proof which was erroneous but much invoked, relying on the societal authority of his author: *The failure by Bohm to achieve a social and cognitive redefinition is not surprising when so much was at stake and considering that he was fighting against the embodiment of the arithmetic ideal, the much worshipped and revered, but little understood, von Neumann proof.*

Contingency

What would have happened if ...? This question, already asked in the previous section, points out to the issue of contingency extensively discussed by Cushing [48, 131, 166], a defender of Bohm's pilot wave. The Cushing thesis is entirely contained in the following quotation [166]: *... it is conventional wisdom that the quantum revolution necessarily required that an objective, causal arena be abandoned for microprocesses — our common sense simply leads us astray. In this essay, I want to focus on a little-known, and even less widely accepted, program that does represent all physical processes — both in the microrealm as well as in the macrorealm — as evolving deterministically in a definite and objectively existing arena. I examine how this largely classical, common-sense view came to be rejected and indicate that the contingent context of the philosophical commitments of the major protagonists was an essential element in this. The present consensus was not, and still is not, the only logically and empirically adequate description possible for quantum phenomena ... neither empirical nor logical considerations necessitate a rejection of Bohm's program ...*

Like Cushing [131], I now recall, once again that, indeed, both quantum theory and Bohm's pilot wave are viable. They are indeed both logically consistent and observationally equivalent. But, once quantum mechanics has become the dominating stream, once the bifurcation has been made at some time and amplified at later times, we are facing some kind of irreversible process which, as in non-linear dynamics and bifurcation theory, and in the Prigogine's spirit [456, 457], participates to a genuine history of physical processes and of the world. The question addressed by Cushing is to understand how the quantum mechanical branch has been initiated and amplified. For Cushing, such considerations *are necessary to block any claim that a causal interpretation is incoherent so that the (clever) founders of quantum mechanics would (surely) have spotted that flaw and hence not bothered pursuing such an interpretation.*

In bifurcation theory, an uncontrollable (or uncontrolled) fluctuation makes the evolving system following one branch of the bifurcation diagram and, with another kind of fluctuations, another branch is followed, and still another branch with still another kind of fluctuations. Whatever the branch followed, any other branch would have been possible. And, therefore, the subsequent development of the history, on this or that branch, is contingent: Whatever happens, it could have been otherwise.

Here, I have used bifurcation theory as a model to explain what contingency would be, and how we could use this concept. But I do not claim that contingency is actually a safe concept. On the contrary, in one sense, it is not. In practice, we cannot make the (macroscopic) history starting again with the same initial conditions (notwithstanding uncontrollable superimposed fluctuations) and, therefore, the effects of contingency, if there is indeed any contingency, cannot be observed. Hence, (macroscopic) contingency is outside of the domain of experience: It is a metaphysical concept in the sense of Popper. Conversely, in microphysics, when for a given Ψ we may observe different outcomes of measurements with different probabilities, we can give a precise meaning to the concept of contingency: Just measure and measure again on the same Ψ, and you will check that the outcomes are indeed contingent. But (i) it is not this kind of contingency which interests us at the present time and (ii) it is the very existence of such a quantum mechanical contingency that is challenged by hidden-variables theories.

This warning being stated, I am returning to Cushing who argued that *historical contingency plays an essential and ineliminable role in the construction and selection of a successful scientific theory from among its observationally equivalent and unrefuted competitors,* and illustrated that *an entire plausible reordering of historical factors could reasonably have resulted in the causal program being chosen over the Copenhagen one.* He then described a quite plausible counterfactual scenario to show that, indeed, things could have developed otherwise. Elements akin to the proposed scenario, such as the behavior of Louis de Broglie, and his consequences, regarding the Solvay Congress of 1927, have been already mentioned in the previous section and are not worthwhile to be repeated here. According to the scenario of Cushing, causal theories could have formed the main stream, instead of Copenhagen. Whether, if this had been the case, they would have eventually resisted to the pressure of various counter-arguments, is a question postponed to the last chapter of this book. A point of view similar to the one of Cushing is defended by Holland [40] when he stated that *the historical triumph of the Copenhagen interpretation seems to be a rather fortuitous affair.*

Bohm and Hiley also pointed out to the issue of contingency [180]: *Perhaps the main objection was that the theory gave exactly the same predictions for all experimental results as does the usual theory ... Indeed, it occurred to me that if de Broglie's ideas had won the day at the Solvay Congress of 1927, they might have become the accepted interpretation; then, if someone had come along to propose the current interpretation, one could equally well have said that since, after all, it gave no new experimental results, there would be no point in considering it seriously. In other words, I felt that the adoption of the current interpretation was a somewhat fortuitous affair, since it was affected not only by the outcome of the Solvay Conference but also by the generally positivistic empiricist attitude that pervaded physics at the time.*

Concerning again contingency, in relation with Mach and Einstein, Klein wrote [458]: *Mach points out clearly that the structure of a developed science like mechanics owes something to historical accident. Mechanics, for example, might look quite different if it had been developed along lines suggested by Huygens, which would have been a logical possibility, rather than along the Newtonian lines actually followed. Statements that now have the status of basic laws would then be derived theorems, and vice versa. Mach goes even further when he emphasizes that one should not think of a principle like that of impossibility of perpetual motion as merely a theorem of mechanics. "Since its correctness was perceived long before the completion of mechanics, and since it even contributed in an essential way to the establishment of mechanics, it is surely probable that it does not really depend on mechanical knowledge, but rather that its roots are to be sought in more general and deeper convictions". This result of Mach's "historical-critical exposition" suggests a flexible approach to physics in which certain general principles are seen as more significant, more reliable, than the logical structures on which they seem to be based. Whether or not Einstein learned to think this way from Mach cannot be said, but learn it he did; nothing is more characteristic of his early work than his ability to select such principles.* This quotation is not only relevant to contingency, but also to related issues such as the Duhem–Quine under-determination thesis or the concept of first principles (and should be kept in mind for further use).

I now would like to discuss the issue of contingency by using a topographical mental image. Assume that we are exploring a huge and complicated landscape with mountains, lakes, valleys, and rivers flowing down to the sea level. Theories are located in this landscape (the landscape of theories) and the lower the altitude the best the theory. In particular, we have not yet reached the sea level of the Theory of Everything. The worst, most unstable, theories are located on summits and the best, most stable, ones in wells. Classical physics is located at a rather high level summit (altitude: 3782 m) and the microphysics crisis made us hurtling down the rocks. Several possibilities were opened, at least two of them. Due to contingent original strokes of fortune, contingency drove us down to the quantum mechanical well (altitude: 1462 m). But it could also have driven us toward another well, the well of pilot wave, and indeed some people (a few of them) rushed down this way. Because quantum theory and pilot wave are mathematically and empirically equivalent, there is no reason (for the time being), other than contingent or historical reasons, to prefer one well or the other. Therefore the pilot wave well is also located at an altitude of 1462 m. If we forget the dynamical history which made the quantum mechanical well the most populated one, and if we also forget the people inhabiting one well or the other, then, focusing on the objective landscape as it is independently of ourselves, we observe that it is symmetrical with respect to quantum mechanics and pilot wave. Let us now make again the landscape populated by physicists. The symmetry has been broken, with many inhabitants in the quantum mechanical well and only a few of them in the pilot wave well. Contingency is the agent which produces the breaking down. At least, this is the situation we are facing just now.

Now, I would like the reader to allow me to make a digression outside of the main subject of this book, to topographically illustrate a frightening possibility. Let us start again from the classical physics summit, at 3782 m. Contingency again (interest in the nature of space and time, and the nature of gravity, rather than in the explanation of microphysical phenomena) made Einstein going down from this summit, using another steep descent, to another well, called: General relativity that we may, for the sake of convenience in the exploitation of the image, and without any havoc, still locate at the altitude of 1462 m. Let us now pretend that we are going to unify general relativity and quantum mechanics. With enough energy, we have to feed quantum mechanics and general relativity, so that they will each of them leave out from their respective wells, find new ways to go on descending, join themselves and mix together in a river flowing down to the sea level. But here is the frightening possibility: How do we know and are we sure that the correct journey to the sea level should start from our two wells. May be there is no real efficient escape from these wells. May be the well of general relativity is not the correct one, and that there is another better well, located at the altitude of 857 m which would provide a better starting point. May be the well of quantum mechanics is not the correct one too. May be the correct starting point should be the pilot wave well, which is determinist as is general relativity, with non-linear equations as proposed by the double solution, or another modified quantum mechanical well located at the lower altitude of 542 m. That is to say may be quantum mechanics and general relativity cannot be unified, as they stand now, without prior significant modifications and rewordings. This hot dish will be later served again with other spices, in a somewhat similar but different context.

But another comment must be made here. According to Popper, relying on the English version of his book on objective knowledge [61], *while we can never have sufficiently good arguments in the empirical sciences for claiming that we have actually reached the truth, we can have strong and reasonably good arguments for claiming that we have made progress towards the truth.* Here, Popper does not say that we can asymptotically reach the truth. It just spoke of progress towards the truth, that is to say towards the sea level. What the above landscape picture strongly suggests is that, due to limitations inherent to our sensory and intellectual structures and equipments, and also due to contingency, we might eventually be blocked in some well, or even simultaneously in separated wells, even far above from the sea level. This is stated as a possibility. Kant, I believe, would have stated that it is a certainty.

Chapter 16

Understanding Quantum Mechanics

Understanding

Before trying to understand quantum mechanics (this un-understandable topic as discussed in the introductory chapter), we need to understand the meaning of the word "understanding". Unfortunately, I believe it is impossible to provide an understanding of the word "understanding" which could be agreed by anyone and everyone, just because this word may receive different meanings. It is effectively terribly polysemic. Nevertheless, I shall do something easier, namely I shall review different meanings relevant to quantum mechanics and comment them. From this, I shall select what, at least for me, could be a real and deep understanding, and afterward devote the rest of this chapter to make one step, rather significant I believe, toward such an understanding. On this trip, we shall learn a few things and, in particular, we shall introduce something, a first principle, that I am calling NSP for the time being. NSP is most relevant to the subject of this book. It will allows us, in the next and last chapter, to conclude on hidden variables. Therefore, with this chapter, we begin to explicitly discuss the second hidden world, namely the one of the possible existence of first principles to help us understanding quantum mechanics, to be coupled, in the next chapter, with the first hidden world, the one of hidden variables.

But, we also have to remember the introduction: We are indeed entering the speculative part of this book. Not everything will be speculative, but there will be a fair amount of speculative material. What is my personal opinion is going to be obvious, I guess. But I do not intend to claim that this opinion is the right one. Indeed, there will be a loophole... So, after having in mind the best that I feel able to convey, the reader will still get his freedom to make up this mind.

Understanding the formalism of quantum mechanics

When a newcomer (usually a student) has to learn quantum mechanics, he may essentially encounter two extreme ways to access to it, possibly mixed together to form different variants, depending on the book studied, or on the lectures attended.

In a first way, axioms (or postulates) are presented straight away and consequences of the axioms are afterward developed and discussed. This axiomatic style is the one of von Neumann [26] who used the word "foundations" (Grundlagen) rather than "axioms", of Dirac [42] who preferred using the word "principles", or of Cohen-Tannoudji *et al.* [43]. This is a way to tell the student that the postulates must be accepted as they are (even von Neumann's postulate on the collapse of a wave packet during a measurement), and that there are no better higher level principles to invoke. It is also a way to acknowledge the extraordinary predictive power of quantum mechanics, anticipating, or better, knowing in advance, that the consequences of the axioms will agree with experimental results.

Actually, from a historical point of view, the principles of quantum mechanics have grown on the breeding ground of the experimental results, although experimental results alone do not make a theory. The exposition of the principles may then be related to these historical developments, using pedagogic and plausibility arguments, and prototypical experiments like the one with two slits. This is the heuristic style represented by Bohm [41], Schiff [459] or Feynman [14], and also by Cohen-Tannoudji *et al.* [43] who supplemented the axiomatic style with a bit of heuristics. The word plausibility, used by Schiff [459], among others, may point out to the fact that experimental results alone do not completely determine the theory (this is the issue of under-determination of theory by experiments to be examined in the last chapter). This is something that we know, indeed, if only because of the existence of the pilot wave. The same word (plausibility) may also be viewed as pointing out to the use of analogies, guesses, and trials.

Once these basic elements are digested, the mastering of quantum mechanics requires the mastering of calculation rules generated by the postulates, and of the usually high-skilled mathematics generated by the calculation rules. Making exercises and problems, i.e. mastering the paradigm [115], the newcomer will learn how to make predictions for such or such experiment and, after a sufficient amount of training, he may even possibly develop some kind of "physical sense", of intuition, adapted to the realm of quantum mechanics. And he may also develop the illusion that, finally, he understands quantum mechanics. But this would genuinely be an illusion: He did not understand quantum mechanics, he just understood the formalism.

Understanding the everyday life

The reason why it is so difficult to understand quantum mechanics is that it is a realm departing much from our everyday life experience, full of roguishnesses and evil intrigues that would drive a magician performing on the stage to the Paradise. Let us just remember the spooky and ghostly features of non-locality and contextuality which are so alien to the common sense that some people, even scientists, detected in quantum mechanics some promises to escape from the intolerable constraints of the rules of the world, and invoked it to explain parapsychology.

Therefore, could we understand quantum mechanics if we succeed to understand how the macroscopic world and our intuition of it is connected with the microscopic world and our non-intuition of it? The early developments of quantum mechanics, even before the Copenhagen age is born in 1927, resorted to such a connection with the use of the Bohr correspondence principle. This principle, basically telling us that, by taking some kind of limit, we must recover classical physics from quantum physics, has been an invaluable tool to open paths in the virgin forest of microphysics. But it did meet limits and did not allow us to reach a consistent enough theory [24, 82].

More generally, the problem of the classical limit is still an enormous problem, and a whole book might be required to do justice to it. More humbly, the student may be told of this piece of dust with a diameter equal to $1\,\mu$m, a mass m of the order of 10^{-15} Kg, and a velocity v of about $1\,$mm/s. The wave-length λ (de Broglie wave-length), given by h/mv, is then equal to typically $6.6.10^{-16}$ m, that is to say $6.6.10^{-6}$ angstroms [43], from which we conclude that the wavy properties of matter cannot manifest themselves at the macroscopic level. We also have the Ehrenfest's theorem relating the classical well-known expressions:

$$\frac{d\mathbf{r}}{dt} = \frac{\mathbf{p}}{m} \tag{16.1}$$

$$\frac{d\mathbf{p}}{dt} = -\boldsymbol{\nabla}V(\mathbf{r}) \tag{16.2}$$

and the corresponding quantum mechanical expressions:

$$\frac{d\langle\mathbf{R}\rangle}{dt} = \frac{\langle\mathbf{P}\rangle}{m} \tag{16.3}$$

$$\frac{d\langle\mathbf{P}\rangle}{dt} = -\langle\boldsymbol{\nabla}V(\mathbf{R})\rangle \tag{16.4}$$

in which $\mathbf{r}$ and $\mathbf{p}$ denote the classical position and momentum, $\mathbf{R}$ and $\mathbf{P}$ the associated quantum mechanical observables, respectively, V is the potential, t the time, m the mass, and $\langle\mathbf{X}\rangle$ is the mean value of $\mathbf{X}$. These expressions associate classical equations for classical quantities with corresponding quantum mechanical equations for quantum mechanical averages. We might find many other examples of that sort but the most elaborated piece of theory is decoherence, already discussed, showing how the classical world could emerge from quantum mechanical measurements.

For some people, such considerations contribute to the understanding of quantum mechanics, as exemplified by the title of a book by Omnès [15] which is (after translation) the same than the headline of the present subsection. But this is to be contested. We do not gain any real understanding of quantum mechanics by such invocations. We are simply reassuring ourselves on the fact that our common sense inherited from our everyday experience is not rubbish, and that we may go on attending to our everyday affairs, without worrying too much (jus a bit!) about quantum mechanics.

Understanding quantum mechanics

There is a certain amount of rationality in the universe, as testified by the efficiency of our mathematically oriented theories to make successful predictions, and to efficiently grasp on our environment, outside of ourselves. For those who believe in God, there is no mystery in that: The rationality of the universe is an emanation of the rationality of God and, us, His creatures, are blessed enough to be allowed to take part in His rationality thanks to the soul which animates us. A part of ourselves is in contact with the transcendence and that part may and can, without referring to any experiment, access to the ultimate truths by the power of the reason. Admittedly, such an access is not easy, obscured as we are by our complementary participation to the degrading world of matter, but it is allowed and, therefore, possible. This was essentially the doctrine of Descartes, this high priest of rationality.

Two millenniums ago, and about four centuries before the advent of Christianity, a somewhat similar approach (with however significant differences) was to make Plato most famous. Instead of God, there was a world of Ideals, certainly dignified enough to deserve to have the word "transcendence" attached to it. Instead of human minds taking part in the mind of God, we have the souls who already contemplated the Ideals before, during their previous peregrinations in the world of perfection, and which can still access to it by reminiscence. And we have this degraded sensible world to which we pertain, and which obscured ourselves, like the prisoners in Plato's cavern.

I believe it is fair to say that most mathematicians (which do not have to rely on experiments to manipulate their concepts and symbols, and to give sense to them) are instinctively Platonists, like the emblematic Erdös with his Big Book [435]. With the mind, we may perceive objective relationships objectively existing between objective abstract quantities. So, why this world should be subjective? After the process of dissolution of metaphysics which followed Descartes and his excessive confidence to the power of mind, I believe it is fair to say that most experimentalists are positivists, or even naively positivists (in the same way that there exist naive realists, and naive metaphysicians). As for theoretical physicists, they form some kind of mixed community with one finger pointing to the sky of Ideals and one hand laying on a measurement device.

One of the best way for theoretical physicists to point the finger to the sky is to use thought experiments, better called Gedanken experiments, to remind us of one of the masters of them, Einstein (although the word is usually better attached to the name of Mach). But, rather than Einstein, I shall invoke Galileo Galilei who, in a few words, could destroy the Aristotelian mechanics, after two millenniums of dominance. Let us consider two objects of masses m and M, with $M > m$. According to Aristotle, the heavier object of mass M must fall down at a velocity v_M larger than the velocity v_m of the lighter object ($v_M > v_m$). Now, let us make a third object of mass $(M + m)$ by assembling the two previous objects. Since we have $(M + m) > M$, the velocity v_{M+m} of the composite object must be larger than the velocity of the object of mass M: $v_{M+m} > v_M$. However, we may take another point of view. When falling down, the lighter object of mass m in the composite system, falling down slowly, must decelerate the heavier object of mass M, falling down quickly, and reciprocally. Therefore, the composite object must fall down with a velocity comprised between v_m and v_M: $v_M > v_{M+m} > v_m$. Hence a contradiction: Exit, Aristotle, and, by using only the mind (this mysterious entity with mysterious properties), Galileo Galilei obtained some kind of knowledge on the laws of nature, and this: *a priori*, in the Kantian sense. Hence, Galileo Galilei did not need to carry out any experiment to confirm this discovery, that all objects should fall down at the same velocity and, in particular he did not need to carry out this famous real experiment, from Pisa tower (that, actually, as far as I know, he did not carry out).

If we reject any transcendence, and decide, by conviction or by method, to confine our search to the immanent realm, then the possibility of mathematics and of its efficiency becomes most problematic, an issue that could be solved, according to Kant, thanks to his synthetics *a priori*. This must also be the issue Einstein had in mind when he stated that the most unintelligible thing in the world is that it is intelligible. But, even if we restrict ourselves to immanence, without any attempt to produce any final and definitive epistemological theory, we may observe a few things which would be very hard to dismiss, even for positivists. First of all, there are regularities in nature (something resulting from our experience and from our experiments, that positivists cannot reject), and we attempt to make laws of nature and theories outside of them. Second, the mind pertains to nature too, a so obvious and stainless "matter of fact" that Descartes, after his methodical exercise of systematic doubt, was to conclude: *cogito, ergo sum.* Now, whatever the solution to the unsolved mind-body problem, we cannot escape, whether dualism or monism, or something else, to the fact that mind and matter interact (or, a monist would say, that the interaction is so strong that mind and matter are actually the same thing). We are made from the stuff of universe, since his beginning. Its rationality is finger-printed in each of our organs, cells, atoms... How then could we imagine that it does not somehow perspire in our brain, in our mind? Furthermore, there has been this long-lasting evolution of species, an evolution during which the information inscribed into our genes had to pour inside us some of the essences underlying the regularities of the world.

Therefore, whether our reason is the result of a transcendent or of an immanent injection, we are right to invoke first principles, in the strong sense discussed in the introductory chapter. Let us dare it, and let us dare to try to find the first principles underlying quantum mechanics, not yet revealed by the postulates. If we succeed and find such principles, comfortable for our intuition, then we could say: Now, I understand quantum mechanics. Actually, I do not possess a set of complete first principles from which I could reconstruct all of quantum mechanics. But I claim that I possess one such principle, the NSP, which is sufficient to break a breach in the wall and whose consequences will be drawn in the rest of this chapter, and in the last chapter.

As a last word of caution, I may escape from any philosophical and metaphysical considerations by taking a societal point of view. NSP is fruitful if we can draw consequences from it, and if it receives a large enough agreement. Therefore, one of the sanctions of NSP has to come from the inter-subjective world, which defines an inter-subjective objectivity. Also, of course, the mind may be obscured and the inter-subjective process inefficient. So, I also have to admit that NSP, like any other first principles of physics, may have to compete in the Darwinian and Popperian process of evolution of science [25, 61] and even that it could be defeated.

Derivations of Schrödinger's equation

Consequences to be drawn from NSP will concern (i) the status of Schrödinger's equation and (ii) the fate of hidden-variables theories. The second item will be examined in the final chapter. The first item is to be discussed in the present chapter. NSP is going to imply a change of the epistemological status of Schrödinger's equation. Therefore, to appreciate this change, we need to have in mind the previous status, that is to say the current status. For this, the present section discusses different ways which have been used to derive (not to demonstrate: Derivation does not mean demonstration) Schrödinger's equation. Actually, in deriving Schrödinger's equation, each author proposes a variant of his own, tracing his path among analogies and plausibility arguments. It is clearly impossible to do justice to all authors from a huge literature, but I shall extract from it what I believe to be a representative sample of derivations.

Schrödinger-like derivation

Schrödinger's equation has been introduced in Refs. [460, 461] under its stationary form and in Ref. [462] under its time-dependent form. English translation is available from Ref. [44] and French translation from Ref. [45]. The derivation relies on an analogy between the Hamilton–Jacobi's formulation (again this Hamilton–Jacobi's formulation!) of classical mechanics and geometrical optics. The analogy is thereafter used to proceed from wave optics to wave mechanics, i.e. to Schrödinger's equation. As rather usual when something new is exposed for the first time, Schrödinger's derivation is more complicated than necessary. For instance, Schrödinger used non-Cartesian coordinates and a non-Euclidean interpretation of the configuration space. This requires the use of covariant and contravariant components of vectors (and more generally tensors), which may be unfamiliar to some readers and which, in any case, can be avoided without any havoc. Feynman [14] even commented that some arguments invoked by Schrödinger are erroneous

Without showing any disrespect to Schrödinger's work, I would prefer to present a more recent exposition extracted from Winogradski [463] who has been my professor in theoretical physics (she produced her thesis, with Louis de Broglie as a supervisor. More precisely, according to her testimony that she conveyed to me, she produced her dissertation by herself and, afterward, when she felt it ready, presented it to Louis de Broglie and asked him whether he would accept to take care of her defence. He did). Winogradski's exposition runs along the same line and in the same spirit than Schrödinger's one and can be viewed as a rewording, with a bit of decantation. Furthermore, while Schrödinger's exposition is available, Winogradski's exposition is unpublished. I am then giving the reader the opportunity of being in contact with a somewhat new material. PEDESTRIANS ($\sim\sim\sim\sim$) might better skip the end of this section, and land again on the beginning of the next section, entitled: The non-singularity principle. They may also try to read in a loose way, picking up the essence of the comments.

We have to start with the Hamilton Jacobi's formulation of classical mechanics. This is easy. I simply have to kindly request the reader to revisit Chapter 2. Once this is done, we can turn to scalar wave optics which relies on a wave equation reading as

$$\frac{\partial^2 \Psi}{\partial x_j^2} - \frac{1}{u^2}\frac{\partial^2 \Psi}{\partial t^2} = 0 \tag{16.5}$$

in which $u = u(x_j, t)$ is the velocity of the wave $\Psi(x_j, t)$. We may also introduce the refractive index n of the medium according to

$$n = \frac{c}{u} \tag{16.6}$$

in which c is the speed of light.

We now specify the situation to the case of a steady medium ($\partial n/\partial t = 0$) which may support monochromatic waves of angular frequency ω, reading as

$$\Psi(x_j, t) = \Psi_0(x_j) \exp(-i\omega t) \tag{16.7}$$

Because Ψ and Ψ_0 are, in general, complex fields, we set

$$\Psi_0 = A \exp(i\phi_0) \quad A, \phi_0 \in \mathcal{R} \tag{16.8}$$

leading to

$$\Psi = A \exp(i\phi) \tag{16.9}$$

with

$$\phi(x_j, t) = \phi_0(x_j) - \omega t \tag{16.10}$$

In these expressions, Ψ_0 is a complex amplitude, A a real amplitude, $\phi(x_j, t)$ and $\phi_0(x_j)$ are phases. We may then introduce the wave-number vector reading as

$$k_j = \frac{\partial \phi}{\partial x_j} = \frac{\partial \phi_0}{\partial x_j} \tag{16.11}$$

The wave-number k is defined as $\sqrt{k_j^2}$ and the wave-length λ is defined by $\lambda = 2\pi/k$. Also, we have

$$\omega = -\frac{\partial \phi}{\partial t} \tag{16.12}$$

Inserting Eq. (16.7) into Eq. (16.5), we obtain

$$\frac{\partial^2 \Psi_0}{\partial x_j^2} + \frac{\omega^2}{u^2} \Psi_0 = 0 \tag{16.13}$$

Next, inserting Eq. (16.8) into Eq. (16.13), we obtain two equations relating the real amplitude A and the phase ϕ_0

$$\frac{1}{A} \frac{\partial^2 A}{\partial x_j^2} - \left(\frac{\partial \phi_0}{\partial x_j}\right)^2 + \frac{\omega^2}{u^2} = 0 \tag{16.14}$$

$$\frac{2}{A} \frac{\partial A}{\partial x_j} \frac{\partial \phi_0}{\partial x_j} + \frac{\partial^2 \phi_0}{\partial x_j^2} = 0 \tag{16.15}$$

If the medium, besides being steady, is homogeneous ($\partial n/\partial x_j = 0$), the wave equation admits plane wave solutions reading as

$$\Psi(x_j, t) = A \exp i(k_j x_j - \omega t) \tag{16.16}$$

in which A, k_j, ω are constant quantities, and λ becomes the spatial period of the wave along the direction of propagation.

We are now equipped enough to turn to a discussion of geometrical optics which is an approximation to wave optics. This approximation is valid whenever the optical wave approximately behaves as a plane wave over a distance of the order of the wave-length λ, that is to say when $A(x_j)$ and $k_j = \partial\phi_0/\partial x_j$ are approximately constant over λ. Equivalently, we may take the limit $\lambda \to 0$.

This may be formally and rigorously introduced in Eqs. (16.14) and (16.15) by examining the relative variations of $\Delta A/A$ and $\Delta k_j/k$ over λ, in the direction $x_{(k)}$, relying on Taylor expansions

$$\frac{\Delta A}{A} = \frac{1}{A}\sum_{n=1}^{\infty}\frac{1}{n!}\frac{\partial^n A}{\partial x_{(k)}^n}\lambda^n \tag{16.17}$$

$$\frac{\Delta k_j}{k} = \frac{1}{k}\sum_{n=1}^{\infty}\frac{1}{n!}\frac{\partial^n k_j}{\partial x_{(k)}^n}\lambda^n \tag{16.18}$$

The insertion of Eqs. (16.17)–(16.18) in Eqs. (16.14)–(16.15), and the investigation of the result to draw consequences, although straightforward, is a bit tedious. I shall rather use heuristic and convincing enough arguments which furthermore lead to the correct results. Because A is approximately a constant, Eq. (16.14) reduces to

$$-\left(\frac{\partial\phi_0}{\partial x_j}\right)^2 + \frac{\omega^2}{u^2} = 0 \tag{16.19}$$

Furthermore, because $k_j = \partial\phi_0/\partial x_j$ is approximately a constant too, Eq. (16.15) reduces to an identity $0 \equiv 0$. Therefore, Eq. (16.19) is the geometrical optics version of the wave optics Eqs. (16.14) and (16.15), i.e. two equations have collapsed into a single one. We observe that Eq. (16.19) contains the phase ϕ_0, but does not contain any more the amplitude A. This means that the concept of amplitude has no meaning, in a strict sense defined by the above derivation, in geometrical optics (this does not prevent to build geometrical optics models using the concept of amplitude).

Also, from Eqs. (16.11) and (16.19), we have

$$k^2 = \frac{\omega^2}{u^2} \tag{16.20}$$

Now, similarly as for S_0 and S, ϕ_0 and ϕ are equiphase surfaces satisfying the following obvious analogous results (refer to Chapter 2). The locus of the points for which ϕ_0 possesses a given value C_0, i.e. $\phi_0(x_j) = C_0$, is a time-independent equiphase surface. There is one surface, and only one, containing a point P of space, given by $C_0 = \phi_0(x_j(P))$. The whole space is therefore filled by a set of motionless surfaces forming the static phase field. The trajectories orthogonal to these surfaces are called rays. The locus of the points for which ϕ possesses a given value C, i.e. $\phi(x_j,t) = C$, is a time-dependent equiphase surface. For a given time t, the moving equiphase surface $\phi = C$ coincides with a motionless equiphase surface $\phi_0 = C_0$. When time goes on, the moving surface $\phi = C$ sweeps over all motionless surfaces $\phi_0 = C_0$.

Assembling the results obtained for the conservative Hamilton–Jacobi's classical mechanics and for geometrical optics, we obtain a remarkable analogy exhibited in Table 1.

This analogy has been discovered by Hamilton, about one century (!) before its use to the discovery of Schrödinger's equations, see Refs. [464, 465], references therein and prior references from Hamilton. Formally, we may express the same structure by using a mechanical language or an optical language. Both languages may be translated, from one to the other, by using a dictionary D exhibited in Table 2,

Table 1. Analogy between Hamilton–Jacobi classical mechanics and geometrical optics.

Classical mechanics	Geometrical optics
$S = S_0 - Et$	$\Phi = \Phi_0 - \omega t$
$S_0 = S_0(x_j)$	$\Phi_0 = \Phi_0(x_j)$
$E = \text{constant}$	$\omega = \text{constant}$
$p_j = \dfrac{\partial S}{\partial x_j} = \dfrac{\partial S_0}{\partial x_j}$	$k_j = \dfrac{\partial \Phi}{\partial x_j} = \dfrac{\partial \Phi_0}{\partial x_j}$
$E = -\dfrac{\partial S}{\partial t}$	$\omega = -\dfrac{\partial \Phi}{\partial t}$
$\left(\dfrac{\partial S_0}{\partial x_j}\right)^2 = \dfrac{E^2}{w^2} = p^2$	$\left(\dfrac{\partial \Phi_0}{\partial x_j}\right)^2 = \dfrac{\omega^2}{u^2} = k^2$
$w = E/p$	$u = \omega/k$
trajectory	ray

Table 2. The dictionary.

$S = G\Phi$	(a)
$S_0 = G\Phi_0$	(b)
$p_j = \frac{\partial S}{\partial x_j} = G\frac{\partial \Phi}{\partial x_j} = Gk_j$	(c)
$E = -\frac{\partial S}{\partial t} = -G\frac{\partial \Phi}{\partial t} = G\omega$	(d)
$w = \frac{E}{p} = \frac{\omega}{k} = u$	(e)
trajectory$\longleftrightarrow$ray	

where the newly introduced constant G has the dimension of an action.

An analogy is not necessarily significant but, although we shall eventually reject that particular one as a way to reach Schrödinger's equation, any analogy should be, at least tentatively taken seriously. If the analogy is fully meaningless, then the value of the constant G does not matter, and any value for G would do. *A contrario*, if the analogy is somehow meaningful, that is to say if the motion of a material point can be somehow associated with the propagation of a certain scalar field (the point of view taken *very seriously* by Louis de Broglie in his double solution), then the constant G should be a new fundamental constant of nature. We now know that the analogy under study, even if it is somewhat misleading as I shall show in the sequel of this section, may be taken seriously enough, and that it eventually leads to $G = \hbar$. Lines (c) and (d) of Table 2 then lead to

$$p_j = \hbar k_j \tag{16.21}$$

$$E = \hbar\omega \tag{16.22}$$

which we call de Broglie, or Einstein-de Broglie relations. Equation (16.21) expresses an equivalence between momentum (mechanical language) and wave-number (optical language), while Eq. (16.22) expresses an equivalence between energy (mechanical language) and angular frequency (optical language). Historically, these relations appeared in connection with the study of light waves, in particular with the interpretation by Einstein of the photoelectric effect, using the concept of photons. Later on, they have been extended by de Broglie to matter waves, with the status of a hypothesis. This

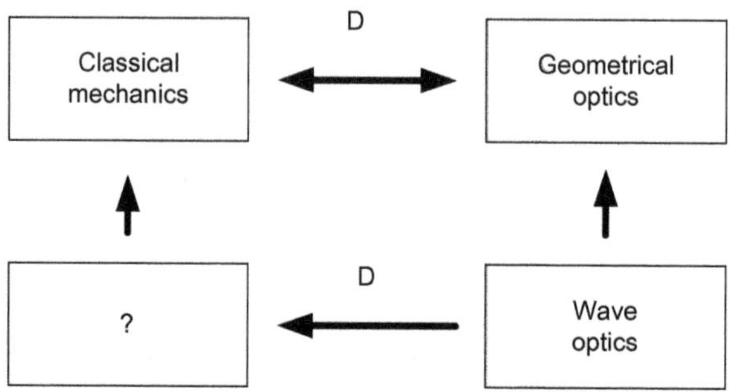

Figure 3. A key to a derivation of Schrödinger's equation.

hypothesis bas been beautifully justified by the experiments of Davisson and Germer concerning the diffraction of electrons. In this book, before refining, we can view Eqs. (16.21)–(16.22) as the result of (i) an analogy or (ii) an hypothesis or (iii) an interpretation of experiments.

The situation we are facing is now sketched in Fig. 3, providing a key to a derivation of Schrödinger's equation. First, we possess an analogy between Hamilton–Jacobi's classical mechanics and geometrical optics, expressed by a dictionary D. Second, geometrical optics is an approximation to scalar wave optics. Figure 3 then exhibits three filled rectangles, and we may feel intuitively but clearly that something is lacking, corresponding to the fourth empty rectangle. To fill this rectangle, we apply the dictionary D to wave optics. From the dictionary of Table 2, with $G = \hbar$, we have

$$\frac{\omega^2}{u^2} = k^2 = \frac{p^2}{\hbar^2} = \frac{2m(E - V)}{\hbar^2} \tag{16.23}$$

We may then translate Eq. (16.13) to

$$\frac{\partial^2 \Psi_0}{\partial x_j^2} + \frac{2m}{\hbar^2}(E - V)\Psi_0 = 0 \tag{16.24}$$

which is exactly the time-independent (stationary) Schrödinger equation.

Therefore, Eq. (16.7) is translated to

$$\Psi = \Psi_0 \exp(-iEt/\hbar) \tag{16.25}$$

and we readily establish that Ψ also satisfies Eq. (16.24) that we better rewrite as

$$-\frac{\hbar^2}{2m}\frac{\partial^2 \Psi}{\partial x_j^2} + V\Psi = E\Psi \tag{16.26}$$

Next, we can eliminate E from Eq. (16.26) by using Eq. (16.25). The simplest way to do it is to write

$$E\Psi = i\hbar\frac{\partial \Psi}{\partial t} \tag{16.27}$$

leading to

$$i\hbar\frac{\partial \Psi}{\partial t} = -\frac{\hbar^2}{2m}\frac{\partial^2 \Psi}{\partial x_j^2} + V\Psi \tag{16.28}$$

which is the general time-dependent Schrödinger's equation. Invoking the "simplest way" to obtain Eq. (16.27) rules out awkward expressions such as the one obtained by deriving Eq. (16.25) twice with respect to time

$$E\Psi = i\hbar\sqrt{\Psi\frac{\partial^2 \Psi}{\partial t^2}} \tag{16.29}$$

Actually, many derivations of Schrödinger's equation invoke a simplicity principle. More generally, we shall refer to this as the use of the Occam razor principle, which is a parsimony principle, usually stated by saying that entities are not to be multiplied beyond necessity (Encyclopaedia Britannica article), but that we summarize by two related precepts (i) always choose the simplest strategy available and, in particular (ii) never add more ingredients than strictly necessary. This principle is more metaphysical than physical but, from a heuristic point of view, it is extremely efficient.

Instead of invoking the Occam razor principle to derive Eq. (16.28) from Eq. (16.26), we may refer to the first postulate in Cohen–Tannoudji list (Chapter III) reworded and commented as follows. At any time t, the state of a system is defined by a wave function $\Psi(t)$. With this statement, in which I am using a wave function instead of a ket (remember, a ket was used to introduce the first postulate), I just get rid, without any havoc for the sequel, of systems requiring more complicated descriptions, such as spinors. Equation (16.28) then implies that, if the state of the system $\Psi(t_0)$ is known at $t = t_0$, it is also known, upon forward (backward) integration in time, at any time $t > t_0$ $(t < t_0)$. In other words, the first postulate is consistent with Schrödinger's equation being of first-order with respect to time. Equation (16.28) is then called an evolution equation.

Variational principles

For the sake of completeness and for further reference, we now examine another analogy, between mechanical and optical variational principles, known as Hamilton, Maupertuis and Fermat principles [39, 466]. For this, we consider a configuration space (X, t) in which a point P designates a space point X and a time t. Hamilton principle tells us that the actual motion of a particle, from $P_1 = (X_1, t_1)$ to $P_2 = (X_2, t_2)$, optimizes a Hamilton action Lt according to

$$\delta \int_{P_1}^{P_2} L\left(x_k, \frac{dx_k}{dt}, t\right) dt = 0 \tag{16.30}$$

in which L is the Lagrangian. For a conservative motion, the Hamilton principle reduces to the Maupertuis principle in which time is no more involved

$$\delta \int_{X_1}^{X_2} p\, dl = 0 \tag{16.31}$$

in which p is the momentum, and dl is an infinitesimal path along the trajectory. From the analogy between classical mechanics and geometrical optics (Tables 1 and 2), and using Eq. (16.6) for the definition of the refractive index n, we have

$$p = Gk = G\frac{\omega}{u} = G\omega\frac{n}{c} \tag{16.32}$$

Therefore, the Maupertuis principle of Eq. (16.31) is translated to the Fermat principle of geometrical optics

$$\delta \int_{X_1}^{X_2} n\, dl = 0 \tag{16.33}$$

Other derivations

Schrödinger's derivation in 1926 [460–462] constituted an extremely reactive and efficient elaboration on de Broglie's works published in 1924, 1925 [157]. Louis de Broglie afterward also reacted quickly and efficiently, providing in 1930 [37] his own derivation of Schrödinger's equation. He explicitly discussed the problem in terms of a wave associated with a particle, his favorite point of view, immediately suggested by Hamilton–Jacobi's formulation in which a field S, or S_0, is associated with particle trajectories (Fig. 1) and which boosted his double solution interpretation of quantum mechanics. Considering a relativistic free particle in the framework of special relativity, Louis de Broglie derived Eqs. (16.21)–(16.22) which are therefore valid in the relativistic domain too. The more general case of a particle in a potential is afterward discussed by using the analogy between Hamilton–Jacobi's theory and geometrical optics, more particularly between the Maupertuis principle and the Fermat principle (actually, as a by-product, Louis de Broglie claimed that it turned the status of Fermat principle from a principle to a theorem). In the next step, a wave equation is derived satisfying the constraint that, in the approximation of geometrical optics, rays should coincide with the trajectories of classical mechanics. Finally, after taking the non relativistic limit ($v/c \to 0$), Louis de Broglie obtained the stationary Schrödinger's equation and thereafter the time-independent Schrödinger's equation. By construction, this derivation used the celebrated correspondence principle of Bohr: Classical mechanics can be obtained as a limit case of quantum mechanics.

Wichmann's derivation [467], as the one by Louis de Broglie, is anchored to the relativistic domain. It starts with the Klein–Gordon's equation, which is Lorentz invariant, and describes the propagation of matter in vacuum (far away from any potential). This equation is satisfied by all plane waves of the form $\Psi = \exp[i(p_j x_j / \hbar - \omega t)]$, associating a mechanical quantity (momentum p_j) and a wave quantity (angular frequency ω), i.e. already somehow incorporating an analogy between mechanics and optics. It is derived by assuming linearity (this is the superposition principle, a postulate as we shall later discuss with Bohm) and maximal simplicity (Occam razor principle). Being linear, the Klein–Gordon's equation is actually valid for any superposition of de Broglie waves. Being relativistic, it is of second order with respect to time, i.e. involving $\partial^2 \Psi / \partial t^2$. This is the consequence of the quadratic relationship between energy and momentum in the relativist domain ($E^2 = p^2 c^2 + m^2 c^4$).

Afterward, Wichmann considered the nonrelativistic limit of the Klein–Gordon's equation and looked for the simplest (Occam razor principle), linear (superposition principle), differential equation satisfied by the nonrelativistic wave equation of quantum mechanics. As an ingredient, he also invoked the interpretation of $\Psi^* \Psi$ as a density of probability of presence (this invocation may seem strange to some readers because, in this case, a prior interpretation of Ψ is used to derive the equation satisfied par Ψ; we might prefer having an equation before interpreting its content). Wichmann then obtained Eq. (16.28), without V however, since the validity of the Klein–Gordon's equation, used as a starting point, was restricted to the free space case. Subsequent plausibility arguments, involving de Broglie hypothesis, lead to Eq. (16.26) and afterward, upon elimination of E, to Eq. (16.28).

Landau and Lifchitz [56] claimed that the wave function Ψ completely determines the state of the system, which is again the first postulate in the list from Cohen-Tannoudji *et al.* [43]. Hence, the wave function equation involves only the first derivative $\partial \Psi / \partial t$ which must depend only on Ψ itself. Invoking the superposition principle, the dependence must be linear, hence $\partial \Psi / \partial t = \mathbf{L} \Psi$ in which $\mathbf{L}$ is a linear operator. The expression of the operator $\mathbf{L}$ is next obtained from the analogy between mechanical and optical variational principles, leading to $i\hbar \partial \Psi / \partial t = \mathbf{H} \Psi$ in which $\mathbf{H}$ is the Hamiltonian operator. The expression of this last operator is derived by considering a classical limit (correspondence principle), eventually leading to Eq. (16.28).

Blokhintsev's approach [38] is very similar to the one of Landau and Lifchitz, invoking our first postulate and the superposition principle, leading again to $\partial \Psi / \partial t = \mathbf{L} \Psi$. The expression of the linear operator $\mathbf{L}$ is subsequently obtained by relying on the examination of the behavior of de Broglie waves of the form $\Psi = \exp[-\frac{i}{\hbar}(Et - p_j x_j)]$.

Schiff's approach [459] relies on de Broglie relations (mechanical/optical analogy) and on the superposition principle (the wave equation must be linear). Schiff also demands that the coefficients of the wave equation must only involve constants such as $\hbar$, the mass and the charge of the particle, that is to say these coefficients should not include any parameter associated with a particular kind of motion of the quantum object, such as E or p. Schiff then established that this last demand rules out any differential equation of the form $\partial^2 \Psi / \partial t^2 = \gamma \partial^2 \Psi / \partial x^2$ (1D-formulation) because, for such equations, γ should depend on E or p. Thereafter, Schiff tentatively examined an equation of the form $\partial \Psi / \partial t = \gamma \partial^2 \Psi / \partial x^2$, established that γ should be equal to $i\hbar/(2m)$, and obtained the 1D-version of Schrödinger's equation for a free particle of mass m. This is followed by a natural generalization to the 3D-case, and also to the case when the particle is not free, eventually leading to Eq. (16.28). This derivation and underlying assumptions give, according to Schiff, a high degree of plausibility to Schrödinger's equation.

Bohm [41] considered the motion of a wave-packet of matter waves (this supposes the validity of the superposition principle). The path of the packet is assumed to coincide with the classical limit of a particle trajectory, i.e. the group velocity of the packet must be equal to the classical particle velocity p/m (this is a form of the correspondence principle). Then, assuming one de Broglie relation, namely $E = \hbar\omega$, he obtained the second one, $p = \hbar k$.

Afterward, Bohm expanded the wave packet Ψ of a free particle as a Fourier integral (superposition principle). This is done first at time $t_0 = 0$ and afterward at any time $t > t_0$. The passage from $t_0 = 0$ to $t > t_0$ invokes the relation $\omega = \hbar k^2/(2m)$ which is obtained by using the de Broglie relation $E = \hbar\omega$ and the classical expression $E = p^2/(2m)$. This step implicitly uses a mechanical/optical analogy, or more explicitly mixes together ingredients borrowed from the quantum world ($E = \hbar\omega$) and from the classical world ($E = p^2/(2m)$) to access to the quantum world, a fairly inconsistent step, we must say. Doing this, Bohm reached Eq. (16.28), in the 1D-case, with $V = 0$ (free space). The fact that the wave equation must be of first order with respect to time is justified by the demand that motions of wave packets may approach the classical limit correctly (correspondence principle) and that there *may exist a conserved probability function with generally sensitive properties*. The last requirement refers to $\Psi^*\Psi$ as a density of probability of presence and, as in Wichmann [467], invokes the interpretation of Ψ as an ingredient to derive the equation for Ψ. Bohm then concluded that, even when forces are present, we must have $i\hbar\partial\Psi/\partial t = H(\Psi)$ where the function $H(\Psi)$ does not involve any time derivative of Ψ and, invoking the superposition principle, has to be linear. Afterward, it is demanded that the average velocity of a wave packet be equal to the classical particle velocity, even when forces are present (this is an invocation of the correspondence principle). This demand allows one to obtain the form of $H(\Psi)$, and leads to Eq. (16.28).

We now comment on the fact that the superposition principle is to be viewed as a postulate (according to Bohm). Indeed, this principle allows one to explain interference effects and, upon summation or integration, to form wave packets. It is actually the simplest one (Occam razor principle) allowing one to do so and it has been efficient to deal with various kinds of waves, such as electromagnetic or acoustic waves. Therefore, at least tentatively, we may try to extend it to quantum mechanical objects. However, as Bohm commented: *Whether this is the only hypothesis which can explain interferences is not known.*

From this sample of derivations (although sketchily reported, for the sake of conciseness), we can see that, although there are many variants, there are also many common points, the most important one being the use of the mechanical/optical analogy, under one form or under another one, including the use of de Broglie relations or of variational principles. Other recurrent ingredients are (i) the first postulate of the list from Cohen–Tannoudji (ii) the superposition principle (iii) some variants of the correspondence principle and (iv) the Occam razor principle. I intend now to show that, motivated by the use of a first principle (NSP), we can reach Schrödinger's equation in a much more economical way, or at least, without using any analogy.

The non-singularity principle

In this section, I am now going to introduce a first principle, called the non-singularity principle (NSP) whose validity will be also discussed after its introduction.

Nature abhors vacuum

As a preliminary training step, I intend to discuss the fact that nature abhors vacuum, before discussing the fact that nature locally abhors infinity. Indeed, any physicist knows about the famous dispute, from the time of Pascal, concerning the question to know whether nature loves or not the vacuum [468]. Actually, there has been a long tradition, beginning at least with the Greek philosophers, including Plato, and Aristotle [7, 33], to refuse the existence of vacuum, conceived as the emptiness of a piece of space from which everything is removed. The argument for this refusal can be more or less paraphrased as follows. Because this piece of space does not contain anything any more, it is nothing. But if it is nothing, it cannot be something. But if it is not something, it does not exist. Hence, nature abhors vacuum. This was contrasting with the proposals of the atomists, Leucippus and Democritus, according to which the matter was made out from atoms surrounded by the vacuum. This problem of the existence or non existence of vacuum went on propagating, in particular all along the Middle Age, where theological arguments have been used such as the idea that where there is nothing, God too is not present, a conception obviously impossible to accept and quite heretical, implying that the vacuum does not exist. To be even-handed, theological arguments may also be used as a proof to the opposite thesis. For instance, theological arguments are used by Locke [469], arguing that the power possessed by God to annihilate a body is a proof for the existence of the vacuum.

According to Descartes (second part in [470]), and following Koyré that I extensively quote below in this paragraph [58], vacuum is not only physically impossible, but it is essentially impossible. An empty space — if something like that could exist-would be a *contradictio in adjectio*, a nothingness which does exist. To imagine two meters of empty space separating two bodies is absurd. If there was vacuum between these bodies, there would not be any separation, and bodies separated by nothingness would be in contact. If there is a separation and a distance between two bodies, this separation and this distance cannot be properties of nothingness, but properties of a subtle matter that we do not perceive (I agree with Descartes on this issue, without any reluctance). More than that, for Descartes, anticipating the spirit of general relativity, there is no space conceived as an entity distinct from matter. Space and matter are identical things. Bodies are not in space, but between other bodies. The space filled by a body is not something different from the body itself.

In contrast, for Torricelli, the top of a mercury barometer was a place where vacuum did exist. However, regarding the affirmation of the existence of vacuum, Pascal is better known. Inspired by the Italian observations, he made his "New experiments regarding the vacuum" [468] and reported on several investigations using *pipes, syringes, bellows and U-bends having various lengths and shapes, with various liquids, like quicksilver, water, wine, oil, air* (sic, for us, air is not a liquid). He concluded, concerning the apparent vacuum observed in such experiments: *After having demonstrated that none of the matters which can affect our senses, and which we know, fills this space seemingly empty, my feeling will be, up to the time where one proved to me the existence of whatever kind of matter filling it, that it is genuinely empty, and devoid of any kind of matter.* With a bit of subtlety, as he wrote: Nature abhors vacuum but accepts it. Descartes who did not believe to the existence of the vacuum, visited Pascal, discussed the issue with him, and returned unconvinced. This gave him the opportunity to wrote that Pascal had *too much vacuum in his head.* Pascal also conceived a famous experiment, carried out by his brother-in-law Périer, on the Puy de Dôme, near Clermont-Ferrand, France, evidencing that the atmospheric pressure decreases which height. Pascal persisted and signed, coming to the conclusion that there was a vacuum above the atmosphere.

Pascal was right to interpret his experiments in terms of physical concepts, such as pressure and static equilibria of fluids. This was a real physical progress, relying on experiments and allowing the development of a quantitative theory (nowadays taught to students when they start learning fluid mechanics). It is physically much more efficient that a simple bold qualitative principle telling us that nature abhors the vacuum (abhors *and* does not accept it). Yet, Pascal was wrong to reject this qualitative principle. The old Greek philosophers were indeed right: Nature abhors and does not accept vacuum.

This, we know from quantum mechanics and further developments, with quantum field theory. If we could remove everything from a piece of space, including the molecules of vapor agitating themselves at the top of a Torricelli mercury barometer, then there would still be quantum fluctuations of fields, virtual particles, bouncing to and fro the wall of nothing. Ironically enough, this piece of space, filled with so many and evanescent entities, is nowadays called the vacuum. But this vacuum is not nothing. It is a quantum vacuum to which the argument of the ancient Greeks does not apply. Nature, abhorring the old vacuum, had to fill it with a new vacuum. Furthermore, even if we confine ourselves to the realm of nonrelativistic quantum mechanics, where particles are not allowed to play a game with nothingness, it is impossible to make a piece of space empty, due to the spatial extensions of wave functions: Where there is nothing, there is a probability to find something.

Nature locally abhors infinity

Our cherished old Greek philosophers were already worried, or even tormented, by the concept of infinity. In the pre-socratic age [3, 4], one of the most prominent characters was Zeno of Elea who, with his paradoxes (only five of them reached us, and many others have been lost), pointed out to some difficulties associated with some infinity oriented concepts, such as infinitely divisible space and infinitely divisible time. All the previous debates on infinity (recurrently perceived as something wild and dangerous) found a first crystallization, a genuine domestication at home of the concept, to remain praised and unchallenged for centuries and centuries, until the nineteenth century, in the work of Aristotle. This philosopher, and genius of sciences, accepted the infinity of time; for him, there was no beginning nor end to the time. In contrast however, he did not accept the infinity of space, claiming that the universe is not infinite, and producing the closed world that we already discussed [7]. Indeed, for him, an infinite body is impossible and, much later on, Kepler, at the frontier between the two systems of the world (the closed one of Aristotle, and the open one from nowadays), was still sharing the same opinion. For Kepler, an infinite body cannot be understood by the mind, and the concept of infinity, such as perceived by the mind, cannot refer to an actual infinity, insofar as an infinite measure cannot be conceived [58]. A bit later on, Galileo Galilei, in a letter to Ingoli, could ask him: Don't you know that it is still not certain (and I believe it will always remain so for human science) whether the world is finite or infinite? [58]. Nowadays, it is fair to state that the issue is still not solved. What we can reach with our sense data, dramatically improved by our modern cosmic probes, is limited (and finite) inside a cosmic horizon inescapably generated by the speed of light. What happens behind this horizon is empirically inaccessible.

The expression "actual infinity" used previously refers to a predominant distinction, introduced by Aristotle, between potential infinity and actual infinity (again this couple potential/actual that we already met in another context, the one of measurement in quantum mechanics). Let us discuss this in the spirit of Aristotle, although a bit modernized. We may define potential infinity by saying that it refers to an object, say a mathematical object or an associated physical object, which cannot be completed but can be repeated, and indefinitely approached. The simplest example is formed by natural numbers. We can start from 1 and add 1 to obtain 2, and add again 1 to obtain 3, and so on. Therefore, it is impossible to say that the series of natural numbers ever finished. There is then an infinity involved in the series of natural numbers, but this infinity is only potential. You cannot face it immediately as a whole, and it is therefore not actual. Potential infinity is also met in mathematics in the concept of limit (which has been used to provide solutions to the paradoxes of Zeno of Elea). You may approach a limit, but never reaches it. You can potentially reach the limit, but it is actually out of grasp.

Predecessors of Aristotle, besides Zeno of Elea, like Archimedes (with the use of recursive methods), or Euclid (with the infinity of points in a line) already used the concept of infinity, but their infinities were potential infinities in the terminology of Aristotle.

But, if such infinities are conceivable in the mind and possess there some kind of existence, they cannot occur in the real concrete physical world. An argument used by Aristotle refers to the existence of bodies in space. Assume a physical object which is infinite. Then, it is boundless, but a body must have bounds. Hence, the associated infinity does not exist (note that the argument, as we stated it, has some general relativity Einsteinian flavour: It uses bodies, not space in which the bodies would be embodied; it makes the body a primary thing, from which we could conclude: No body, no space, in contrast with the Newtonian vision of the absolute space as a fundamental prior arena). Such an infinity in the physical world, an infinity that you could imagine to grasp on, and touch, is an actual infinity. But you cannot actually grasp on an actual infinity.

The distinction between potential and actual infinities may then be used to state a theorem (or may be better said: A thesis): We may accept the use of potential infinity but must reject the use of actual infinity. Infinity cannot be a property of a real thing. It can in principle be examined in the mind, where it may be potential, but never observed in practice in the world, where it should have to be actual, an impossible happening.

Years after years, the centuries went on, up to the time when Galileo Galilei noticed something strange with mathematical infinities [34]. He considered the set of natural numbers $\mathbf{N} = \{1, 2, 3, 4, \ldots\}$ and the set of the squares of natural numbers $\mathbf{N}_S = \{1, 4, 9, 16, \ldots\}$. Each element of the set $\mathbf{N}$ may be put in an $1 - 1$ correspondence with each element of the set $\mathbf{N}_S$, namely 1 with 1, 2 with 4, $\ldots$ and, relying on the definition of the equality of cardinals for finite sets (the one used by ancient shepherds to count their sheeps, a stone or *calculus* for each sheep), extending it to infinite sets, we have to state that the sets $\mathbf{N}$ and $\mathbf{N}_S$ have the same size. Yet, the set $\mathbf{N}$ possesses elements $(3, 5, \ldots)$ that the set $\mathbf{N}_S$ does not possess. Then *infinity should obey a different arithmetic than finite numbers.*

About 250 years later, it has been the fate and fame of Cantor to expand the remark of Galileo Galilei to its *logical conclusion*, and to put infinity on a firm logical foundation, introducing concepts and using words such as: Aleph numbers, continuum hypothesis, transfinite set theory, the infinite number of the kinds of infinity, or cardinal arithmetic. In particular, cardinal arithmetic has been used to demonstrate that the number of points in the real line is equal to the number of points in any segment of the line, and also to the number of points in any space $\mathbf{R}^n$, whatever n, an integer larger than or equal to 1, with the case $n = 1$ being a tautology (this is actually the content of the Cantor-Bernstein theorem that we already had the opportunity to use).

The views of Cantor (a deeply religious Lutheran) on infinity were not independent from the Christian philosophy nor from the Christian theology, influenced as they have been by Aristotle. The word *infinite* should be attributed to God alone. The way to escape from the contradiction between the infinity of God, inaccessible to the human mind, and the infinities perceived by this same mind, can be solved by introducing two kinds of infinity, as done by Thomas Aquinas, between God which is absolutely infinite, and things other than God, which can be relatively infinite. This terminology can be translated in the language of Aristotle. A quantity can be relatively infinite if it is simply unbounded, corresponding to the potential infinity of Aristotle. In contrast, saying that God is absolutely infinite, means that it is actually infinite, or even better said: More than actually infinite.

Before Cantor, the mathematicians viewed the infinities they were dealing with as potential infinities. With Cantor, another huge step is made with the introduction of the notion of a completed set. For instance, by considering the set of natural numbers as a completed set, no more as a set generated by successive additions of 1 to the previous result, the status of this set is dramatically modified, although apparently in an innocent way. It is now a completed set by itself, a whole entity, allowing Cantor to define the first transfinite number, denoted by ω instead of the usual symbol ∞ for unbounded infinity. And, because it is a whole entity, that you can grasp with a simple wink, it is actual. The quantity ω does not denote any more a potential infinity. It has, with a kind of *fiat*, become an actual infinity.

Even now, after Cantor, some mathematicians remain uncomfortable with the concept of infinity. Some of them still consider the concept of infinity as of no real value, arguing that even mathematics using infinities should get rid of them, by using inexhaustible but finite quantities (this is the potential infinity of Aristotle with other clothes). In particular, we know that Kronecker (a contemporary fervent opponent to Cantor) rejected the notion of infinity, even in mathematics, opening the development of a school of thought, known as finitism.

Well, I am writing a book on physics, and although, as a physicist, I may need to use mathematics, I do not feel so much interested by the disputes to determine whether infinities in mathematics are potential or actual, or whatever. But I am much concerned with the possibilities of infinities in the phenomenal world studied by my discipline. To state it in another way, I can imagine the infinity of God (whether I believe in God or not) and that there are infinite things, just as may be the universe, or concepts, in the mind of God. They are for Him, not for me, a poor creature, even if I am allowed to get some divine light pouring from above down to me, such as by the revelation of transfinite numbers. And, indeed, I know that there are all kinds of infinities in my mind (*in abstracto*), such as these mathematical infinities from the world of Plato's Ideals. But can I find infinities in the phenomenal world outside of me (*in concreto*)?

We indeed have examples of possible infinities *in concreto*. For instance, we may believe that time has no beginning nor end, and similarly for space. There is no difficulty for the human mind to accept such ideas even if they are the source of breathtaking discomforts. The model of the Big Bang, in the framework of general relativity, may very well contradict such views: There could have been an origin of time at the moment of the original singularity, and space can be both finite and unbounded. But this is not the issue. First of all, the Big Bang is a model, in the framework of general relativity which, we know, is ultimately erroneous. There could has been a pre-Big Bang time to our universe. The spatial infinity of this very universe does not contradict general relativity: The expansion of space is compatible with an infinite original universe. And, furthermore, there could be an infinity of universes, generated by an infinity of Big Bangs, as suggested by some scenarii. And more important, this is the second point, I have not been talking of infinities *in concreto*. I have been talking about *possible* infinities *in concreto*. And, as far as we can see, even if such infinities do not exist, they are *possible*.

To avoid dealing with such difficulties, not useful for my purpose, let us reformulate our question as follows: Can I find *local* infinities in the phenomenal world outside of me? Here, the word "local" refers to both space and time, or, in a more elaborated vision, to space-time. The answer is a first principle, that would have been accepted by Aristotle, that there is no local actual infinity, or in other words that nature locally abhors infinity. I call it the non-singularity principle (NSP) and state it as follows:

Non-singularity principle:

LOCAL INFINITY IN PHYSICS IS NOT ADMISSIBLE.

This should be easy to accept for physicists who most usually would not accept an infinite result when measuring something. For example, a local body with infinite mass or containing an infinite amount of usable energy would be implacably frightening. Yes, we use the notion of infinite plane waves, but we are well aware that it is an idealization which simplifies our theories and facilitates our computations, but that nothing of that sort is to be found in nature. We also use the Dirac function, or better said the Dirac distribution δ, but we are well aware too that it pertains to the mathematical world and that it an idealization of a narrow function with great amplitude. We can find the manifestation of such narrow functions in nature, but not the limit δ of a series of such functions.

Effectively, each time we encounter local infinity in physics, we immediately recognize that something is going deeply wrong, or that we are dealing with convenient idealizations which cannot be actual, and possibly that we have to think more. The non-singularity principle is indeed something, at least implicitly, anchored in the mind of physicists. For them, an infinite outcome from any computation relevant to the physical world, as would be an infinite outcome from any experiment, is something which cannot be thought, a real non sense. What we are doing here is to state it explicitly just now, before drawing some consequences of it, relevant to quantum mechanics.

The applications of NSP, from its definition, in particular due to the use of locality in space and time, have to be confined to space-time representations. Therefore, according to its definition, it should not be applied to quantum mechanics which works outside of space and time. Nevertheless, the occurrence of infinite quantities in quantum mechanics, and beyond, has always been considered as a disease too. The best emblematic example is to be found in the infinities arising in quantum field theory, and that have been eventually eliminated by using a process called renormalization, in order to obtain physically meaningful results. This suggests that NSP, as stated above, did not receive yet his best and most general formulation. But I do not dare making any further step at the present time toward such a most general formulation. Furthermore, NSP in space and time, or space-time, as it stands now, is sufficient for the purpose of this book, and any search for a generalized formulation can be postponed without any damage.

Why should we consider NSP as a first principle, in the sense already discussed in the introductory chapter? We might rely on logical arguments (*à la Aristotle*), unfortunately possibly biased, or theological arguments (no more receivable nowadays). We might invoke our contact with the mind of God, or with the world of Platonic Ideals, however not enough secure as amply demonstrated by the history of sciences and philosophy. We should better invoke the fact that it matches our intuition and therefore, if this intuition is correct, *whatever the origin of this intuition*, we should, at least tentatively, examine its consequences. Indeed, when the intuition is satisfied, we have found a possible entry to the world of understanding. Furthermore, the reluctance of physicists to the acceptance of local actual infinity means that NSP can receive a large enough inter-subjective agreement, another reason to draw its consequences.

And there is also an inescapable argument to assert that we indeed need first principles. In the words of Selleri [380], *in fact, explaining a word (or an idea) means reducing its meaning to other words (or other ideas). But since the total number of words ever said (or ideas ever had) by human beings is finite, one cannot explain everything without circular reasoning. Some ideas must be taken as self-evident without the need for any explanation. These are the a priori ideas.* I dare to say that NSP should be taken as such an idea.

Therefore, with this new weapon (not a new weapon indeed, but a weapon newly again exhibited, and ready to be used), we are going to try to make some progress toward an understanding of quantum mechanics. By using this weapon, we shall be able to claim that quantum mechanics may be viewed as the *a priori* consequence of a rational demand. In other words, it is rationally necessary. Telling that a certain thing is necessary, and could not be otherwise, and understanding why it is necessary, transitively provides an understanding of the thing itself (by transitivity from the understanding of the necessity of the thing to the thing). We shall also be able to draw some negative consequences concerning causal theories.

Nevertheless, although I shall from now on, as a methodological decision, hold fast on this principle (NSP) in most of the sequel, and draw its consequences, it has to be admitted that there is may be no universal intuition, that any statement is in principle revisable, even any logical statement [138, 471, 472] and that the acceptance of the NSP may be the consequence of a prejudice or, in a more well-tempered way, that it is simply a hypothesis. This gives, from now on, to a significant amount of this chapter, and of the next one, a speculative character, and will eventually oblige me to end with a loophole, in the last section of the next chapter. Note however that there is no harm to speculate: Speculations and thoughts about speculation may be a source of inspiration. In any case, I hope that the reader will go on following me in this journey, to the end of the book, before making up his mind.

Before entering the next section, I however cannot resist to the pleasure of making the reader contemplating a beautiful and genial anticipation of the second postulate of the relativity, due to Epicurus [473]. For him, abhorring infinity like the nature, there cannot be any infinite speed. Hence, he concluded, there must be a finite speed which is the largest speed possible. This speed is the speed of thought. But it is also, he convincingly enough argued, the speed of simulacra, some kinds of images, for instance from a mountain, entering the eye to excite the vision (let us say, the speed of light...). This gives to the second postulate some flavour of a first principle, relying on NSP.

Preliminary trainings

Fluid mechanics and atomism

Before applying NSP to quantum mechanics, I intend to provide a few examples of its application to more conventional fields of science, starting with fluid mechanics. Let us consider a shock wave in fluid mechanics. Across the shock wave (which occurs on a surface in the Newtonian 3D-space), the specific mass changes abruptly from a value ρ_1 to a value $\rho_2 \neq \rho_1$. Therefore, the gradient of specific mass is infinite, a result which is forbidden by NSP. We all know the dramatic effects that a real shock wave can produce. But, just imagine what could be the effect of an actual idealized shock wave with an actual infinite gradient of specific mass. Such unbelievable things do not happen because nature locally abhors infinity. We have really to conclude that our idealized shock wave is indeed an idealization, and have to wonder how nature protected ourselves from such an awful happening, or, in other words, we have to wonder where is the faulty premise in our shock wave theory which produces such an awful happening. This is easy to identify. The origin of the singularity exhibited by our idealized shock wave theory is the assumption of continuity of the medium supporting the wave, which is stated at the very beginning of nearly any exposition of fluid mechanics (with however, I must admit, the usual careful statement that it is not a continuity *stricto sensu*, but rather a quasi-continuity). Anyway, NSP then implies that we must reject the idea that matter is continuous. From this simple fact, we may conclude to the existence of atoms.

Atoms and ancient time

The introduction of the atoms from NSP gives us an opportunity to revisit this concept as it occurred in the ancient times. Indeed, as we know, the idea that the matter is constituted from atoms may already be found in the old Greeks. It is usually granted to Leucippus from which however we know little, and which had been made essentially known by his disciple Democritus [3, 4]. According to Diogène Laërce [474], Leucippus believed that things could mutually transform each in other, and that the universe was simultaneously empty and filled with bodies. This is a discontinuous conception of matter such as the one we just above deduced as necessary by applying NSP. The differences between bodies would result from specific combinations between the atoms they are made of. Each thing receives a particular configuration due to the properties of these atoms, their weights, their shapes, their velocities (in the vectorial sense), their locations ... This theory can be related to the theory of elements of Empedocles and others. The elements (fire, air, water, earth) can be perceived at a macroscopic level. What is introduced with Leucippus is a sub-level, a sub-macroscopic level, similar to our sub-thermodynamical level, made out from things which cannot be seen, touched... i.e. not perceived or observed by any means (a crime for positivist people).

From Democritus, the disciple, we know more, in part because we have access to at least some of his writings (although only through a few fragments), even catalogued such as by Thrasyle and listed by Diogene Laërce, and also because he had enough commentators. For him, as for Leucippus, there are only two things: The atoms and the vacuum between the atoms. These atoms, as for Leucippus, are the sub-elements of the elements. Actually, we could repeat with Democritus everything we already have said with Leucippus because it seems clear that the disciple essentially reproduced the teachings of the master, in such a way that it is virtually impossible to distinguish between what is due to Leucippus and what is due to Democritus.

Aristotle has been an efficient opponent to atomism and, due to its authority and influence, he succeeded to make atomism relegated to the cemetery of premature theories, to be forgotten, before they could be possibly resuscitated later on, as indeed has been the case, for this particular example of atomism, with the chemist Dalton. One reason attested for the rejection of atomism by Aristotle is determinism. Indeed, the atomist theory which is born at the time of Leucippus and Democritus was a deterministic theory, something which has been accepted by the post-Newtonian classical physicists, but that was reluctant to Aristotle. Indeed, for Aristotle, who developed a theory of causes and a classification of them, there was no room in the atomist doctrine for final causes. Therefore, determinism at this time was already a source of debates.

Atomism was deterministic (in contrast with our atomist quantum physics which exhibits an intrinsic indeterminacy) and missed the indeterministic part of the nature, represented by final causes, associated in particular with the aim of the things, and with an underlying free will (that we have ejected from our modern physics). The determinism implied by this ancient atomism is effectively so radical that it could allow one to perceive the physical world as the consequence of a causal development, without any reference to God, or better said, to the gods invoked at that time. Some fundamental questions, still raised today were therefore already raised at that time, although in a somewhat different context: Determinism versus indeterminism, free will or not free will, God (gods) or no God (no gods).

Nowadays, we are often arrogant and I guess that this is not a specificity of our time. But, as far as we are concerned, we may believe that we, lucky men living in a lucky time for thinking and science, are enlightened by the wisdom of Reason which would just have appeared a few centuries ago, or even a few years ago, after we are born, and after the dark age represented by unlucky people represented by... our parents. *Actually*, the essential questions have already be discussed by very clever people, much more clever than most of us, a long time ago, questions still under question, and it is not losing time to examine how they debated and settled them.

And therefore, not surprisingly, the debate concerning determinism versus indeterminism went on further, even at the time of the ancient atomists. Indeed, Epicurus was a follower of Democritus, but he could not accept the determinism of the atomism of Democritus because it did not leave room for the free will. Therefore, atomism had to be modified. This was done by introducing the concept of clinamen which is a spontaneous deviation of the motion of the atoms. This concept is also introduced for physical reasons. For Epicurus, the atoms are moving in the vacuum between them in a free fall way, and therefore all in the same direction, from up to down (and with the same velocity!). The clinamen disturbs these motions and allows the atoms to interact. For many philosophers, the concept of clinamen has been unintelligible but the kind of unintelligibility involved is actually of the same nature of, or at least similar to, another one that we encountered in quantum mechanics: With the clinamen, the physics of atoms is equipped with an intrinsic indeterminacy. In *De rerum natura*, Lucretius described the world, as a follower of Epicurus, and in particular accepted and described the concept of clinamen, insisting on its necessity to understand the free will. The intrinsic indeterminacy of the ancient atomism was therefore explicitly justified by a philosophical necessity.

Fractals and atomism again

Fractals have been efficiently popularized and advertised by Mandelbrot [475, 476], and the reader very likely already possesses some sufficient knowledge of the topic, available not only from Mandelbrot's books but also from many other books and from a huge list of scientific papers. Let us however recall for convenience the construction of one of the most famous fractal objects, namely the Koch snowflake, exhibiting a curve which is everywhere continuous but nowhere differentiable. To build this monster curve, start with an equilateral triangle, and on each line segment of the triangle, apply *ad infinitum* a certain very simple transformation. Let us just describe this transformation applied to any one of the three segments of the triangle, whose length is taken equal to 1. In this unit segment, distinguish three equal segments (each one having then a length equal to 1/3), a segment on the left, one in the middle, and the last one on the right. Replace the middle segment with a pair of line segments, each segment in the pair having a length equal to 1/3, forming a kind of equilateral protuberance. After this first step, the original segment of length 1 has been transformed to a continuous curve made out from four segments of length 1/3, with 3 points where the curve is not differentiable. The total length has been changed from 1 to 4/3. For the second step, apply again the same transformation (with the removal of the middle third and its replacement by a new pair of line segments) to each of the four differentiable segments obtained at the first step. The result is a new continuous curve now made out from 16 differentiable segments, each one having a length equal to 1/9, therefore with a total length equal to $16/9 = (4/3)^2$, with 15 points where the curve is not differentiable. After the Nth step, the total length is $(4/3)^N$ and the number of singular points has considerably increased (according to a formula that we may avoid to consider). When the *ad infinitum* process is achieved, we obtain a continuous curve of infinite length which is nowhere differentiable. Nevertheless, this curve has been obtained by working with only one of the three line segments of the original equilateral triangle. Let us repeat it for the two other line segments. The result is a closed continuous curve, nowhere differentiable, of infinite length, however bounding a finite surface. This monster curve is the Koch snowflake.

The kind of infinity exhibited in this mathematical example is what Aristotle would have called a potential infinity. Instead of the potential infinity generated by 1, then 2 obtained by adding 1 to 1, then 3 obtained by adding 1 to 2, and so on, we have quite a similar potential infinity obtained by making a transformation numbered 1, then a transformation numbered 2 obtained by applying the transformation numbered 1 to the result of the transformation numbered 1, then a transformation numbered 3 obtained by applying the transformation numbered 1 to the result of the transformation numbered 2, and so on. Very likely, Cantor would have contemplated the Koch snowflake as a whole and claimed that it possesses a real existence in the mathematical world. We may imagine, us, poor plain human beings, that we have to construct it with a potential infinite number of steps. But, for Cantor, it is there, that is to say, actual. Whether the infinity exhibited in the Koch snowflake is potential or actual, actually should not bother physicists: It is only a potential problem.

But fractals (the Koch snowflake, and many others like the Cantor sets, the boundary of the Mandelbrot set, the Sierpinski triangle and carpet, space filling curves, strange or chaotic attractors, and whatsoever) are not only attached to the mathematical world but have been used for the description of many physical objects: Clouds, mountains, lightning bolts, snowflakes, soot aggregates, non-linear dynamical systems, and many others. An interesting example is furnished by Mandelbrot who, building on an earlier work of Richardson, asked the question to know how long is the coast of Brittany, a problem already perceived about fifty years before by Perrin in his famous book on atoms [477], as acknowledged by Mandelbrot in one of his books [476]. The reader, superposing in mind the structure of the Koch snowflake and the highly irregular structure of the Brittany coastline, can anticipate the result: Infinity.

This can be experimentally tested. You may follow the coastline by plane, crossing over details, and only managing with the larger scales of the landscape and, as a first approximation, you will measure a length equal to $L_{\text{plane}}(s_{\text{plane}})$ in which s_{plane} is a large scale associated with the plane motion (for instance the plane flies along straight lines, changing only its direction every $10\,\text{km}$). If you follow the coastline by car, you will measure another length $L_{\text{car}}(s_{\text{car}})$. You may probe a still smaller scale by walking along the coastline, following way marked foot pathways, exploring gulfs, and bays inside gulfs, and creeks inside bays, and find $L_{\text{foot}}(s_{\text{foot}})$. For the dog accompanying you, allowed to leave the pathways and to better approach the coastline, running around rocks, the length is $L_{\text{dog}}(s_{\text{dog}})$. For an ant having to deal and play with very fine details, such as sand grains, it is $L_{\text{ant}}(s_{\text{ant}})$. Obviously: $L_{\text{plane}} < L_{\text{car}} < L_{\text{foot}} < L_{\text{dog}} < L_{\text{ant}}$. Furthermore, surprisingly enough, the function $L(s)$ expressing the length versus the scale is very regular, allowing one to characterize the coastline by a quantity called the fractal dimension of the coastline. The function $L(s)$ can be in practice conveniently determined by using maps with more and more refined scales. When extrapolating to $s = 0$, it is then established that the length of the Brittany coastline is infinite.

When I heard of this result for the first time, I could not believe it. Instinctively, such a result, making infinity actual in the physical world, was impossible and ever funny. For me, it was just another kind of games that mathematicians like to play to tease physicists. And, indeed, it is an impossible result as NSP tells us. The flaw is very easy to identify. We cannot reject the law $L(s)$ from which the result was obtained, since it is an empirical result. But we can, and should, reject its extrapolation to vanishingly small scales, which, surely, must be undue. Therefore, there must be a small cut-off scale, preventing the divergence to the impossible actual infinity, below which one assumption in the fractal model of the Britanny coastline must lose any meaning. Identifying this assumption, we must conclude that, around this small cut-off scale, the boundary between the sea and the earth of Britanny, not only loses its differentiability but, more importantly, must lose its continuity too. In other words, at a small refined enough scale, there should not be any more any physical boundary. If Democritus had been aware of the existence of fractals, and of its silly consequence concerning the infinity of the coastline of Britanny (more generally of any coastline), he would have correctly concluded to the existence of atoms breaking the faulty continuity erroneously assumed. Next, if clever enough (but we are unfortunately never clever enough), he would have remarked that atoms imply vacuum between the atoms, and that this is contradiction with the fact that nature abhors vacuum. Holding fast on the existence of atoms, and trying to understand how the contradiction so raised implied by the existence of the vacuum between atoms can be eliminated, he could have concluded that the vacuum is not the void. There must be something in this vacuum, and, today, we know what: Quantum probabilities.

Black holes

With black holes, we are facing one of the most impressive consequences of our picture of the physical world, generated by the general relativity which is our more elaborated theory concerning events taking place in the space-time. Usually, a stellar or galactic black hole is described as an object whose gravity is so strong that not even light can escape from it. Under some simplifying assumptions such as spherical symmetry, it possesses a characteristic radius which defines a critical distance from the center, and some kind of immaterial sphere which is called the horizon of events. Above the horizon, you still have some chance to escape to normal space-time. Below the horizon, you are immediately trapped. This is the place where the ordinary laws of physics break down: Everything collapses to a one-point singularity, with an infinite density of energy. At least, this is what the equations of general relativity tell us. There is strong indirect experimental evidences that black holes do exist, but there is no experimental evidence concerning what actually happens below the horizon. The question is: Do you believe to the existence of such a singularity?

If we apply NSP as it has been stated above, we should say: No, and conclude that the equations of general relativity become incorrect when the singularity is approached. Going on like that, we must afterward conclude that there must exist a more general theory which would wash out, in some way, the singularity. We know what such a theory could be. It could be a theory unifying general relativity working in the space-time, and quantum mechanics allowing processes outside of space and time. So, the situation seems in principle comfortable. "Just build this theory", and you will indeed find that the singularity disappears, replaced by something likely very strange and weird (after all, a black hole is already indeed a very weird object) but without any conflict with NSP.

But, the situation is not so simple. There is what is called a cosmic censorship, which is the consequence of the existence of an horizon of events. Indeed, the singularity, if it exists, is hidden behind the horizon. You are charitably protected from it: It is not for your eyes. Well, I am not particularly frightened by a monster that I cannot see. So, after all, why not accept the singularity and just modify a bit NSP. Instead of saying: Local infinity in physics is not admissible, let us refine a bit and say: Local observable infinity in physics is not admissible. Furthermore, both versions may be used in the sequel without influencing the conclusions we are going to draw later.

But wait, again the situation is still not so simple. Something more frightening might happen: The monster could possibly be not hidden.

Some scientists claim that naked singularities could exist. I am not really equipped, at the present time, to discuss this issue and I do not intend to tell something that I do not clearly enough understand. Let us simply say that naked singularities could exist in the case of spinning black holes according to certain computations. Intuitively, I am not reluctant to the idea that a black hole with a big enough angular momentum could have a sufficient effect on the horizon of events to disrupt it, but I would not like to say more. Let us just accept the idea and examine what we should do with a singularity observable from the outside. We are then returning to a previous step, the one in which we can again invoke NSP, and, applying again the NSP, we could claim: This is not possible. And we would have further good and strong arguments to support the idea that, indeed, this is not possible: Just because we know that general relativity is not the ultimate theory. It is only a theory of space-time and it should be clear enough now that space and time are not the ultimate ingredients of the universe. A quantum gravity theory, or something of that sort, should be able to kill all monsters generated by general relativity. If we hold fast on NSP, the conclusion is: General relativity is incorrect. I am ready to hold fast on it. Furthermore, as a fallback position, it might be possible to argue that NSP has to be improved to account for such extreme conditions like the ones possibly producing naked singularities. And, furthermore again, I shall not use NSP under such extreme conditions. But, let us say it frankly: I do not believe that such a fallback position should be useful here. "Just build the correct theory", and naked singularities will evaporate. Indeed: Even non naked singularities disappear, as far as we know, when quantum gravity is accounted for, producing a foam of probabilities, in which seemingly time is destroyed and space becomes unshaped. For a comprehensive and popularized exposition on black holes, and associated singularities, the reader may refer to Thorne [478].

Big Bang

Astronomical observations, in agreement with general relativity, indicate that our universe (not *the* universe) began with a singularity called the Big Bang. There are some similarities between the Big

Bang and the singularities exhibited by the black holes, the most important one being that they all result from the equations of general relativity. Accordingly, the Big Bang is often viewed as a naked singularity. But there is a drastic difference. In the case of black holes, the singularities are below the horizon of events, which would be reached in the future of an observer approaching this horizon. In the case of the Big Bang, the singularity happened in our past and, instead of being located outside of the horizon, we are located inside the horizon. Furthermore, inside this horizon, we are facing a space-time expansion (not a shrinking process), an expansion which could possibly have experienced a dramatic phase of inflation, a long time ago, and which is now in a phase of acceleration.

Should we believe to the existence of such a singularity? If we apply NSP, we must answer: No. But the Big Bang has something very special. Some theologians could affirm that this universe is The Universe (the only one), that it has been created by God in a *fiat*, that God being absolutely infinite, there would not be any problem for Him to achieve His creation, *ex nihilo*, under the form of a singularity. Furthermore, they could add, the Big Bang is the very origin of time, of space, and more precisely of space-time. It is easy to admit, as some do, that God is seating somewhere, comfortably located in a place having no words for it, insofar as this place is actually not a place, being outside of space-time. God does not know time but His creatures had to know it, the origin of the universe being the very origin of the time. From this point of view, what is very special with the Big Bang is that it would be one exception, the only exception, to NSP. This principle then could be refined as follows: Local observable infinity in physics is not admissible, excepted at the beginning of time.

But I am still ready to hold fast on NSP as it has been stated before. Then, we have to conclude that the equations of general relativity which are the ones which produced the Big Bang monster are erroneous. And we indeed know, once again, that they are. A theory like quantum gravity, the Theory of Everything, or whatever of that sort, would allow us to get rid of the Big Bang.

It is now from within this universe, our universe, under much less extreme and exotic conditions, that I am going to use NSP to show the rational inadmissibility of classical mechanics, and the rational necessity of quantum mechanics.

The rational necessity of quantum mechanics

The optical rainbow

One of the most beautiful phenomenon in nature is the optical rainbow in the sky [479–481]. Furthermore, the use of the optical rainbow for optical particle characterization in industry and in the laboratory (sizes, refractive indices, temperatures, chemical species...) nowadays constitutes an active field of research, e.g. [482–484], references therein, and many other references elsewhere.

The simplest way to understand the basic features of the optical rainbow (we only consider the primary rainbow) is to start with geometrical optics, more precisely with ray tracing, an approach usually granted to Descartes, although there are precursors. The existence of the rainbow then comes from the fact that the deviation of the once internally reflected ray passes through a minimum when the angle of incidence is varied (this is called a stationary ray). The concentration of rays, near the stationary ray, generates a singularity which is a real caustic, separating a bright side from a dark side. The singularity corresponds to an abrupt transition between both sides of the stationary ray, one associated with an infinite intensity (divergence at the rainbow angle), the other with vanishing intensity. We therefore have actually two kinds of singularities, one that we may call a longitudinal singularity associated with the divergence at the rainbow angle, and the other one that we may call a transverse singularity associated with the abrupt transition perpendicularly to the emerging rays. More generally, infinite intensities are predicted by geometrical optics at focal points, lines and caustics. The word "caustics" can be viewed as a generic word to refer to any kind of singularity produced by ray families filling regions of space.

Invoking NSP, we then may state that the existence of singularities in geometrical optics is *a priori* (in the Kantian sense, that is to say without referring to experiments) inadmissible. This points out to the fact that something is going wrong and that actually geometrical optics must be an approximation to a more general theory which is a wave theory (waves remove singularities). Indeed, such is the fact. We know how nature solves this problem: Light waves are described by the vectorial Maxwell's equations. The exact theory of the rainbow is provided by the Lorenz-Mie theory [485–487] which describes the interaction between a sphere and an illuminating electromagnetic plane wave, or by the generalized Lorenz-Mie theory [488] in the case of laser illumination.

The mechanical rainbow

We now consider classical mechanics. A sub-topic of classical mechanics is classical scattering, e.g. [489, 490] which, in contrast with electromagnetic scattering, is a scalar scattering instead of being a vectorial scattering. Following Nussenzveig [490], we discuss the scattering of a nonrelativistic particle of mass m by a central potential $V(r)$. The trajectory of the incoming particle is deflected by an angle $\bar{\theta}$ called the deflection angle which depends on the impact parameter b. For a repulsive interaction, $\bar{\theta}$ ranges from 0 to π but, for an attractive interaction, it can take arbitrary large values (in modulus). There is a simple relationship between the deflection angle $\bar{\theta}$ and the more usual scattering angle θ, namely: $\bar{\theta} + 2n\pi = \pm\theta$, $n = 0, 1, 2, \ldots$ Relying on the conservation laws of angular momentum and energy, it is then established that the differential cross-section in the direction θ reads as

$$\sigma_{sca}(\theta) = \sum_j \frac{b_j(\theta)}{\sin\theta} \left| \frac{d\theta}{db_j} \right|^{-1} \tag{16.34}$$

where the summation over j arises from the fact that, in general, there exist several impact parameters b_j that lead to a same scattering angle θ.

This expression implies the occurrence of several singularities. In particular, we have a divergence whenever $d\theta/db = 0$, for at least one j, and for some angle $\theta_R = \theta$, that is to say when the deflection function, say $\theta(b)$, goes through an extremum. We then have a stationary trajectory and the singularity is a rainbow singularity occurring at the rainbow angle θ_R. Other singularities involved in the expression are glory, forward scattering, orbiting, but we do not need to discuss them here.

Invoking again NSP, we conclude, without referring to any experiment, that the divergence exhibited by the mechanical rainbow (still having both a longitudinal and a transverse character) is inadmissible and that classical mechanics, which therefore contained the germ of its own destruction, is an approximation to a more general theory. This more general theory must be a wave theory, say a wave mechanics, in order to smooth out the rainbow singularity, and any other singularity, as well. We therefore also conclude to the rational necessity of something which will eventually become quantum mechanics, or even of something more general (as we shall see). While modern physicists are usually satisfied when they can answer the question: "how?", this rational necessity tries to point to another forgotten question: "why?".

The wave mechanics we have to build must then exhibit a wave, or more specifically a quantum wave, that we shall denote Ψ. We shall soon establish a set of possible differential equations, by relying on another principle. But before, I would like to make a few comments.

Ultra-violet catastrophe

Historically, at least if we rely on some textbooks and lectures on quantum mechanics, the quantum physics arose from experiments and associated features which could not be satisfactorily explained in the framework of classical physics. The failure of classical physics was not viewed as the consequence of its internal inconsistency (although it indeed contained the germ of its own destruction), but as the consequence of its inability to deal with the matters of fact. Among these matters of fact, we might select the most famous ones (but not actually the only ones), namely (i) the stability of atoms and their spectra (ii) photo-electric effect and (iii) black-body radiation. These three examples all involve and require a discussion of the properties of radiation.

The stability of atoms is intimately connected with the properties of the classical description of radiation, in the framework of classical electrodynamics, for, in this classical framework, an electron is vectorially accelerated in its motion around the nucleus (even if it orbits with a constant speed). Therefore, it must lose energy by emitting radiation, and spiral down to the nucleus, and eventually crash on it (this should happen in a very short time). Hence, this result of classical physics is in contradiction with the very existence of stable atoms. Furthermore, the radiation emitted by the electron during its fast journey to the nucleus should be continuous, in contrast with the observed fact that atom spectra are made out from a set of discrete lines. Hence, we have still another contradiction between classical physics and matters of fact. For the photo-electric effect, its connection with electromagnetism is even faster to report: It only found an explanation by invoking light quanta. As for black-body radiation, it is still simpler: The word radiation is explicitly mentioned in the denomination of the issue.

It is on this black-body problem that I would like to make an important comment, relevant to NSP. As is well known, even to students, a prediction of classical physics concerning an ideal black body at thermal equilibrium is that it emits radiation with infinite power. The infinite power divergence obtained is the result of the contributions from the high frequency region of the electromagnetic spectrum, justifying the denomination of ultra-violet catastrophe (coined in 1911 by Ehrenfest). Another denomination sometimes used is Rayleigh-Jeans catastrophe.

The ultra-violet catastrophe may be viewed as an ingredient of a more general feature: The properties of the black-body radiation, as calculated in the framework of classical physics, do not agree with experiments. Therefore, even without referring to the ultra-violet catastrophe, classical physics was falsified. This falsification may be viewed as being *a posteriori* (referring to experiments). To reconcile theory and experiments, Planck, in 1900, made the founding step of introducing the indivisible quantum of action $\hbar$. But it is important to remark that Planck did not feel really concerned with the ultra-violet catastrophe which therefore has not been, historically, a decisive feature for the emergence of quantum physics [458, 491].

The relevance of the ultra-violet catastrophe to the validity of classical physics has been actually put forward later, in particular by

Einstein, in 1905. What is very important with an argument using the ultra-violet catastrophe is that it is *a priori*, that is to say, once again, without referring to any experiment. A theoretical formula resulting from a theoretical framework is immediately perceived as leading to an impossibility. We could say that NSP implies the failure of classical physics.

But, this invocation of NSP actually does not concern all of classical physics. It is rather specific of this part of classical physics which treated of radiation, already sustained at this time by Maxwell's equations. It did not say anything on classical mechanics, something which is easy to see since there was a deep contradiction between classical mechanics allowing infinite velocities, associated with the Galilean composition of velocities, and the electromagnetism of Maxwell which exhibits the constancy of the speed of light in vacuum and therefore conflicts with the Galilean composition of velocities (a problem soon to be solved thanks to relativity). Hence, it is clear that, with our invocation of NSP in the framework of classical mechanics, we have done something else. We have actually demonstrated the inadmissibility of classical mechanics (without referring to the properties of radiation which is not relevant to mechanics). Therefore, consistently, we are now going to focus on matter waves.

A set of generalized Schrödinger equations

A demonstration for the set

PEDESTRIANS ($\sim\sim\sim\sim$) should better jump to the next subsection, being content of trying to get a bit of information from the comments associated with the equations.

To proceed further, the reader is kindly but firmly invited to forget everything he knows about quantum mechanics. We are indeed supposed to know only a few things as follows (i) we have a good enough knowledge of classical mechanics but (ii) we have concluded, from NSP, that classical mechanics is inadmissible. On this minimal basis, we now have to search for a differential equation satisfied by the wave Ψ of the wave mechanics to be built and we are trying to do this without any analogy, guess, or trial, as far as possible.

Our starting point is classical mechanics which, although not perfect and even inadmissible, is the best model we have up to now (quantum mechanics being assumed to be forgotten, and relativistic effects being neglected) to describe the world. More particularly, we shall rely on the Hamilton–Jacobi's formulation of classical mechanics which is entirely contained in a real field, the action $S(x_j, t)$, and its properties. In face of this, we know that the wave mechanics must rely on a wave $\Psi(x_j, t)$ which will have the virtue of washing out the singularities exhibited by classical mechanics. The most general form for a wave reads as

$$\Psi = e^{iT} \tag{16.35}$$

in which $T = T(x_j, t)$ is a complex dimensionless phase.

At this stage, we possess one field $S(x_j, t)$ for classical mechanics and two fields $\Psi(x_j, t)$ and $T(x_j, t)$ for wave mechanics. These fields are the only quantities involved in the problem. Therefore, we have to search for a relationship between Ψ and S (first option), or between T and S (second option). Because T and S possess the same nature, I mean they are not proper waves, I preferably choose the second option. Of course, the first option is valid too, but it would lead to more complicated derivations and equations.

For the relationship between T and S, we could look for $T(S)$ or for $S(T)$. Because wave mechanics (T) is more general than classical mechanics (S), it is apparent that we better have to try to determine $T(S)$ rather than the inverse version $S(T)$. We therefore have to explicitly consider $T(x_j, t) = T(S(x_j, t))$. However, this is to be slightly corrected. Indeed, T is dimensionless while S is an action (the action). This will require us to introduce a new constant, that will be denoted g.

Now, I invoke another principle, that I call the lifting principle (later to be commented a bit more when the demonstration is completed). This principle tells us something very simple, even looking a bit like tautological, as follows.

Lifting principle:

Classical Mechanics is an Approximation to Wave Mechanics.

Rather than simply using the argument S in $T(S)$, we then have to look for a function $T(\overline{S})$ in which the functional argument $\overline{S} = \overline{S}(x_j, t)$ reads as

$$\overline{S} = \frac{1}{g}(S + i\epsilon S_1) \tag{16.36}$$

in which g is a constant having the dimension of an action, S_1 is a correcting function, and ϵ is a small parameter. To recover classical mechanics from wave mechanics, we shall have to take the limit $\epsilon \to 0$. However, we shall discover that this condition is necessary but not sufficient, because an extra-condition on g will be required. Also, we can take $\epsilon \in \mathcal{R}$. Indeed, if ϵ were complex, it would exhibit a phase factor which could be absorbed in S_1. We shall later comment, when appropriate, on the reality of the constant g which, at least provisionally, may be viewed as a complex constant.

The function $T(\overline{S})$ may be explicitly written as

$$T = T_\epsilon \left(\frac{S + i\epsilon S_1}{g} \right) \tag{16.37}$$

in which we used a subscript ϵ to insist on the fact that T depends on ϵ. Eq. (16.37) may give the feeling that we are dealing with a restricted first-order perturbation approach. However, instead of Eq. (16.36), let us assume

$$\overline{S} = \frac{1}{g}\left(S + i\epsilon \overline{S_1} + (i\epsilon)^2 \overline{S_2} + \cdots \right) \tag{16.38}$$

This can be rewritten as

$$\overline{S} = \frac{1}{g}\left[S + i\epsilon \left(\overline{S_1} + i\epsilon \overline{S_2} + \cdots \right) \right] \tag{16.39}$$

which, relabelling, identifies with Eq. (16.36).

We are now looking for a differential equation satisfied by the wave Ψ, involving partial derivatives with respect to x_j and t. This equation must be fundamental, that is to say it must contain lowest-order derivatives compatible with the constraints imposed by the problem under study. Once the fundamental equation is obtained, we can of course generate other equations by further differentiating with respect to x_j and t, but such extra-equations are said to be non-fundamental.

We begin with the assumption that, besides derivatives with respect to x_j, the wave equation only contains the first derivative $\partial\Psi/\partial t$ with respect to time. We shall afterward comment on the use of higher-order derivatives with respect to time.

The derivative $\partial\Psi/\partial t$ may always be written as

$$\frac{\partial\Psi}{\partial t} = f_\epsilon(K, \{\partial\Psi\}) \tag{16.40}$$

in which we again use a subscript ϵ to insist on the dependence on ϵ. Also, K is an extra-field (i.e. a function of time and space, but not a dynamical field possessing its own differential equation), possibly a constant, and $\{\partial\Psi\}$ represents a set of arguments formed from various derivatives of Ψ with respect to x_j

$$\Psi_{i_1 i_2 i_3 \ldots i_r} = \frac{\partial}{\partial x_{i_1}} \frac{\partial}{\partial x_{i_2}} \frac{\partial}{\partial x_{i_3}} \cdots \frac{\partial}{\partial x_{i_r}} \Psi \tag{16.41}$$

The set $\{\partial\Psi\}$ is infinite and there is a systematic way to generate all arguments of the set. For instance, the subset generated by Ψ_{ijk} contains $\Psi_{ijk}\Psi_i\Psi_j\Psi_k$, $\Psi_{ijk}\Psi_{ij}\Psi_k$, $\ldots$, and other arguments obtained by using complex conjugations.

We may also express the derivative $\partial\Psi/\partial t$ from Eqs. (16.35) and (16.37), so that we obtain

$$\frac{\partial\Psi}{\partial t} = i\frac{dT_\epsilon}{d\bar{S}}\frac{1}{g}\left(\frac{\partial S}{\partial t} + i\epsilon\frac{\partial S_1}{\partial t}\right)\Psi \tag{16.42}$$

We rewrite Eq. (16.42) as

$$-\frac{\partial S}{\partial t} = i\epsilon\frac{\partial S_1}{\partial t} - \frac{g}{i\frac{dT_\epsilon}{d\bar{S}}\Psi}\frac{\partial\Psi}{\partial t}, \quad \Psi \neq 0 \tag{16.43}$$

or, invoking Eq. (16.40)

$$-\frac{\partial S}{\partial t} = i\epsilon\frac{\partial S_1}{\partial t} - \frac{g}{i\frac{dT_\epsilon}{d\bar{S}}\Psi}f_\epsilon(K, \{\partial\Psi\}) \tag{16.44}$$

But, Hamilton–Jacobi's equation (and the lifting principle) implies that the right-hand-side of Eq. (16.44) must contain a term with no derivative associated with V in Eq. (2.1), and a term involving $(\partial S/\partial x_j)^2$, associated with the first term in the right-hand-side of Eq. (2.1). These terms have to be involved in the function f_ϵ. Upon investigation, we find that the term involving $(\partial S/\partial x_j)^2$ can only be generated by Ψ_{jj} which indeed is found to be

$$
\Psi_{jj} = \frac{i\Psi}{g} \left\{ \frac{\mathcal{T}}{g} \left(\frac{\partial S}{\partial x_j} \right)^2 + \frac{2i\epsilon\mathcal{T}}{g} \frac{\partial S}{\partial x_j} \frac{\partial S_1}{\partial x_j} \right.
$$
$$
\left. - \frac{\epsilon^2 \mathcal{T}}{g} \left(\frac{\partial S_1}{\partial x_j} \right)^2 + \frac{dT_\epsilon}{d\overline{S}} \left(\frac{\partial^2 S}{\partial x_j^2} + i\epsilon \frac{\partial^2 S_1}{\partial x_j^2} \right) \right\} \quad (16.45)
$$

in which

$$
\mathcal{T} = i \left(\frac{dT_\epsilon}{d\overline{S}} \right)^2 + \frac{d^2 T_\epsilon}{d\overline{S}^2} \quad (16.46)
$$

We therefore set, without any loss of generality

$$
f_\epsilon(K, \{\partial\Psi\}) = a \frac{\partial^2 \Psi}{\partial x_j^2} + b\Psi + h_\epsilon(K, \{\partial\Psi\}) \quad (16.47)
$$

in which h_ϵ is a complementary function, possibly including non-linear terms, and which also could possibly annihilate the terms $a\partial^2\Psi/\partial x_j^2$ and $b\Psi$ if, eventually, we would find that they should be zero.

The evolution equation (16.40) then takes the form

$$
\frac{\partial \Psi}{\partial t} = a \frac{\partial^2 \Psi}{\partial x_j^2} + b\Psi + h_\epsilon(K, \{\partial\Psi\}) \quad (16.48)
$$

and our next task is to evaluate a and b.

To this purpose, we now return to Eq. (16.44) and insert in it Eqs. (16.47) and (16.45), leading to

$$
-\frac{\partial S}{\partial t} + \frac{a}{g} \left(\frac{\partial S}{\partial x_j} \right)^2 \left(i \frac{dT_\epsilon}{d\overline{S}} + \frac{d^2 T_\epsilon/d\overline{S}^2}{dT_\epsilon/d\overline{S}} \right) + \frac{gb}{i\frac{dT_\epsilon}{d\overline{S}}} = \mathcal{A} + \mathcal{B} + \mathcal{C} \quad (16.49)
$$

with

$$\mathcal{A} = -a\frac{\partial^2 S}{\partial x_j^2} - g\frac{h_\epsilon}{i\frac{dT_\epsilon}{d\overline{S}}}\Psi$$

$$\mathcal{B} = \epsilon\left[i\frac{\partial S_1}{\partial t} - \frac{2ia}{g}\left(i\frac{dT_\epsilon}{d\overline{S}} + \frac{d^2 T_\epsilon/d\overline{S}^2}{dT_\epsilon/d\overline{S}}\right)\frac{\partial S}{\partial x_j}\frac{\partial S_1}{\partial x_j} - ia\frac{\partial^2 S_1}{\partial x_j^2}\right]$$

$$\mathcal{C} = \epsilon^2\frac{a}{g}\left(\frac{\partial S_1}{\partial x_j}\right)^2\left(i\frac{dT_\epsilon}{d\overline{S}} + \frac{d^2 T_\epsilon/d\overline{S}^2}{dT_\epsilon/d\overline{S}}\right)$$

In the classical limit ($\epsilon \to 0$), Eq. (16.49) simplifies to

$$-\frac{\partial S}{\partial t} + \frac{a}{g}\left(\frac{\partial S}{\partial x_j}\right)^2\left(i\frac{dT_0}{d\overline{S}} + \frac{d^2 T_0/d\overline{S}^2}{dT_0/d\overline{S}}\right) + \frac{gb}{i\frac{dT_0}{d\overline{S}}}$$

$$= -a\frac{\partial^2 S}{\partial x_j^2} - g\frac{h_0}{i\frac{dT_0}{d\overline{S}}}\Psi \tag{16.50}$$

which must identify with Hamilton–Jacobi's equation. Under the proviso that the right-hand side of Eq. (16.50) must be vanishingly small, we then obtain, from the left-hand side:

$$\frac{gb}{i\frac{dT_0}{d\overline{S}}} = -V \tag{16.51}$$

$$\frac{a}{g}\left(i\frac{dT_0}{d\overline{S}} + \frac{d^2 T_0/d\overline{S}^2}{dT_0/d\overline{S}}\right) = -\frac{1}{2m} \tag{16.52}$$

in which $T_0 = T_0(S/g)$ and $\overline{S}$ therefore reduces to S/g. Eq. (16.51) implies

$$b = -\frac{iV\frac{dT_0}{d\overline{S}}}{g} \tag{16.53}$$

We must now recall that the coefficient b has been actually set as a function $b(x_j, t)$, and Eq. (16.48) shows that it must pertain to the wave mechanical level. In other words, it does not pertain to the classical mechanical level, that is to say, as a rational demand, we would not like it to depend on S. Therefore, $dT_0/d\overline{S} = dT_0/d(S/g)$

must be a constant that we denote as C_1. Then, upon integration:

$$T_0 = C_1 \frac{S}{g} + C_2 \tag{16.54}$$

From Eq. (16.53), we then have

$$b = \frac{-iV}{g} C_1 \tag{16.55}$$

With $d^2 T_0/d\overline{S}^2 = 0$ (since the first derivative is a constant), Eq. (16.52) then implies

$$a = \frac{ig}{2mC_1} \tag{16.56}$$

Inserting Eqs. (16.55) and (16.56) into Eq. (16.48), we then obtain

$$ig \frac{\partial \Psi}{\partial t} = -\frac{g^2}{2mC_1} \frac{\partial^2 \Psi}{\partial x_j^2} + VC_1 \Psi + igh_\epsilon \tag{16.57}$$

Concerning the constant C_2, it does not appear in Eq. (16.57) and can therefore be viewed as irrelevant (we may set $C_2 = 0$). Concerning the constant C_1, I have (at least at the present time) no theoretical reason to assign a value to it.

Let R denote the right-hand side of Eq. (16.50). We still have to check that it is vanishingly small. With Eq. (16.56), we obtain

$$R = -\frac{g}{C_1} \left(\frac{i}{2m} \frac{\partial^2 S}{\partial x_j^2} + \frac{h_0}{i\Psi} \right) \tag{16.58}$$

which is indeed 0 in the limit $g \to 0$. This implies that g is a small action, actually so small that it could not be detected in a classical framework.

Equation (16.57) is the main result of this subsection. It provides a set of generalized Schrödinger equations, being admitted that they are evolution equations (first derivative with respect to time), obtained by a deformation of Hamilton–Jacobi's equation, according to the lifting principle. The process used to obtain this set may be viewed as a demonstration.

The function h_ϵ in Eq. (16.57) may be significant because it allows one to introduce non-linear wave equations. Nonlinear Schrödinger equations in quantum theory (this restriction means that I am not considering soliton theory) are considered in the literature in many papers. For example, they are comprehensively discussed by Doebner and Goldin in Ref. [492], and in many references therein. We also already encountered such equations such as in the Bohm-Bub hidden-variables theory [297] or with the GRW's equation for spontaneous collapse of the wave function [92]. More generally, as we have seen, non-linear equations may provide a solution to the measurement problem (recall that linear equations, in utmost rigor, do not allow one to get rid of quantum superpositions). This fact has been recently emphasized by Penrose in one of his books [493].

It is however the right place now, after all the previous comments made on non-linearities, to give a word of caution. According to Gisin [494], *the Schrödinger evolution is the only quantum evolution that is deterministic and compatible with relativity.* Hence, *the fact that a deterministic evolution compatible with relativity must be linear puts heavy doubts on the possibility to solve the measurement problem ... by adding non linear terms to the Schrödinger equation.* However, Gisin's result relies on the use of the quantum mechanical formalism that, at the present time, we are not supposed to know. In particular, Gisin used EPR-like correlations that we are not allowed to use since we are currently dealing with an one-body problem.

Returning to such a problem, setting $h_\epsilon = 0$ in Eq. (16.57), and $C_1 = 1$, we obtain the simplest equation among the set of generalized equations, simplest in the sense that, in Eq. (16.47), we retain only those sufficient terms $a\partial^2\Psi/\partial x_j^2$ and $b\Psi$ required to match Hamilton–Jacobi's equation in the classical limit, and dismiss all the other unnecessary terms (notwithstanding the value of C_1 which has to be considered, in the absence of any theoretical argument, as an empirical value). The result is the time-dependent Schrödinger's equation:

$$ig\frac{\partial\Psi}{\partial t} = -\frac{g^2}{2m}\frac{\partial^2\Psi}{\partial x_j^2} + V\Psi \qquad (16.59)$$

in which the constant g identifies with Planck constant $\hbar$. This means in particular that g, as C_1, also receives the status of an empirical constant and that, eventually, it is real. As is well known, the wave mechanics associated with Schrödinger's equation is equivalent to the matrix mechanics of Heisenberg, Born and Jordan, and, therefore, the above approach to a wave mechanics is simultaneously an approach to quantum mechanics.

The fact that g in Schrödinger's equation is real may also be given a theoretical justification. In the limit $\epsilon \to 0$, from Eq. (16.54), with $C_1 = 1$ and $C_2 = 0$, we have

$$T_0 = S/g \tag{16.60}$$

For the time being, g is complex and may therefore be rewritten as $g = |g|e^{i\theta}$ in which $|g|$ is the modulus and θ is a real phase. We also write T_0 as $R + iI$. Equation (16.60) then implies

$$(R + iI)e^{i\theta} = S/|g| \tag{16.61}$$

in which the r.h.s is real, implying that the left-hand side is real too. This left-hand side may be rewritten as

$$L = R\cos\theta - I\sin\theta + i(R\sin\theta + I\cos\theta) \tag{16.62}$$

leading to the apparent solution

$$\tan\theta = -I/R \tag{16.63}$$

In the left-hand side of Eq. (16.63), the phase angle θ is a constant because g is a constant. Eq. (16.63) then would rely the imaginary part I of T and the real part R of T, in such a way that $R = \alpha I$, with α a constant. Invoking Eq. (16.35), we then would obtain

$$\Psi = e^{-I}e^{i\alpha I} \tag{16.64}$$

This is to be dismissed because it relies the amplitude of Ψ and its phase which, conversely, in general, have to be independent. From the reality of L in Eq. (16.62), we must then have independently $I = 0$, and $\theta = 0$ or $\theta = \pi$ (hence T_0 is a real field). For $\theta = 0$, $g = |g|$ and for $\theta = \pi$, $g = -|g|$. Therefore, in both cases, g is a real constant.

Now, taking the classical limit of Schrödinger's equation (16.59), we readily recover Hamilton–Jacobi's equation, as described for instance by Blotkhintsev [38]. This is a case when the correspondence principle is satisfied. Conversely, the lifting principle, consistently elevates Hamilton–Jacobi's equation to the generalized Eq. (16.57) which contains Schrödinger's equation plus additional terms, with an additional constant C_1. However, now that we have obtained Eq. (16.57) by using the lifting principle, we may get rid of it and only rely on the correspondence principle, then providing another demonstration. This is done by using an Ansatz, namely

$$ig\frac{\partial \Psi}{\partial t} = A(x_j, t)\frac{\partial^2 \Psi}{\partial x_j^2} + B(x_j, t)\Psi + H_\epsilon \tag{16.65}$$

in which H_ϵ is a supplementary function (similar to h_ϵ above) which guarantees the validity of the Ansatz. In the classical limit $\epsilon \to 0$, this equation becomes

$$ig\left(\frac{\partial \Psi}{\partial t}\right)_{\epsilon \to 0} = A\left(\frac{\partial^2 \Psi}{\partial x_j^2}\right)_{\epsilon \to 0} + B\Psi_{\epsilon \to 0} + H_0 \tag{16.66}$$

Now, from Eqs. (16.35) and (16.37), we have

$$\Psi_{\epsilon \to 0} = e^{iT_0(S/g)} \tag{16.67}$$

Also, from Eqs. (16.39), (16.42), and using Eq. (16.67):

$$\left(\frac{\partial \Psi}{\partial t}\right)_{\epsilon \to 0} = i\frac{dT_0}{dS}\frac{\partial S}{\partial t}e^{iT_0(S/g)} \tag{16.68}$$

Furthermore, from Eqs. (16.45), (16.39), and using again Eq. (16.67), we obtain

$$\left(\frac{\partial^2 \Psi}{\partial x_j^2}\right)_{\epsilon \to 0} = ie^{iT_0(S/g)}\left\{\mathcal{T}_0\left(\frac{\partial S}{\partial x_j}\right)^2 + \frac{dT_0}{dS}\frac{\partial^2 S}{\partial x_j^2}\right\} \tag{16.69}$$

in which

$$\mathcal{T}_0 = i\left(\frac{dT_0}{dS}\right)^2 + \frac{d^2T_0}{dS^2} \tag{16.70}$$

We now set

$$H_0 = K_0 e^{iT_0(S/g)} \tag{16.71}$$

and insert Eqs. (16.67)–(16.69), and (16.71) in (16.66), leading to

$$-\frac{\partial S}{\partial t} = \frac{A}{g}\left(\frac{\partial S}{\partial x_j}\right)^2 \left[i\frac{d^2T_0/dS^2}{dT_0/dS} - \frac{dT_0}{dS}\right]$$
$$+ i\frac{A}{g}\frac{\partial^2 S}{\partial x_j^2} + \frac{B+K_0}{gdT_0/dS} \tag{16.72}$$

Comparing this equation and Hamilton–Jacobi's equation (2.1), we obtain

$$\frac{B}{gdT_0/dS} = V \tag{16.73}$$

$$\frac{A}{g}\left[i\frac{d^2T_0/dS^2}{dT_0/dS} - \frac{dT_0}{dS}\right] = \frac{1}{2m} \tag{16.74}$$

$$\frac{A}{g}\frac{\partial^2 S}{\partial x_j^2} = 0 \tag{16.75}$$

$$\frac{K_0}{gdT_0/dS} = 0 \tag{16.76}$$

From Eq. (16.73), similarly as for our previous discussion of the coefficient $b(x_j,t)$, we recover Eq. (16.54), and, using this equation, we obtain

$$B = C_1 V \tag{16.77}$$

From Eq. (16.74), using again Eq. (16.54), we then readily obtain

$$A = -\frac{g^2}{2mC_1} \tag{16.78}$$

Next, from Eqs. (16.75) and (16.78), we have

$$-\frac{g}{2mC_1}\frac{\partial^2 S}{\partial x_j^2} = 0 \tag{16.79}$$

which tells us that g must be small, while Eq. (16.76) implies that $K_0 = H_0 = 0$.

Finally, inserting Eqs. (16.78) and (16.79) into the Ansatz, we obtain

$$ig\frac{\partial\Psi}{\partial t} = -\frac{g^2}{2mC_1}\frac{\partial^2\Psi}{\partial x_j^2} + C_1 V\Psi + H_\epsilon \tag{16.80}$$

which identifies with Eq. (16.57).

Equations (16.57) and (16.59) have been obtained with the assumption that the variation of the field Ψ with respect to time is completely defined by the first order derivative $\partial\Psi/\partial t$. Let us now examine the possibility of occurrence of a second-order derivative $\partial^2\Psi/\partial t^2$. A possible attack could be to write

$$C\frac{\partial^2\Psi}{\partial t^2} = q\left(\frac{\partial\Psi}{\partial t}, K, \{\partial\Psi\}\right) \tag{16.81}$$

in which C is a constant. From Eq. (16.42), we have

$$\frac{\partial^2\Psi}{\partial t^2} = \frac{i\Psi}{g}\frac{dT_\epsilon}{d\overline{S}}\left[\frac{\partial^2 S}{\partial t^2} + i\epsilon\frac{\partial^2 S_1}{\partial t^2} + \frac{1}{g}\left(\frac{\partial S}{\partial t} + i\epsilon\frac{\partial S_1}{\partial t}\right)^2\right.$$
$$\left.\times\left(i\frac{dT_\epsilon}{d\overline{S}} + \frac{d^2 T_\epsilon/d\overline{S}^2}{dT_\epsilon/d\overline{S}}\right)\right] \tag{16.82}$$

Therefore, in the classical limit $\epsilon \to 0$:

$$\frac{\partial^2\Psi}{\partial t^2} = \frac{i\Psi}{g}\frac{dT_0}{d\overline{S}}\left[\frac{\partial^2 S}{\partial t^2} + \frac{1}{g}\left(\frac{\partial S}{\partial t}\right)^2\left(i\frac{dT_0}{d\overline{S}} + \frac{d^2 T_0/d\overline{S}^2}{dT_0/d\overline{S}}\right)\right] \tag{16.83}$$

in which $\overline{S}$ is to be understood as $\overline{S}$ taken for $\epsilon = 0$.

But the terms $\partial^2 S/\partial t^2$ and $(\partial S/\partial t)^2$ do not appear in Hamilton–Jacobi's equation. Therefore, $\partial^2\Psi/\partial t^2$ would not be admissible and, in Eq. (16.81), we would have to set $C = 0$, leading to an implicit equation for $\partial\Psi/\partial t$, having the explicit form of Eq. (16.40). Similarly, higher-order derivatives $\partial^p\Psi/\partial t^p$, $p > 2$, would be forbidden. However, the argument is faulty. Indeed, rather than being a constant, C could be a function $C(\epsilon)$ with $C(\epsilon) \to 0$ when $\epsilon \to 0$. Then the limit $\epsilon \to 0$ would not tell us anything on the admissibility of second-order (and higher-order) derivatives with respect to time. Therefore, having, at least at the present time, no formal argument to dismiss higher-order derivatives with respect to time, I shall rely on physical considerations.

In the Hamilton–Jacobi's formulation of classical mechanics, we may observe that the state of the system is defined by the field S, satisfying a derivative equation which is of first-order with respect to time (i.e. an evolution equation). Knowing the field S at time $t = 0$, we may then obtain the field S at any time $t > 0$ ($t < 0$), by using a forward (backward) time-marching procedure.

In wave mechanics, the field S has been deformed to a wave Ψ and we may then assume, in a natural way, that the state of the system is defined by Ψ, this being true at any time (this point of view is obviously reminiscent of the first postulate in the list of Cohen–Tannoudji *et al.*). For this reason, we may demand that the equation for Ψ be an evolution equation too. Then, knowing Ψ at time $t = 0$, we may obtain Ψ at any time $t > 0$ ($t < 0$), by using a forward (backward) time-marching procedure, as for S. In contrast, the use of a second-order derivative with respect to time would require, for integration, to have the state defined by Ψ and by $\partial\Psi/\partial t$ at some initial time (similar considerations hold for higher-order derivatives with respect to time).

Then, in utmost rigor, what we have demonstrated is that Schrödinger's equation is the simplest evolution equation satisfying the lifting principle. We have not ruled out, in the framework of the theoretical approach proposed, that higher-order derivatives with respect to time should necessarily be dismissed, although this would be an unpleasant prospect. Also, we have to accept the theoretical possibility of the occurrence of extra-nonlinear terms in Schrödinger's equation and this, considering their relevance for the measurement problem, is not necessarily an unpleasant prospect, although it would require further investigations that are outside of the scope of the book.

A demonstration for the set: The **PEDESTRIAN** ($\sim\sim\sim\sim$) path

We now summarize the previous subsection to the address of PEDESTRIANS ($\sim\sim\sim\sim$). The approach is as follows (i) acknowledge, relying on NSP, that classical mechanics is inadmissible, and that, therefore, we need a wave mechanics because waves wash out singularities (ii) afterward, assume the validity of the lifting principle telling us that classical mechanics is an approximation to wave mechanics. Then, look for a first-order differential equation satisfied by the wave mechanics to be built. The result obtained, after a bit of mathematics, is a set of generalized Schrödinger equations. I already mention that a nightmare of theoreticians is to commit computational errors. The interesting point however is that, even in the presence of computational mistakes, the conclusion "we obtain a set of generalized Schrödinger equations" is a robust conclusion, and even an obvious one. Indeed, classical mechanics can be obtained by taking some limit on Schrödinger's equation. Hence, conversely, by deforming classical mechanics, it is just normal that we obtained a set of generalized Schrödinger's equations. The interesting point then is that Schrödinger's equation itself is the simplest one (in some sense) inside the set.

The lifting principle

The lifting principle should not be confused with the correspondence principle. To see this, let us consider two theories, denoted T_G and T_A, with the subscript G for "General" and the subscript A for "Approximate". The theory T_G (for instance quantum mechanics to be specific), is a general enough theory, and the other theory T_A (for instance classical mechanics to be specific) is an approximation to T_G.

Let us assume that T_G is known (or, alternatively, it could be a candidate theory). Then, by taking some kind of limit on T_G, we must recover T_A, something that is denoted as $T_G \to T_A$. We then say that T_G satisfies a correspondence principle (with respect to T_A). In the specific examples taken above, where T_G is quantum mechanics and T_A is classical mechanics, this correspondence principle provides a possible formulation for the correspondence principle of Bohr. If T_G is unknown and under construction, any valid candidate, say $T_{G_1}, T_{G_2}, \ldots$, must satisfy the correspondence principle: $T_{G_1} \to T_A$, $T_{G_2} \to T_A, \ldots$ If it does not, it is not valid and must be rejected. If several valid candidates are retained, then the discrimination between the candidates may need to rely on other considerations, such as analogies, guesses, or trials (as in the processes previously used for the derivations of Schrödinger's equation) or simply experiments.

This is what I shall call an up→down ($\downarrow$) strategy, from T_G above to T_A below.

The lifting principle tells us that classical mechanics is an approximation to quantum mechanics. Therefore quantum mechanics must satisfy a correspondence principle: (quantum mechanics)→(classical mechanics), that is to say the correspondence principle is contained in the lifting principle. But it does not identify with it. Actually, it is more general and somewhat more flexible, as apparent from the use we made of it. Indeed what we have done is to start from T_A below and to elevate ourselves to something generalizing classical mechanics. It is not an up $\rightarrow$ down process ($\downarrow$), but a down $\rightarrow$ up process ($\uparrow$). This is the justification for the name given to the principle.

However, the word "lifting" may have other meanings, in other contexts, for example in mathematics. In particular, in the theory of nonlinear dynamical systems, you may have to study a low-dimensional dynamical system and find that it is easier to understand it by rather studying a higher-dimensional system [240, 495]. In such a case, you have the demand that the low-dimensional system must be recovered from the higher-dimensional system, by some kind of projection ($\downarrow$). This is a kind of mathematical "correspondence principle". Furthermore, mathematicians say that the low-dimensional system has been conversely ($\uparrow$) lifted to the higher-dimensional system. We could also say that the higher-dimensional system is the result of a lifting of the low-dimensional system. The word "lifting" I have used for the lifting principle presents obvious analogies with this example from the theory of nonlinear dynamical systems, and it is actually from these analogies that I picked up the word. Yet, the contexts are different. Those who believe that using the same word for different contexts may be dangerous might prefer to refer to another terminology like: Elevation principle. The terminology "deformation principle" is also acceptable, or even more precise, because what we have done is somehow to deform the Hamilton–Jacobi's equation of classical mechanics to a set of generalized Schrödinger equations for a family of variants of wave mechanics. But although alternative terminologies are possible, or even have been suggested to me, I still prefer to keep on with my original choice: Lifting principle.

Clearly, with the lifting principle, we should not expect to elevate classical mechanics to an unique theory. The situation is easily explained by using a metaphor due to Feynman in his famous lectures [14]: The correspondence principle proceeds from one object to its shadow (and there is one shadow for one object) while the lifting principle proceeds from a shadow to objects (and there are several possible objects for a given shadow). Our result agrees with this expectation. We did not reach Schrödinger's equation, but a set of generalized Schrödinger equations. This is much more significant that it could seem at first sight. The derivation of Schrödinger, and all other Schrödinger-like derivations, reach a single result because they used analogies, guesses, and trials, with many more or less explicit or implicit assumptions. Conversely, NSP telling us that classical mechanics is inadmissible, and the lifting principle allowing us to turn this qualitative statement to a mathematical formulation, simultaneously provided the whole set of admissible possibilities still available. We have reached all candidates.

It is then afterward still required to discriminate between candidates. Concerning the fact that we retained an evolution equation, that is to say that we rejected all derivatives of Ψ with respect to time higher than the first-order derivative, we relied on physical considerations which seem natural, but we did not exhibit any theoretical necessity to reach this result. Once an evolution equation is obtained, it took the form of a set of generalized Schrödinger equations, but there is no known theoretical reason, at least known to me, to reject the complementary function h_ϵ involved in Eq. (16.57). It might even happen that this function could be vital for a solution of the measurement problem, definitely getting rid of quantum superpositions.

Schrödinger's equation is the simplest equation among the set and this happens as the result of the rejection of h_ϵ described in the previous paragraph. For this rejection, it might be possible to invoke the Occam razor principle which has often been found efficient in the history of sciences. But this principle is a metaphysical principle rather than a physical principle. Whatever its heuristic virtue is, it does not offer any definitive guarantee. Indeed, a first point is that the history of sciences provides many examples of the efficiency of the Occam razor principle, but also provides many counter-examples. To give one example, we may refer to fluid mechanics, where, when deriving Navier–Stokes equations, the simplest relation between stress and strain tensors is assumed to be linear [496]. This Occam razor assumption indeed encompasses all Newtonian fluids, but there are many non-Newtonian fluids. Second, there is the difficult question to accurately define the concept of simplicity (and correlatively of complexity), as testified for instance by the existence of a huge field of research concerning algorithmic complexity. We know the witticism of Einstein telling that things should be as simple as possible, but not simpler. In practice, it is not necessarily simple to use this motto on simplicity.

Well, could we say. It does not matter since we have a good judge for a final discrimination: Experiments. But there are several points to raise concerning this opinion (i) Schrödinger's equation is indeed corroborated by experiments, and not falsified, but this does not prove the validity of the equation, as Popper would have remarked (ii) more importantly, we are concerned in this chapter with the understanding of quantum mechanics, but experiments reveal, they do not explain and (iii) finally, we surely might be able to find equations, different from Schrödinger's equation, pertaining to the set of generalized equations, whose consequences would also agree with experiments. It would suffice to make modifications sufficiently small to be observationally undetectable. Such equations might be of a limited practical interest, but could have an interest to solve some problems, such as the measurement problem. This is in the spirit of GRW's equation [92], even if this equation has been criticized. In any case, we must preserve our freedom of thinking and speculating: Imagination, besides experimental unexpected revelations, is the main source of progress in sciences.

To end this subsection, let us remark that there is an implicit assumption that we have made and that is better to make explicit. The assumption is that classical mechanics is a correct starting structure to reach quantum mechanics by using the lifting principle, and in particular, that it is a correct final structure to be reached by using the descending correspondence principle. This assumption was already made by Bohr, and others, when classical mechanics was used as a valid limit theory of quantum mechanics. I call it the faithfulness assumption. It means that we can trust classical mechanics. Whether we like it or not, how could we do if we did not rely on this lucid faith?

Revisiting the non-singularity principle

As we already discussed, we define words by using other words and, the number of words we used being finite, any definition is logically spoiled by the logical sin of circularity. We may escape definitions by simply showing: Look, this is a stone, look, this is a tree, and so on. But we cannot use concepts by just showing particular things, such as a special stone or a special tree. Concepts must be abstracted and have to circulate in the third world. There, to avoid circular circulation, we need basic principles such as the ones formed by the postulates of quantum mechanics. If we want to get a feeling of understanding, basic principles should better satisfy the intuition, and then can be called first principles.

There is a first principle on which I intend to hold fast (at least up to the time when I accept a loophole, near the end of this book), namely that the rainbow singularity, and more generally any singularity, in classical mechanics are inadmissible, from which I concluded to the necessary existence of a wave mechanics. I presented this as a consequence of a more general non-singularity principle. I also intend to hold fast on the existence of such a NSP, but I do not intend to hold fast on the particular general formulation I used to present it. It might possibly be provisional. But, assuming that this provisional formulation is indeed correct, I would like here to tentatively draw a few complementary consequences of it.

In a preliminary training, we discussed the Koch curve and its infinite length, acceptable in a mathematical framework, but inadmissible in a physical framework. Instead of considering the whole Koch curve, let us concentrate on one of the original line segments (one side of a triangle) used to built it. This line segment has a finite length that we have taken equal to 1 but it contains an infinite number of points. The existence of such an infinity has been a source of many problems all along the history of mathematics, already starting with the discovery of irrationals at the time of the Pythagoreans, a real and dramatic scandal for them. For a Pythagorean who would understand that there are infinitely many irrationals on our line segment, the scandal would be infinite and could only end with a heart attack. This was also an important issue for Aristotle which discussed the fact that infinity is absolutely present in a continuum and wondered whether there can exist an infinite sensible (physical) body, answering in particular that an infinite body cannot exist [33].

But we, at least many of us, are ready to accept actual infinities in mathematics (we have been trained by Cantor). Many of us, yes, but may be not all of us, because we must recall the existence of a mathematical school which produced a doctrine called finitism, initiated by Kronecker (a bad-tempered opponent to Cantor), who rejected the notion of infinity even in mathematics. But, in the sensible world of physics, if we insist on using NSP, the line segment must not contain an infinite number of points because we have to reject actual infinity in physics, a feature which points out again to the existence of atoms. Indeed, when we draw a so-called continuous line on a so-called continuous sheet of paper, we actually deposit "black" atoms on "white" atoms.

We may then argue that, nevertheless, in the process of drawing the line on the sheet of paper, the pencil has to move from the initial point of the line segment to a final point. This happens through the continuous space and there must be an infinite number of space points between the initial and final points of the line segment. If we still insist on using NSP, we must conclude that the concept of space used above is a convenient but, at a deeper fundamental level, inadequate abstraction. We might for instance state that space does not actually exist as an independent background in which objects would be immersed but that the existing things are the objects (locally in finite quantity), and that space is only a convenient way to speak of a particular kind of relationship between objects (this is a point of view which has actually be shared by many philosophers, such as Descartes, when he said, as we have seen, that space and matter are identical). Alternatively, if we decide to abstract space from matter, then we should have to conclude that the structure of space, at some more fundamental level, does not resemble to a continuum, at least not in the way we are accustomed to deal with. Similar considerations are valid for time.

And now, as a last exercise, what about the existence of an infinite velocity of bodies or of the propagation of effects? This has been a most important and tremendous effective issue since Newton published his theory of gravitation. Newton himself did not like the idea of an instantaneous interaction between two bodies, and many opponents used the *a priori* impossibility of such an instantaneous influence, *passion-at-a-distance*, to reject the newly born Newtonian theory. Let us for instance, to be concrete, consider the Sun and the Earth related by gravitation, instantaneously, across a space devoid of matter. How could this be possible? Nowadays, we know what is the solution to this question, thanks to general relativity: Attraction is not a force, but the result of the space-time curvature, with bodies freely running along space-time geodesics; and any perturbation of the structure of space-time propagates, not instantaneously, but at the speed of light. The existence of the speed of light therefore solves the problem raised by the instantaneous propagation of Newtonian gravitational forces (let us here recall the extraordinary anticipation of Epicurus concerning the speed of thought).

Let us take this issue from another point of view, the one of NSP. Then, any interaction between two bodies separated in space by a finite distance requires a finite time because an actual infinite velocity is forbidden. This means that we have to face the existence of a limit maximal velocity. Let us now relate this conclusion to the first principle of relativity, that laws of nature should not depend on the observer or more objectively, that is to say without using the word "observer", should be covariantly invariant with respect to any frame of reference. Then, we have to conclude that the limit maximal velocity should be unique. Therefore, instead of basing relativity on a first principle (covariance) and a basic principle (constancy of the speed of light), we may build it on two first principles (covariance and NSP). Quantum non-locality, another kind of phenomenon, which apparently leads us to the consideration of instantaneous influences, is however not submitted to NSP because, as we have discussed, this does not happen in a space-time framework.

As we see, discussing NSP leads us to the consideration of some vertiginous issues which are among the most fundamental and difficult ones ever considered by the human mind, to questions asked since the beginning and that possibly we shall not be able to answer definitely at the end. For this reason, although we may tentatively attempt to set down a non-singularity principle expressed under a general formulation, possibly to be later revised, we may feel more comfortable in considering specific, better controlled, and less ambitious particular formulations of a most general NSP, may be not yet correctly formulated. The non admissibility of singularities in classical mechanics is one of these restricted particular formulations, that I believe should fit the intuition of most physicists. Other cases are provided below.

A detector is not allowed to receive an infinite amount of energy (it would be destroyed). In a correlated way, physicists agree that the result of a measurement outcome cannot be infinite. For example, the mass of a local body cannot be infinite. Any such body can only contain a finite amount of energy. Infinities arising in quantum field theory had to be eliminated (this is done by a renormalization procedure) in order to provide meaningful results. Infinities are allowed to be used in physics, such as for plane wave of infinite lateral extensions, Dirac deltas of infinite heights, infinite straight motions of classical particles in free space, and so on, but although such permissions may provide efficient and convenient mathematical idealizations, they should always be considered as physically meaningless (not present in our actual world).

But, recall: You are allowed to claim that the NSP is revisable, that it is a prejudice, and that all its consequences are therefore speculative. But, even so, you have to admit that we may speculate on these matters, at the frontiers...

On the classical nature of measurement devices

As we already have seen, Bohr (and others) insisted on the compulsory classical nature of measurement devices. More generally, for him, the formulation of the postulates (basic or fundamental principles) of quantum mechanics is impossible to state without referring to classical mechanics although, at the same time, classical mechanics is to be viewed as an approximation to quantum mechanics. Bohr's motivation for such a standpoint is that the aim of physics is to establish some correlations between outcomes of measurements which have to be the same for any one. This is again a principle of mediocrity similar to the one of Einstein telling us that the laws of physics must be expressed in such a way that they are the same for any one. Indeed, with Bohr, not only the laws of nature must be the same for any one, but the outcomes generated by the laws of nature must be the same for any one too. As a result, Bohr insisted that we then have to communicate the outcomes of the observations to each other and for this, we need to employ a language which can be understood by any one. This language can only be our everyday language, augmented with technical words. And this language is a classical language, implying that the measurement devices must also be described with this language and, therefore, must be given a classical nature. However, if such a point of view is effective for many, if not all, practical purposes, it cannot be accepted in utmost rigor. Classical physics being rejected by NSP, this rejection must extend to the measurement devices which, in utmost rigor, must be viewed as quantum systems, and only approximately as classical systems. Any measurement process, viewed as an interaction process, requires at least the exchange of one quantum of action. This implies a kind of non-separability between a quantum system and the measuring instrument with which he has to interact, to be however opposed to the non-separability of elementary entities having interacted in the past, as done by d'Espagnat [497]. Prosperi [498] discussed a point of view intermediary between the one of Bohr (the measuring instrument is a classical instrument) and the one of von Neumann (the measuring instrument is a quantum instrument).

Epistemological comments

In this subsection, I intend to provide a few comments concerning the relations between the two ways we have exposed to reach Schrödinger's equation, namely (i) derivations relying on analogies and (ii) extraction in an Occam-style from a demonstrated set of generalized equations.

In mathematics, the word "demonstration" possesses a clear enough meaning. In essence, we consider a set of axioms generating a mathematical framework. Then, a demonstration proceeds from a proposition P (possibly formed from a set of elementary propositions), pertaining to the set of axioms or already demonstrated previously, by using small steps with well defined rules, to another proposition Q called the result of the demonstration, or: A theorem. It is interesting to remark that we can make the steps as elementary and as small as possible but, eventually, the passage from a sub-proposition to another sub-proposition by an indivisible step, a quantum of demonstration, is not analyzable and is to be grasped immediately by something which can be called intuition, and gives us a feeling of understanding (a feeling that a Turing machine, simply applying the rules, cannot perceive).

In physics, the meaning of the word "demonstration" cannot be the same, just because we do not possess a set of axioms generating the laws of nature. But the word may be, or has been used, in physics, for instance by Louis de Broglie when he claimed having turned Fermat's principle from a principle to a theorem, as already mentioned above in this chapter under the headline: Other derivations.

Instead of using axioms in physics, we may use basic principles, like the postulates of quantum mechanics. But we have seen that basic principles are not necessarily first principles, and that the lack of first principles constitutes a barrier to understanding. Hence, more ideally, a demonstration in physics should rely on first principles. These first principles must be convincing enough, looking like inescapable rational demands. For instance, the statement that the laws of physics should not depend on the observer (or, more objectively, on the frame of reference) is a first principle of relativity: It is a principle of mediocrity that we would not like to question. Conversely, the relativity postulate of the constancy of the speed of light in vacuum is a basic principle of relativity but, at least up to now, is not a first principle. Also, the postulate concerning the wave function collapse in quantum mechanics (postulate 5 in our list of postulates) is a basic principle of (orthodox) quantum mechanics, but it is not a first principle.

In the same way that a good set of axioms must be minimal in mathematics (no redundancy between the axioms), a good set of first principles must be minimal in physics. This no-redundancy demand is also a parsimony demand. Before exhibiting the minimal set of our demonstration to a set of generalized Schrödinger equations, let us examine what has not been used. The most important issue is certainly that I did not use the mechanical/optical analogy under any form, that is to say, in particular, I did not use the de Broglie relations, nor any analogy between variational principles. *A contrario*, I proceeded directly from classical mechanics to wave mechanics.

The use of the mechanical/optical analogy, with Schrödinger as the beginner, has been, in my opinion, may be, or even certainly, a fortunate idea, if not a *felix culpa*, but also something like a psychologically deterministic historical accident. First, light has been described as being made out of waves (after Newton, proposing a corpuscular interpretation, being defeated by Huyghens) and, afterward, it had to be viewed as made out from quanta, in particular with the Einstein interpretation of the photo-electric effect. Then it was clear that light could exhibit both a wave-like behavior and a particle-like behavior. Second, matter has been described as being made out from particles and, after the de Broglie hypothesis, and its confirmation by Davisson and Germer experiments, the particles had to simultaneously receive a wave-like behavior, hence the idea of a so-called wave/particle duality. It was then very natural, and almost compulsory (from a psychological point of view), to design the wave mechanics by exploiting the well-known Hamilton analogy between mechanics and optics, at least if you are aware of the existence of this analogy (although Heisenberg followed another track).

The fact that we did not use the mechanical/optical analogy established that it is unnecessary (to build a wave mechanics) but, furthermore, I make the claim that this analogy is misleading and even unpleasant. Light waves are actually not scalar but vectorial, because they result from the vectorial Maxwell's equations. Analogies between electromagnetic wave scattering and quantum wave scattering have recently been established [499, 500], but they are of limited scope in the context under interest for us here. What is important is that the Hamilton analogy does not in fact consider vectorial electromagnetic waves but the scalar approximation, and it is therefore an approximation which is used to reach the wave equation of quantum mechanics, supposed not to be an approximation, a rather strange conjuring trick indeed. Conversely, the lifting principle starts from a scalar field S and then automatically generates a scalar wave Ψ. We do not need to make any implicit assumption in which vectorial electromagnetic waves would be approximated by scalar waves, before proceeding further.

Rather than relying on experiments to recognize that classical physics is unsatisfactory, as has been done in the historical process, we obtained our set of generalized Schrödinger equations by relying on *a priori* theoretical considerations. We first used NSP to conclude that classical mechanics is inadmissible and thereafter the lifting principle to deform the Hamiton-Jacobi's equation for S to a set of equations for Ψ. The two first principles then involved form the minimal set of principles used to reach the result. Occam razor principle is afterward required to reach Schrödinger's equation *stricto sensu* but, as already commented, other possibilities, not rejected by the Occam razor principle, may be worth to be considered. The "demonstration" we have used therefore provides a theoretical legitimation for the examination of such more general opportunities.

Of course, the result obtained is only valid in the physical framework generated by the way the minimal set of first principles has been used. From this point of view, there is nothing more to say concerning NSP, but the lifting principle has been applied to classical mechanics, more specifically to non relativistic classical mechanics. Therefore, the wave equation obtained is non relativistic too. Obviously, other interesting prospects would concern the lifting of relativistic classical mechanics, starting from the formulation of special relativity, or even from the formulation of general relativity. I have no idea concerning what the results could be, at least no idea that I would be ready to confess, but these prospects are worth to be written on a list of decent problems.

Another reason why Schrödinger's equation is not the ultimate equation originates from the phenomenological status of the potential V. Schrödinger's equation describes the behavior of one quantum object, but the source of V is actually one or several other quantum objects. Therefore, all quantum objects are not treated on the same footing. Solving this problem requires a more elaborated theory (e.g. quantum field theory).

And, to end this subsection, once more, what about the concept of understanding? We cannot *understand* a postulate. Hence, Schrödinger's equation, viewed as a postulate, even if it is motivated by derivations, cannot be understood. Conversely, our generalized set of evolution equations, viewed as the result of a "demonstration", can be understood as the consequence of a rational necessity. Schrödinger's equation itself, pertaining to the necessary set, does not possess any more its postulational charge. Wave mechanics, whatever its eventual wave equation or more generally its eventual formulation, is necessary as the result of the *a priori* inadmissibility of classical mechanics. This points out to a possibility of understanding, even if only partially, quantum mechanics which, most often, is reputed as being non intelligible. And this is because rational necessity implies intelligibility.

Yes, Aristotle has been a genius of sciences, even if his physics has been erroneous enough to block further advances for so many centuries, and even if this physics is certainly very old-fashioned for modern scientists. But Aristotle can still be a source of inspiration. For instance, he made very little difference, in physics, between "provide a demonstration" and "exhibit the cause". The cause in question may be a rational cause. With this point of view, "exhibit the cause" may be interpreted as follows: Starting from first principles, explain why the thing is so and not anything else. We may then state that the cause of the quantum mechanics is the rainbow (and its singularity). This establishes a very beautiful relationship between the rainbow in the sky and the structure of atoms, say: In the sky and on earth. Such a remark possesses a poetic flavour which may be not scientific, in a strict sense, and may be a bit out of place in a book like this one. I hope the reader will accept me to end the present subsection with such a license.

A few anticipations and more speculations: On the necessity of experiments in physics

Is it possible to find enough first principles which would allow one to *a priori* construct quantum mechanics, and more generally to recover the laws of nature of our universe? This is a kind of dream or even of nightmare, which is in deep contradiction with the philosophical mood progressively developed during, say, the last three centuries, including positivism, and the deep anchored idea that physics is an experimental science. Before the progressive rejection of such a possibility (finding enough first principles) historically happened, let us however recall that this was essentially the vision of Descartes, e.g. [470, 501]. For instance, in his preface to the principles of philosophy, Descartes stated that we need to start with the search of the primary causes, that is to say of the principles, and that these principles must satisfy two conditions, one is that they must be so clear and obvious that the human kind could not question their truth... and the other one is that the knowledge of the other things must depend on them. The importance of what Descartes called clear and distinct ideas is also acknowledged by Spinoza when he stated that ideas which are clear and distinct can never be false [502], or that the true principles of science... must be so clear and distinct that they would not need any demonstration, that they are beyond any questioning, and that nothing can be demonstrated without them [503].

This definition from Descartes perfectly agrees with what we have presented as first principles (in opposition to basic principles). But, unfortunately, we know what eventually happened with the physics of Descartes, based on his first principles, particularly with his theory of eddies [112] defeated by Newtonian mechanics. The failure of

Descartes (dogmatic metaphysics and dogmatic physics) has been followed by an intense epistemological effort toward a dissolution of metaphysics, and to a rejection of the idea of first principles. My claim is that we should not throw overboard the baby with the water of the bath. We must go on looking for first principles, because they provide us with a genuine feeling of understanding and also, because, when they are good ones, they provide us with a splendid opportunity to organize our knowledge. This does not mean that the process is not risky (and Descartes has certainly been arrogant in believing that he found the ultimate and unique method), but looking for some kind of truth has been a risky unachieved affair since the origin of the human kind. Also, we should not been afraid of the fact that first principles are reminiscent of the divine nature, and of the possibility for the human mind, even obscured as it is, to participate to this nature. Indeed, there are genuine atheists among the scientists. And these atheists, just because they are making science, assume, at least implicitly, that there is some amount of rationality in the universe. The belief to the existence of such a rationality may be viewed as a positivistic attitude (we experimentally observe it), or as a matter of methodology (if you want to make science, you have to assume it), but it is somehow a kind of metaphysical faith. If you assume a certain amount of rationality, whatever your reasons for doing so, why not assuming also that this rationality may be grasped by first principles, even if they are difficult to discover? I have here another claim, namely that we must not renunciate, and that we must try hard.

Let us also mention that there still exist a certain amount of modern physicists who somehow believe that the universe is what it is because it could not be otherwise, that is to say, more or less implicitly, that the things are ruled out by a rational necessity so pregnant that, if we could grasp on it, we would have a Theory of Everything explaining everything. Then, concerning the idea that physics is an experimental science, we may argue in somewhat the same mood. Surely, physics has been an experimental science. But, from this historical observation, we should not logically conclude that it should be an experimental science for ever, nor that, ideally, it had to be an experimental science. To support this iconoclast doctrine, or at least to discuss it, we may very well return to Descartes and argue that he was basically right, that the power of reason sustained by God should reveal us the secrets of physics, that we have simply to apply his method (the one of Descartes), proceeding careful step after careful step, being sure at each step that we only keep on with clear and distinct ideas.

An opponent to Descartes would argue that Descartes himself did not succeed with his method because, although he made significant mathematical and physical discoveries, his physics, as a whole, collapsed, in particular his theory of eddies (allowing for instance Descartes to explain how a fixed star can become a comet or a planet!) did not resist to the attack of the Newtonian school. An opponent to the opponent, that is to say a defender of the method of Descartes, could simply retort that still the method of Descartes is correct, that Descartes simply did not succeed to have clear enough and distinct enough ideas, that he was not clever enough (and more generally, that we are never clever enough), that the use of experiments in physics is not in principle necessary, but that it is simply necessary in practice to compensate the deficiencies of our mind. And an opponent to the opponent of the opponent could then insist that ... and, I am afraid, the story would never end.

As an attempt to provide a definitive answer to the issue, let us accept some assumptions, possibly to be criticized by many philosophers, but which are consistent with the scientific venture, or even necessary to legitimate it. The first assumption relies on the fact that we observe regularities in nature, that we extract laws of nature from them, that is to say that we postulate a legal rationality in nature. The second assumption is that this rationality may be expressed in a mathematical way. These two assumptions are justified by the practical success of science. Let us note that it is not necessary to postulate that everything in nature is legally rational, nor that any rationality is of a mathematical nature. It is sufficient to postulate that there is some amount of mathematical legal rationality. As far as I am concerned, I use to think that physics describes the plumbing of the universe, but plumbing does not exhaust everything in a house.

Now there exist many incompatible mathematical systems generated by many incompatible systems of axioms. To know which system is associated with our universe, we therefore need to make experiments. May be the best known typical example concerns the structure of Riemannian spaces, more specifically of the 3D-space, or better of the 4D-space-time in which we are embedded. Let us consider the square ds^2 of the infinitesimal element of length ds. Depending on the structure of the metric tensor, the space may possess a definite positive metric ($ds^2 \geq 0$), a definite negative metric ($ds^2 \leq 0$) or an indefinite metric ($ds^2 \lesseqgtr 0$). If the metric is definite positive, one may define the cosine of an angle and more generally an usual geometry. And one may also distinguish between an Euclidean space (Riemannian space with a definite positive metric, and a Riemann tensor equal to a zero tensor), and curved spaces (for which the Riemann tensor is not zero), with in the last case another subdivision depending on the kind of curvature (spaces of the sphere-kind or of the saddle-kind) [504]. To know in which kind of space we are embedded, we must rely on measurements, that is to say on experiments. There is no smallest clue to indicate that this could be *a priori* known.

But, if experiments are required, meaning that physics is indeed an experimental science, it remains that we are may be making too much experiments, just because we are not clever enough. Since we use our mind, and have to use it, even if we do not know exactly what is mind, and how much we could be confident with it, or how much we should be suspicious of it, and how it works, with many other questions of that sort, it is much reasonable to use our reason. And, without believing that we could be able to *a priori* reconstruct quantum mechanics from nothing, it is worthwhile to try as far as possible, just because, surely, we shall learn something from this effort. In the rest of this subsection, I intend to provide a bit more of information concerning such a kind of attack, may be simply the entry of a blind-alley, may be in contrast something which, with much time and many efforts, could give us illuminating insights.

Because this rest will involve a certain amount of mathematics, and that it may be viewed as some kind of digression,

PEDESTRIANS ($\sim\sim\sim\sim$) may avoid it and prefer to go on with the last section of this chapter, which is a simple qualitative one, however essential before proceeding to the last chapter.

On the concept of energy

Let us recall that we are supposed having forgotten everything we knew on quantum mechanics. We are only allowed to use what we have established: The inadmissibility of classical mechanics from NSP, and a set of generalized Schrödinger equations from the lifting principle. Furthermore, let us thereafter decide to focus on Schrödinger's equation itself, the simplest equation inside the set. For the time being, that is all we have in hand, and should have in mind.

Equation 2.2 defines the energy W for a classical system, denoted E if the system is conservative. We however have no concept of energy in our skeleton of wave mechanics because Eq. (16.59) is silent on the concept of energy in quantum mechanics. However, Eq. (16.59) possesses special solutions of the form:

$$\Psi(x_j, t) = \Psi_0(x_j) \exp\left(-iE_q t/g\right) \tag{16.84}$$

in which E_q, with the subscript q standing for "quantum" is a constant. We have preserved the notation g for the elementary quantum of action (not $\hbar$) to insist on the fact that we are working in the framework of our embryonic wave mechanics, not in the framework of the well developed and established quantum mechanics, that we are supposed to ignore. Inserting Eq. (16.84) into Eq. (16.59), we obtain

$$\frac{\partial^2 \Psi_0}{\partial x_j^2} + \frac{2m}{g^2}(E_q - V)\Psi_0 = 0 \tag{16.85}$$

which identifies with Eq. (16.24) and that, not reluctantly, we can call the time-independent, or stationary, Schrödinger's equation. Because V is an energy, Eq. (16.85) shows that E_q is an energy too. By definition, we call it the energy associated with the quantum object of Eq. (16.84). In other words, we can define the concept of energy for quantum waves of the form given by Eq. (16.84). A superposition of such quantum waves, with different values of E_q, still satisfies Eq. (16.59), but such a more general wave does not possess any well-defined value of energy.

Therefore, we have defined the concept of energy in a quantum context without any reference to classical mechanics, at least as far as this definition is concerned. It is however interesting to examine what happens to this concept in a classical framework. Let us then consider the classical limit for a conservative motion. From Eqs. (16.35), (16.54) (again with $C_2 = 0$ and, since we are dealing with Schrödinger's equation, $C_1 = 1$) and 2.4, we have

$$\Psi = e^{iT_0} = e^{iS/g} = e^{iS_0/g}e^{-iEt/g} \tag{16.86}$$

Comparing Eqs. (16.84) and (16.86), we have

$$\Psi_0(x_j) = e^{iS_0/g} \tag{16.87}$$

$$E_q = E \tag{16.88}$$

that is to say the quantum energy E_q, defined in a quantum mechanical context for solutions of the form given by Eq. (16.84), identifies with the classical energy E of a conservative motion.

On the concept of momentum

Similarly, we can directly define the concept of momentum in a quantum mechanical context. Let us first consider a free quantum object ($V = 0$). Schrödinger's equation reduces to

$$ig\frac{\partial \Psi}{\partial t} = -\frac{g^2}{2m}\frac{\partial^2 \Psi}{\partial x_j^2} \tag{16.89}$$

This equation possesses solutions of the form

$$\Psi = A\exp\left[\frac{i}{g}(p_{qj}x_j - E_q t)\right] \tag{16.90}$$

in which the subscript q in p_{qj} is not a tensor subscript but stands for "quantum". Inserting Eq. (16.90) in Eq. (16.89), we readily obtain

$$E_q = \frac{p_{qj}^2}{2m} \tag{16.91}$$

to be compared with the classical expression

$$E = \frac{p_j^2}{2m} \tag{16.92}$$

Since the quantum energy E_q identifies with the classical energy E, we see that p_{qj} may be called the quantum momentum, identifying with the classical momentum p_j in the classical limit.

If V is a constant, but different from 0, we then obtain, instead of Eq. (16.92), with $E_q = E$ and $p_{qj} = p_j$

$$(E - V) = \frac{p_j^2}{2m} \tag{16.93}$$

which exactly identifies with Eq. (16.92) if we decide to measure the energies with respect to the constant value of V.

Einstein-de Broglie relations

We have directly defined the energy E_q and the momentum p_{qj} in a quantum mechanical context. Relying on dimensional analysis, we may then introduce an angular frequency ω_q and a wave-number k_{qj} according to

$$p_{qj} = g k_{qj} \tag{16.94}$$

$$E_q = g \omega_q \tag{16.95}$$

which identify with Einstein-de Broglie relations. However, since a quantum object is an entity in its own right, the interpretation of these relations should not drive us any more to a so-called wave/particle duality. It is just a matter of taste, allowed by the existence of the quantum of action g, whether we prefer to discuss energies with E_q or ω_q, and momenta with p_{qj} or k_{qj}.

This remark is valid for matter waves since we have been dealing with matter waves. To extend it to photons, we should first find a way to introduce the existence of photons. This could be done by solving Schrödinger's equation (16.85) for the hydrogen atom (or for hydrogenoid atoms). For bound states, we would mathematically determine that the atom can be found in one or in another state pertaining to a discrete spectrum. If we assume that transitions between these states are allowed, and hold fast on the conservation of energy, we should postulate that, during a transition, the energy is transferred to another quantum object, that we may call photon, or whatsoever. Again, there is no trace any more of the so-called wave/particle duality. It is again just a matter of taste whether we prefer to discuss the energy of the photon in terms of E_q or of ω_q, and its momentum in terms of p_{qj} or of k_{qj}.

Other liftings?

What is so special with the Hamilton–Jacobi's formulation of classical mechanics that we gave it such a privilege? There is obviously its historical importance, in particular it has been a key formulation for the mechanical/optical analogy which allowed Schrödinger to derive his equation. And therefore, because it provided us with a field $S(x_j, t)$, it has been a natural formulation to successively apply the lifting principle. But we have to wonder what would have happened if we started with the other equivalent formulations of classical mechanics. This is the issue to be discussed in this section.

Let us then consider the Lagrangian formulation, and first of all recall it. We already alluded to it, but without any development, when we discussed variational principles. We now need to learn a bit more. In this formulation, the motion of a Newtonian particle is determined by the variational principle of Hamilton, reading as

$$\delta \int_{P_1}^{P_2} ds = 0 \tag{16.96}$$

in which P_1 and P_2 are two fixed points of the configuration space (x_k, t), according to

$$P_1 = (x_k^{(1)}, t^{(1)}) \tag{16.97}$$

$$P_2 = (x_k^{(2)}, t^{(2)}) \tag{16.98}$$

In the configuration space, there are an infinity of motions going from $x_k^{(1)}$ at time $t^{(1)}$ to $x_k^{(2)}$ at time $t^{(2)}$, but all these motions are virtual, the actual motion being given by Eq. (16.96). The quantity s is called the action of Hamilton.

We write

$$ds = Ldt \qquad (16.99)$$

in which

$$L = L(x_k, dx_k/dt, t) \qquad (16.100)$$

is the Lagrangian. The variables x_k and dx_k/dt are independent variables called Lagrange variables.

The solutions of Eq. (16.96) are geodesic equations, called Lagrange–Euler equations, reading as

$$\frac{\partial L}{\partial x_k} - \frac{d}{dt}\frac{\partial L}{\partial(dx_k/dt)} = 0 \qquad (16.101)$$

In order to recover Newtonian mechanics, we have to set

$$L = E_c - V \qquad (16.102)$$

in which E_c is the kinetic energy. Indeed, we have

$$E_c = \frac{1}{2}m\left(\frac{dx_k}{dt}\right)^2 \qquad (16.103)$$

Inserting this equation into Eq. (16.101), we readily obtain Newton's equation

$$\frac{d}{dt}\left(m\frac{dx_k}{dt}\right) = -\frac{\partial V}{\partial x_k} \qquad (16.104)$$

establishing the equivalence between Newton's and Lagrange's formulations.

We therefore have

$$ds = Ldt = (E_c - V)dt \qquad (16.105)$$

Now, we also have

$$\left.\begin{array}{l} L = E_c - V \\ W = E_c + V \end{array}\right\} \qquad (16.106)$$

leading to

$$L = 2E_c - W = m\left(\frac{dx_k}{dt}\right)^2 - W \qquad (16.107)$$

Hence,

$$ds = Ldt = m\frac{dx_k}{dt}dx_k - Wdt \tag{16.108}$$

which may also be rewritten as

$$ds = mv_k dx_k - Wdt = mvdl - Wdt \tag{16.109}$$

in which we have used $v_k = dx_k/dt$. The action s involved in Eq. (16.96) may therefore be expressed as

$$s = \int_{P_1}^{P_2} ds = \int_{x_k^{(1)},t^{(1)}}^{x_k^{(2)},t^{(2)}} \left(m\frac{dx_k}{dt}dx_k - Wdt\right) \tag{16.110}$$

But the result of this integration is a number, not a field, and therefore does not allow one to carry out any lifting. However, instead of considering the action s, we may consider an action s_0 in which the point $P_1 = (x_k^{(1)}, t^{(1)})$ is still fixed, but in which the point P_2 is no more fixed, according to

$$s_0 = \int_{x_k^{(1)},t^{(1)}}^{x_k,t} \left(m\frac{dx_k}{dt}dx_k - Wdt\right) \tag{16.111}$$

in which, to avoid complicating the notations, we denote by x_k and t both the quantities in the integrand, and the quantities in the upper limit of the integrand.

We now return to the Hamilton–Jacobi's formulation and to its field $S = S(x_k, t)$, in which x_k and t are the arguments of the field (the Lagrange variable dx_k/dt is not involved). We have

$$dS = \frac{\partial S}{\partial x_k}dx_k + \frac{\partial S}{\partial t}dt = m\frac{dx_k}{dt}dx_k - Wdt \tag{16.112}$$

in which we have used Eqs. 2.2 and 2.3. Hence,

$$s_0 = \int_{x_k^{(1)},t^{(1)}}^{x_k,t} dS \tag{16.113}$$

leading to

$$s_0 = S + \alpha \tag{16.114}$$

in which α is a constant. Since the real motions from $P_1(x_k^{(1)}, t^{(1)})$ to $P(x_k, t)$ are given by $\delta s_0 = 0$, the constant is irrelevant, and we may set it to zero. Hence,

$$s_0 = S \tag{16.115}$$

We therefore conclude that the Lagrangian formulation can be lifted too, starting from the action s_0. The demonstration starting from s_0 would mimic the one starting from S, and it is not worthwhile to repeat it.

Let us now consider Newton's formulation. There is no direct way, in this formulation, to obtain a field which could possibly be lifted. We essentially just have x_k depending on t, that is to say trajectories. In the Hamilton–Hacobi's formulation, these trajectories are dressed by a pilot field $S(x_k, t)$ and they have to be orthogonal to this pilot field. In the Lagrangian formulation, the trajectories are dressed by a swarm of virtual trajectories. But, in Newtonian formulation, we just have naked trajectories. And this is consistent with a very important fact which will allow us soon to return to hidden variables: Naked trajectories cannot be seen. They simply do not exist.

Newtonian trajectories of matter points do not exist

At the beginning of this book, the reader has been introduced to two figures, namely Figs. 1 and 2. Fig. 1 has been already commented, but Fig. 2 is still waiting. Time has come to explain what it means. But, before, let us return to Figs. 1(a) and 1(b), and comment them again. We are all used to the concept of trajectory, so easy to extract from classical mechanics, in deep agreement with our intuition, and with our sense data: Trajectories of cars, of balls,..., or even of planets. There is no apparent difficulty to consider trajectories of matter points. They at least constitute efficient models for the behavior of so many objects around us. Furthermore, physicists, when they are students, have been trained, when learning classical mechanics, to the study of the mechanical behavior of matter points, before entering the realm of the mechanics of extended solids. There is also this famous theorem telling us *that the force between two uniform spheres of masses M_1, M_2 is equal to the force between two point masses M_1, M_2 at their respective centers* [505]. As a specific example, the motion of a satellite around the earth can be accurately enough (but not exactly because the earth is not spherical) computed as if all the mass of the earth was condensed to a single point (this physically would produce some kind of black hole, but does not invalidate the theorem).

We feel therefore very comfortable with Fig. 1(a) and, as a consequence, we may feel very uncomfortable with Fig. 1(b). What could physically be this field S, which extends in the whole space around the object under motion, that no one has ever seen and which is effectively unobservable? At the best, it is a convenient intermediary for computations and, at the worst, it is some kind of *metaphysical artificial superstructure*. The two figures provide different visions of the world and, from this point of view, they do not correspond to two different interpretations of the same theory (classical mechanics), but they genuinely correspond to different theories. Although these two different theories are mathematically equivalent (we may pass from one to the other, and reciprocally, by mathematical manipulations) and also experimentally equivalent (the trajectories computed are the same), we could believe that one of them, at least is erroneous. Going on thinking that way, noting that we indeed can observe trajectories, and cannot observe the field S, we immediately and instinctively conclude that the vision of the world suggested by the Hamilton–Jacobi's formulation is the one which is erroneous, and that the Newtonian vision of the world is the correct one. This point of view would be indeed supported, I guess, by all positivists of the universe. Then, indeed, S must be a fictitious field, convenient to mathematically solve some classical mechanical problems, but without any physical significance, i.e. not pertaining to its ontology, or to the *furniture of the world*, to borrow an expression from Cushing [131].

But, actually, although this way of thinking seems very reasonable, it is dramatically misleading. Let us effectively recall NSP and its application to classical scattering. We have rejected classical mechanics, as being inadmissible, and therefore claimed the necessity of the wave mechanics, on the basis that, in some cases, a classical problem may produce singularities, in particular the rainbow singularity when the differential cross-section becomes infinite. I intend, up to the end of this book (notwithstanding a final loophole), to hold fast on this conclusion and draw all its consequences, in particular that one: Newtonian trajectories of matter points do not exist (and cannot exist).

Therefore, let us return to Fig. 1(a). We must remove the trajectory depicted in this figure, and let is blank. Let us say that the trajectory has to vanish in the BLUE of the sky, and this leads us to Fig. 2(a). We also have to remove the trajectories depicted in Fig. 1(b). leading us to Fig. 2(b), with the field remaining, but which does not pilot anything any more. Next, the lifting leads us from Fig. 2(b) for S to Fig. 2(c) for Ψ.

All along Fig. 2, from Figs. 2(a) to 2(b), we are dealing with unobservable quantities. In Fig. 2(a), trajectories cannot be observed because they simply do not exist. And effectively, no one ever observed the trajectory of a matter point. Simply try to imagine, at the most fundamental level of understanding, what could be the trajectory of a matter point, that is to say the trajectory of an object of mass m, confined to a vanishingly small volume. Such an object would have an infinite density, and Newtonian mechanics would have to be merged with the most extraordinary predictions of general relativity. The more lax intuition is not ready to accept the existence of such an object, excepted may be if it is a matter of life or death, with the gun pushing the temple.

A special kind of trajectory is the motionless trajectory of a particle at rest. Then, we have the particle (the matter point) standing still (in some frame of reference). From this, we can see that the concept of a matter point, independently of the concept of trajectory, cannot be accepted. Besides what we have already said, including the idea of a collapse of a mass m to a geometrical point to generate a matter point, just think, for instance, of the unbelievable behavior of the electrostatic potential of a point charge varying as $1/r$, and therefore diverging when the point charge is indefinitely approached.

Obviously, from a pure logical point of view, we might start to refute first the existence of matter points invoking NSP, and thereafter the existence of trajectories of matter points, since something which does not exist cannot have any trajectory. But I preferred the reverse way, not only a matter of taste however, but because it is easier to draw consequences, invoking, as we are doing at the present time, Newton's and Hamilton–Jacobi's formulations.

Next, in Fig. 2(b), the dressing field S, which does not dress anything any more, is unobservable too. And this unobservability is transferred to the wave Ψ of Fig. 2(c). But the unobservabilities of the trajectories and of Ψ (in orthodoxy) are not of the same nature. Trajectories are unobservable because they do not exist. The wave function Ψ is unobservable, but it is the most fundamental object of one of our most fundamental theories available to us, at the present time: Usual quantum mechanics.

When teaching students, it is often heavily pointed out that trajectories do not exist in quantum mechanics. This may be viewed as a consequence of Heisenberg's uncertainty relations, forbidding the simultaneous measurements of locations and momenta, these quantities which are so vital for the very possibility of defining a trajectory. And the students are supposed to be surprised by such a rupture. But there should not be any surprise here. If trajectories of matter points already do not exist in classical mechanics, why should they exist in quantum mechanics?

But, what about trajectories of cars, balls... and planets? No problems with them. Cars, balls,..., planets are not matter points that we cannot observe, but macroscopic objects that we do observe. Their existence, their properties, the fact that we can observe them, and by which processes we observe them, must emerge as the consequences of a quantum theory, whatever its ultimate formulation will be. No U-turn allowed.

Chapter 17

Coup de Grâce (with however a loophole)

After the Big Book classification, we were left with only two physical PHV-theories: The double solution and the pilot wave. To these, we must also add stochastic quantum mechanics, sharing a sufficient amount of ingredients with the causal theories of Louis de Broglie and of Bohm to be allowed to enter the same family. To end this book, we have to definitely decide, if possible, whether the family is valuable or not, whether we still want to live with it or not, or whether we must definitely leave away or not. I shall mainly focus on Bohm's pilot wave, for convenience and for being specific, but the arguments we shall meet concerning this causal theory could be applied, *mutatis mutandis*, to the double solution and to stochastic quantum mechanics as well. To be fair, I shall begin to defend Bohm's pilot wave but, at the end, with clarity, but charity and a bit of sadness, I shall have to blow a coup de grâce (although offering a loophole).

Naughty reasons to reject Bohm's pilot wave

The positivistic accusation of metaphysicism

The word "metaphysics", and even the concept without the word, is so old that it had time, too much time, to receive many meanings, and to appear under many avatars. In its golden age, metaphysics was much dignified. It has been the science of the knowledge of Being, as far as Being is Being, probing the most ultimate questions. It was a science above all sciences, the culminating summit, the most important intellectual affair of thinkers. Excesses of metaphysical considerations in treating philosophical, religious, and later on scientific questions (Descartes again, may be the last metaphysician among the metaphysicians), provoked a reaction, and a process of dissolution of metaphysics which has a venerable and long history. Kant is one of the predominant milestones for the story of this dissolution. He attacked the dogmatic metaphysics, threw it away from the house by the door, but let it enter back in a disguised form by the window, restoring the dignity of God and of ethics for practical reasons [506]. As he wrote, at the beginning of his master opus [103], he had to abolish the knowledge to obtain a room for the belief, so announcing its most intimate research program in philosophy.

I do not intend to report more on or even to summarize this story but let us point out that, as a result, it also produced its own excesses which, by a counter-reaction, will generate, I dare to predict, an opposite movement of dissolution of the dissolution of metaphysics. This process of dissolution of the dissolution seems to be required if we just accept to watch some inescapable features which are in front of us, do not pertain to the empirical world but rather to the third world of Popper: Our ability to reach the mathematical realm and to make journeys in its landscapes, the fact that nature is at least partly intelligible, the power of reason (called by Galileo Galilei the power of understanding [34]) even if it has limitations, and also our use of logic and our confidence in it, more generally all our ability of reasoning and playing successfully with ideas.

Let us now, once again, mention Popper who introduced his famous demarcation line between scientific theories which can be falsified, through a virtuous Darwinian process of objective scientific progress, and metaphysical doctrines which are immune to falsification (I am excluding mathematics from the discussion). From this point of view, we can say that metaphysics discuss subjects that are outside of the physical empirical world. Although Popper nevertheless did not aim to the destruction of metaphysics (for him metaphysical statements are not meaningless), but only to a clarification of its domain of validity and of an appraisal of its pretensions, the use of the word has become, even before Popper, pejorative or even insulting, particularly in the mouth of positivist scientists. There has been indeed a whole efficient process in which metaphysics experienced a dramatic loss of credibility, sustained by the prejudice that, since now a long time, it is impossible any more to deal with metaphysics, so that the accusation of metaphysicism became something violent, uttered *ex cathedra*, pointing out with the finger the irrational madness of the unfortunate ones who dared to wander in no-think areas. For the defenders of the orthodox Copenhagen interpretation, at least for those who believe that they stand firmly on an orthodoxy forever, the arrogant accusation of metaphysicism helped to push away hidden variables toward the darkness of a medieval heretical way of thinking.

We recall that indeed Bohm's pilot wave has been accused of metaphysicism, sometimes with a spiteful anger. This was often motivated by the fact that the introduction of hidden variables did not provide any new prediction that could be tested experimentally. Therefore, the introduction of hidden variables has been viewed not as physical, but as metaphysical. For Bitbol [128], dealing with hidden variables is a kind of intellectual game which pretends to describe some processes which are outside of any experimental investigation, and therefore, for many physicists, it is a sufficient reason to disregard them.

Obviously, Bohm was well aware of the fact that his theory could be attacked on such grounds. In 1953 [261], he provided a rebuttal on the question to know whether the introduction of unobserved entities is forbidden: *...Past experience in a wide variety of fields would, however, suggest the falsity of such a conclusion; for it has very often turned out to be very fruitful indeed to postulate the existence of things before any one knew, even in principle, how to go about trying to verify their existence. In fact, the means of verification are very often discovered precisely because of researches springing from the hypothesis themselves. To require that the means of verification already be known when a new hypothesis is suggested would therefore often make almost impossible the suggestion of new concepts.*

In 1957 [60], he reiterated a similar rebuttal, and added: *The proponents of the usual interpretation of the quantum theory however, have adopted the hypothesis of Heisenberg ... They do not assume that the indeterminacy principle is just a consequence of the quantum theory in its current stage of development, which could therefore turn out to have only a limited range of validity if, as seems likely it is later discovered that the present form of the theory must be modified, corrected, or extended. Instead they assume that this principle represents an absolute and final limitation of our ability to define the state of things by means of measurement of any kind that are now possible or that ever will be possible.*

If one makes the assumption described above, then one comes to a conclusion having far-reaching consequences. For even if a sub-quantum mechanical level containing hidden variables of the type described previously should exist, these variables would then never play any role in the prediction of any possible kind of experimental results. In fact, if this hypothesis is true, the future behavior of a system would, at least as we are concerned, be predictable to just that degree of accuracy corresponding to the limits set by the indeterminacy principle, and to no higher degree.

Thus, it is concluded that the present general form of the quantum theory is able to deal with every kind of measurement that we could possibly carry out. Any theory (such as one involving hidden variables) which claim to deal with more than this would then be just a metaphysical exercise of the imagination because nothing in physics would be different from what it would be if these hidden variables did not exist.

More generally, what should we say today concerning the fate of things which are, not only unobserved, but also unobservable? How should we deal today with the concept of metaphysics, possibly in a somewhat renovated and more balanced way? There are certainly several issues relevant to this issue. But the most important one is that, even if only implicitly, we (at least scientists, possibly omitting provocative philosophers) adhere to considerations which, in one sense or many, are of a metaphysical nature. We believe that we may rely on logic and reason, that there are laws of nature which can be expressed with mathematics, that these mathematics themselves, at least in our mind, can be constructed independently of any experiment and that they tell us something relevant to veracity, and that our theories, so far away from the sense data even if corroborated by them, are not completely rubbish.

Therefore, the idea that Bohm's pilot wave should be rejected, under the accusation of metaphysicism, because it uses unobserved, or even unobservable, entities, is not convincing enough. All our elaborated theories indeed use unobservable entities, that is to say concepts which may be used for making experimental predictions, possibly to provide us with a vision of the world, or also to help us gaining some understanding concerning what is going on, but which pertain to the third world, in the sense of Popper, and certainly not to the world of physical events surrounding us. The incommensurability between the physical word around us and the mental world inside us definitively colors what we are (thinking subjects) with metaphysical tonalities. And even our most positivistic theory, quantum theory, is based on the most unobservable and metaphysical entity: Ψ.

If we refer to the Popper criterion of demarcation, both theories (pilot wave and quantum mechanics), which simultaneously agree with experiments, are actually not metaphysical since they can be (simultaneously) falsified by experiments and, being (simultaneously) in agreement with experiment up to now, they are (simultaneously) corroborated. Things would be easier if one theory were on one side of the demarcation line, and the other theory on the other side. But such is not the case: They are on the same non metaphysical side, and the real problem is a problem of discrimination between two theories on the same right side of the demarcation line.

If we insist in saying that hidden variables are metaphysical, relying on the unobservability of the supplementing parameters, we have to say that it is true too for quantum mechanics, relying on the unobservability of Ψ and of its evolution between observations. We could then argue that hidden variables are to be rejected because they are more metaphysical than quantum mechanics. Indeed they cumulate the unobservability of hidden variables and of Ψ while, in quantum mechanics we only have the unobservability of Ψ. But, besides being a bit lame and pernickety, this is not quite correct: In the pilot wave theory, Ψ is a consequence of the sub-level of hidden variables and, therefore, it would not be fair to count both the hidden variables and Ψ as two independent sources of unobservability.

Hence, both theories are simultaneously non-metaphysical, or metaphysical, depending on the point of view taken. This, we may call it the metaphysical symmetry. This symmetry has been historically broken, may be contingently. Once an orthodoxy is established, it is socially easy for it to declare any heterodoxy as heretical. From the point of view of the Copenhagen interpretation, the dominant orthodox one, it is then easy to accuse the pilot wave of metaphysicism, without the risk of any serious rebuttal. If the objective pilot wave succeeded in becoming the dominant one, it would have been easy from it to accuse the Copenhagen interpretation of metaphysicism, or even of mysticism (a criticism which has actually be addressed to Bohr), arguing that it relies on a subjective and unintelligible fuzzy description of nature.

Therefore, if we really and fairly think of it, we cannot reject the pilot wave with such an accusation. We have to find better reasons, may be its lack of simplicity?

Lack of simplicity

We have to deal again with the Occam razor principle, already previously discussed in many occasions. Referring to it, at least partly, Belinfante [1] stated that *physicists call a theory satisfactory if (1) it agrees with the experimental facts (2) it is logically consistent (3) it is simple as compared to other explanations.* As we already know, the pilot wave satisfies the two first demands. The question is to evaluate whether it also satisfies the last one. For many physicists, in particular for the defenders of the Copenhagen interpretation, the answer is negative. Their point of view is well represented by London and Bauer [27] when they stated that, in quantum mechanics, *the knowledge of the representative of the state at a given instant is necessary and sufficient for an unambiguous calculation, with the aid of the dynamical law, of the representative of the state at every subsequent moment* (and therefore, I recall, for an unambiguous calculation of experimental predictions). Hence, for him, we cannot *add to them supplementary data without introducing useless tautologies.*

Bohm [31] was well aware of this possible weakness of his theory, as we can see from the fact that he remarked that, according to Heisenberg, it is desirable to formulate the physical laws in terms of a minimal set of notions, implying that hidden variables should not be considered because they add nothing to the physical predictions of the theory, while they complicate the formalism in an irrelevant way. In his 1993-book with Hiley [68], his last book indeed, Bohm defended himself by stating that, although his interpretation has the additional assumption of particles, this is balanced by the fact that it does not require the usual assumption of probability. So, he said, at least on the basis of a formal count of the number of assumptions, it cannot be concluded that either interpretation is favoured over by the principle of Occam's razor. I already used and commented just above a variant of the formal count. Whether it is correct or not, it is in any case, a very poor argument, far from being able to convince any orthodox physicist. According to Margenau [116]: *Logically, it is extremely difficult to state the conditions under which a set of axioms may be regarded as simple or even as simpler than another, and this situation is likely to remain until theoretical physics has been penetrated completely by the methods of symbolic logic. Only when the number of independent fundamental axioms involved in a theory can be counted will simplicity become a quantitative concept.* It seems to me that Bohm was actually embarrassed by the problem, but he missed the point to defend him efficiently enough. Let us try to defend him in a better way.

We could again, as we have done previously, argue on the complexity of the notion of simplicity, and on the heuristic status of the Occam razor principle. The reader might also refer to Popper [25] who, in a short but penetrating chapter, discussed the difficulties involved in the definition of what is simple, or simpler. The demand for simplicity cannot be a first principle of physics, and there is no reason to believe that it would be correct to apply it whatever the circumstances, and also no reason to believe that the laws of nature or our representations of the laws of nature should satisfy it. Indeed, it can be misleading. The best way to convince the reader of this fact is to provide examples: Two of them will pertain to the history of sciences, and a third one, the most striking one, at least in the framework of this book, is new and unnoticed, as far as I know.

The first example concerns the history of the development of the System of the World [8]. Already my cherished pre-socratic thinkers participated to this history with, sometimes, striking and genial anticipations. But (and this was not a genial anticipation), the idea arose that planets should follow circular motions. The reason why is easy to grasp: The circle is the most perfect (and the simplest) curve. We know the end of the story, actually a very long and complex story, from Ptolemy with his epicycles (circles within circles, in the language of Bohm [294]) and his deferents, to Kepler with his laws and his ellipses, Newton with his principles of mechanics and his $(1/r^2)$-law of gravity, and to Poincaré who, working on the long-standing problem of the orbit of the Moon introduced what would be later named deterministic chaos. There are two kinds of simplicities involved in this example: The conceptual simplicity of the principle generating the circle (a principle of symmetry for the circle, simpler than the laws of Newton) and a computational simplicity. What is simpler to draw than a circle, and what is more complicated to calculate than a chaotic orbit? Well, actually, there is something as simple as the circular line, namely a straight line. For Aristotle, the circle pertains to the heavens, and the straight line to the Earth, upward (e.g. for fire) or downward (e.g. for earth) [7]. However, in a famous dialogue by Galileo Galielo [34], the honest character Sagredo wondered why a cylindrical helix, which is everywhere similar to itself, should not be admitted inside the set of simple lines.

The second example concerns the phenomenological thermodynamics (upper macro-level) and the existence of atoms (lower micro-level), already discussed. For Bohm, this is an important example used to justify hidden variables (lower micro-level) as an underlying substrate for quantum mechanics (upper quantum-level). In the same way than phenomenological thermodynamics alone is simpler than this thermodynamics supplemented with hidden atoms (they have indeed been hidden for a significant amount of time), quantum mechanics alone is simpler than quantum mechanics supplemented by hidden variables. Yet, the complexity involved by the existence of atoms generated dramatic advances in our understanding and mastering of the world. Bohm, very often, expressed the idea that it could be the same with hidden variables. This does not only concern a failure of the search for a conceptual simplicity. It is also related to the issue of computational simplicity. The engineer wanting to design heat exchangers, fluidized beds, mixing reactors, or whatsoever, should not loose his time with statistical computations at the atomic level. Concerning quantum mechanical computations however, the issue is a bit more subtle. Any one used to such computations with the usual quantum mechanics would prefer to keep on with them. But why not computing Bohmian trajectories as many people have done? This requires computer programming. It may need a lot of computations which will be carried out by the computer. But, from the algorithmic point of view, you essentially just have to make a computer program to deal with the guidance formula. Is this so difficult?

But, here is now the third example, likely the best one in the context of this book. To explain it, let us return once again to Figs. 1(a) and 1(b). In Fig. 1(a) sketching the Newtonian formulation, we just have trajectories. In Fig. 1(b) sketching the Hamilton–Jacobi's formulation, we have trajectories supplemented by the field S. Both formulations agree with experimental facts and are logically consistent. They are indeed observationally and mathematically equivalent, therefore satisfying the two first criteria expressed by Belinfante. But, according to the third criterion (the one of simplicity), we should reject the Hamilton–Jacobi's formulation and retain the Newtonian formulation. This is exactly the contrary of what we have done. It is the simplest formulation (theory) that we have rejected as inadmissible, and it is the more complicated formulation (theory) that we have retained. And it is that one which opened the way to a set of generalized Schrödinger's equation. The situation may abstractly and formally be expressed as follows. We are facing two theories T_1 and T_2, which are observationally and mathematically equivalent, with T_1 simpler than T_2. This abstract formulation provides us with a logical relationship between T_1 and T_2. Now, replace T_1 by "quantum mechanics", relying on Ψ only, and T_2 by "pilot wave", relying on Ψ *and* on supplementary hidden variables. After this translation, we obtain the following logically equivalent statement: Quantum mechanics and pilot wave are observationally and mathematically equivalent, with quantum mechanics simpler than pilot wave. But, if in the previous case, we retained the most complicated theory (Hamilton–Jacobi), why, should we, in the present case, reject it (pilot wave) on the basis of Occam razor principle?

Another remark, not on conceptual simplicity, but on computational simplicity is to be made. When one starts learning the Hamiton–Jacobi's formulation of classical mechanics, and train ourselves to manipulate it with various exercises, the most immediate (and simplest) problem to solve concerns the motion of a free particle. With the Newton's formulation, the result is most easily obtained with just a snap of fingers. You only need to integrate twice the equation of motion $d^2 x_k / dt^2 = 0$ to readily obtain: $x_k = \alpha_k t + \beta_k$, in which α_k and β_k are constants determined by the initial location and initial velocity, say at time $t = 0$. But, the same problem is a bit more complicated to solve with the Hamilton–Jacobi's formulation. The student might then wonder why it would be good to use a more complicated formulation leading to more complicated computations. The answer is that the Hamilton–Jacobi's formulation has been designed to deal with complicated problems or, at least, to add a new weapon to the mathematical task force. It may be more complicated for simpler problems and simpler for complicated problems. This result may also be applied to the issue of conceptual simplicity: Things which look complicated at the beginning may appear simpler later, and things which are simpler at the beginning may afterward become unwieldy complicated.

I would like to end this subsection with an anecdote on the topic of metaphysical simplicity. I remember, when being a child, how much I felt surprised by the existence of the world around me, something which looked to me like quite incongruous. Surely, this world should not be there. I later learnt that my questioning was in phase with the famous question of Leibniz wondering why there is something rather than nothing. If we try to raise the issue in a rational way, we may find that it has something to do with the concept of simplicity: Nothing is infinitely much simpler than something. With nothing, there is no question, no answer, no problem, and if you feel like that, "nothing" is more perfect than "something". Therefore *there should be nothing*! But just open the eyes and look: There is something. Well, nothingness may be the simplest but, on the basis of experimental evidence, we must conclude that the simplest is too simple. This is an opportunity to recall the witticism from Einstein, saying that things have to be simple, but not too simple. Even today I am still puzzled by such an incoherence lying in the heart of all these things which *should not be there*. Well, clearly, the concept of simplicity, even if it can be a good guide for making physics (or metaphysics!) is not so easy to manipulate. By the way, Leibniz himself was well aware of this issue concerning simplicity. Indeed, here is a complete quotation from him: "Why is there something rather than nothing? Since nothing is simpler and easier than something" [507].

Therefore, as a conclusion concerning the issue of simplicity, we still need a good reason to reject Bohm's pilot wave, may be incompleteness?

Incompleteness

The words completeness, completude, completion, may be used in different contexts, in relation with the fact of being not complete or not completed. We have for instance the question to know whether quantum mechanics is complete or not (could it be or should it be supplemented by hidden variables?) and, in another context, we have the completude theorems of Gödel. In the present subsection, we use the word incompleteness to denote the fact of being unfinished, unachieved.

In this sense, the double solution is incomplete, and remains a far from being achieved research program. Concerning the pilot wave, it is incomplete too. Many authors indeed noticed that the pilot wave theory has not been extended successfully to the relativistic domain, particularly to the high energy physics when particles can be created and annihilated, in contrast with quantum mechanics which, extended to the relativistic domain and to quantum field theory, successfully deals with such phenomena.

Quantum mechanics, with all its extensions, is however incomplete too, in the precise sense used here, since its marriage with general relativity is not yet consummated. Therefore, it becomes a bit less obvious to reject the pilot wave (and the double solution) on the basis of their incompletenesses. Surely, we might insist and state that, if both quantum mechanics and pilot wave are indeed incomplete, it happens that the pilot wave is more incomplete, much more incomplete, than quantum mechanics. But this cannot be a fully convincing and definitive argument.

Let us begin to illustrate this last statement with a mental picture, recalling, as we recently mentioned, that things which are complicated at the beginning may become simpler afterward, after we succèeded getting over some pass. Assume that we are exploring a massif of mountains, on search for the Lost Mount. Some people decide to follow an easy path, walking along a torrent upstream to its source which, presumably, is fed by one of the Lost Mount glaciers. And, suddenly, here they are: Impassable stiff and steep cliffs, that no one is able to climb up. Others choose another way, the uneasy one, fighting against the brambles, step by step, meter by meter. Eventually, discouraged, they give up, excepted one who goes on with perseverance and ... dies of exhaustion. But, fifty meters later, there were no brambles any more, and here it is: The Lost Mount.

And also, furthermore, how could we compare the huge army who attacked the orthodox way and the small group who braved, may be contingently, the heterodox one? Likely, the small group would have been defeated too if it opted for the orthodox path. But, may be, the army making an assault on the heterodox way could have succeeded. No, definitely, we cannot use the accusation of incompleteness to get rid of Bohm.

Bitbol's list

Bitbol [128] provided five reasons, expressed by various authors, to motivate the rejection of Bohm's pilot wave. Let us examine all of them to decide whether they are naughty or not. The first reason is the accusation of metaphysicism. As discussed above, it is a naughty reason. The second reason concerns the issues of undecidability and under-determination, to be discussed below. It is also a naughty reason. If you cannot decide, you cannot reject. The third reason is that there exist other less adventurous means to preserve a realist conception of quantum mechanics, such as the one of the veiled reality. This reason is naughty too because it is evidently too much subjective. The fourth reason is that the associated research program exhibits a regressive character. This is another naughty reason. The question is not necessarily to know whether the pilot wave theory is regressive when compared with quantum mechanics taken as a standard, but it could be to know whether quantum mechanics is not unduly hazardous with respect to the classical trio. The fifth reason is that the early Bohm's theory even did not succeed to keep on with its own spirit which, initially, would have been of an atomist nature. In agreement with this spirit, it starts with a view of the world made out from particles which are individual entities on their own, equipped with properties having a classical flavour, but has afterward to accept highly non classical features (such as non-locality and contextuality). This is also a naughty reason. Indeed, it happens very often that a researcher starts a work with some preconceptions and prejudices indicating a certain direction and that, eventually, the landscape reached does not look like the landscape expected. This is a matter of fact and a matter of life, not a basis for rejection. If we add to Bitbol list two more reasons, discussed above, and which are naughty too (lack of simplicity and incompleteness), we arrive to seven naughty reasons. But a set of naughty reasons should never, from a logical point of view, constitutes a good reason.

Duhem–Quine under-determination thesis

Undecidability

We did not find any good reason to definitively and convincingly enough, excepted for those who are already faithfully convinced, reject Bohm's pilot wave. And the summation of all more or less subjective insufficient reasons does not make any good final objective reason. Obviously, researchers involved in practical experiments and/or computations, well trained as they are in quantum mechanics, may very well feel poorly interested by such a debate. After all, quantum mechanics still go on working, and that is enough to go on walking with it. Very understandably, they possibly do not intend to take the risk of losing time with such a risky affair and, furthermore, you know: My career (I objectively need money to support my family and educate my children), my reputation (I do not want to become isolated from, or despised by my community of friends and colleagues) ... It might be that such a volcanic-hot topic should be reserved to young researchers, rich enough thanks to generous fairies having surrounded the cradle, or to well-established retired emeriti, or to ... crazy outsiders. But we need people going on dealing with the issue, because it is not closed.

And it is furthermore not closed in an embarrassing manner: We have no way to decide between quantum mechanics and pilot wave. We have to set a diagnosis of undecidability. It has been argued that the pilot wave is simply quantum mechanics recasted in another language and, if this were true, the issue of undecidability could be of little significance. We could say: Just use the language you have learnt, excepted if you are inquisitive enough to feel the desire to learn another language. But the fact is that, when you learn another language (for instance English when you are French, or more convincingly Japanese or Chinese when you are English), you really enter into another world, facing a quite different way of thinking. And, if this is true for natural languages, it is still true, or even more, when we compare quantum mechanics and pilot wave. These two theories offer different visions and interpretations of the world. The undecidability between both theories is therefore also an undecidability between two different visions of the world. This undecidability may also be expressed in the Popper's framework of objective knowledge because, let us recall it, both theories may be simultaneously corroborated or falsified.

This is a severe blow against realism, in some sense, that is to say against the very idea that science could pretend to really tell us something on the world around us. This might drive us toward relativism, a philosophical position that is, nowadays, praised by many philosophers. But we have to be careful and avoid to draw hastily erroneous conclusions. What our case study does imply is not an absolute relativism. It precisely tells us the following: In some cases , we may be facing some undecidable statements because they pertain to undecidable and even contradictory theories, which have however to be simultaneously accepted. It does not tell us that undecidability is the ultimate fate in *all cases.* In balanced terms, the consequence of any undecidability between two theories concerning realism is that we must abandon the naive scientist project to know the world in all aspects, although some of the aspects of the veiled reality could be unveiled.

The most significant undecidable issue concerning quantum mechanics and pilot wave is, for some people, the one of determinism, actually an issue which boosted the search for hidden variables. Indeed, if we cannot decide between quantum mechanics and pilot wave, we cannot decide between the intrinsic indeterminacy of quantum mechanics and the determinism postulated and afterward constructed by causal theories. Both Einstein (the opponent to quantum mechanics) and Born (the defender of quantum mechanics) even agreed on the undecidability between determinism and indeterminism. In a letter to Einstein, Born wrote: *You are absolutely right that an assertion about the possible future acceptance or rejection of determinism cannot be logically justified. For there can always be an interpretation which lies one layer deeper than the one we know (as your example of the kinetic theory as against the macroscopic theory shows).* Bohm used a similar argument and explicitly extended it to defend the undecidability under question. For him, starting from a deterministic (indeterministic) level of description, we can always imagine and construct a sub-level which would be indeterministic (deterministic), and this *ad infinitum.* For instance, determinism at a certain upper level may be the result of indeterministic, or possibly stochastic, processes at the next lower level, e.g. macroscopic determinism resulting from averages over quantum indeterminist processes or atomic classical stochastic processes, at a microscopic level. And indeterminism at a certain upper level may be the result of deterministic processes at the next lower level, e.g. quantum indeterminacy underlain by deterministic hidden variables. As stated by Bitbol [133], in a complementary way, phenomena which are strictly predictable, seemingly determined, may be underlain by statistical regularities emerging from a background of random events as well as by genuine deterministic laws. And, reciprocally, unpredictable phenomena, seemingly undetermined, may be underlain by deterministic chaotic processes as well as by genuine stochastic laws. In the words of Bohm himself [60], ... *one sees that the possibility of treating causal laws as statistical approximation to laws of chance is balanced by a corresponding possibility of treating laws of probability as statistical approximations to the effects of causal laws... The assumption that any particular kind of fluctuations are arbitrary and lawless relative to all possible contexts, like the similar assumption that there exists an absolute and final determinate law, is therefore evidently not capable of being based on any experimental or theoretical developments arising out of specific scientific problems, but it is instead a purely philosophical assumption.*

The above argument from Bohm may be invoked to settle down an ultimate undeterminacy between determinism and indeterminism, relying on the possible existence of a descent to deeper and deeper levels, and may look appealing. Although the argument from Bohm is correct, I however intuitively do not believe it is correct to use it in a descent process to call for an ultimate impossibility of closing the issue. It looks to me that any step from up to down, from an upper level to the next lower level, implies new formal constraints, and that the addition of such formal constraints, if we intend to pursue the down-process *ad infinitum*, should eventually block us somewhere. But I am not able to prove this point.

Inevitably, when we discuss the issue of undecidability, we have to say a few more words concerning Gödel. Indeed, when we say that we cannot decide between two theories, we also say that we cannot state on the truth or falsehood of either theory. This is obviously reminiscent of something similar demonstrated by Gödel in mathematics. More specifically, Gödel demonstrated that, given any finite set of axioms, it is possible to find meaningful questions that cannot be answered in the system generated by the axioms. Such questions are said to be undecidable. You may generate a larger set of axioms (for instance by going from the set of axioms generating arithmetic to the set of axioms generating the real line) and then possibly decide about the undecidable propositions, but you also simultaneously create new undecidables. On one hand, this has the disappointing consequence that one of the problems on the famous list of Hilbert, namely the "Entscheidungproblem" to know whether there exists an algorithm to decide on the truth of any proposition, was negatively answered. On the other hand, this has the exciting and vertiginous consequence that the world of mathematics cannot be exhaustively explored by the human mind. This is a new kind of infinity that Cantor himself, likely, did not dream of, as far as I know. Thanks to God and Gödel, there will always be more room for the imagination of mathematicians to play around.

We should certainly resist to the temptation to invoke Gödel whenever we see the faintest opportunity of doing so, but, if we believe that the physical world is some kind of incarnation of a mathematical structure, it may seem likely that the inexhaustibility of the mathematical world could be conveyed to the physical world. At least, it is not ridiculous to expect that we should have to face undecidability in physics too and to believe that, may be, the undecidability between quantum mechanics and pilot wave could be the manifestation of such a possibility. This idea is in agreement with Nicolescu [508] who remarked that the mathematics used by the physicists include arithmetic and therefore, logically, physical theories should be submitted to Gödel's theorem according to which a system of consistent axioms, rich enough, is necessarily open.

Interestingly enough, Bohm and Hiley [68] took advantage of the issue of undecidability to advocate a peaceful relationship between Bohm's interpretation and quantum mechanics. For them, and also for us at the present time, *there does not seem to be any valid reason ... to decide finally what would be the accepted interpretation,* and they asked: *Is there a valid reason why we need to make such a decision at all?* So, they went on, *would it not be better to keep all options open and to consider the meaning of the interpretations on its own merits, as well as in comparison with others? This implies that there should be a kind of dialogue between different interpretations rather than a struggle to establish the primacy of any of them.* If we observe that these quotations are contained in the last book of Bohm (with Hiley), published one year after his death, we can view the above statements as forming a kind of final testamentary auto-appraisal to add to the file of Bohm's drama. A similar opinion is shared by Ord [509], with however a different and interesting wording, when he stated that *the contrasting views of Bohr and Bohm may well be complementary.*

Allusions to or invocations of Gödel may also be found in Belinfante [1] or Peres and Zurek [510], and certainly in many other places. In particular, Peres and Zurek stated that the inability to completely describe the measurement process in quantum mechanics is not a flaw of the theory but a logical necessity and is analogous to Gödel's undecidability theorem.

The undecidability between quantum mechanics and Bohm's pilot wave also implies that we may use both of them, at our convenience. For Dias and Prata [511], there is a practical interest in each individual formulation: *... in some cases, it may be advantageous to perform a specific calculation in one formulation, rather than another.* Also, asking the question to know whether particles are all real, Goldstein *et al.* [512] discussed Bohm-type theories, always feasible they said, without any possibility to decide by relying on experiments.

The thesis tentatively stated

To avoid any confusion, I have now to state an important remark concerning the terminology: Duhem-Quine thesis, recurrently to be used from now on. There exists a Duhem-Quine thesis according to which it is impossible to test a single hypothesis inside a theoretical framework. Conversely, what can be tested is a theory as a whole, so that the word "holism" is sometimes used to denote the thesis. From now on, I shall use the terminology: Duhem–Quine thesis to denote another thesis, the one of under-determination of theories by experiments.

According to this thesis, theories are under-determined by experiments or, in other words, we may have two contradictory theories simultaneously agreeing with experiments (although Quine himself eventually hesitated about what would be a correct formulation of the thesis). In other more general words, given a certain physical world (our world to be specific), there exists a class of equivalent descriptions. Each one is true, in some sense, and the question to know which one is absolutely true is meaningless. Let us remark that this would have sound very strange to Kant. He would not have believed that the same categories applied to the same empirical data could produce two different systems, particularly if they were contradictory (see Léna Soler discussions in Ref. [117]). For Kant, this would have implied an intolerable undecidability originating from the mind himself, an inherent imperfection of his system of categories.

Nevertheless, Duhem and Quine had predecessors and precursors that are worthwhile to visit to gain a better understanding of the issue. I am pretty sure that a careful enough reading of my cherished pre-socratic thinkers would show that they were already aware of it, but I have to confess that I did not succeed to find any report on it (most likely because I did not read carefully enough). Concerning antiquity however, we might possibly, but very charitably, refer to Epicurus as a precursor of precursors when, in a letter to Pythoclès, he associated to the experimental fact of an eclipse of the sun or of the moon two possible theoretical explanations, namely (i) main extinction or (ii) interposition of other bodies [473]. But this is very charitable indeed because, as we shall see, we will be reluctant to consider these possible theoretical explanations as theories in a more dignified sense of the term.

Next, as far as I have been able to find, the first relevant historical notice is to be found in Osiander (laconically mentioned by Harré [118], but without referring to the Duhem–Quine thesis). Osiander has been in charge of the printing of the famous book by Copernicus named: *De Revolutionibus Orbium Coelestium*. In a rather contested famous preface [8], he wrote that "Copernicus hypotheses did not need to be true or even probable. It would be sufficient if they *saved the appearances*". Answering a letter from Copernicus, wondering whether he should publish his book, Osiander replied that "as far as he is concerned, he always believed that hypotheses are not articles of faith but starting points for computations, so that, even if they are false, it does not matter, if they simply exactly represent the phenomena". In another letter, he wrote that "Aristotelians and theologians will be easily pacified if they are told that several hypotheses can be used to explain the same apparent movements; and that the present hypotheses are not proposed because they are actually true, but because they are the most convenient to compute the apparent combined motions."

Later on, in the same context (geocentrism versus heliocentrism), but regarding the famous Galileo Galilei's affair, the Pope Urbain VIII stated that an hypothesis which gives results does not necessarily represent the reality, for there may be several ways to explain how the Lord produces phenomena. This statement from Urbain VIII is reflected in a statement from Simplicio, in a famous book by Galileo Galilei [34], concerning the ebb and flow of tides: ... God ... can give to water the alternated motion that we observe ... by using various ways, unthinkable to our mind. Let us however mention that this statement might have been motivated by a political and strategic desire to pacify the catholic hierarchy.

Afterward, we have to mention Descartes, quoted by both Duhem [513] and Van Fraassen [21]. In the third part of his principles of philosophy [112], Descartes indeed stated that what he wrote should be taken as an hypothesis, possibly far remote from truth. But, even in this case, Descartes believed he would have done much if only all the things which can be deduced from the causes he exhibited would entirely agree with the experiments. Furthermore, such a false hypothesis, agreeing with experiments, would be as useful to the life as a correct one, because it can be used in the same way than the correct one to produce the effects that we want to produce.

There are actually many places in the principles of philosophy of Descartes where such considerations are put forward. Let us simply mention another one, taken from the fourth part of the principles [35], in which Descartes wrote that, ... in the same way than an industrious watchmaker can produce two watches which indicate the same time, without any difference which can be perceived for the external eye ... it is certain that God possesses an infinity of various means, allowing him to make that all the things of this world look as they now look, without however any possibility for the human mind to know, among all the means it could use, which one he actually used. And Descartes reiterated the idea that, in some sense, it does not matter: ..and I would believe having done enough if the causes I explained are such that all the effects they can produce are like the ones that we see in the world, without informing myself whether it is by them or by others that they are produced. I even believe that it is equivalently useful for the life to know the causes we have imagined or to know the true ones.

Just after, Descartes, referring to Aristotle, also wrote that regarding the things which our senses do not perceive, it is sufficient to explain what they could be, though perhaps they are not as we describe them. For Van Fraassen [21], Descartes here *voiced a thoroughly empirical sentiment. Under a given interpretation, a physical theory describes how things are — one way they might indeed be — but the story is not unique, and for an empiricist it need not be unique.* The idea that we may have several stories is indeed in the spirit of the Duhem-Quine thesis. There is only a small ambiguity if we understand that things which our senses do not perceive drive us away from experimental data required by the Duhem-Quine thesis but, let us remark, an actual experiment in physics indeed involves many things that our senses do not perceive, if only the internal functioning of our computer used to record the data. Having therefore quoted the French Descartes, it is fair enough to also refer to the British (Scottish) Hume who, with a critique of induction, was to conclude that empirical evidence cannot determine the choice of a theory.

In most cases, the history of the Duhem–Quine thesis, as has been reported up to now, heavily relies on the invocation of the Supreme Being, namely God. I believe that this is very far from being stupid, even from a logical point of view. Indeed, we may at least translate the Divine Argument in other words, closer to the framework of a more modern and familiar theory of knowledge as follows. If, for the theologian, God knows everything and the creatures only a small amount of the things, we can nowadays, in a naturalist and immanent way of thinking, still argue that the human mind is too fragile to completely grasp on the mysteries of the universe. Then, one explanation could be enough if it is good, but why should we be arrogant to believe that, when we possess one good explanation, there would not exist other good explanations?

The process from a theological understanding of the Duhem-Quine thesis to its secular interpretation is already achieved with Pascal. In a letter to the Révérend Père Noël [514], he wrote: "In the same way than a same cause may produce several different effects, a same effect can be produced by several different causes. Therefore, when one discusses, with our human abilities, of the motion and of the stability of the Earth, all phenomena of movements and retrogradations of the planets may perfectly be derived from the hypotheses of Ptolemy, of Tycho, of Copernicus and of many others, although from all of them only one can be correct".

With Duhem, and his famous book on the structure of physical theories [513], we arrive to one of the two thinkers who gave his name to the thesis under discussion. For a correct appraisal of the work of Duhem, we need to know that the date of death of Duhem is: 1916, and that his book was first published in 1906. Therefore, Duhem could have been aware of the physical structure of Newton's mechanics (and of various formulations of it) concerning particles, and of the physical structure of the electromagnetic theory of Maxwell, which has been modernized by Hertz, concerning fields, but it could not forecast the huge developments of relativity and quantum mechanics. When reading Duhem, it is extraordinarily obvious that we are facing a classical mind formed by classical physics. This is not to express any disrespect, but just to mention a fact of life and of contingency.

Duhem expressed the issue under discussion, in general terms, by comparing, and opposing, the French and the English styles of making physics. As he stated, for geometricians, like Laplace or Ampère (on the French side, but forgetting Descartes, a theological precursor of Duhem), it would be absurd to give to a law of physics two different theoretical explanations, and to express the idea that such two different explanations can be simultaneously valid. But, in contrast, for scientists building models, like Thomson or Maxwell (on the English side), there is no contradiction in the fact that a given law could be apprehended by two different models.

Lifting the issue to still more general terms, beyond a plain comparison between different national ways of thinking and working, Duhem raised the question to know whether it is allowed to symbolize several distinct groups of experimental laws, or even a single group of laws, by using several theories, each one relying on hypotheses which cannot be reconciled with the hypotheses underlying the other ones. For Duhem, the answer is that, *if we restrain ourselves to the use of pure logical reasons* (emphasized because Duhem himself emphasized), then one cannot prevent a physicist to use several irreconcilable theories to represent various sets of laws, or even a single set of laws. And, he concluded, one cannot condemn incoherence in the physical theory. This is a point of view which agrees with the opinion of Poincaré, quoted by Duhem, according to which two contradictory theories, if we do not combine them, and if we do not search with them to do things thoroughly, can both of them be useful tools for research.

These observations are likely sufficient to justify that the name of Duhem can be attached to the thesis under discussion. But it is not clear, in his famous book, whether Duhem made a significant distinction between an experimental law (like what is called a correlation in fluid mechanics, or in other disciplines), a model (with all its heuristic virtues), and a theory (with all its internal consistency and demands). This is particularly obvious when he leaves his lucid general considerations about the issue of undecidability to go down to examples. I am then hesitating, beyond the written words, to decide what Duhem really had in mind and what he had in mind looks to me somewhat and sometimes ambiguous.

For instance, Duhem also made the general statement that an infinity of different theoretical facts can be taken as a translation of the same practical fact. This general statement, to which we may agree, is illustrated by a disappointing *too cheap* example. Here it is.

We may express several contradictory theoretical facts such as: The length of a line is 1 cm, or 0.999 cm, or 0.993 cm, or 1.002 cm, or 1.003 cm. All these theoretical facts may agree with only one experimental fact, if we simply account for measurement uncertainties. Hence, such theoretical facts are under-determined by experiments.

The example taken here by Duhem has nothing to do with a formally and consistently structured theory such as relativity or quantum mechanics (that Duhem could not know) or Newtonian mechanics and Maxwell's electromagnetism (that Duhem did know). It concerns a rather trivial example with theoretical facts which are just atomic propositions and cannot be assimilated to proper theories. Other examples are available from Duhem's book, all related to the existence of measurement uncertainties, and disappointingly illustrating a Duhem–Quine thesis.

We may better approach the concept of theory by discussing the concept of physical laws derived from experiments. Even if you assume that there is no measurement uncertainties, you may have recorded many experimental data concerning the variation of one quantity Q versus another one R. Since the number of measurements made is countable, you are then facing a finite set of points on your sheet, and you need to deduce from this the physical law $Q(R)$. This is an interpolation problem and, logically, there are an infinite number of physical laws which are in agreement with the experimental data. Of course, the researcher will try to build a simple law $Q(R)$, from the data, using a straight line, or a parabola, or an exponential or whatsoever. But such an invocation of simplicity (recall Occam razor) is may be a methodology or parsimony invocation, but it is not of a logical nature. According to Duhem, any physical law is an approximate law and therefore, for a strict logician, it cannot be either true or false. Any other law which represents the same experiments with the same approximation may pretend, as well as the first one, to the name of true law or more exactly to the name of acceptable law. So was speaking Duhem. Hence, physical laws are under-determined by experiments.

But a physical law is not yet a theory. To do better justice to Duhem, it is necessary however to mention that he actually tackled the problem of under-determination of theories, although in an unsatisfactory way. For this, he compared Newton's theory and other hypothetical (not built) theories of gravitation (general relativity was not yet available). The predictions of these different theories, he observed, could become significantly different after ten millions of years. But, can we know what will actually happen after ten millions of years? So no theory among the different theories could pretend to be the best one, because they are practically equivalent. This is better viewed as concerning the issue of Popper's falsification, although we may accept this example, but reluctantly, to illustrate the fact that, hence, theories are under-determined by experiments. However, in the present case, they are under-determined by *available* experiments, but it is not a case of contradictory theories making the same experimental predictions because experimental predictions are indeed different after ten millions of years. Another example, but no more convincing for similar reasons, concerns the possibility to construct (something easy, according to Duhem) a new thermodynamics which would predict that the entropy of universe must increase, and then decrease, and so on, alternatively, by periods of hundred millions of years.

So, up to now, we possess some general formulations of the Duhem–Quine thesis but without any convincing concrete example. Could possibly Quine himself be more convincing than Duhem? The reader might refer to Word and Object, regarded as the most important book from Quine [515] or to the Pursuit of Truth [516], his last book. The French reading reader may also find a synthetic account of the work of Quine in Ref. [517]. But a specific and specialized report is available in a famous paper by Quine [518] to which the reader might prefer to refer.

Quine's thesis of under-determination of theory by experiments is related to his other famous thesis concerning the impossibility of a radical translation. Assume that you are learning a foreign language from nothing, without in particular any dictionary nor teacher, just by sharing the life of a native individual. Suddenly, you see a white rabbit jumping across the trees, and the native pronounces the word "gavagai". What does "gavagai" means? Does it means an object (the rabbit), a property (being white), an action (jumping), or something else like the fact of perceiving something on the right moving to the left? It is impossible to affirm with certainty that the equation *gavagai = rabbit* is the correct one. Indeed, we have several contradictory theoretical choices like *gavagai = rabbit, gavagai = white* or *gavagai = jumping*, for an unique experimental fact: Gavagai. For Quine, the situation is similar with theories, in particular with physical theories. The thesis of under-determination of theories by experiments says that experimental facts do not determine an unique theory for describing the reality. Furthermore, the truth depends immanently on the conceptual scheme of our language and on the entities that this language allows us to manipulate (this insistence on the fact that we have to use a language, with its inherent limitations, is also typical of Wittgenstein [517, 519]).

In his famous paper, Quine [518] stated what could be the essence of the thesis as follows: *If all observables can be accounted for in one comprehensive scientific theory — one system of the world, to echo Duhem's echo of Newton- then we may expect that they can all be accounted for equally in another, conflicting system of the world.* Such an expectation may rely on the examination of *how scientists work. For they do not rest with mere inductive generalizations of their observations: Mere extrapolation to observable events from similar observed events. Scientists invent hypotheses that talk of things beyond the reach of observation. These hypotheses are related to observation only by a kind of one-way implication, namely, the events we observe are what a belief in the hypotheses would have led us to expect. These observable consequences of the hypotheses do not, conversely, imply the hypotheses.* In the words of Quine, this defines a doctrine rather than a thesis, a doctrine from which we can say that it *is plausible insofar as it is intelligible, but it is less readily intelligible that it may seem.* The doctrine is afterward exposed in general terms (again without any convincing example) which alone do not allow one to be clearly convinced, something which is implicitly acknowledged by Quine when, instead of using the word "truthfulness", he is content of using the word "plausibility". Redefining more carefully the thesis, Quine arrived to the following formulation: *Under-determination says that for any one theory formulation there is another that is empirically equivalent to it but logically incompatible with it, and cannot be rendered logically equivalent to it by any reconstrual of predicates.* Whether the thesis, under this form, is true or not, this remains for Quine *an open question.* However Quine, as he said, believes to it. A mathematician would roar in face of such an incongruous object: An "objective" thesis which cannot be demonstrated but must be a matter of faith. At this stage, we might certainly agree with Harré [118] when he spoke of the *myth of the underdetermination of theories by data.*

One of the most difficult things to correctly state the Duhem–Quine thesis is to precisely define what is a theory. Consider for instance the old theory of Bohr correctly explaining the energy levels of the hydrogen atom. The energy levels predicted by Bohr are in perfect agreement with the energy levels predicted by Schrödinger from his equation. Should we consider that we have here a simple example of the application of Duhem–Quine thesis? Certainly not, and this for two reasons. First, Bohr's theory is more a model than a theory and Bohr himself was aware of the fact. Second, all predictions of Bohr's model do not agree with all predictions of Schrödinger's wave mechanics. Therefore, we here have two theories (let us accept the name of theory) which are not experimentally equivalent in all aspects. But, unfortunately, there is a loophole to these arguments. We necessarily have to attach a domain of validity to any theory (this will not be true if one day we possess *the ultimate theory* encompassing all domains of validity). If we affirm, by a *fiat*, that we are just interested with energy levels, therefore defining in this way a domain of validity, then we could take Bohr–Schrödinger as a decent example of the Duhem–Quine thesis, something that we are however intuitively very reluctant to do. Therefore, the problem is not only to define exactly what can be named a theory, but also what can be named a domain of validity. There is indeed a quasi-continuum of theoretical constructions, from a simple experimental law *à la Duhem,* to a consistent theory like relativity, and there is also a quasi-continuum of domains of validity, from a certain domain, like the quantum mechanical domain, to sub-domains, and sub-sub-domains, like the one concerning only the predictions of energy levels of the hydrogen atom.

A pertinent analysis of Quine's anthropology is available from Laugier-Rabaté [520]. According to this author, it appears that Quine did not provide convincing examples for his thesis of the under-determination of the radical translation. And, as with Duhem, the lack of convincing examples is also characterizing the thesis of the under-determination of theories by experiments. Quine himself would certainly agree with this statement if we refer to the mood expressed in his Erkentniss-paper [518].

Another author which is often quoted as relevant to the issue of the Duhem-Quine thesis is Van Fraassen, in particular in reference to his book *Laws and Symmetries* [57]. Van Fraassen distinguished between theoretical equivalence and empirical equivalence. For him, if two statements are logically (theoretically) equivalent, then they are saying the same thing. However, if we consider a theory as a kind of statement, say a meta-statement, we have at least one example to demonstrate that the affirmation of Van Fraassen is erroneous. This example concerns the Newton's and Hamilton–Jacobi's formulations of classical mechanics. Both theories may be viewed as logically, theoretically, equivalent by the simple fact that they are mathematically equivalent. But they are not saying the same thing. Contrarily, they are proposing two drastically different visions of the world (two ontologies), one with local trajectories, the other with local trajectories embedded into a non-local field.

The status of empirical equivalence alone is however simpler to discuss without any flaw. We just have to admit the possibility of theories saying the same thing, not on the nature of the world, but on the predictions of experimental facts. But the above example of Newton's and Hamilton–Jacobi's formulations demonstrate that two theories which are logically equivalent, and saying the same thing as far as empirical facts are concerned, may be deeply contradictory if we interpret them in a larger framework, a framework in which we are concerned with the visions of the world they provide. I must insist however on the fact that Van Fraassen did not refer to this example taken from classical mechanics. Actually, he discussed the issue in general terms without, never (as far as I know), referring to explicit examples.

In the same spirit, Van Fraassen believed (as everyone he said!) that there exist numerous theories, may be not yet formulated, but which are in agreement with the experimental facts which are up to now known, and these theories are at least as explanatory as the best theories that we possess nowadays. However, he added, the big majority of these theories must be erroneous because they may disagree, in many different manners, with experimental facts not yet known. He concluded that our best theory is very unlikely to be correct. I would like to express this idea with a probabilistic flavour: If there is only one truth, we are sure to be in error.

Elsewhere, Van Fraassen [21] discussed the lack of unicity asserted by the Duhem-Quine thesis in association with the existence of different interpretations of quantum mechanics. As he wrote, *why then be interested in interpretation at all? If we are not interested in the metaphysical question of what the world is really like, what need is there to look into these issues? Well, we should still be interested in the question of how the world could be the way quantum mechanics — in its metaphysical vagueness but empirical audacity — says it is. That is the real question of understanding. To understand a scientific theory, we need to see how the world could be the way that the theory says it is. An interpretation tells us that.* After this quotation concerning the meaning of the word understanding, the main issue of the previous chapter, Van Fraassen immediately went on with something else, logically connected, but relevant to the present chapter. But, as he wrote, *the answer is not unique, because the question "How could the world be the way the theory says it is?" is not the sort of question to call for a unique answer.*

We may refine our discussion by considering two versions of the Duhem–Quine thesis, the weak version, and the strong version. In the weak version, there might exist two (or several) contradictory theories explaining the experimental facts known up to now (or a set of experimental facts known up to now but restricted to a certain well defined domain of validity, for instance the domain of validity of classical mechanics with $v/c \to 0$ and $\hbar \to 0$, or the domain of validity of non relativistic quantum mechanics with $v/c \to 0$). In the strong version, there might exist two (or several) contradictory theories explaining all experimental facts, the ones we know, and the ones we do not know, encompassing all domains of validity. To tell it frankly, I do not believe to the strong version (not only one truth, but several of them), although obviously without any proof. But we should not bother too much about this strong version which is definitely outside of any human possibility of falsification, if only because experimental facts are in practice (and may be also in principle if we are allowed to apply Gödel to physics) inexhaustible.

Let us therefore focus on the weak version. But even for this weak version, the philosophical literature lacks of convincing examples which are vital not only for making the skeptical one convinced, but for a proper understanding of the philosophical thesis to which the names of Duhem and Quine have been attached. The aim of the next subsection is to provide such examples.

Before, I shall end this subsection by giving the point of view of Cushing dated 1992 [48]: *Very loosely, the formalism refers to the equations and calculational rules that prove empirically adequate (i.e. getting the numbers right) and the interpretation refers to the accompanying representation the theory gives us about the physical universe (i.e. the picture story that goes with the equations of what our theory "really" tells us about the world. Since a (successful) formalism does not uniquely determine its interpretation, there may be two radically different interpretations (and ontologies) corresponding equally well to one adequate formalism. This can be taken as an instantiation of the Duhem–Quine thesis of undetermination of theory by an empirical base. Even if one wants to restrict (and, arguably, that would be a mistake) the Duhem–Quine thesis to different formalisms each handling equally well a given body of empirical information, there nevertheless remains the interesting and important point of opposing ontologies equally well supported by a common empirical base.*

The thesis actually exemplified

The possibility of an under-determination of theories by experiments is usually much disliked by physicists (when they are aware of it!), and the vagueness which remains attached to this concept after the somewhat fuzzy efforts of philosophers does not help. Squires [13] expressed this point of view with a bit of humor or even irony by saying that *the statement that two theories, both of which fit the data, are equally good can be seen to be unreasonable if we note that a theory in which the sun always turns into cream cheese as it disappears over the horizon, and turns back again later, gives a perfectly adequate account of my observation.* To be convinced, the physicist demands convincing examples. I am going to provide some of them, concerning the weak version of the Duhem–Quine thesis, restricted to specific domains of validity.

The first example concerns classical electromagnetism, and particularly light scattering theory. In this field of research, the most famous theory is likely to be a theory published by Mie in 1908 [487]. This theory describes the quasi-elastic interaction (no change of frequency except that one due to the Doppler effect, and the other singular one from a finite frequency to a "null" frequency due to absorption) between an illuminating electromagnetic plane wave and a homogeneous non-magnetic sphere defined by its diameter and its complex refractive index, embedded in a non-absorbing material. It allows one to calculate scattered fields outside of the sphere, internal fields, phase relations, various cross-sections (for scattering, absorption, and extinction), and radiation pressure forces. The theory is built by using Maxwell's electromagnetism.

However, there is a bit of injustice in granting Mie only for this theory. Actually somewhat twenty years before, a similar theory had been produced by Lorenz [485, 486], with however an important memoir lost. Hence, rather than speaking of Mie theory, I always preferred to give the name of Lorenz-Mie to the theory. Several historical papers are available concerning the formulation of Lorenz, and its relationship with the one of Mie [521–524]. I master very well Mie's version but did not deeply go into Lorenz's version but, according to the previous references, a most important thing is that they lead to the same results, that is to say, they are empirically equivalent. The work of Lorenz has been overlooked, certainly in part because it has been written in Danish. But, more relevant to our subject, it did not rely on Maxwell's equations. Indeed, it did rely on the old theory of ether that, after Einstein, we would not like to consider any more. From this, I conclude that the whole of the classical radiation theory could be exposed either following Maxwell or using the ether. When I heard of this for the first time, I felt much surprised and impressed. Now, I consider this story as a beautiful example of the Duhem-Quine thesis because Maxwell's electromagnetic theory and the electromagnetic theory of ether are to be viewed as contradictory: They do not tell the same thing on the world. Not saying the same thing on the world means that they do not share the same ontology, i.e. furniture of the world (electromagnetic fields with Maxwell, mechanical ether with Lorenz). Such a situation (conflicting because proposing different ontologies, yet being empirically equivalent) defines what I have called an ontological under-determination, a variant of Duhem–Quine thesis or of its terminology [525, 526].

If we may have two contradictory theories, empirically equivalent, to explain the classical behavior of light and other electromagnetic radiation (when no quantum jumps are involved), we may also have two contradictory theories, empirically equivalent, to explain the classical behavior of particles. We already well know a corresponding example: Just go back again to Newton's and Hamilton–Jacobi's formulations of classical mechanics (and forget that we used NSP to discriminate between them). We also commented on this example (the second example) above, when discussing Van Fraassen (although Van Fraassen did not refer to it). In this case, the conflict between ontologies is as follows: Particles following trajectories with Newton, and particles following trajectories dressed by fields with Hamilton–Jacobi.

And, finally, let us now consider something that we also already know: Just consider usual quantum mechanics (in the non relativistic domain) and Bohm's pilot wave. If the Duhem–Quine thesis is correct, and we now know that it is correct, in some sense, from the previous two examples, then the equivalence between quantum mechanics and pilot wave would possibly simply provides us with a third example. In this case, the conflict between ontologies is as follows: Underlying particles following trajectories, dressed by a quantum field Ψ, with the pilot wave, and the quantum field Ψ alone with quantum mechanics.

These three examples have something particular which must be emphasized : All of them assume theories expressing what we call laws of nature. But we may also pretend that there is no law of nature. Indeed, following Popper [25], let us assume that we could examine the universe on a very long period of time, much longer that the period we can observe (if only because we have no real access to the pre-Big-Bang age nor to the "end" of the future). If we assume that the world is governed by randomness rather than by laws, we can imagine that we are facing an era in which the law of gravitation seems to be experimentally valid, although actually it is not, but merely the result of probabilistic chances. This kind of explanation may be applied to any regularity we observe in our laboratories and in the world. That is to say, we may explain in this way all our so-called laws of nature as being valid for a long period of time, embedded in the infinity of the time ruled out by accidental chaos. Such a speculation is strictly metaphysical, without any interest for the scientist, and presumably for the human kind. But it shows that the choice between two theories, one telling us that there are laws of nature, and the other telling us that there are no law of nature, is under-determined by experiments.

Harré [118] provided another point of view on this issue by referring to the famous problem of induction (irrevocably attached to the name of Hume). For him, *it has also been pointed out that the methodological principle "Reject that which has been refuted" can be used categorically only under the inductive assumption that the world will not change so that what was not true today is true to-morrow and for evermore. This version of the principle of the uniformity of nature is no more acceptable than the traditional uniformitarian assumption that would support positive inductive reasoning from the truth of a prediction to the acceptance of the hypothesis from which it was derived.*

There is also a softer version of this example according to which the laws of nature we have in hands and in minds (this assuming that we have accepted the idea that there are laws of nature) are not reliable but only the consequences of the fact that we can only observe the world for a ridiculously small, or even vanishingly small, period of time. The recent discovery of the unexpected accelerating expansion of the universe gives some credence to the softer version.

The relevance of the Duhem-Quine thesis for a discussion of Bohm's pilot wave has been noticed and discussed by Cushing [131, 166]. His point of view is that [166] *one formalism, with two different interpretations, counts as two different theories* and that [131] *the physical interpretation refers to what the theory tells us about the underlying structure of ... phenomena, i.e. the corresponding story about the furniture of the world.* Let us here remember that a "furniture of the world" is an "ontology". Therefore, quantum mechanics and Bohm's pilot wave, although empirically equivalent (implying that both of them *should* be accepted by positivists, in contrast with their usual attitude), provided two different ontologies. Bohm himself was aware of the issue at least implicitly [60]. Espagnat [85], without however explicitly referring to the Duhem-Quine thesis, mentions three mathematically and empirically equivalent theories, namely : The theory of Dirac sea, Feynman diagrams and quantum field theory, which may be or not considered as a valid example, as for the Bohr/Schrödinger example mentioned previously, depending whether we are lax or not in the interpretation of the thesis. For Bitbol [133], referring to Quine, scientists invent hypotheses which are out of reach of observational possibility. But hypotheses are linked to observation only in an one-way manner, without any reciprocal. The observable consequences of the hypotheses do not imply the hypotheses (a nice way to state Duhem–Quine thesis).

Also, when discussing contingency, we expressed the possibility of the frightening idea that quantum mechanics and general relativity could not be unified because, possibly, they are not the right starting points for such an unification. This possibility may be served again by referring to the Duhem–Quine thesis. Let us denote the usual quantum mechanics as QM and imagine that there exists another Duhem–Quine equivalent theory denoted as QMDQ (Bohm's pilot wave, or another one still unknown). Similarly, let us denote the general relativity as GR and imagine that there exists another Duhem-Quine equivalent theory denoted as GRDQ. Imagine that QM and GR are, in principle, correct starting points for an unification (this is not guaranteed), then due to the Duhem-Quine equivalence between QM and QMDQ, and between GR and GRDQ, we could transitively believe that QMDQ and GRDQ, or even QM and GRDQ, QMDQ and GR, also form correct starting couples for the unification. But we are facing couples of theories which are not conceptually equivalent and even conceptually contradictory, even if they are logically consistent, mathematically and empirically equivalent. If the way to the unification requires us to rely on concepts, in order to guide our intuition, it may be that the concepts underlying QM and QR are not the right ones to consider. Therefore, even if QM and QR are, in principle, correct starting points, it could not be the case in practice. Furthermore, the situation could be worst because it remains also the possibility, due to contingency, in particular due to experimental facts that we still, may be contingently, ignore, that QM and QR are not correct starting points, even in principle.

After this digression, returning to the main stream of the book, we may ask : So, are we done? Should we now proceed simply to the conclusion of this book, just saying that we cannot in utmost rigor choose between quantum mechanics and Bohm's pilot wave, and that this undecidability is a manifestation of the no any more surprising Duhem–Quine thesis? Not yet, we still may use ampliative arguments.

Ampliative arguments

According to Harré [118], *a theory is plausible if it is both empirically adequate and framed within the constraints of the current communally approved ontology*. Discussing more extensively the concept of plausibility (and the one of implausibility), Harré exposed five conditions for plausibility. Rather than following Harré, I shall simplify my exposition by referring to the rewording of Van Fraassen [57]. Following Van Fraassen, a theory is definitely acceptable for Harré if it satisfies two conditions (1) it must be in agreement with empirical facts (2) it must be plausible. In the mind of Harré, plausibility means that the theory must imply mechanisms and entities pertaining to the unique hierarchy of ontological types underlying the history of the scientific enterprise. I am basically accepting these two criteria but for the fact that I shall discuss the plausibility demand in a way different from the one of Harré and Van Fraassen. Actually the explanation of the word "plausibility" as given by Harré and Van Fraassen does not really satisfy me. As stated and as I understand it, it looks to me too much conservative insofar as it refers to an approved ontology or to an hierarchy of ontological types related to the past of the history of sciences. For instance, in the actual enterprise of sciences, new entities may have to be built and old entities may have to be destroyed. Such new entities do not pertain to ontological types already used previously in the history of sciences. Old ontologies may have to be destroyed and new ontologies may have to be built.

The first demand of Harré, in Harré's list (reworded by Van Fraassen), concerning the agreement with empirical facts, is the same than the first demand of Belinfante, in the Belinfante's list that has been previously exposed in this chapter when we discuss the issue of lack of simplicity. We can then decompose the second demand of Harré into sub-demands. The first sub-demand (denote it as 2a), is taken as the second demand of Belinfante, the one on logical consistency. It is indeed plausible (even more than plausible!) that a satisfactory theory must be logically consistent. The third demand of Belinfante on simplicity may be used as a heuristic guide for the search of a satisfactory theory but, because it is insecure, as we have seen, we should reject it as a demand. With the demands we have retained up to now, namely demand 1 for agreement with experimental facts and demand 2a for logical consistency, we have enough to state a possible formulation of the Duhem–Quine thesis: Two theories which are logically consistent, and observationally equivalent, may be contradictory. We may now introduce a second sub-demand in the plausibility criterion of Harré, let us call it demand 2b. The demand 2b states that, besides being logically consistent and making predictions agreeing with experiments, a satisfactory theory must also satisfy other demands. We shall soon discuss what could be such demands but, for the time being, let us call them ampliative arguments. The word "ampliative" means that these arguments enlarge our possibilities of choosing between several theories, the enlargement being made with respect to the Duhem–Quine thesis as it has been just stated above. In other words, ampliative arguments allow us to decide between undecidables. Therefore, we may now build a new list of criteria as follows. A theory is satisfactory if (1) it agrees with experimental facts (2) it is logically consistent (3) it satisfies ampliative arguments.

I am not taking the risk of listing what could be acceptable ampliative arguments (this might be postponed to future episte-mological researches). But some thinkers are ready to accept the invocation of simplicity (not to be accepted, as we have seen, because we know too much counter-examples), or of beauty (too much subjective, and insecure), or even the acknowledgment of societal influences. More prudently, I shall be content in using one very specific ampliative argument, as an example, but also as our tool to proceed further, beyond what was allowed to us by the mere use of the Duhem–Quine thesis. For this, let us return once again to Newton's and Hamilton–Jacobi's formulations of classical mechanics, which are Duhem–Quine equivalent. But, using a first principle, the non-singularity principle (NSP), we have been able to discriminate between both theories. We have been able to understand that Newtonian trajectories of matter points do not exist, and find a way, relying on the lifting principle, from S to Ψ. Invoking NSP is an ampliative argument.

It is interesting to remark that Einstein had something to tell us concerning the Duhem–Quine thesis (without referring to it) and ampliative arguments (without using this terminology). Indeed, in an address he delivered in 1918 before the Physical Society of Berlin on the occasion of Planck sixtieth birthday, he expressed himself as follows [48]: *There is no logical paths in these laws; only intuition, resting on sympathetic understanding of experience, can reach them. In this methodological uncertainty, one might suppose that there were any number of possible systems of theoretical physics all equally well justified; and this opinion is no doubt correct, theoretically. But the development of physics has shown that at any given moment, out of all conceivable constructions, a single one has always proved itself decidedly superior to all the rest. Nobody who has really gone deeply into the matter will deny that in practice the world of phenomena uniquely determines the theoretical system, in spite of the fact that there is no logical bridge between phenomena and their theoretical principles.*

Furthermore, let us remark that, if we return to the (strong) metaphysical version of the Duhem–Quine thesis, we may discriminate between the two theories involved (whether there are laws of nature or not), by using an ampliative argument. We may believe, us, scientists, that there are laws of nature, although others, some philosophers, do not believe in them. But faith is a dangerous ampliative argument. We may instead take the decision that there are laws of nature [25], not as a matter of faith, not as the result of free will, but as a methodological rule. If you do not like this decision, better stop making science.

Coup de grâce

You know how a novel is written. You have to expose the characters, and this takes a bit of time. You have to tell the story, and this requires a bit more of time. And you really know the end, only at the very end. This is particularly true for detective novels, when you understand everything (if you are lucky enough), although the name of the murderer is only known at the last page. It is similar for a detective film when the end extends only during the five last minutes. So will be the case for this book. Furthermore, giving a coup de grâce is fast. After the coup de grâce, we shall however make a few comments (and provide a loophole). You know, after such a terrible thing like a coup de grâce, comments are required.

For the coup de grâce, we just need to remark that the same ampliative argument (NSP) which has been used to discriminate between Newton's and Hamilton-Jacobi's formulations may be used to discriminate between Bohm's pilot wave and quantum mechanics. If objective, deterministic, Newtonian trajectories of matter points (and matter points themselves) do not exist, it is a non sense to introduce them in quantum mechanics as done by Bohm. The attempt to propel the classical concepts of matter points and matter points trajectories in the quantum domain leads definitively to a failure. The coup de grâce is given. More now with comments.

First, the issue of contingency, as discussed by Cushing is, and remains, an important issue relevant to the history of sciences. But, in the particular case we are studying, it is not relevant any more. The same thing can be said from other kinds of influences, even the irrational ones, which influenced or could have influenced the story of the causal theories. If the pilot wave had been successful at the Solvay Congress of 1927, and attracted a huge army of researchers, even if it became the dominant stream, even if Bohm, having driven it to its final logical conclusion, would have been largely accepted, eventually, it would have been defeated.

Next, I do not mean that Bohm's pilot wave is not an useful model. Indeed it can be much useful as is still very useful the classical Newtonian mechanics of matter points. In particular, the fact that singularities in the behavior of Newtonian trajectories are rare, occurring only occasionally, demonstrate that the concept of trajectory still remains useful for many practical purposes, in the same way that optical ray computing and tracing will forever remain invaluable tools. Trajectories are then useful idealizations, and we would make fool of ourselves if we absolutely and definitely repudiate the concept of trajectory and its so convenient use: We just need to keep in mind that it possesses a finite, although significant, range of validity. In the same mood, but more prosaically, we shall go on saying that the sun rises, as Aristotelians would say, without any practical consequence: This reflects a very convenient "model" to speak of everyday life, in the same way that trajectories provide a very convenient "model" to speak of mechanics. But, NSP tells us that a perfect physics should avoid singularities. Obviously, this does not mean that we shall be able, one day, to build a perfect physics.

Therefore, we should not be arrogant, nor unduly optimistic: For a very long time, may be forever, we shall have to manage with "models" and idealizations, being unable to reach the gist of the things. But, some idealizations may be better than others and, if we need to dig deeper into the mysteries of the world, it is definitely of good advice, and even compulsory, to abandon the concept of trajectories because, already at the classical level, it has been identified as faulty. The idea, expressed by Bohm in its auto-appraisals, and by other authors, that the pilot wave could provide complementary insights to a better, more thorough, and deeper, understanding of quantum mechanics, then appears as erroneous and even dangerous, for there could not be complementary insights in erroneous ideas, just only an opportunity to spoil the clarity of the mind. Any theoretician in quantum field theory, for whom the description of reality in terms of particles being permanent entities with fixed numbers is more than naive, would agree with this statement.

Classical Newtonian trajectories of extended objects are still to be accepted too, as a convenient model, but such extended objects, rather than being considered as a collection of matter points, in the Newton's style, must ultimately emerge from a more fundamental description of nature, e.g. from the underlying quantum mechanical level. Hence, the rejection of matter points, and of their trajectories, does not offend our everyday intuition.

In a previous subsection on undecidability, we referred to Dias and Prata [511] stating that, after all, we can use Bohm's pilot wave or quantum mechanics, at our convenience, for computational purposes. However, there is something I have to add to this. Dias and Prata, in the same reference, also expressed the idea that we might use one theory or the other for conceptual purposes because they, both of them, provide insights in the interpretation of quantum mechanics. This idea is well founded if we stop thinking at the level of undecidability or at the one of the Duhem–Quine thesis. But, if we use our ampliative argument for discrimination, this idea is no more acceptable. Indeed, much more advances in physics are required and, if we accept, on the same footing, contradictory conceptual solutions which can actually be discriminated, then we are spoiling our mind with inadmissible lines of research, which might prevent us to progress forward to a completely satisfactory solution. We just accept possibilities which are misleading, erroneous, and flawed from the very beginning.

The ampliative argument we have used against Bohm's theory simultaneously works for the double solution and for stochastic quantum mechanics. The fundamental intuition of Louis de Broglie that both waves and particles must be preserved in the new mechanics, and that the aim of wave mechanics is to achieve a synthesis between the dynamics of matter points and the Fresnel-like theory of waves [62] has died, and all his effort to implement this intuition in a healthy-defined formalism can be viewed as a courageous but desperate rearguard action. Also, incidentally, the original proposition of Louis de Broglie, that a corpuscle in the double solution would be represented by a mathematical singularity becoming infinite at the location of the corpuscle, however afterward corrected by Louis de Broglie, is in immediate conflict with NSP. Concerning now stochastic quantum mechanics, let us note that the idea of a sub-quantum mechanical level is not rejected (there is indeed such a level taking the form of the quantum vacuum, a quantum ether). What is rejected is any explanation of this level in terms of Newtonian trajectories.

The ampliative argument therefore works for all physical hidden-variables theories that we have been able to identify. In particular, a basic germ for all causal theories, since Louis de Broglie, has been to introduce trajectories orthogonal to the equiphase surfaces of Ψ in the same way that, in the Hamilton–Jacobi's formulation, trajectories are orthogonal to the equivalue surfaces of S. Since the classical trajectories had vanished in the blue, the basic germ has been destroyed, and the whole causal enterprise, as we can see now, was of an infinite fragility.

Furthermore, the ampliative argument is not a downstream argument, to be used once these theories are completed, and we therefore have not to worry whether they are indeed completed or not, or whether they can be completed or not. We have not to invoke the many successes of Bohm's theory, such as its solution to the measurement problem which is solved in a way that a realist oriented mind would likely like, to claim that the theory is a good one. Similarly, we have not any more to go on discussing the naughty arguments used to reject Bohm's theory, or other existing physical-PHV theories, to claim that they are bad ones. Indeed the ampliative argument is an upstream argument which dries up the flow at the source. It is indeed much more comfortable to possess an upstream final objection, which dries the flow at the source, than having permanently to fight downstream with non convincing objections, facing a growing flow flooding all objections and counter-objections. Causal theories were like the Lernean Hydra with cut heads growing again. An upstream argument cuts all the heads in one stroke. And this upstream argument just tells us that everything was basically flawed from the very beginning.

Furthermore, artificial hidden-variables theories, because they are not physical, have been rejected. They have been useful, and others may be useful too, to gain a better understanding of the world if only by motivating beautiful experiments in quantum mechanics. But, because they do not offer new interpretations, new physical visions of the world, and therefore no improvement in our understanding of quantum mechanics, I do not consider them as relevant to the future progress of a genuine intrinsic understanding of the world.

The rejection of artificial hidden variables may also be viewed as the result of an ampliative argument (although not so inescapable as is NSP). Indeed, according to Bitbol [128], the decision of rejecting such theories may be understood as "pertaining to the very large set of fundamental choices often easy to choose to orient ourselves, allowing the whole scientific enterprise to avoid getting lost in shortcuts". These fundamental choices are given the name of ampliative criteria by Bitbol. His quotation however applies to all hidden-variables theories, and not only to artificial hidden-variables theories. NSP used to reject the pilot wave, the double solution, and stochastic quantum mechanics, is indeed also an ampliative criterion in the strict sense that it allows one to escape from the undecidability of the Duhem–Quine thesis. But I also consider it as a rational hard demand, which must be satisfied in a compulsory way. The ampliative argument used to reject artificial hidden-variables theories is softer. It just says that if they are acceptable (they are acceptable), they add nothing to our physical understanding. Someone not worrying about understanding, but solely in making predictions, is not obliged to accept such an argument. This, ironically enough, means that a positivist of strict obedience should accept artificial hidden-variables theories.

The rejection of all hidden variables implies that Ψ is complete (it does need to be supplemented with hidden variables). This is indeed also pictorially and vividly illustrated by Figs. 2(b) and 2(c). In Fig. 2(b), we are left with S (trajectories removed) and, in Fig. 2(c), S has been lifted to Ψ. So, clearly, Ψ alone tells all the story. However, we have to be careful in interpreting these figures. These figures alone do not imply that Ψ is the only quantity relevant in quantum mechanics. We could imagine that Fig. 2(c) could also contain other fields or quantities Λ, Σ, Ξ, ... which would vanish in the classical limit. Indeed, such quantities are allowed by the lifting process and, for evolution equations, they have to be contained in the supplementary function h_ϵ of Eq. (16.57). Therefore, they have to be related to Ψ, possibly through non-linear terms, or more generally, if we accept equations which are not evolution equations, through higher time derivatives of Ψ. But the rejection of hidden variables implies that there should not be other quantities than these, consistently with the use we have made of the lifting principle. This consistency has a conceptual cause. One one hand, the lifting principle has been used *after* the invocation of NSP, so that trajectories are removed from Fig. 2(b). On the other hand, following another route, the invocation of NSP implied the rejection of hidden-variables theories relying on trajectories.

From these discussions and conclusions, can we say that the issue on hidden variables is definitely closed? No, in utmost logical rigor, we cannot go so far. There remains always the possibility of a *lack of imagination*, that someone, very inspired, could propose a satisfactory alternative. As stated fairly recently by Laloë [335], *in fact, it is not impossible that the fundamental theory of microscopic processes will, one day, include a (nonlocal) deterministic mechanism behind what we now call the orthodox quantum theory of measurement.* But this seems now, at least to me, very unlikely. However, at least, is it possible to say that the issue is closed, as far as the currently existing hidden-variables theories are concerned? To this question, my answer is: Yes. But this is a personal opinion. People which do not agree with it will soon be given a way to escape.

Next, considering the alternative between hidden variables and quantum mechanics, having rejected hidden variables, can we say that quantum mechanics is right. There is no logical reason to draw this conclusion, on the contrary. As a matter of fact, the response of Popper would be no: You can corroborate a theory, you can falsify it, but you can never prove the validity of a theory. Those who believe that the measurement problem in quantum mechanics is still unsolved, irritating and puzzling, could possibly even consider this feature as providing a falsification, if not of an experimental nature, at least of a conceptual nature, some kind of logical inconsistency. Whether the measurement problem can be solved inside quantum mechanics (some may uphold that this is achieved by the decoherence theory) or will require a no-cheap alteration of it, may be a matter of debate. Furthermore, we indeed know that quantum mechanics is incorrect, that something more general is still waiting and hopefully coming. And unexpected new disruptions are to be expected.

So, now, indeed, at least concerning the purpose of this book, we are done. But, surely, we are not done forever. And furthermore, I have to add, we are not done for everybody: There is a loophole.

The loophole

Bohmian people still do exist: I encountered a few of them. In particular, one of them, namely Bricmont already quoted [208], in a private communication, did not accept my rejection of Bohmian mechanics. The essence of the disagreement is as follows. I started from the NSP, from which I came to the conclusion that Newtonian trajectories of matter points (and matter points themselves) do not exist, this being used as an ampliative argument to discriminate between the Duhem–Quine equivalent quantum mechanics and pilot wave. The loophole is then easy to identify: If you reject the NSP, you may reject my rejection of Bohmian mechanics. The existence of such a loophole illustrates the difficulties we may encounter when we try to think at the frontiers of science, in a fuzzy region containing simultaneously the Popper demarcation line, and speculations. But the reader now possesses in hand a book, and in mind, the contents of a book which, hopefully, may help him to decide whether he prefers Bohmian mechanics rather than quantum mechanics, or quantum mechanics rather than Bohmian mechanics, or whether, like a Pyrrhonian skeptical man, he would prefer to defer his judgment (or belief), waiting for something more satisfactory... this being for the ones who are unsatisfied by both Bohmian and quantum mechanics. Other loopholes, pertaining to the theory of knowledge, are discussed in [527].

Conclusion

A long time ago, something very special happened: The Big Bang of Mind, when someone, somehow, somewhere, could see the world with new eyes, and an impressive (but far from being perfect) brain behind. Questions, problems, stupefactions and incredulities, resignations and naive beliefs, began to populate the third world. Step after step, invention after invention, idea after idea, theory after theory, Homo Sapiens Sapiens (and may be others forgotten or overlooked) started the process of creating an objective knowledge. It has been a very slow and difficult process, over may be one hundred thousands years, without virtually any significant remnant, excepted in the last five thousands years or so. But it also has been an incredibly fast process, because one hundred thousands years only make one thousand hundred-year-old individuals, and five thousands years only make fifty (yes: Fifty) hundred-year-old individuals.

During these last centuries, since, let us say Galileo Galilei, the scientific enterprise began its cosmic age of inflation. This only required 4 hundred-years-old individuals. And the twentieth century, with Planck, Einstein, Louis de Broglie, Schrödinger, Heisenberg, Born, Jordan, Dirac, and many others ... and Bohm, only required

1 hundred-years-old individual, like one of my uncles who is born with the first light of relativity and died with the world wide web.

The landscape to explore is more than huge, with blind-paths, traps, pits, abysses and cliffs, deceitful demons with the face of angels, and deceptive mirages. Some explorers are running ahead, neck and neck together (we need them) while others, solitary, are wandering around, watching the inside of a dark cavern or the strange foliations of a flowering plant (we need them too). We actually really need the ones who are running and the ones who are wandering, because no clue should be overlooked and, when there is a clue, we do not know whether it will be found in a populated town or in an isolated chalet. Although science is made by individuals (which, admittedly, usually, do not have one hundred years at their disposal), the scientific enterprise is the result of a collective effort.

Louis de Broglie has been a solitary wanderer, and his Nobel Prize seemingly did not protect him from anxieties, and may be from despair. The objective situation has likely be worst for Bohm, but he grasped fast on his commitment and, for me, he is a hero. I feel sad having blown a coup de grâce to such a character. But it had to be done. Many possible ways may be open to the explorer. And, if possible, all of them have to be explored. But, if you choose the wrong way, you can walk a very long distance before perceiving that things become hopeless. And later on, there might a guy who just examine the travel diaries, and conclude that the courageous explorer, unfortunately, did embark himself on a rotting boat, without any destination. Well, this is the life.

Regularly, the explorers are facing many-way decisions. It has been so at the beginning of quantum mechanics. One way has been the imperial highway leaving from Copenhaguen and Göttingen. The other way has been a small old road running nearly parallely to the highway. Only one hundred meters to cross, and you can pass from the highway to the small road, and reciprocally. The highway has been preserved in good shape, due to the huger and huger number of scientists driving on it. The small road, followed only by a few outsiders, deteriorated and now, here it is: The small road, for most people, is blocked. It has become a no-drive road.

May be another outsider, with a particularly sharp mind, will find another path, another physical contextualist and non-local theory, different from the one of Bohm, which would make again the debate wide open. For me, it seems unlikely today. And this because, once classical particles and classical waves have been pushed away from the stage, it is difficult to imagine which kind of more or less physical objects could be reintroduced to support the existence of Ψ. But the possibility is still present. May be it will not be to-morrow. May be it will need another 20 hundred-years-old individuals. May be longer. Dazzled by our recent successes, we might possibly avoid to be arrogant, and to think that the thing is here. I believe that many blind-alleys will still have to be explored before reaching the unreachable of an universe which could very well, even conceptually, be inexhaustible (thanks again to God and Gödel).

But it has been the merit of Louis de Broglie and of Bohm to have perseveringly explored dry rivulets outside of the main stream, to have reported on their findings, speculations, and hopes, allowing us with only a bit of effort to make a conclusion, or to make up our mind. They have to be respected and admired for that. In particular, among the many references from Bohm, if there is only one book to read, I would recommend his last opus [68], showing how much his approach and his mind are attractive. And I must confess that giving a coup de grâce to such an elegant and consistent commitment has been painful to me, at least this is my feeling. But we had to get rid of it, actually avoiding to impede future developments. In any case, I hope that the reader found the journey in hidden worlds pleasant, even if the end of the journey may have been a bit bloody (a coup de grâce is always bloody).

Ah! I was going to forget to tell a last thing. For those who cannot or will not, whatever the reasons, get rid of Bohm: Remember, there is a loophole...

Complementary reference: Since the first publication of this book, Jean Bricmont, who is the preface writer of the present book, published a defence of Bohm's theory under the title "Making Sense of Quantum Mechanics", published by Springer in 2016. I obviously disagree but this is already known to the reader since the preface he wrote for this book.

References

[1] F.J. Belinfante. *A Survey of Hidden-Variable Theories*. Pergamon, Oxford, 1973.

[2] A. Forsee. *Albert Einstein, Theoretical Physicist*. MacMillan, 1963.

[3] Jean-Paul Dumont, editor. *Les écoles pré-socratiques*. Gallimard, 1991.

[4] Flammarion Garnier Frères, *Les penseurs grecs avant Socrate, de Thalès de Milet à Prodicos, traduction, introduction et notes, par J. Voilquin*. Gallimard, 1964.

[5] C. Rovelli. *Anaximandre de Milet ou la naissance de la pensée scientifique*. Dunod, Paris, 2007.

[6] Plato. *Timée, et Critias, A New Translation, with Introduction and Notes*, by Luc Brisson. Flammarion, 2001.

[7] Aristotle. *Traité du ciel (De Caelo), Presentation by P. Pellegrin*. Flammarion, 2004.

[8] A. Koestler. *Les somnambules, French translation of "The sleepwalkers"*. Calmann-Lévy, 1960.

[9] L. Smolin. *Rien ne va plus en physique, l'échec de la théorie des cordes, French translation of "The Trouble with Physics — The Rise of String Theory, the Fall of a Science, and what comes Next"*, 2006. Dunod, 2007.

[10] W. Heisenberg. *La partie et le tout, le monde de la physique atomique, French translation of: der Teil und das Ganze, Gespräche im Umkreis der Atomphysik*. Editions Albin Michel, 1972.

[11] W. Heisenberg. *La nature dans la physique contemporaine, translated from German*. Gallimard, collection Idées, 1962.

[12] W. Heisenberg. *Umkreis der Kunst, Eine Festschrift für Emil Preetorius*, chapter: Platons Vorstellungen von den kleinstein Bausteinen der Materie und die Elementarteilchen der modernen Physik, pp. 137–140. Wiesbaden, 1953.

[13] E. Squires. *The Mystery of the Quantum World.* Adam Hilger, Bristol, 1986.

[14] R.P. Feynman. *Le cours de physique de Feynman, mécanique quantique, French Translation of The Feynman Lectures on Physics, 1965.* Dunod, 2000.

[15] R. Omnès. *Comprendre la mécanique quantique.* EDP Sciences, Paris, 2000.

[16] M. Le Bellac. *Physique Quantique.* CNRS, Paris, 2007.

[17] G. Charpak. *Mémoires d'un déraciné.* Odile Jacob, Paris, 2008.

[18] R.P. Feyman. *La nature de la physique, French translation of "The nature of physical laws".* Seuil, Paris, 1980.

[19] M. Gell-Mann. *The Nature of Matter, Wolfson College Lectures 1980, edited by J.H. Mulvey,* chapter: Questions for the future, pp. 169–186. Clarendon Press, Oxford, 1981.

[20] M. Gell-Mann. *Le quark et le jaguar, French translation of "The quark and the jaguar, adventures in the simple and the complex".* Editions Albin Michel, 1995.

[21] B. Van Fraassen. *Quantum Mechanics: An Empiricist View.* Clarendon Press, Oxford, 2000.

[22] P. Duhem. *Le système du monde.* Paris, 1913.

[23] W.M. Elsasser. *Louis de Broglie, physicien et penseur, collection dirigée par André George,* chapter: Les mesures et la réalité en mécanique quantique, pp. 87–108. Editions Albin Michel, 1953.

[24] M. Jammer. *The Philosophy of Quantum Mechanics. The Interpretations of Quantum Mechanics in Historical Perspective.* Wiley, New York, 1974.

[25] K.R. Popper. *La logique de la découverte scientifique, French translation of "The logic of scientific discovery", Hutchinson, 1959.* Payot, 1973.

[26] J. von Neumann. *Les fondements de la mécanique quantique, French translation of "Mathematische Grundlagen der Quantenmechanik", 1932. English version: Mathematical foundations of quantum mechanics, Princeton University Press, Princeton, 1955.* Librairie Félix Alcan, 1946.

[27] F. London and E. Bauer. *Actualités scientifiques et industrielles: exposés de physique générale, number 775, edited by P. Langevin, English translation in Quantum Theory and Measurement, edited by J.A. Wheeler and W.H. Zurek, Princeton University Press, 1983,* chapter: The theory of observation in quantum mechanics. Hermann et Compagnie, 1939.

[28] S. Kochen and E.P. Specker. The problem of hidden-variables in quantum mechanics. *Journal of Mathematics and Mechanics,* 17:59–67, 1967.

[29] D. Bohm. A suggested interpretation of the quantum theory in terms of "hidden" variables. part 1. *Physical Review*, 85:166–179, 1952.

[30] D. Bohm. A suggested interpretation of the quantum theory in terms of "hidden" variables, part 2. *Physical Review*, 85:180–193, 1952.

[31] D. Bohm. *Wholeness and the Implicate Order*. Routledge and Paul Kegan, 1980.

[32] J.A. Wheeler. *Quantum Theory and Measurement*, edited by J.A. Wheeler and W.H. Zurek, chapter: Law without law, pp. 182–213. Princeton Series in Physics, Princeton University Press, Princeton, New Jersey, 1983.

[33] Aristotle. *Physique, traduction et présentation par Pierre Pellegrin.* Flammarion, 2002.

[34] Galileo Galilei. *Dialogue sur les deux grands systèmes du monde, French translation of: Dialogo sopra i due massimi sistemi del mondo tolemaico e copernicano.* Editions du Seuil, 1992.

[35] R. Descartes. *Les principes de la philosophie, quatrième partie: la Terre et son histoire.* Paleo, France, 1999.

[36] L.E. Ballentine. A survey of hidden-variables theories. *Physics Today*, october:53–55, 1974.

[37] Louis de Broglie. *Introduction à l'étude de la mécanique ondulatoire.* Hermann et Cie, Paris, 1930.

[38] D.I. Blokhintsev. *Mécanique quantique.* Masson et Cie, 1967.

[39] L.D. Landau and E.V. Lifchitz. *Mécanique, English translation: Mechanics, Pergamon Press, Oxford, 1969.* Editions Mir, Moscou, 1966.

[40] P.R. Holland. *The Quantum Theory of Motion, An Account of the de Broglie-Bohm Causal Interpretation of Quantum Mechanics.* Cambridge University Press, 1993.

[41] D. Bohm. *Quantum Theory.* Prentice-Hall, New York, 1951.

[42] P.A.M. Dirac. *The Principles of Quantum Mechanics (first published, 1930).* The International Series of Monographs on Physics, 27. Fourth Edition. Oxford Science Publications, Clarendon Press, Oxford, 1958.

[43] C. Cohen-Tannoudji, B. Diu, and F. Laloë. *Mécanique quantique, tomes 1 et 2.* Hermann et Cie, Paris, 1973.

[44] E. Schrödinger. *Collected Papers on Wave Mechanics.* Blackie and Sons, Ltd, Glasgow, 1928.

[45] E. Schrödinger. *Mémoires sur la mécanique quantique.* Edition Jacques Gabay, 1988.

[46] *Electrons et photons, Rapports et discussions du cinquième congrès de physique tenu à Bruxelles du 24 au 29 octobre 1927.* Gauthier-Villars, 1928.

[47] P. Marage and G. Wallenborn. *Les conseils Solvay et les débuts de la physique moderne.* Université libre de Bruxelles, 1995.

[48] J.T. Cushing. *Wave-Particle Duality*, edited by F. Selleri, chapter Causal quantum theory, why a nonstarter? pp. 37–63. Plenum Press, New York and London, 1992.

[49] R. Penrose. *Quantum Implications*, Essays in honour of David Bohm, edited by B.J. Hiley and F.D. Peat, chapter: Quantum physics and conscious thought, pp. 105–120. Routledge and Kegan Paul, 1987.

[50] J.P. Vigier. *Structure des micro-objets dans l'interprétation causale de la théorie des quanta.* Gauthier-Villars, 1956.

[51] A. Shimony. *Foundations of quantum mechanics, Proceedings of the International School of Physics "Enrico Fermi", course IL, edited by B. d'Espagnat,* chapter: Philosophical comments on quantum mechanics, pp. 470–480. Academic Press, New York, 1971.

[52] N. Bohr. *Physique atomique et connaissance humaine, French version of "Atomphysics and human knowledge", 1958.* Gallimard, 1991.

[53] W. Heisenberg. *Physique et philosophie, translated from "Physics and Philosophy, the Revolution of Modern Physics", Harper and Brothers, New York, 1958.* Editions Albin Michel, 1971.

[54] A. Shimony. Réflexions sur la philosophie de Bohr, Heisenberg et Schrödinger. *Journal de Physique,* Colloque C2, supplément au numéro 3, 42:81–98, 1981.

[55] H.P. Stapp. The Copenhagen interpretation. *American Journal of Physics,* 40:1098–1116, 1972.

[56] L. Landau and E. Lifchitz. *Mécanique quantique, théorie non relativiste, English translation: quantum mechanics, Pergamon Press, Oxford, 1977.* Editions Mir, Moscou, 1967.

[57] B. Van Fraassen. *Lois et symétries, French translation of "Laws and symmetry", Oxford University Press, 1989.* Librairie philosophique J. Vrin, 1994.

[58] A. Koyré. *Du monde clos à l'univers infini, French translation of: From the closed world to the infinite universe.* Gallimard, 1973.

[59] G. Gouesbet, S. Meunier-Guttin-Cluzel, and O.Ménard, editors. *Chaos and its reconstruction.* Novascience Publishers, New York, 2003.

[60] D. Bohm. *Causality and Chance in Modern Physics.* Routledge and Paul Kegan, 1957.

[61] K.R. Popper. *La connaissance objective, French translation of "Objective knowledge",* Oxford University Press, 1979. Flammarion, 1991.

[62] Louis de Broglie. *La physique quantique restera-t-elle indéterministe?* Gauthier-Villars, 1953.

[63] Louis de Broglie. *Une tentative d'interprétation causale et non linéaire de la mécanique ondulatoire (la théorie de la double solution).*

English translation: Nonlinear Wave Mechanics, Elsevier, Amsterdam, 1960. Gauthier-Villars, 1956.

[64] Louis de Broglie. *La théorie de la mesure en mécanique ondulatoire (interprétation usuelle et interprétation causale).* Gauthier-Villars, 1957.

[65] Louis de Broglie. *Etude critique des bases de l'interprétation actuelle de la mécanique ondulatoire.* Gauthier-Villars, 1963.

[66] E. Klein. *Petit voyage dans le monde des quanta.* Flammarion, Paris, 2004.

[67] J.A. Wheeler. *Some Strangeness in the Proportion. A Centennial Symposium to Celebrate the Achievements of Albert Einstein,* edited by H. Woolf, chapter: Beyond the black hole, pp. 341–386. Addison-Wesley, Reading, Mass., 1980.

[68] D. Bohm and B.J. Hiley. *The Undivided Universe, An Ontological Interpretation of Quantum Theory.* Routledge and Paul Kegan, 1993.

[69] B. d'Espagnat. Nonseparability and the tentative descriptions of reality. *Physics Reports,* 110, 4:201–264, 1984.

[70] B. Greene. *La magie du cosmos (French translation of "The fabric of cosmos", 2004).* Laffont, 2005.

[71] W.C. Wickes, Carroll O.Alley, and O. Jakubowicz. *Quantum Theory and Measurement,* edited by J.A. Wheeler and W.H. Zurek, chapter: A "delayed-choice" quantum mechanics experiment, pp. 457–461. Princeton Series in Physics, Princeton University Press, Princeton, New Jersey, 1983.

[72] V. Jacques, E. Wu, F. Grosshans, F. Treussart, P. Grangier, A. Aspect, and J.F. Roch. Experimental realization of Wheeler's delayed-choice Gedankenexperiment. *Science,* 315:966–968, 2007.

[73] E. Schrödinger. *Louis de Broglie, physicien et penseur, collection dirigée par André George,* chapter: La signification de la mécanique ondulatoire, pp. 17–32. Editions Albin Michel, 1953.

[74] E.P. Wigner. *Quantum Theory and Measurement,* edited by J.A. Wheeler and W.H. Zurek, chapter: Interpretation of quantum mechanics, pp. 260–314. Princeton University Press, Princeton, New Jersey, 1983.

[75] J.P. Jarrett. *Philosophical Consequences of Quantum Theory,* edited by J.T. Cushing and E. McMullin, chapter: Bell's theorem: a guide to the implications, pp. 60–79. University of Notre Dame Press, 1989.

[76] Louis de Broglie. *Louis de Broglie, physicien et penseur, collection dirigée par André George.* Editions Albin Michel, 1953.

[77] J.S. Bell. *Quantum Gravity 2,* edited by C. Isham, R. Penrose and D. Sciama, chapter: Quantum mechanics for cosmologists, reprinted

in "Speakable and Unspeakable in Quantum Mechanics", pp. 611–637. Clarendon Press, Oxford, 1981.

[78] H. Everett. Relative state formulation of quantum mechanics. *Review of Modern Physics*, 29:454–462, 1957.

[79] J.S. Bell. *Quantum Mechanics, Determinism, Causality, and Particles*, edited by M. Flato, Z. Maric, A. Milojevic, D. Sternheimer and J.P. Vigier, chapter: The measurement theory of Everett and de Broglie's pilot wave, reprinted in "Speakable and Unspeakable in Quantum Mechanics", pp. 11–17. D. Reidel, Dordrecht-Holland, 1976.

[80] B.S. de Witt. *Foundations of Quantum Mechanics, Proceedings of the International School of Physics "Enrico Fermi", course IL*, edited by B. d'Espagnat, chapter: The many-universes interpretation of quantum mechanics, pp. 211–262. Academic Press, New York, 1971.

[81] P. Byrne. Les nombreux univers de Hugh Everett. *Pour la science*, March 2008:26–31, 2008.

[82] M. Jammer. *The Conceptual Development of Quantum Mechanics.* The History of Modern Physics, 1800–1950, Volume 12, Tomash Publishers, 1989.

[83] D. Home and M.A.B. Whitaker. Ensemble interpretations of quantum mechanics. a modern perspective. *Physics Reports (Review Section of Physics Letters)*, 210, 4:223–317, 1992.

[84] B. d'Espagnat. *Le réel voilé, analyse des concepts quantiques.* Fayard, 1994.

[85] B. d'Espagnat. *Traité de physique et philosophie.* Fayard, 2002.

[86] A. Bassi and G. Ghirardi. Dynamical reduction models. *Physics Reports*, 379:257–426, 2003.

[87] J.S. Bell. Against "Measurement", Chapter 23, in: *Speakable and Unspeakable in Quantum Mechanics*, Cambridge University Press. 2004.

[88] R. Omnés. *Les indispensables de la mécanique quantique.* Odile Jacob, 2006.

[89] E. Wigner. *Foundations of Quantum Mechanics, Proceedings of the International School of Physics "Enrico Fermi", course IL*, edited by B. d'Espagnat, chapter: The subject of our discussion, pp. 1–19. Academic Press, New York, 1971.

[90] H.P. Stapp. Whiteheadian approach to quantum theory and the generalized Bell's theorem. *Foundations of Physics*, 9, 1-2:1–25, 1979.

[91] J.M. Jauch. *Proceedings of the International School of Physics "Enrico Fermi", course IL*, edited by B. d'Espagnat, chapter: Foundations of quantum mechnics, pp. 20–55. Academic Press, New York, 1971.

[92] G.C. Ghirardi, A. Rimini, and T. Weber. Unified dynamics for microscopic and macroscopic systems. *Physical Review D*, 34, 2: 470–491, 1986.

[93] A. Shimony. *Philosophical Consequences of Quantum Theory, Reflections on Bell's Theorem*, edited by J.T. Cushing and E. McMullin, chapter: Search for a worldview which can accommodate our knowledge of microphysics, pp. 25–37. University of Notre Dame Press, Notre Dame, Indiana, 1989.

[94] E.J. Squires. Continuous spontaneous localisation without a stochastic field. *Physics Letters A*, 157, number 8, 9:453–455, 1991.

[95] G.C. Ghirardi. *Quantum [Un]Speakables, From Bell to Quantum Information*, edited by R.A. Bertlmann and A. Zeilinger, chapter: John Stewart Bell and the dynamical reduction program, pp. 287–305. Springer, 2002.

[96] D.Z. Albert and L. Vaidman. *On a Theory of the Collapse of the Wave Function*, Chapter 1, pp. 1–6. Kluwer Academic Publishers, edited by Menas Kafatos, 1989.

[97] R. Omnès. *Alors l'un devint deux, la question du réalisme en physique et en philosophie des mathématiques*. Flammarion, 2002.

[98] B. d'Espagnat and C. Saliceti. *Candide et le physicien*. Fayard, Paris, 2008.

[99] S. Haroche, J.M. Raimond, and M. Brune. Le chat de Schrödinger se prête à l'expérience. *La Recherche*, 301:50–55, 1997.

[100] J.A. Wheeler and W.H. Zurek, editors. *Quantum Theory and Measurement*. Princeton series in physics, Princeton University Press, Princeton, New Jersey, 1983.

[101] L. Rosenfeld. *Louis de Broglie, physicien et penseur, collection dirigée par André George*, chapter: L'évidence de la complémentarité, pp. 43–65. Editions Albin Michel, 1953.

[102] C.M. Patton and J.A. Wheeler. *Quantum Gravity, an Oxford symposium*, edited by C.J. Isham, R. Penrose and D.W. Sciama, chapter: Is physics legislated by cosmogony? pp. 538–605. Clarendon Press, Oxford, 1975.

[103] E. Kant. *Critique de la Raison Pure, French translation of "Kritik der reinen Vernunft"*. Presses universitaires de France, Quadrige, 1944.

[104] B. d'Espagnat. The quantum theory and reality, French version: théorie quantique et réalité, in: Pour la science, 72–87, January 1980. *Scientific American*, 241:158–181, 1979.

[105] A. Schopenhauer. *Le monde comme volonté et représentation, critique de la philosophie kantienne*, pp. 517–668. Presses universitaires de France, Quadrige, 1966.

[106] J. Sebestik and A. Soulez. *Le Cercle de Vienne, doctrines et controverses.* L' Harmattan, 2001.

[107] Teilhard de Chardin. *Le phénomène humain.* Editions du Seuil, 1955.

[108] Aristote. *Métaphysique.* Librairie philosophique J.Vrin, 2000.

[109] Pierre Cassou-Nogués. *Les démons de Gödel, logique et folie.* Science ouverte, Seuil, 2007.

[110] J.C. Pecker. *L'univers expliqué, peu à peu exploré.* Odile Jacob, 2003.

[111] S. Hawking. *Une brève histoire du temps, du big bang aux trous noirs, translated from English.* Flammarion, 1989.

[112] R. Descartes. *Les principes de la philosophie, troisiéme partie: pirouettes et tourbillons des cieux.* Paleo, France, 2000.

[113] Spinoza. *Pensées métaphysiques.* Flammarion, 1964.

[114] P. Pellegrin. *Traité du ciel, by Aristotle, Introduction,* pp. 7–62. Flammarion, 2004.

[115] S. Kuhn. *La structure des révolutions scientifiques, French translation of: the structure of scientific revolutions.* Flammarion, 1983.

[116] H. Margenau. *Albert Einstein: Philosopher-Scientist (First edition, 1949),* edited by P.A. Schilpp, chapter: Einstein's conception of reality, pp. 245–268. The Library of Living Philosophers, Volume VII, 1970.

[117] Grete Hermann. *Les fondements philosophiques de la mécanique quantique. Introduction, présentation et postface critique par Lena Soler (First published, 1935).* Librairie philosophique J. Vrin, 1996.

[118] R. Harré. *Varieties of realism, a rationale for the natural sciences.* Basil Blackwell, 1986.

[119] O. Spengler. *Le déclin de l'Occident.* Editions Gallimard, 1976.

[120] J.S. Bell. Bertlmann's socks and the nature of reality, reprinted in: *Speakable and Unspeakable in Quantum Mechanics. Journal de Physique,* Colloque C2, supplément au numéro 3:C2–41, 1981.

[121] A. Einstein. *Albert Einstein: Philosopher-Scientist (First edition, 1949),* edited by P.A. Schilpp, chapter: Reply to criticisms, pp. 665–688. The Library of Living Philosophers, Volume VII, 1970.

[122] P. Frank. *Albert Einstein: Philosopher-Scientist (First edition, 1949),* edited by P.A. Schilpp, chapter: Einstein, Mach and logical positivism, pp. 271–286. The Library of Living Philosophers, Volume VII, 1970.

[123] P.A. Schilpp, editor. *Albert Einstein: Philosopher-Scientist.* The Library of Living Philosophers, volume VII, 1949.

[124] M. Born. *Albert Einstein: Philosopher-Scientist (First edition, 1949), edited by P.A. Schilpp,* chapter: Einstein's statistical theories, pp. 163–177. The Library of Living Philosophers, Volume VII, 1970.

[125] E. Nelson. *Dynamical theories of Brownian motion.* Princeton University Press, Princeton, New-Jersey, 1967.

[126] B. d'Espagnat. *A la recherche du réel.* Gauthier-Villars, 1979.

[127] A. Einstein. *Albert Einstein: Philosopher-Scientist,* edited by P.A. Schilpp, chapter: Autobiographical notes, pp. 2–94. The Librairy of Living Philosophers, Volume VII (First edtion, 1949), 1970.

[128] M. Bitbol. *L'aveuglante proximité du réel.* Flammarion, 1998.

[129] Spinoza. *Court traité.* Flammarion, 1964.

[130] Spinoza. *Ethique.* Flammarion, 1965.

[131] J.T. Cushing. *Quantum Mechanics, Historical Contingency and the Copenhagen Hegemony.* The University of Chicago Press, 1994.

[132] A. Sommerfeld. *Albert Einstein: Philosopher-Scientist (First edition, 1949),* edited by P.A. Schilpp, chapter: To Albert Einstein's seventieth anniversary, pp. 99–105. The Library of Living Philosophers, Volume VII, 1970.

[133] M. Bitbol. *Mécanique quantique. Une introduction philosophique.* Flammarion, 1996.

[134] H. Reichenbach. *Philosophical Foundations of Quantum Mechanics,* edited by H. Reichenbach, chapter: Three-valued logic and the interpretation of quantum mechanics. University of California Press, Los Angeles, 1944.

[135] H. Reichenbach. *Louis de Broglie, Physicien et penseur, collection dirigée par André George,* chapter: La signification philosophique du dualisme ondes-corpuscules, pp. 117–134. Editions Albin Michel, 1953.

[136] C.A. Hooker, editor. *The Logico-Algebraic Approach to Quantum Mechanics, Historical Evolution.* D. Reidel Publishing Company, 1975.

[137] R.J. Greechie and S.P. Gudder. *Contemporary Research in the Foundations and Philosophy of Quantum Theory,* edited by C.A. Hooker, chapter: Quantum logics. D. Reidel Publishing Company, Dordrecht-Holland, 1974.

[138] W.V. Quine. *Philosophie de la logique, présentation par Denis Bonnay and Sandra Laugier.* Aubier-Montaigne, 1975.

[139] P. Feyerabend. *Contre la méthode, esquisse d'une théorie anarchiste de la connaissance, French translation of "Against Method", New left books, 1975.* Editions du Seuil, 1979.

[140] R. Thom. *Prédire n'est pas expliquer.* Champs Flammarion, 1991.

[141] E. Nagel. *Some Strangeness in the Proportion, a Centennial Symposium to Celebrate the Achievements of Albert Einstein,* chapter:

Relativity and twentieth-century intellectual life, pp. 38–45. Addison-Wesley, Reading, Mass., 1980.

[142] J.S. Bell. On the imposible pilot wave, reprinted in *Speakable and Unspeakable in Quantum Mechanics. Foundations of Physics*, 12,10:989–999, 1982.

[143] A. Einstein, N. Podolsky, and N. Rosen. Can quantum mechanical description of physical reality be considered complete? *Physical Review*, 47:777–780, 1935.

[144] N. Bohr. Can quantum-mechanical description of physical reality be considered complete? *Physical Review*, 48:696–702, 1935.

[145] N. Bohr. Quantum mechanics and physical reality. *Nature*, 136:65, 1935.

[146] W. Heisenberg. *Niels Bohr and the Development of Physics, Essays Dedicated to Niels Bohr on the Occasion of his Seventieth Birthday*, edited by W. Pauli, with the assistance of L. Rosenfeld and V. Weisskopf, chapter: The development of the interpretation of the quantum theory, pp. 12–29. Pergamon Press Ltd, 1955.

[147] A. Pais. *Some Strangeness in the Proportion, a Centennial Symposium to Celebrate the Achievements of Albert Einstein*, edited by H. Woolf, chapter: Einstein on particles, fields, and the quantum theory, pp. 197–268. Addison-Wesley, Reading, Mass., 1980.

[148] E. Klein. *Dictionnaire de l'ignorance, sous la direction de M. Cazenave*, chapter: La complémentarité quantique, pp. 99–107. Bibliothéque Albin Michel Sciences, 1998.

[149] W. Pauli. *Albert Einstein: Philosopher-Scientist (First edition, 1949)*, edited by P.A. Schilpp, chapter: Einstein's contributions to quantum theory, pp. 149–160. The Library of Living Philosophers, Volume VII, 1970.

[150] A. Einstein. *Louis de Broglie, physicien et penseur, collection dirigée par André George*, chapter: Remarques préliminaires sur les concepts fondamentaux, pp. 5–15. Editions Albin Michel, 1953.

[151] N. Bohr. *Albert Einstein: Philosopher-Scientist (First edition, 1949)*, edited by P.A. Schilpp, chapter: Discussion with Einstein on epistemological problems in atomic physics, pp. 201–241. The Library of Living Philosophers, Volume VII, 1970.

[152] Louis de Broglie. Sur la possibilité d'une interprétation causale et objective de la mécanique ondulatoire. *Comptes-Rendus de l'Académie des Sciences*, t 234:265–268, 1952.

[153] Louis de Broglie. *Louis de Broglie, physicien et penseur*, chapter: Vue d'ensemble sur mes travaux scientifiques, pp. 457–486. Editions Albin Michel, Collection dirigée par André George, 1953.

[154] T.J. Pinch. *Social Production of Scientific Knowledge, Sociology of the Sciences, Vol 1*, edited by E. Mendelson, P. Weingart and R. Whitley, chapter: What does a proof do if it does not prove, a study of the social conditions and metaphysical divisions leading to David Bohm and John von Neumann failing to communicate in quantum physics, pp. 171–215. D. Reidel, Dordrecht, 1977.

[155] H. Freistadt. The causal formulation of the quantum mechanics of particles (the theory of de Broglie, Bohm and Takabayasi). *Supplemento al Nuovo Cimento*, Ser. 10, 5:1–70, 1957.

[156] F.M. Pipkin. *Advances in Atomic and Molecular Physics*, edited by D.R. Bates and B. Bederson, chapter: Atomic physics tests of the basic concepts in quantum mechanics, pp. 281–340. Academic Press, New York, 1978.

[157] Louis de Broglie. Recherches sur la théorie des quanta, thése de doctorat de Louis de Broglie, soutenue à Paris, le 25 november 1924. *Annales de Physique*, Série 10:22–128, 1925.

[158] Louis de Broglie. *Electrons et photons, Rapports et Discussions du cinquiéme conseil de physique tenu à Bruxelles du 24 au 29 octobre 1927*, chapter: Nouvelle dynamique des quanta, pp. 105–141. Gauthier-Villars, 1928.

[159] N. Bohr. *Electrons et Photons, Rapports et Discussions du cinquième conseil de physique tenu à Bruxelles du 24 au 29 octobre 1927*, chapter: Le postulat des quanta et le nouveau développement de l'atomistique, pp. 215–287. Gauthier-Villars, 1928.

[160] E. Schrödinger. *Electrons et Photons, Rapports et Discussions du cinquième conseil de physique tenu à Bruxelles du 24 au 29 ocotbre 1927*, chapter: La mécanique des ondes, pp. 185–213. Gauthier-Villars, 1928.

[161] E.P. Wigner. *Some Strangeness in the Proportion*, edited by H. Woolf, chapter: Thirty years of knowing Einstein, pp. 461–472. Addison-Wesley, Reading, Mass., 1980.

[162] D.W. Belousek. Einstein's unpublished hidden-variable theory: Its background, context and significance. *Stud. Hist. Phil.Mod. Phys.*, 27,4:437–461, 1996.

[163] M. Born. *Louis de Broglie, physicien et penseur, collection dirigée par André George*, chapter: La grande synthèse, pp. 165–170. Editions Albin Michel, 1953.

[164] W. Pauli. *Louis de Broglie, physicien et penseur, collection dirigée par André George*, chapter: Remarques sur le problème des paramètres cachés dans la mécanique quantique et sur la théorie de l'onde pilote, pp. 33–42. Editions Albin Michel, 1953.

[165] E. Madelung. Quantentheorie in hydrodynamischer Form. *Zeitschrift für Physik*, 40:322–326, 1926.

[166] J.T. Cushing. Bohm's theory: Common sense dismissed. *Studies in History and Philosophy of Sciences*, 24, 5:815–842, 1993.

[167] T. Takabayasi. On the formulation of quantum mechanics associated with classical pictures. *Progress in Theoretical Physics*, 8, 2:143–182, 1952.

[168] T. Takabayasi. Remarks on the formulation of quantum mechanics with classical pictures and on relations between linear scalar fields and hydrodynamical fields. *Progress in Theoretical Physics*, 9, 3:187–222, 1953.

[169] M. Schönberg. A non-linear generalization of the Schrödinger and Dirac equations. *Nuovo Cimento*, 11:674–682, 1954.

[170] M. Schönberg. On the hydrodynamical model of the quantum mechanics. *Nuovo Cimento*, 12:103–133, 1954.

[171] J. Bass. *Wave-Particle Duality*, edited by F. Selleri, chapter: Probability, pseudoprobability, mean values, pp. 1–16. Plenum Press, New York and London, 1992.

[172] Louis de Broglie. *Foundations of Quantum Mechanics, Proceedings of the International School of Physics "Enrico Fermi", course IL*, edited by B. d'Espagnat, chapter: L'interprétation de la mécanique ondulatoire par la théorie de la double solution, pp. 346–367. Academic Press, New York, 1971.

[173] Louis de Broglie. La mécanique ondulatoire et la structure atomique de la matière et du rayonnement. *J. Phys. Radium*, 8:225–241, 1927.

[174] J. Andrade e Silva. *Foundations of Quantum Mechanics, Proceedings of the International School of Physics "Enrico Fermi", course IL*, edited by B. d'Espagnat, chapter: Une formulation causale de la théorie quantique de la mesure, pp. 368–397. Academic Press, New York, 1971.

[175] J.P. Vigier. Introduction géométrique de l'onde pilote en théorie unitaire affine. *Comptes-Rendus de l'Académie des Sciences*, 233:1010–1012, 1951.

[176] J.P. Vigier. Particular solutions of a non-linear Schrödinger equation carrying particle-like singularities represent possible models of de Broglie's double solution theory. *Physics Letters A*, 135, 2:99–105, 1989.

[177] T. Bonk. Why has de Broglie's theory been rejected? *Studies in History and Philosophy of Sciences*, 25, 3:375–396, 1994.

[178] Louis de Broglie. The reinterpretation of wave mechanics. *Foundation of Physics*, 1, 1:5–15, 1970.

[179] D. Bohm and B.J. Hiley. Measurement understood through the quantum potential approach. *Foundations of Physics*, 14, 3:255–274, 1984.

[180] D. Bohm and B.J. Hiley. *Quantum Implications, Essays in Honour of David Bohm*, edited by B.J. Hiley and F. David Peat, Chapter 2, Hidden variables and the implicate order, pp. 33–45. Routledge and Kegan Paul, 1987.

[181] J.S. Bell. Free variables and local causality, reprinted in *Speakable and Unspeakable in Quantum Mechanics. Epistemological Letters*, February, 1977.

[182] J.S. Bell. Beables for quantum field theory, reprinted in *Speakable and Unspeakable in Quantum Mechanics*. Technical report, CERN-TH, 4035/84, August 2, 1984.

[183] J.S. Bell. *Beables for Quantum Field Theory. In Quantum Implications, Essays in Honour of David Bohm*, edited by B.J. Hiley and F. David, Chapter 12, pp. 227–234. Routledge and Kegan Paul, 1987.

[184] J. Bub and R. Clifton. A uniqueness theorem for 'no collapse' interpretations of quantum mechanics. *Studies in History and Philosophy of Modern Physics*, 27, 2:181–219, 1996.

[185] Y. Aharonov, B.G. Englert, and M.O. Scully. Protective measurements and Bohm trajectories. *Physics Letters A*, 263:137–146, 1999.

[186] Y. Aharonov, B.G. Englert, and M.O. Scully. Erratum to "protective measurements and Bohm trajectories". *Physics Letters A*, 266:216–217, 2000.

[187] A. Drezet. Comment on "protective measurements and Bohm trajectories". *Physics Letters A*, 350:416–418, 2006.

[188] J.S. Bell. On the problem of hidden variables in quantum mechanics, reprinted in Speakable and Unspeakable in Quantum Mechanics. *Review of Modern Physics*, 38:447–452, 1966.

[189] J.S. Bell. On the Einstein-Podolsky-Rosen paradox, reprinted in "Speakable and Unspeakle in Quantum Mechanics". *Physics*, 1:195–200, 1964.

[190] D. Bohm and B.J. Hiley. Non-locality and locality in the stochastic interpretation of quantum mechanics. *Physics Reports (Review Section of Physics Letters)*, 172, 3:93–122, 1989.

[191] J. Finkelstein. How to measure a beable. *Physics Letters A*, 216:1–6, 1996.

[192] C.W. Rietdijk. *On Waves, Particles and Hidden Variables*. Van Gorcum and Company, 1971.

[193] L. Hardy. On the existence of empty waves in quantum theory. *Physics Letters A*, 167:11–16, 1992.

[194] B. Bläsi and L. Hardy. Realism and time symmetry in quantum mechanics. *Physics Letters A*, 207:119–125, 1995.

[195] M. Schmidt. *Wave-Particle Duality*, edited by F. Selleri, chapter: Wind effect of empty quantum waves in a Pfleegor-Mandel-type experiment for electrons, pp. 253–266. Plenum Press, New York and London, 1992.

[196] F. Selleri. *Wave-Particle Duality*, edited by F. Selleri, chapter: Two-photon interference and the question of empty waves, pp. 277–285. Plenum Press, New York and London, 1992.

[197] N.D. Mermin. Hidden variables and the two theorems of John Bell. *Review of Modern Physics*, 65:803–815, 1993.

[198] J.S. Bell. De Broglie-Bohm, delayed choice, double-slit experiment, and density matrix, reprinted in "Speakable and Unspeakable in Quantum Mechanics". *International Journal of Quantum Chemistry: a symposium*, 14:155–159, 1980.

[199] C. Dewdney and B.J. Hiley. A quantum potential description of one-dimensional time-dependent scattering from square barriers and square wells. *Foundations of Physics*, 12, 1:27–48, 1982.

[200] R.P. Feynman and A.R. Hibbs. *Quantum Mechanics and Path Integrals*. Mc Graw-Hill, New York, 1965.

[201] M. Born. *The Born-Einstein Letters, Correspondence Between Albert Einstein and Max and Hedwig Born, from 1916 to 1955 with Commentaries by Max Born*, edited by M.Born. MacMillan, 1971.

[202] J.L. Destouches. *Louis de Broglie, physicien et penseur, collection dirigée par André George*, chapter: Retour sur le passé, pp. 67–85. Editions Albin Michel, 1953.

[203] *Louis de Broglie, physicien et penseur, collection dirigée par André George*, chapter: Remarques sur le problème des paramètres cachés dans la mécanique quantique et sur la théorie de l'onde pilote, pp. 33–42. Editions Albin Michel, 1953.

[204] J.F. Clauser. *Quantum [Un]Speakables, from Bell to Quantum Information*, edited by R.A. Bertlmann and A. Zeilinger, chapter: Early history of Bell's theorem, pp. 61–98. Springer, 2002.

[205] J.S Bell. *Speakable and Unspeakable in Quantum Mechanics*. Cambridge, second edition with a new introduction by Alain Aspect, 2004.

[206] J.S. Bell. The theory of local beables, reprinted in "Speakable and Unspeakable in Quantum Mechanics". *Epistemological Letters*, 9:11, originally appeared as CERN preprint TH–2053, 1975, 1976.

[207] D. Dürr, S. Goldstein, and N. Zanghi. Quantum physics without quantum philosophy. *Studies in History and Philosophy of Modern Sciences*, 26, 2:137–149, 1995.

[208] J. Bricmont and H. Zwin. *Philosophie de la mècanique quantique.* Vuibert, Paris, 2009.

[209] C. Philippidis, C. Dewdney, and B.J. Hiley. Quantum interference and the quantum potential. *Nuovo Cimento*, 52 B, 1:15–28, 1979.

[210] C. Philippidis, D. Bohm, and R.D. Kaye. The Aharonov-Bohm effect and the quantum potential. *Nuovo Cimento*, 71B, 1:75–88, 1982.

[211] D. Home and F. Selleri. *Wave-Particle Duality,* edited by F. Selleri, chapter: The Aharonov-Bohm effect from the point of view of local realism, pp. 127–133. Plenum Press, New York and London, 1992.

[212] A. Tonomura. *Wave-Particle Duality,* edited by F. Selleri, chapter: Experiments on the Aharonov-Bohm effect, pp. 291–298. Plenum Press, New York and London, 1992.

[213] C. Dewdney. Particle trajectories and interference in a time-dependent model of neutron single crystal interferometry. *Physics Letters A*, 109, 8:377–384, 1985.

[214] J.P. Vigier, C. Dewdney, P.R. Holland, and A. Kyprianidis. *Quantum implications, Essays in honour of David Bohm,* edited by B.J. Hiley and F. David Pear, chapter: Causal particle trajectories and the interpretation of quantum mechanics, pp. 169–204. Routledge and Kegan Paul, 1987.

[215] B.J. Hiley. *Bell's Theorem, Quantum Theory and Conceptions of the Universe,* edited by M. Kafatos, chapter: Cosmology, EPR correlations and separability, pp. 181–190. Kluwer Academic Publishers, 1989.

[216] M.M. Lam and C. Dewdney. Locality and nonlocality in correlated two-particle interferometry. *Physics Letters A*, 150, 3/4:127–135, 1990.

[217] M. Bozic and Z. Maric. Probabilities of de Broglie's trajectories in Mach-Zehnder and in neutron interferometers. *Physics Letters A*, 158:33–42, 1991.

[218] H. Wu and D.W.L. Sprung. Ballistic transport: A view from the quantum theory of motion. *Physics Letters A*, 196:229–236, 1994.

[219] L. Shifren, R. Akis, and D.K. Ferry. Correspondence between quantum and classical motion: Comparing Bohmian mechanics with a smoothed effective potential approach. *Physics Letters A*, 274:75–83, 2000.

[220] C.R. Courtney and L.J. Frasinski. Coulomb imaging of h2plus nuclear wavefunctions using the pilot-wave theory of quantum mechanics. *Physics Letters A*, 318:30–35, 2003.

[221] A. Datta, P. Ghose, and M.K. Samal. Bohmian picture of Rydberg atoms. *Physics Letters A*, 322:277–281, 2004.

[222] A. Matzkin. Rydberg wavepackets in terms of hidden variables: de Broglie-Bohm trajectories. *Physics Letters A*, 345:31–37, 2005.

[223] J. Gonzalez, M.F. Gonzalez, J.M. Bofill, and X. Gimenez. Deterministic resonance formation and decay under Bohmian mechanics. *Journal of Molecular Structure: Theochem*, 727:205–212, 2005.

[224] Y. Zheng and D.H. Kobe. Momentum diffusion of the quantum kicked rotor: Comparison of Bohmian and standard quantum mechanics. *Chaos, solitons and fractals*, 34, 4:1105–1113, 2007.

[225] D. Nerukh and J.H. Frederick. Multidimensional quantum dynamics with trajectories: A novel numerical implementation of Bohmian mechanics. *Chemical Physics Letters*, 332:145–153, 2000.

[226] D. Bohm, R. Schiller, and J. Tiomno. A causal interpretation of the Pauli equation. *Supplemento al Nuovo Cimento*, I, Série X:48–66, 1955.

[227] C. Dewdney, P.R. Holland, and A.Kyprianidis. What happens in a spin measurement. *Physics Letters A*, 119, 6:259–267, 1986.

[228] C. Dewdney, P.R. Holland, and A. Kyprianidis. A quantum potential approach to spin superposition in neutron interferometry. *Physics Letters A*, 121, 3:105–110, 1987.

[229] C. Colijn and E.R. Vrscay. Spin-dependent Bohm trajectories for hydrogen eigenstates. *Physics Letters A*, 300:334–340, 2002.

[230] C. Colijn and E.R. Vrscay. Erratum to "spin-dependent Bohm trajectories for hydrogen eigenstates". *Physics Letters A*, 316:424, 2003.

[231] H. Rafii-Tabar. Real Feynman-like stochastic paths in the Bohm-Vigier causal stochastic model of quantum mechanics. *Physics Letters A*, 138, 8:353–358, 1989.

[232] R.H. Parmenter and R.W. Valentine. Deterministic chaos and the causal interpretation of quantum mechanics. *Physics Letters A*, 201: 1–8, 1995.

[233] R.H. Parmenter and R.W. Valentine. Erratum, deterministic chaos and the causal interpretation of quantum mechanics. *Physics Letters A*, 213:319, 1996.

[234] R.H. Parmenter and R.W. Valentine. Properties of the geometric phase of a de Broglie-Bohm causal quantum mechanical trajectory. *Physics Letters A*, 219:7–14, 1996.

[235] S. Sengupta and P.K. Chattaraj. The quantum theory of motion and signatures of chaos in the quantum behaviour of a classically chaotic system. *Physics Letters A*, 215:119–127, 1996.

[236] S. Konkel and A.J. Makowski. Regular and causal trajectories for the Bohm potential in a restricted space. *Physics Letters A*, 238:95–100, 1998.

[237] H. Wu and D.W.L. Sprung. Quantum chaos in terms of Bohm trajectories. *Physics Letters A*, 261:150–157, 1999.

[238] O.F. de Alcantara Bonfim, J. Florencio, and F.C. Sa Barreto. Chaotic Bohm's trajectories in a quantum circular billiard. *Physics Letters A*, 277:129–134, 2000.

[239] A.J. Makowski, P. Peploswski, and S.T. Dembinski. Chaotic causal trajectories: The role of the phase of stationary states. *Physics Letters A*, 266:241–248, 2000.

[240] J. Guckenheimer and P. Holmes. *Nonlinear Oscillations, Dynamical Systems, and Bifurcations of Vector Fields, Applied Mathematical Sciences, 42*. Springer-Verlag, 1983.

[241] P. Falsaperla and G. Fonte. On the motion of a single particle near a nodal line in the de Broglie-Bohm interpretation of quantum mechanics. *Physics Letters A*, 316:382–390, 2003.

[242] P.K. Chattaraj and S. Sengupta. Quantum fluid dynamics of a classically chaotic oscillator. *Physics Letters A*, 181:225–231, 1993.

[243] P.K. Chattaraj and S. Sengupta. Quantum chaos in Rydberg atoms: A quantum potential approach. *Current Science*, 71(2): 134–139, 1996.

[244] P.K. Chattaraj, S. Sengupta, and A. Poddar. Chaotic dynamics of some quantum anharmonic oscillators. *Current Science*, 74(9):758–764, 1998.

[245] P.K. Chattaraj, S. Sengupta, and A. Poddar. Quantum signature of the classical chaos in the field-induced barrier crossing in a quartic potential. *Current Science*, 76(10):1371–1381, 1999.

[246] P.K. Chattaraj and S. Sengupta. Chemical hardness as a possible diagnosis of the chaotic dynamics of Rydberg atoms in an external field. *Journal of Physical Chemistry A*, 103:6122–6126, 1999.

[247] P.K. CHattaraj, B. Maiti, and S. Sengupta. Quantum analogue of the Kolmogorov–Arnold–Moser transition in different quantum anharmonic oscillators. *International Journal of Quantum Chemistry*, 100:254–276, 2004.

[248] P.K. Chattaraj, S. Sengupta, and S. Giri. Quantum-classical correspondence of a field induced KAM-type transition: A QTM approach. *Journal of Chemical Science (Bangalore, India)*, 120(1): 33–37, 2008.

[249] C. Levit and J. Sarfatti. Are the Bader Laplacian and the Bohm quantum potential equivalent? *Chemical Physics Letters*, 281:157–160, 1997.

[250] O.V. Prezhdo and C. Brooksby. Non-adiabatic molecular dynamics with quantum solvent effects. *Journal of Molecular Structure (Theochem)*, 630:45–58, 2003.

[251] E. Gindensperger, C. Meier, and J.A. Beswick. Hybrid quantum/classical dynamics using Bohmian trajectories. *Advances in Quantum Chemistry*, 47:331–346, 2004.

[252] R. Guantes, A.S. Sanz, J. Margalef-Roig, and S. Miret-Artes. Atom-surface diffraction: A trajectory description. *Surface Science Reports*, 53:199–330, 2004.

[253] C. Dewdney, P.R. Holland, A. Kyprianidis, and J.P. Vigier. Cosmology and the causal interpretation of quantum mechanics. *Physics Letters A*, 114, 7:365–370, 1986.

[254] E.J. Squires. A quantum solution to a cosmological mystery. *Physics Letters A*, 162:35–36, 1992.

[255] N. Pinto-Neto, A.F. Velasco, and R. Colistete. Quantum isotropization of the universe. *Physics Letters A*, 277:194–204, 2000.

[256] P. Pinto-Neto and R. Colistete. Graceful exit from inflation using quantum cosmology. *Physics Letters A*, 290:219–226, 2001.

[257] N. Pinto-Neto and E.S. Santini. The accelerated expansion of the universe as a quantum cosmological effect. *Physics Letters A*, 315: 36–50, 2003.

[258] S.P. Kim. Quantum potential and cosmological singularities. *Physics Letters A*, 236:11–15, 1997.

[259] J. Acacio de Barros and N. Pinto-Neto. The causal interpretation of quantum mechanics and the singularity problem in quantum cosmology. *Nuclear Physics B (Proc.Suppl.)*, 57:247–250, 1997.

[260] J. Acacio de Barros, G. Oliveira-Neto, and T.B. Vale. Bohmian trajectories for an evaporating black hole. *Physics Letters A*, 336:324–330, 2005.

[261] D. Bohm. Comments on an article of Takabayasi concerning the formulation of quantum mechanics with classical pictures. *Progress in Theoretical Physics*, 9,3:273–287, 1953.

[262] D. Bohm. Proof that probability density approaches. $|\psi|^2$ in causal interpretation of the quantum theory. *Physical Review*, 89,2:458–466, 1953.

[263] D. Bohm and J.P. Vigier. Model of the causal interpretation of quantum theory in terms of a fluid with irregular fluctuations. *Physical Review*, 96, 1:208–216, 1954.

[264] E. Nelson. Derivation of Schrödinger equation from Newtonian mechanics. *Physical Review*, 150:1079–1085, 1966.

[265] A. Valentini. Signal-locality, uncertainty, and the subquantum H-theorem, I. *Physics Letters A*, 156, number 1,2:5–11, 1991.

[266] A. Valentini. Signal-locality, uncertainty, and the subquantum H-theorem, II. *Physics Letters A*, 158:1–8, 1991.

[267] A. Valentini. Signal-locality in hidden-variables theories. *Physics Letters A*, 297:273–278, 2002.

[268] G. Potel, M. Munoz-Alenar, F. Barranco, and E. Vigezzi. Stability properties of $|\psi|^2$ in Bohmian mechanics. *Physics Letters A*, 299:125–130, 2002.

[269] C. Colijn and E.R. Vrscay. Quantum relaxation in hydrogen eigenstates and two-state transitions. *Physics Letters A*, 327:13–122, 2004.

[270] J.P. Vigier. Possible consequences of an extended charged particle model in electromagnetic theory. *Physics Letters A*, 235:419–431, 1997.

[271] W.A. Hofer. Internal structures of electrons and photons: The concept of extended particles revisited. *Physica A*, 256:178–196, 1998.

[272] J.P. Vigier. Photon mass and Heaviside force. *Physics Letters A*, 270:221–231, 2000.

[273] D. Bohm and B.J. Hiley. An ontological basis for the quantum theory. Non-relativistic particle systems. *Physics Reports*, 144, 6:321–348, 1987.

[274] D. Bohm, B.J. Hiley, and P.N. Kaloyerou. An ontological basis for the quantum theory. A causal interpretation of quantum fields. *Physics Reports*, 144, 6:350–375, 1987.

[275] J.S. Bell. Quantum field theory without observers. *Physics Reports (Review Section of Physics Letters)*, 137,1:49–54, 1986.

[276] S. Colin. A deterministic Bell model. *Physics Letters A*, 317:349–358, 2003.

[277] C. Fenech and J.P. Vigier. Thermodynamical properties/description of the de Broglie-Bohm pilot-waves? *Physics Letters A*, 182:37–43, 1993.

[278] W. Struyve, W. De Baere, J. De Neve, and S. De Weirdt. On the uniqueness of paths for spin-0 and spin-1 quantum mechanics. *Physics Letters A*, 322:84–95, 2004.

[279] W. Weizel. Ableitung der Quantentheorie aus einem klassischen, kausal determinierten Modell. *Zeitschrift für Physik*, 134:264–285, 1953.

[280] W. Weizel. Ableitung der Quantentheorie aus einem klassichen Modell. *Zeitschrift für Physik*, 135:270–273, 1953.

[281] W. Weizel. Abteilung der quantenmechanischen Wellengleichung des Mehrteilchensystems aus einem klassischen Modell. *Zeitschrift für Physik*, 136:582–604, 1954.

[282] S.M. Roy and V. Singh. Maximally realistic causal quantum mechanics. *Physics Letters A*, 255:201–208, 1999.

[283] A.R. Plastino, M. Casas, and A. Plastino. Bohmian quantum theory of motion for particles with position-dependent effective mass. *Physics Letters A*, 281:297–304, 2001.

[284] P.R. Holland. Computing the wavefunction from trajectories: Particle and wave pictures in quantum mechanics and their relation. *Annals of Physics*, 315:505–531, 2005.

[285] C.D. Yang. Quantum dynamics of hydrogen atom in complex space. *Annals of Physics*, 319:399–443, 2005.

[286] A. Khrennikov. Classical and quantum dynamics on p-adic trees of ideas. *Biosystems*, 56:95–120, 2000.

[287] P.R. Holland. The de Broglie–Bohm theory of motion and quantum field theory. *Physics Reports*, 224, 3:95–150, 1993.

[288] M. Dickson. Essay review, antidote or theory? *Studies in History and Philosophy of Modern Sciences*, 27, 2:229–238, 1996.

[289] C. Papaliolios. Experimental test of a hidden-variable quantum theory. *Physical Review Letters*, 18, 15:622–625, 1967.

[290] D. Bohm and Y. Aharonov. Discussion of experimental proof for the paradox of Einstein, Rosen and Podolsky. *Physical Review*, 108, 4:1070–1076, 1957.

[291] M. Mugur-Schächter. *Etude du caractère complet de la mécanique quantique.* Gauthier-Villars, 1964.

[292] N.F. Mott. The wave mechanics of alpha-ray tracks. *Proceedings of the Royal Society, London*, A126:79–84, 1929.

[293] A. Messiah. *Mécanique quantique, nouvelle édition.* Dunod, 2003.

[294] D. Bohm. *Foundations of Quantum Mechanics, Proceedings of the International School of Physics "Enrico Fermi", course IL*, edited by B. d'Espagnat, chapter: Quantum theory as an indication of a new order in physics, pp. 412–469. Academic Press, New York, 1971.

[295] B. d'Espagnat. *Conceptual Foundations of Quantum Mechanics.* Benjamin, Reading, Mass., 1971.

[296] B.J. Hiley. Essay review, foundations of quantum mechanics. *Contemporary Physics*, 18, 4:411–414, 1977.

[297] D. Bohm and J. Bub. A proposed solution of the measurement problem in quantum mechanics by a hidden variable theory. *Review of Modern Physics*, 38, 3:453–469, 1966.

[298] N. Wiener and A. Siegel. A new form for the statistical postulate of quantum mechanics. *Physical Review*, 91, 6:1551–1560, 1953.

[299] N. Wiener and A. Siegel. The differential-space theory of quantum systems. *Supplemento al Nuovo Cimento*, 2, 4:982–1003, 1955.

[300] S.J. Freedman and R.A. Holt. *Quantum mechanics, determinism, causality, and particles, an International Collection of Contributions in Honor of Louis de Broglie on the Occasion of the Jubilee of His celebrated Thesis*, chapter: Experimental status of hidden variable theories, pp. 43–59. D. Reidel Publishing Company, Dordrecht-Holland/Boston-USA, 1976.

[301] J.H. Tutsch. Collapse time for the Bohm-Bub hidden variable theory. *Review of Modern Physics*, january:232–234, 1968.

[302] A. Valentini. Universal signature of non-quantum systems. *Physics Letters A*, 332:187–193, 2004.

[303] A. Siegel. Operational aspects of hidden-variable quantum theories. *Synthèse*, 14:171–188, 1962.

[304] L.E. Ballentine. The statistical interpretation of quantum mechanics. *Review of Modern Physics*, 42,4:358–380, 1970.

[305] M. Flato, Z. Maric, A. Milojevic, D. Sternheimer, and J.P. Vigier, editors. *Quantum mechanics, determinism, causality and particles.* D. Reidel Publishing Company, Dordrecht-Holland, 1976.

[306] E.P. Wigner. On hidden variables and quantum mechanical probabilities. *American Journal of Physics*, 38, 8:1005–1009, 1970.

[307] J. Winogradzki. *Les méthodes tensorielles de la physique, calcul tensoriel dans un continuum amorphe.* Masson et Cie, 1979.

[308] J. Solomon. Sur l'indéterminisme de la mécanique quantique. *Journal de Physique*, tome IV, série VII, 1:34–37, 1933.

[309] P. Destouches-Février. Une nouvelle preuve du caractère essentiel de l'indéterminisme quantique. *Comptes-Rendus de l'Académie des Sciences*, t 220:553–555, 1945.

[310] P. Destouches-Février. Sur l'impossibilité d'un retour au déterminisme en microphysique. *Comptes-Rendus de l'Académie des Sciences*, t 220:587–589, 1945.

[311] M. Strauss. *The Logico-Algebraic Approach to Quantum Mechanics, Historical Evolution*, edited by C.A. Hooker, chapter: The logic of complementarity and the foundation of quantum mechanics, pp. 27–44. D.Reidel Publishing Company, 1975.

[312] F. Kamber. The structure of the propositional calculus of a physical theory. *Nachrichten der Akademie der Wissenchaften Mathematisch-Physikalische Klasse*, 10:103–124, 1964.

[313] A.M. Gleason. Measures on the closed subspaces of a Hilbert space. *Journal of Mathematics and Mechanics*, 6, 6:885–893, 1957.

[314] J.M. Jauch and C. Piron. Can hidden variables be excluded in quantum mechanics? *Helvetica Physica Acta*, 36:827–837, 1963.

[315] D. Bohm and J. Bub. A refutation of the proof by Jauch and Piron that hidden variables can be excluded in quantum mechanics. *Review of Modern Physics*, 38, 3:470–475, 1966.

[316] D. Bohm and J. Bub. On hidden variables — a reply to comments by Jauch and Piron and by Gudder. *Review of Modern Physics. Letters to the editor*, January:235–236, 1968.

[317] S.P. Gudder. Hidden variables in quantum mechanics reconsidered. *Review of Modern Physics. Letters to the Editor*, January:229–231, 1968.

[318] J.M. Jauch and C. Piron. Hidden variables revisited. *Reviews of Modern Physics*, 40, 1:228–229, 1968.

[319] J.M. Jauch. *Foundations of quantum theory*. Addison-Wesley, Reading, Mass., 1968.

[320] S.P. Gudder. Dispersion-free states and the exclusion of hidden-variables. *Proceedings of the American Mathematical Society*, 19:319–324, 1968.

[321] S.P. Gudder. On hidden-variable theories. *Journal of mathematical physics*, 11,2:431–436, 1970.

[322] J. Bub. What is a hidden variable theory of quantum phenomena? *International Journal of Theoretical Physics*, 2, 2:101–123, 1969.

[323] M. Flato, Z. Maric, A. Milojevic, D. Sternheimer, and J.P. Vigier, editors. *Quantum Mechanics, Determinism, Causality, and Particles, an International Collection of Contributions in Honor of Louis de Broglie on the Occasion of the Jubilee of His celebrated Thesis*, chapter: Quantum mechanism and determinism, pp. 19–31. D. Reidel Publishing Company, Dordrecht-Holland/Boston-USA, 1976.

[324] J. Gréa. *Quantum Mechanics, Determinism, Causality, and Particles*, edited by M. Flato, Z. Maric, A. Milojevic, D. Sternheimer and J.P. Vigier, chapter: Pre-quantum mechanics. Introduction to models with hidden-variables, pp. 61–103. D. Reidel Publishing Company, 1976.

[325] A. Fine and P. Teller. Algebraic constraints on hidden variables. *Foundations of Physics*, 8, 7/8:629–636, 1978.

[326] A. Peres. *Bell's Theorem, Quantum Theory and Conceptions of the Universe*, edited by M. Kafatos, chapter: The logic of quantum nonseparability, pp. 51–60. Kluwer Academic Publishers, Dordrecht/Boston/London, 1989.

[327] J.S. Bell. *Foundations of Quantum Mechanics, Proceedings of the International School of Physics "Enrico Fermi", course IL*, edited by B. d'Espagnat, chapter: Introduction to the hidden-variable question, reprinted in "Speakable and Unspeakable in Quantum Mechanics", pp. 171–181. Academic Press, New York, 1971.

[328] J.S. Bell. Einstein-Podolsky-Rosen experiments, reprinted in "Speakable and Unspeakable in Quantum Mechanics". In *Proceedings of the symposium on frontier problems in high energy physics, Pisa, 33–45*, June 1976.

[329] A. Fine. Hidden variables, joint probability, and the Bell inequalities. *Physical Review Letters*, 48:291–295, 1982.

[330] N.D. Mermin. Simple unified form for the major no-hidden-variables theorems. *Physical Review Letters*, 65, 27:3373–3376, 1990.

[331] D. Bohm and B.J. Hiley. On the intuitive understanding of nonlocality as implied by quantum theory. *Foundations of Physics*, 5,1:93–109, 1975.

[332] D. Bohm and B.J. Hiley. Nonlocality and polarization correlation of annihilation quanta. *Nuovo Cimento*, 35 B, 1:137–144, 1976.

[333] N.D. Mermin. Bringing home the atomic world: Quantum mysteries for anybody. *American Journal of Physics*, 49, 10:940–943, 1981.

[334] D. Howard. *Philosophical Consequences of Quantum Theory*, edited by J.T. Cushing and E. McMullin, chapter: Holism, separability, and the metaphysical implications of the Bell experiments, pp. 224–253. University of Notre Dame Press, 1989.

[335] F. Laloë. Correlating more than two particles in quantum mechanics. *Current Science*, 68, 10:1026–1035, 1995.

[336] D. Bohm. *The Paradox of Einstein, Rosen, and Podolsky*, chapter 22, pp. 611–614. Prentice Hall, New York, 1951.

[337] D. Bohm and Y. Aharonov. Further discussion of possible experimental tests for the paradox of Einstein, Podolsky and Rosen. *Nuovo Cimento*, 17, 6:964–976, 1960.

[338] C. Dewdney, P.R. Holland, and A. Kyprianidis. A causal account of non-local Einstein–Podolsky–Rosen spin correlations. *Journal of Physics A: Math. Gen.*, 20:4717–4732, 1987.

[339] P. Teller. *Philosophical Consequences of Notre Dame Press*, edited by J.T. Cushing and E. McMullin, chapter: Relativity, relational holism, and the Bell inequalities, pp. 208–223. University of Notre Dame Press, 1989.

[340] W.H. Furry. Note on the quantum-mechanical theory of measurement. *Physical Review*, 49:393–399, 1936.

[341] W.H. Furry. Remarks on measurements in quantum theory. *Physical Review*, 49:476, 1936.

[342] Costa de Beauregard. Discussion on temporal asymmetry in thermodynamics and cosmology. In *Proceedings of the International Conference on Thermodynamics*, edited by P.T.Landsberg, Butterworths, London, 539–548, 1970.

[343] Costa de Beauregard. Time symmetry and interpretation of quantum mechanics. *Foundations of Physics*, 6,5:539–559, 1976.

[344] Costa de Beauregard. Time symmetry and the Einstein paradox, i. *Il Nuovo Cimento B*, 42,1:41–64, 1977.

[345] Costa de Beauregard. S-matrix, Feynman zigzag and Einstein correlation. *Physics Letters A*, 67,3:171–178, 1978.

[346] Costa de Beauregard. Time symmetry and the Einstein paradox, ii. *Il Nuovo Cimento*, 51B,2:267–279, 1979.

[347] F. Selleri and G. Tarozzi. Quantum mechanics reality and separability. *La Rivista del Nuovo Cimento*, 4, 2:1–53, 1981.

[348] H.P. Stapp. Comments on quantum theory with shadow states. *Physical Review D*, 13:947–949, 1976.

[349] H.P. Stapp. Theory of reality. *Foundations of Physics*, 7:313–323, 1977.

[350] C.W. Rietdijk. Proof of a retroactive influence. *Foundations of Physics*, 8:615–628, 1978.

[351] D.M. Greenberger, M.A. Horn, A. Shimony, and A. Zeilinger. Bell's theorem without inequalities. *American Journal of Physics*, 58, 12:1131–1143, 1990.

[352] S.J. Freedman and R.A. Holt. Tests of local hidden-variable theories in atomic physics. *Comments Atom. Molec. Phys.*, 5, 2:55–62, 1975.

[353] J.S. Bell. *Between Science and Technologie*, edited by A. Sarlemijn and P. Kroes, reprinted in "Speakable and Unspeakable in Quantum Mechanics", chapter: La nouvelle cuisine. Elsevier Science Publishers, 1990.

[354] J.F. Clauser, M.A. Horne, A. Shimony, and R.A. Holt. Proposed experiment to test local hidden-variable theories. *Physical Review Letters*, 23:880–884, 1969.

[355] A. Shimony. *Foundations of Quantum Mechanics, Proceedings of the International School of Physics "Enrico Fermi", course IL*, edited by B. d'Espagnat, chapter: Experimental test of local hidden-variable theories, pp. 182–194. Academic Press, New York, 1971.

[356] J.F. Clauser and M.A. Horne. Experimental consequence of objective local theories. *Physical Review D*, 10, 2:526–535, 1974.

[357] J.F. Clauser. Experimental investigation of a polarization correlation anomaly. *Physical Review Letters*, 36, 21:1223–1226, 1976.

[358] J.F. Clauser. Measurement of the circular-polarization correlation in photons from an atomic cascade. *Nuovo Cimento*, B 33, 2:740–746, 1976.

[359] J.F. Clauser and A. Shimony. Bell's theorem. Experimental tests and implications. *Rep. Prog. Phys.*, 41:1881–1927, 1978.

[360] J.H. McGuire and E.S. Fry. Restrictions on nonlocal hidden-variable theory. *Physical Review D*, 7, 2:555–557, 1973.

[361] J. Edwards. Sufficient conditions for objective local theories. *Nuovo Cimento*, 29 B, 1:100–104, 1975.

[362] J. Edwards and F. Ballentine. Sufficient conditions for objective local theories. *Nuovo Cimento*, 34 B, 1:91–94, 1976.

[363] A. Garuccio and F. Selleri. Nonlocal interactions and Bell's inequality. *Nuovo Cimento*, B 36, 2:176–185, 1976.

[364] A.J. Leggett. Nonlocal hidden-variable theories and quantum mechanics: an incompatibility theorem. *Foundations of Physics*, 33:1469–1493, 2003.

[365] S. Gröblacher, T. Paterek, R. Kaltenbaeck, C. Brukner, M. Zukowski, M. Aspelmeyer, and A. Zeilinger. An experimental test of non-local realism. *Nature*, 446:871–875, 2007.

[366] D. Bohm. *Quantum Mechanics, Determinism, Causality, and Particles, an International Collection of Contributions in Honor of Louis de Broglie on the Occasion of the Jubilee of his Celebrated Thesis*, edited by M. Flato, Z. Maric, A. Milojevic, D. Sternheimer and J.P. Vigier, Chapter 1, On the creation of a deeper insight into what may underlie quantum physical law, pp. 1–10. D. Reidel Publishing Company, Dordrecht-Holland, Boston-USA, 1976.

[367] H.P. Stapp. Are superluminal connections necessary? *Nuovo Cimento*, 40 B, 1:191–204, 1977.

[368] P.H. Eberhard. Bell's theorem and the different concepts of locality. *Nuovo Cimento*, 46 B, 2:392–419, 1978.

[369] J.T. Cushing and E.McMullin, editors. *Philosophical Consequences of Quantum Theory*. Notre Dame, IN, 1989.

[370] L. Hardy. Quantum mechanics, local realistic theories, and Lorentz-invariant realistic theories. *Physical Review Letters*, 68, 20:2981–2984, 1992.

[371] N.D. Mermin. Quantum mysteries for anyone. *The Journal of Philosophy*, 78:397–408, 1981.

[372] N.D. Mermin. *Philosophical Consequences of Quantum Theory*, edited by J.T. Cushing and E. McMullin, chapter: Can you help your team tonight by watching on TV? More experimental metaphysics from Einstein, Podolsky and Rosen, pp. 38–59. University of Notre Dame Press, 1989.

[373] N.D. Mermin. Quantum mysteries revisited. *American Journal of Physics*, 58:731–734, 1990.

[374] N.D. Mermin. Quantum mysteries refined. *American Journal of Physics*, 62:880–887, 1994.

[375] A. Aspect. *Quantum [Un]speakables, from Bell to Quantum Information*, edited by R.A. Bertlmann and A. Zeilinger, chapter: Bell's theorem: the naive view of an experimentalist, pp. 119–153. Springer, 2002.

[376] M. Kafatos, editor. *Bell's Theorem, Quantum Theory and Conceptions of the Universe*. Kluwer Academic Publishers, 1989.

[377] M. Kafatos. *Bell's Theorem, Quantum Universe and Conceptions of Universe*, edited by M. Kafatos, chapter: Horizons of knowledge

in cosmology, pp. 195–210. Kluwer Academic Publishers, Dordrecht/Boston/London, 1989.

[378] C.A. Hooker. *Bell's Theorem, Quantum Theory and Conceptions of the Universe*, edited by M. Kafatos, chapter: Bell, book and candle: The limning of a mystery, pp. 235–249. Kluwer Academic Publishers, Dordrecht/Boston/London, 1989.

[379] R.A. Bertlmann and A. Zeilinger, editors. *Quantum [Un]speakables, from Bell to Quantum Information*. Springer, 2002.

[380] F. Selleri. *Foundations of Quantum Mechanics, Proceedings of the International School of Physics "Enrico Fermi", course IL*, edited by B. d'Espagnat, chapter: Realism and the wave-function of quantum mechanics, pp. 398–406. Academic Press, New York, 1971.

[381] J.A. Wheeler. Polyelectrons. *Annals New York Academy of Sciences*, 48:219–238, 1946.

[382] M.H.L. Pryce and J.C. Ward. Angular correlation effects with annihilation radiation. *Nature*, 160:435, 1947.

[383] H.S. Snyder, S. Pasternak, and J. Hornbostel. Angular correlation of scattered annihilation radiation. *Physical Review*, 73, 5:440–448, 1948.

[384] E. Bleuler and H.L. Bradt. Correlation between the states of polarization of the two quanta of annihilation radiation. *Physical Review*, 73:1398, 1948.

[385] R.C. Hanna. Polarization of annihilation radiation. *Nature*, 162:332, 1948.

[386] C.S. Wu and I. Shaknov. The angular correlation of scattered annihilation radiation. *Physical Review*, 77, 1:136, 1950.

[387] F. Hereford. The angular correlation of photoelectrons ejected by annihilation quanta. *Physical Review, Letters to the Editor*, 81:482, 1951.

[388] G. Bertolini, M. Bettoni, and E. Lazzarini. Angular correlations of scattered annihilation radiation. *Nuovo Cimento*, 2,3: 661–662, 1955.

[389] H. Langhoff. Die Linearpolarisation der Vernichtungsstrahlung von Positronen. *Zeitschrift für Physik*, 160:186–193, 1960.

[390] L.R. Kasday, J. Ullman, and C.S. Wu. The Einstein-Podolsky-Rosen argument: positron annihilation experiment. *Bulletin. Am. Phys. Soc.*, 15,4:586, 1970.

[391] L. Kasday. *Foundations of Quantum Mechanics, Proceedings of the International School of Physics "Enrico Fermi", course IL*, edited by B. d'Espagnat, chapter: Experimental test of quantum predictions for widely separated photons, pp. 195–210. Academic Press, New York, 1971.

[392] L.R. Kasday, J.D. Ullman, and C.S. Wu. Angular correlation of Compton-scattered annihilation photons of hidden-variables. *Nuovo Cimento*, B 25, 2:633–661, 1975.

[393] G. Faraci, D. Gutowski, S. Notarrigo, and A.R. Pennisi. An experimental test of the EPR paradox. *Lett. Nuovo Cimento*, 9, 15:607–611, 1974.

[394] A.R. Wilson, J. Lowe, and D.K. Butt. Measurement of the relative planes of annihilation quanta as a function of separation distances. *J. Phys. G: Nucl. Phys.*, 2, 9:613–624, 1976.

[395] M. Bruno, M. d'Agostino, and C. Maroni. Measurement of linear polarization of positron annihilation photons. *Nuovo Cimento*, 40 B, 1:143–152, 1977.

[396] C.A. Kocher and E.D. Commins. Polarization correlation of photons emitted in an atomic cascade. *Physical Review Letters*, 18, 15:575–577, 1967.

[397] S.J. Freedman and J.F. Clauser. Experimental test of local hidden-variable theories. *Physical Review Letters*, 28:938–941, 1972.

[398] R.A. Holt. (unpublished). PhD thesis, Harvard University, Cambridge, Massachusetts, 1973.

[399] R.A. Holt and F.M. Pipkin. Quantum mechanics vs hidden variables: Polarization correlation measurement on an atomic mercury cascade. Harvard University Preprint, 1974.

[400] E.S. Fry. Two-photon correlations in atomic transitions. *Physical Review A*, 8, 3:1219–1232, 1973.

[401] E.S. Fry and R.C. Thompson. Experimental test of local hidden-variable theories. *Physical Review Letters*, 37:465–468, 1976.

[402] M. Lamehi-Rachti and W. Mittig. Quantum mechanics and hidden variables: A test of Bell's inequality by the measurement of the spin correlation in low-energy proton-proton scattering. *Physical Review D*, 14:2543–2555, 1976.

[403] A. Aspect. Proposed experiment to test the nonseparability of quantum mechanics. *Physical Review D*, 14:1944–1951, 1976. Reprinted in "Quantum theory and measurement", J.A. Wheeler and W.H.Zurek (ed). Princeton series in Physics, Princeton University Press, Princeton, New Jersey, 1983.

[404] A. Aspect. Expériences basées sur les inégalités de Bell. *Journal de Physique*, Colloque C2, supplément au numéro 3, 42:C2–63, 1981.

[405] B. d'Espagnat. Use of inequalities for the experimental test of a general conception of the foundation of microphysics. *Physical Review D*, 11, 6:1424–1435, 1975.

[406] J.T. Cushing and E. McMullin, (editors). *Philosophical Consequences of Quantum Theory, Reflections on Bell's Theorem*. University of Notre Dame Press, Notre Dame, Indiana, 1989.

[407] H.P. Stapp. *Philosophical Consequences of Quantum Theory*, J.T. Cushing and E. McMullin, editors, chapter: Quantum nonlocality and the description of nature, pp. 154–174. University of Notre Dame, IN, 1989.

[408] R.I.G. Hughes. *Philosophical Consequences of Notre Dame*, edited by J.T. Cushing and E. McMullin, chapter: Bell's theorem, ideology, and structural explanation, pp. 195–207. University of Notre Dame Press, 1989.

[409] A. Aspect. Proposed experiment to test separable hidden-variable theories. *Physics Letters A*, 54,2:117–118, 1975.

[410] Y. Aharonov and D. Bohm. Discussion of experimental proof for the paradox of Einstein, Rosen and Podolsky. *Physical Review*, 108:1070–1076, 1957.

[411] A. Aspect, P. Grangier, and G. Roger. Experimental tests of realistic local theories via Bell's theorem. *Physical Review Letters*, 47,7:460–463, 1981.

[412] A. Aspect, P. Grangier, and G. Roger. Experimental realization of Einstein-Podolsky-Rosen-Bohm Gedankenexperiment: A new violation of Bell's inequalities. *Physical Review Letters*, 49,2:91–94, 1982.

[413] A. Aspect, J. Dalibard, and G. Roger. Experimental test of Bell's inequalities using time-varying analyzers. *Physical Review Letters*, 49,25:1804–1807, 1982.

[414] G. Weihs, T. Jennewein, C. Simon, H. Weinfurter, and A. Zeilinger. Violation of Bell's inequality under strict Einstein locality conditions. *Physical Review Letters*, 81:5039–5043, 1998.

[415] G. Weihs. *Quantum [Un]Speakables, from Bell to Quantum Information*, edited by R.A. Bertlmann and A. Zeilinger, chapter: Bell's theorem for space-like separation, pp. 155–162. Springer, 2002.

[416] P.M. Pearle. Hidden-variable example based upon data rejection. *Physical Review D*, 2, 8:1418–1425, 1970.

[417] M. Genovese. Research on hidden variable theories: A review of recent progresses. *Physics Reports*, 413:319–396, 2005.

[418] T.W. Marshall, E. Santos, and F. Selleri. Local realism has not been refuted by atomic cascade experiments. *Physics Letters 98A*, 1, 2:5–9, 1983.

[419] W. Tittel, J. Brendel, B. Gisin, T. Herzog, H. Zbinden, and N. Gisin. Experimental demonstration of quantum correlations over more than 10 km. *Physical Review A*, 57:3229–3232, 1998.

[420] R. Ursin, F. Tiefenbacher, T. Schmitt-Manderbach, H. Weier, T. Scheidl, M. Liedenthal, B. Blauensteiner, T. Jennewein, J. Perdigues, P. Trojek, B. Ömer, M. Fürst, M. Meyenburg, J. Rarity, Z. Sodnik, C. Barbieri, H. Weinfurter, and A. Zeilinger.

Entanglement-based quantum communication over 144 km. *Nature Physics*, 3:481–486, 2007.

[421] C.H. Bennett, G. Brassard, C. Crépeau, R. Jozsa, A. Peres, and W.K. Wootters. Teleporting an unknown quantum state via dual classical and Einstein-Podolsky-Rosen channels. *Physical Review Letters*, 70, 13:1895–1899, 1993.

[422] S. Popescu. Bell's inequalities versus teleportation: What is nonlocality? *Physical Review Letters*, 72, 6:797–799, 1994.

[423] M.L.G. Redhead. *Philosophical Consequences of Quantum Theory*, edited by J.T. Cushing and E. McMullin, chapter: Nonfactorizability, stochastic causality, and passion-at-a-distance, pp. 145–153. University of Notre Dame, IN, 1989.

[424] J. Butterfield. *Philosophical Consequences of Quantum Theory*, edited by J.T. Cushing and E. McMullin, chapter: A space-time approach to the Bell inequality, pp. 114–144. University of Notre Dame Press, 1989.

[425] I. Fenyes. Eine wahrscheinlichkeitstheoretische Begründung und Interpretation der Quantenmechanik. *Zeitschrift für Physik*, 132:81–106, 1952.

[426] D. Bohm and B.J. Hiley. On the intuitive understanding of nonlocality as implied by quantum theory. *Foundations of Physics*, 5, 1:93–109, 1975.

[427] H.J. Folse. *Philosophical Consequences of Quantum Theory*, edited by J.T. Cushing and E. McMullin, chapter: Bohr on Bell, pp. 254–271. University of Notre Dame, 1989.

[428] D. Greenberger, M. Horne, and A. Zeilinger. *Bell's Theorem, Quantum Theory and Conceptions of the Universe*, edited by M. Kafatos, chapter: Going beyond Bell's theorem, pp. 69–72. Kluwer Academic Publishers, Dordrecht/Boston/London, 1989.

[429] D. Greenberger. *Quantum [Un]Speakables, from Bell to Quantum Information*, edited by R.A. Bertlmann and A. Zeilinger, chapter: The history of the GHZ paper, pp. 281–286. Springer, 2002.

[430] S. Goldstein. Nonlocality without inequalities for almost all entangled states for two particles. *Physical Review Letters*, 72, 13:1951, 1994.

[431] D. Boschi, S. Nranca, F. Fe Martini, and L. Hardy. Ladder proof of nonlocality without inequalities: Theoretical and experimental results. *Physical Review Letters*, 79, 15:2755–2758, 1997.

[432] A. Peres. Incompatible results of quantum measurements. *Physics Letters*, 151A, 3, 4:107–108, 1990.

[433] L. Hardy. Non-locality for two particles without inequalities for almost all entangled states. *Physical Review Letters.*, 71,11:1665–1668, 1993.

[434] H.R. Brown and G. Svetlichny. Nonlocality and Gleason's lemma. deterministic theories. *Foundations of Physics*, 20, 11:1379–1387, 1990.

[435] P. Hoffman. *Paul Erdös: l'homme qui n'aimait que les nombres.* Librairie Lavoisier, 2000.

[436] A. Einstein and L. Hopf. Über einen Satz der Wahrscheinlichtkeitsrechnung und seine Anwendung in der Strahlungtheorie. *Annalen der Physik*, 33:1096–1105, 1910.

[437] A. Einstein and O. Stern. Einige Argumente für die Annahme einer molekularen Agitation beim absoluten Nullpunkt. *Annalen der Physik*, 40:551–560, 1913.

[438] S. Bergia, P. Lugli, and N. Zamboni. Zero-point energy, Planck's law, and the prehistory of stochastic electrodynamics, part ii: Einstein and Stern's paper of 1913. *Ann. Fond. L. de Broglie*, 5:39–62, 1980.

[439] E. Schrödinger. Über die Umkehrung der Naturgesetze. *Berliner Sitzungsberichte*, pp. 144–153, 1931.

[440] R. Fürth. Über einige Beziehungen zwischen klassischer Statistik und Quantenmechanik. *Zeitschrift für Physik*, 81:143–162, 1933.

[441] N.S. Kalitsin. On the interaction of an electron with the fluctuating electromagnetic field of the vacuum, in Russian. *Soviet Phys. JETP*, 25:407–409, 1953.

[442] L. de La Pena-Auerbach. New formulation of stochastic theory and quantum mechanics. *Journal of Mathematical Physics*, 10, 9:1620–1630, 1969.

[443] L. de La Pena-Auerbach and A.M. Cetto. A new formulation of stochastic theory and quantum mechanics. Lagrangian formulation and electromagnetic coupling. *Revista Mexicana de Fisica*, 18, 3:253–264, 1969.

[444] L. de La Pena-Auerbach. Stochastic theory of quantum mechanics for particles with spin. *Journal of Mathematical Physics*, 12, 3:453–461, 1971.

[445] Z. Maric and DJ. Zivanovic. *Quantum Mechanics, Determinism, Causality, and Particles*, edited by M. Flato, Z. Maric, A. Milojevic, D. Sternheimer and J.P. Vigier, chapter : The stochastic interpretation of quantum mechanics and the theory of measurement, pp. 137–146. D. Reidel Publishing Company, Dordrecht-Holland/ Boston-USA, 1976.

[446] M. Davidson. A generalization of the Fenyes–Nelson stochastic model of quantum mechanics. *Letters in Mathematical Physics*, 3:271–277, 1979.

[447] L. Fritsche and M. Haugk. A new look at the derivation of the Schrödinger equation from Newtonian mechanics. *Annalen der Physik*, 12,6:371–402, 2003.

[448] J.P. Vigier. Model of quantum statistics in terms of a fluid with irregular stochastic fluctuations propagating at the velocity of light: A derivation of Nelson's equations. *Lett. Nuovo Cimento*, 24, 8:265–272, 1979.

[449] B.G. Sidharth. Space-time as a random heap. *Chaos, Solitons and Fractals*, 12:173–178, 2001.

[450] L.F. Santos and C.O. Escobar. A proposed solution to the tail problem of dynamical reduction models. *Physics Letters A*, 278:315–318, 2001.

[451] E. Santos. On a heuristic point of view concerning the motion of matter: From random metric to Schrödinger's equation. *Physics Letters A*, 352:49–54, 2006.

[452] F. Nietzsche. *Lettres choisies*. Gallimard, 1950.

[453] F. Nietzsche. *Le crépuscule des idoles*. Denoël-Gonthier, 1976.

[454] P. Forman. *Historical Studies in Physical Science, 3*, edited by R. Mc Cormak, chapter: Weimar Culture, Causality, and quantum theory, 1918–1927: Adaptation by German physicists and mathematicians to a hostile intellectual environment, pp. 1–115. University of Pennsylvania Press, Philadelphia, 1971.

[455] W. Heitler. *Albert Einstein: Philosopher-Scientist (First edition, 1949), edited by P.A. Schilpp*, chapter: The departure from classical thought in modern physics, pp. 181–198. The Library of Living Philosophers, Volume VII, 1970.

[456] I. Prigogine and I. Stengers. *La nouvelle alliance*. Gallimard, 1979.

[457] I. Prigogine. *La fin des certitudes*. Editions Odile Jacob, 1996.

[458] M.J. Klein. *Some Strangeness in the Proportion*, edited by H. Woolf, chapter: No firm foundation: Einstein and the early quantum theory, pp. 161–185. Addison-Wesley, Reading, Mass., 1980.

[459] L.I. Schiff. *Quantum Mechanics*. Mc Graw-Hill, New York, 1968.

[460] E. Schrödinger. Quantisierung als Eigenwertproblem. *Annalen der Physik*, 79,4:361–376, 1926.

[461] E. Schrödinger. Quantisierung als Eigenwertproblem. *Annalen der Physik*, 79,6:489–527, 1926.

[462] E. Schrödinger. Quantisierung als Eigenwertproblem. *Annalen der Physik*, 81,18:109–139, 1926.

[463] J. Winogradski. Lectures notes and: An imaginary history of the making of nonrelativistic quantum mechanics. Unpublished personal communication.

[464] W.R. Hamilton. On a general method of expressing the paths of light and of the planets by the coefficients of a characteristic function. *Dublin University Review and Quarterly Magazine*, 1: 795–826, 1833.

[465] W.R. Hamilton. On a general method in dynamics. *Philosophical Transactions of the Royal Society, Part II*, pp. 247–308, 1834.

[466] L.D. Landau and E.V. Lifchitz. *Théorie des champs, English translation: Field theory, 1970*. Editions Mir, Moscow, 1966.

[467] E.H. Wichmann. *Cours de physique de Berkeley, tome 4, physique quantique*. Librairie Armand Colin, Quantum Physics, Education Development Center Inc., Newton, Massachussets, 1967.

[468] B. Pascal. *Oeuvres complètes*, chapter: Physique, pp. 355–543. Gallimard, collection de la Pléiade, 1998.

[469] J. Locke. *Essai sur l'entendement humain. French translation of "An Essay Concerning Human Understanding"*. Librairie générale française, Paris, 2009.

[470] R. Descartes. *Les principes de la philosophie, première et seconde parties, Tome 1, Principes*. Paleo, France, 2007.

[471] W.V. Quine. *Relativité de l'ontologie et autres essais, présentation par Sandra Laugier*. Aubier-Montaigne, Paris, 1977.

[472] W.V. Quine. *Du point de vue logique, neuf essais logico-philosophiques, French translation of "From a logical point of view"*. Librairie philosophique J. Vrin, Paris, 2003.

[473] Epicure. *Lettres, maximes, sentences*. Librairie générale française, Paris, 1994.

[474] D. Laërce. *Vie, doctrines et sentences des philosophes illustres*. Garnier-Flammarion, 1965.

[475] B. Mandelbrot. *The Fractal Geometry of Nature*. W.H.Freeman, 1983.

[476] B. Mandelbrot. *Les Objets Fractals*. Flammarion, 3ème édition, 1994.

[477] J. Perrin. *Les atomes*. Flammarion, 1991.

[478] K.S. Thorne. *Black Holes and Times Warps*. Norton, New York, 1994.

[479] Kerker. M. *The Scattering of Light and Other Electromagnetic Radiation*. Academic Press, 1969.

[480] H.C. van de Hulst. *Ligth Scattering by Small Particles*. Dover Publications, Inc., 1981.

[481] J.A. Adam. The mathematical physics of rainbows and glories. *Physics Reports*, 356:229–365, 2002.

[482] M.R. Vetrano, J.P.A.J. van Beeck, and M.L. Riethmuller. Global rainbow thermometry: Improvement in the data inversion algorithm and validation technique in liquid-liquid suspension. *Applied Optics*, 43,18:3600–3607, 2004.

[483] J.P.A.J. van Beeck, D. Giannoulis, L. Zimmer, and M.L. Riethmuller. Global rainbow refractometry for droplet temperature measurement. *Optics Letters*, 24:1696–1698, 1999.

[484] S. Saengkaew, T. Charinpanitkul, H. Vanisri, W. Tantha-panichakoon, L. Mées, G. Gouesbet, and G. Gréhan. Rainbow refractometry: On the validity domain of Airy's and Nussenzveig theories. *Optics Communications*, 259:7–13, 2006.

[485] L. Lorenz. Lysbevaegelsen i og uden for en af plane lysbolger belyst kulge. *Vidensk. Selk. Skr.*, 6:1–62, 1890.

[486] L. Lorenz. *Sur la lumière réfléchie et réfractée par une sphère transparente*. Lib. Lehmann et Stage, Oeuvres scientifiques de L.Lorenz, revues et annotées par H.Valentiner, 1898.

[487] G. Mie. Beiträge zur Optik trüben Medien speziell kolloidaler Metalösungen. *Annalen der Physik*, 25:377–452, 1908.

[488] G. Gouesbet and G. Gréhan. *Generalized Lorenz-Mie theories*. Springer, Berlin, 2011.

[489] R.G. Newton. *Scattering Theory of Waves and Particles*. Dover Publications, Inc., 2002.

[490] H.M. Nussenzveig. *Diffraction Effects in Semiclassical Scattering*. Cambridge University Press, 1992.

[491] E. Klein. *Le monde quantique*, chapter: La naissance de la physique quantique, pp. 6–12. Les dossiers de la recherche, 2007.

[492] H.D. Doebner and G.A.Goldin. Introducing nonlinear gauge transformations in a family of nonlinear Schrödinger equations. *Physical Review A*, 54,5:3764–3771, 1996.

[493] R. Penrose. *A la découverte des lois de l'univers, la prodigieuse histoire des mathématiques et de la physique, French translation of "The road to reality, a complete guide to the laws of the universe"*. Odile Jacob, Paris, 2007.

[494] N. Gisin. Stochastic quantum dynamics and relativity. *Helvetica Physica Acta*, 62:363–371, 1989.

[495] J.M.T. Thompson and H.B. Stewart. *Nonlinear Dynamics and Chaos*. John Wiley and Sons, 1986.

[496] J.O. Hinze. *Turbulence, Introduction to Its Mechanism and Theory*. McGraw-Hill Book Company, 2nd edition, 1975.

[497] B. d'Espagnat. *Foundations of Quantum Mechanics, Proceedings of the International School of Physics "Enrico Fermi", Course IL, edited by B. d'Espagnat*, chapter: Mesure et non séparabilité (revue sommaire), pp. 84–96. Academic Press, New York, 1971.

[498] G.M. Prosperi. *Foundations of Quantum Mechanics, Proceedings of the International School of Physics "Enrico Fermi", course IL*, edited by B. d'Espagnat, chapter: Macroscopic physics and the problem of measurement in quantum mechanics, pp. 97–126. Academic Press, New York, 1971.

[499] G. Gouesbet. Electromagnetic scattering by an absorbing macroscopic sphere is cross-sectionnally equivalent to a superposition of two effective quantum processes. *Journal of Optics A: Pure and Applied Optics*, 9:369–375, 2007.

[500] G. Gouesbet. On the optical theorem and non plane wave scattering in quantum mechanics. *Journal of Mathematical Physics*, 50(112302): 1–5, 2009.

[501] R. Descartes. *Discours de la méthode*. J. Vrin, 1938.

[502] Spinoza. *Traité de la réforme de l'entendement*. Flammarion, 1964.

[503] Spinoza. *Les principes de la philosophie de Descartes*. Flammarion, 1964.

[504] J. Winogradski. *Les méthodes tensorielles de la physique, calcul tensoriel dans un continuum structuré*. Masson et Cie, 1997.

[505] C. Kittel, W.D. Knight, and M.A. Ruderman. *Mechanics*. McGraw-Hill Book Company, 1962.

[506] E. Kant. *Critique de la raison pratique*. Presses universitaires de France, Quadrige, 1943.

[507] G.W. Leibniz. *Principes de la nature et de la grâce fondés en raison: principes de la philosophie ou monadologie*. Presses Universitaires de France, Paris, 1954.

[508] B. Nicolescu. *Dictionnaire de l'ignorance, sous la direction de M. Cazenave*, chapter: Relativité et physique quantique, pp. 108–120. Bibliothèque Albin Michel Sciences, 1998.

[509] G.N. Ord. Bohr, Bohm and entwined paths. *Chaos, Solitons and Fractals*, 25:769–774, 2005.

[510] A. Peres and W.H. Zurek. Is quantum theory universally valid? *American Journal of Physics*, 50, 9:807–810, 1982.

[511] Nuno Costa Dias and Joao Nuno Prata. Bohmian trajectories and quantum phase space distributions. *Physics Letters A*, 302:261–272, 2002.

[512] S. Goldstein, J. Taylor, R. Tumulka, and N. Zanghi. Are all particles real? *Studies in History and Philosophy of Modern Physics*, 36:103–112, 2005.

[513] P. Duhem. *La théorie physique, son objet, sa structure. English version: The Aim and Structure of Physical Theory*. Librairie philosophique J. Vrin, 1997.

[514] B. Pascal. *Oeuvres complètes, tome 1*, chapter: Lettre de Blaise Pascal au très Révérend Père Noël, pp. 377–386. Gallimard, 1998.

[515] W.V. Quine. *Le mot et la chose, French translation of "Word and Object"*, MIT Press, 1960. Flammarion, 1977.

[516] W.V. Quine. *La poursuite de la vérité, French Translation of "Pursuit of Truth"*, Harvard University Press, 1990. Editions du Seuil, Paris, 1993.

[517] Sous la direction de Pierre Wagner, editor. *Les philosophes et la science*. Gallimard, 2002.

[518] W.V. Quine. On empirically equivalent systems of the world. *Erkenntnis*, 9:313–328, 1975.

[519] L. Wittgenstein. *Tractatus Logico-Philosophicus*. Gallimard, 1993.

[520] S. Laugier-Rabaté. *L'anthropologie logique de Quine, l'apprentissage de l'obvie*. Librairie philosophique J. Vrin, 1992.

[521] N.A. Logan. Survey of some early studies of the scattering of plane waves by a sphere, reprinted in selected papers on light scattering by M.Kerker. *SPIE*, p. 951, 1965.

[522] N.A. Logan. Survey of some early studies of the scattering of plane waves by a sphere. In *Proceedings of the Second international symposium on optical particle sizing*, pp. 7–15, edited by D. Hirleman, Arizona State University, 1990.

[523] H. Kragh. Ludwig Lorenz: His contributions to optical theory and light scattering by spheres. In *Proceedings of the Second International Symposium on Optical Particle Sizing*, pp. 1–6, edited by D. Hirleman, Arizona State University, 1990.

[524] H. Kragh. Ludwig Lorenz and the nineteenth century optical theory. the work of a great Danish scientist. *Applied Optics*, 30,33:4688–4695, 1991.

[525] G. Gouesbet. From theories by Lorenz and Mie to ontological underdetermination of theories by experiments. In *Mie theory, 1908–2008*, pp. 73–100, 2012.

[526] G. Gouesbet. On empirically equivalent systems of the world with conflicting ontologies: Three case-studies. *International Journal of Philosophy and Theology*, 4(1):22–41, 2016.

[527] G. Gouesbet. Hypotheses on the a priori rational necessity of quantum mechanics. *Principia — An International Journal of Epistemology*, 14(3):393–404, 2010.

9 789819 806485